Bioelectrochemistry: Ions, Surfaces, Membranes

Bioelectrochemistry: Ions, Surfaces, Membranes

Martin Blank, EDITOR
Columbia University

Based on a symposium sponsored by the Division of Colloid and Surface Chemistry at the 176th Meeting of the American Chemical Society in Miami Beach, Florida, September 11–14, 1978.

ADVANCES IN CHEMISTRY SERIES **188**

AMERICAN CHEMICAL SOCIETY
WASHINGTON, D. C. 1980

Library of Congress CIP Data

Bioelectrochemistry.
(Advances in chemistry series; 188 ISSN 0065-2393)

Includes bibliographies and index.

1. Membranes (Biology)—Electric properties—Congresses. 2. Electrophysiology—Congresses.
I. Blank, Martin, 1933- . II. American Chemical Society. Division of Colloid and Surface Chemistry. III. Series.
[DNLM: 1. Biochemistry—Congresses. 2. Electrochemistry—Congresses. 3. Ions—Congresses. 4. Cell membranes—Congresses. 5. Surface properties—Congresses. QD273 B615 1978]

QD1.A355 No. 188 [QH601] 540s [574.19'283] 80-18001
ISBN 0-8412-0473-X ADCSAJ 188 1-527 1980

PRINTED IN THE UNITED STATES OF AMERICA

Advances in Chemistry Series

M. Joan Comstock, *Series Editor*

FOREWORD

ADVANCES IN CHEMISTRY SERIES was founded in 1949 by the American Chemical Society as an outlet for symposia and collections of data in special areas of topical interest that could not be accommodated in the Society's journals. It provides a medium for symposia that would otherwise be fragmented, their papers distributed among several journals or not published at all. Papers are reviewed critically according to ACS editorial standards and receive the careful attention and processing characteristic of ACS publications. Volumes in the ADVANCES IN CHEMISTRY SERIES maintain the integrity of the symposia on which they are based; however, verbatim reproductions of previously published papers are not accepted. Papers may include reports of research as well as reviews since symposia may embrace both types of presentation.

CONTENTS

MEMBRANE PROCESSES

CELLS AND TISSUES

PREFACE

As scientists, we pride ourselves on our independence—i.e. our freedom to seek answers to the problems that arouse our curiosity. Yet, when pressed to support this view of ourselves, we reluctantly must admit that our independence is largely a myth. Scientists, like other groups in society, always have been constrained in a social environment as products of a specific time, education, and professional milieu. And now, in the era of government support, of dependence on sophisticated instruments, and of greater specialization, the constraints have increased many fold.

In the face of the growing organization and specialization of scientific activity, scientists have been quite resourceful in devising ways of enhancing their freedom and creativity. One of the techniques which has evolved in recent years is the formation of interdisciplinary fields to broaden the scope and applicability of research. (Most activities of this type are multidisciplinary, i.e. the teaming up of scientists from two disciplines, generally to apply the techniques of one to the systems of the other. Ideal interdisciplinary research involves scientists with different and frequently opposing views, attacking the same problem.) One of the new interdisciplinary areas that appears to offer many possibilities for fresh approaches to problems in biology and medicine is bioelectrochemistry. This book is based on the symposium, "Surface Chemistry in Biology and Medicine: Bioelectrochemistry," which took place on September 13–14, 1978 in Miami Beach, FL.

My active interest in bioelectrochemistry was stimulated when, as physiologist for the Office of Naval Research (London) during 1974–1975, I became aware of rapid developments in Europe in this recently formulated interdisciplinary research area. During the past few years I have been trying to foster this area on this side of the Atlantic largely through symposia, e.g. the Fourth International Symposium on Bioelectrochemistry (the first one held in the United States), Electrochemical Processes at Biological Membranes (part of the Electrochemical Society Meeting), and the present symposium. The papers at these meetings have been in a variety of areas, but in all cases an attempt has been made to bring together the biological and electrochemical thinking, rather than merely applying electrochemical techniques to biological systems. The framework of this meeting, colloid and surface chemistry, provides a traditional interdisciplinary background and a particular emphasis on the ability of interfacial electrochemistry to elucidate a number of important problems in biology and medicine.

The 28 chapters included here have been grouped into four categories based on the areas of major emphasis. As can be expected in a rapidly growing field, there is overlap between the categories and some chapters would be at home in several categories. This probably is illustrated best by our keynote address on red cell aggregation, which contains elements of all of the major topics and shows clearly how it is possible to understand the interactions of cell membranes when they are treated as charged surfaces.

There are five major section headings in this volume and the prevailing point of view in each of the sections is what can be learned about the biological problem as a result of bringing electrochemical and surface chemical intuition to bear on the system. Thus, we can think of cell and membrane interactions as resulting from the contact of electrically charged plates, of dissolved macromolecules as surface-stabilized suspension droplets; of electrical changes at excitable membrane surfaces owing to electrochemical interactions in arrays of charged molecules, and of complex natural membranes being comprised of compartments that contribute to the overall responses of the system, etc. Without mentioning each chapter individually, we have attempted to break new ground through fresh approaches to important problems.

I wish to thank the authors and the many others who were involved in the organization of the meeting and the production of this volume. I also wish to extend special thanks to the Office of Naval Research for their continued support and to the Division of Colloid and Surface Chemistry of the American Chemical Society for their willingness to use the proceeds of this volume to support further meetings and symposia devoted to the same theme.

Department of Physiology
College of Physicians and Surgeons
Columbia University
630 W. 168th St.
New York, NY 10032

MARTIN BLANK

April 30, 1979

KEYNOTE ADDRESS

1

Aggregation of Red Blood Cells: An Electrochemical and Colloid Chemical Problem

SHU CHIEN

Division of Circulatory Physiology and Biophysics, Department of Physiology, Columbia University, College of Physicians and Surgeons, New York, NY 10032

The aggregation of red blood cells by macromolecular bridging involves energy balance at the cell surface. The energy of macromolecular binding to the cell membrane provides the primary aggregating energy. The disaggregating energies result from electrostatic repulsion between the sialic acids on cell surfaces and mechanical shearing stress. A portion of the net aggregation energy (aggregating energy minus disaggregating energies) is stored in the cell membrane as a change in strain energy. Experimental manipulations of these various forms of energies have provided evidence supporting the concept of energy balance at cell surfaces. The results indicate that an understanding of red cell aggregation requires an interdisciplinary approach by which principles in electrochemistry and colloid chemistry are applied to this biological system.

Cell aggregation is an important phenomenon in biology and medicine. The aggregation of like cells gives rise to the structural basis of tissues and organs. The aggregation process involves interactions between membranes of adjacent cells. Since biological cell surfaces possess charged groups and the body fluids are composed of electrolyte solutions containing macromolecules, cell interactions entail electrochemical and colloid chemical factors. An understanding of the role of these factors in controlling cell aggregation would help to elucidate the physicochemi-

0-8412-0473-X/80/33-188-003$09.00/1

cal principles underlying many important physiological and pathological processes. These include the development, growth, and differentiation of cells, the functional behaviors of cells including secretion, contraction, and signal transmission, and the growth and metastasis of tumor cells.

Among the various types of cells, the red blood cells (RBCs) are most easily obtainable and amenable to experimental manipulations, and the electrochemical properties of the RBC membrane are most clearly understood. Therefore the aggregation of RBCs provides an excellent model for studying cell-to-cell interactions in general. Of course, applying the knowledge gained from studies of RBC aggregation to other cell systems necessitates appropriate modifications by taking into account the structural and functional differences among different cell types.

The whole blood of normal human subjects consists of a concentrated suspension of RBCs (approximately 40–45% by volume, with white blood cells (WBCs) and platelets contributing only less than 1%) in plasma (containing 7 g/dL proteins, i.e. albumin, globulins, and fibrinogen). Under normal conditions, the blood flowing through most parts of the circulatory system is subject to sufficiently high shear stresses to disperse the RBCs during flow. In regions of flow stagnation or when blood is examined in vitro in a stationary condition, extensive RBC aggregation occurs to form rouleaux. This RBC aggregation necessitates the presence of appropriate macromolecules in the suspending medium, such as fibrinogen and some globulin fractions present in the normal blood plasma or other polymers (e.g. dextrans). In the absence of such macromolecules, RBCs do not aggregate even in the absence of flow. The physicochemical properties and the concentration of the macromolecules play a significant role in affecting RBC aggregation, suggesting that RBC aggregation results from an interaction of the macromolecules with the RBC membrane surface (*1*). Our current model is that the energy which causes RBC aggregation results from the bridging of the surfaces of two adjacent cells by the adsorbed macromolecule (*2*). In order to attain a stable aggregation, the aggregating energy owing to macromolecular bridging (E_b) must overcome the disaggregating energies (*3*). Thus, the electrostatic repulsive energy (E_e) between two negatively charged cell surfaces brought into close range would cause disaggregation. As mentioned above, mechanical shearing also causes RBC disaggregation by supplying dispersing stress. RBC aggregation leads to an alteration in the shape of the RBCs, and such cell deformation involves a change in membrane strain energy (E_m) supplied by the aggregation energy.

The purpose of this chapter is to review the influences of electrochemical and colloid chemical factors on RBC aggregation from the viewpoint of the balance of these aggregating and disaggregating energies

Table I. Molecular Weight of Dextrans

Dextran Fractions	$\overline{M}_w$	$\overline{M}_w/\overline{M}_n$ [a]
Dx 20	20×10^3	1.24
Dx 40	42×10^3	1.63
Dx 70 or 80 [b]	75×10^3	1.54
Dx 150	141×10^3	1.64
Dx 500	450×10^3	2.54
Dx 2000	2000×10^3	ND [c]

[a] The ratio of $\overline{M}_w$ to $\overline{M}_n$ gives an indication of the dispersion of molecular weight distribution. $\overline{M}_w/\overline{M}_n = 1$ for monodispersion.

[b] Dx 70 and Dx 80 have been used interchangeably in this chapter for the same dextran fraction with a $\overline{M}_w$ of 75,000.

[c] Not determined.

at the cell surface. Most of the studies to be presented involve RBC aggregation by dextrans, which are neutral polymers of glucose synthesized by bacterial action (*4*). The availability of various dextran fractions with controlled molecular size (*see* Table I) renders them particularly suited for investigations on RBC aggregation.

Aggregation of Normal RBCs in Dextrans

Quantitative Relationship Between RBC Aggregation and the Molecular Weight and Concentration of Dextrans. The aggregation of normal RBCs in suspending media containing dextrans with various molecular weights and concentrations has been quantified by microscopic counting (*5*), determination of erythrocyte sedimentation rate (*6, 7, 8*), viscometry at low shear rates (*8, 9*) and measurements by optical methods (*10, 11, 12*). In Figure 1 the various aggregation indices are plotted against the logarithm of the molar concentration of dextrans (*3*). No aggregation occurs with Dx 20. RBCs can be aggregated by Dx 40, but only to a slight degree; furthermore, it occurs only in a narrow range of Dx 40 concentrations. When higher-molecular-weight dextrans are used, the degree of cell aggregation becomes more prominent. The molar concentration of dextrans required to induce aggregation varies inversely with the molecular size (*see* Figure 1). For each dextran molecular size, RBC aggregation and dextran concentration have a bell-shaped relationship, i.e., with progressive increases in dextran concentration, there is first an aggregation phase to reach a peak aggregation and then a disaggregation phase at high dextran concentrations. The cell aggregation curves in different dextran fractions resemble one another more closely when plotted against the weight concentration of dextrans (e.g. in grams per deciliter). For all fractions studied the maximum aggregation occurs at a dextran concentration of approximately 4 g/dL (*see* Figure 2). As

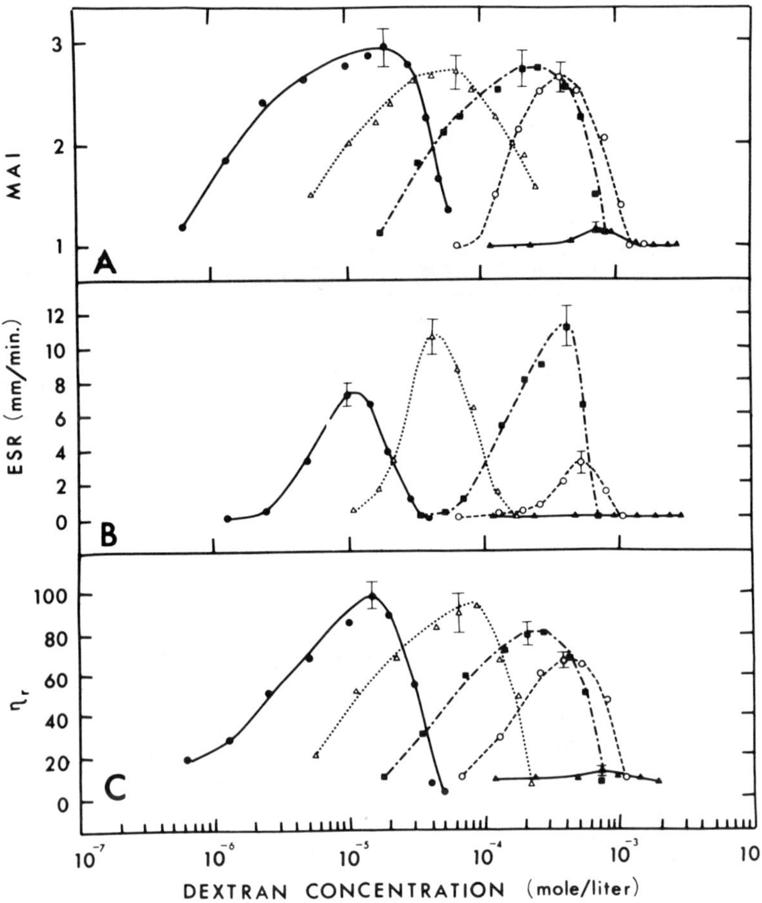

Figure 1. Effects of the concentration of dextrans with various molecular weights on three indices of RBC aggregation (16). MAI indicates the average number of RBCs in each aggregation unit counted under the microscope. ESR is the maximum rate of sedimentation of erythrocytes in a calibrated tube, with corrections made for changes in viscosity and density of the suspending medium following the addition of dextrans. The relative viscosity (η_r) is the ratio of the viscosity of RBC suspension to that of the suspending medium at a shear rate fo 0.1 sec^{-1}. The RBC concentration of the suspension was 1% for MAI and 45% for ESR and η_r measurements. The vertical bars represent SEM. (▲), Dx 40; (○), Dx 80; (■), Dx 150; (△), Dx 500; (●), Dx 2000.

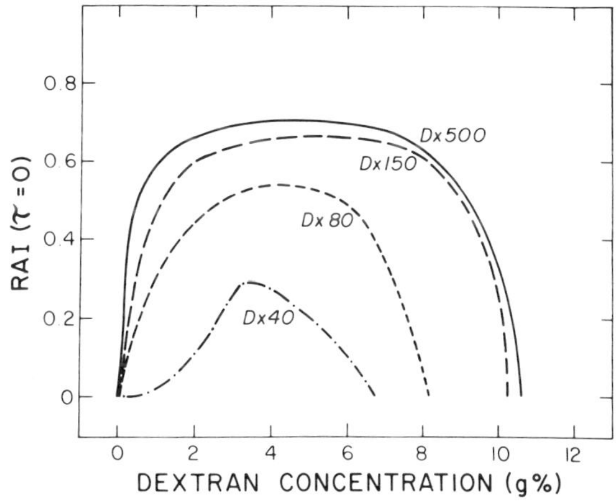

*Figure 2. Variation in RAI with the concentration of dextrans with different molecular weights. RAI is calculated as one minus the ratio of light reflection reading at stationary state to that at a shear rate of 200 sec^{-1}. Cell concentration = 45%. I = 150m*M.

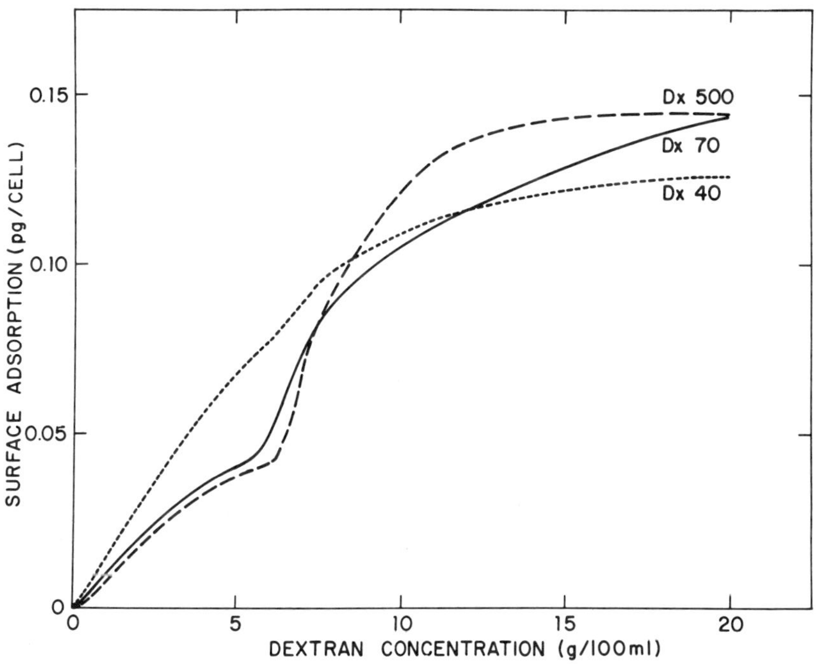

Figure 3. Adsorption isotherms of Dx 40, Dx 70, and Dx 500 on the surfaces of normal RBCs. Abscissa is bulk concentration.

shall be discussed later, the disaggregation phase in high dextran concentrations results from an enhancement of electrostatic repulsion.

Adsorption of Dextrans with Various Molecular Weights on the Surface of Normal RBC Membranes. The adsorption isotherms of dextrans on normal RBC surfaces have been determined with ^{3}H-labeled dextrans (*13*) and unlabeled dextrans (*14*). In the latter study, ^{14}C-carboxyldextran (mean molecular weight 16,000) was used as a tracer to correct for the trapped fluid following centrifugal packing of RBCs. This correction of fluid trapping amounted to less than 1% of the packed-cell volume and agreed with electron microscopic observation of packed cells. Therefore, this negatively charged low-molecular-weight dextran apparently does not have significant adsorption to RBC surfaces. The results obtained on Dx 40, Dx 70, and Dx 500 are shown in Figure 3.

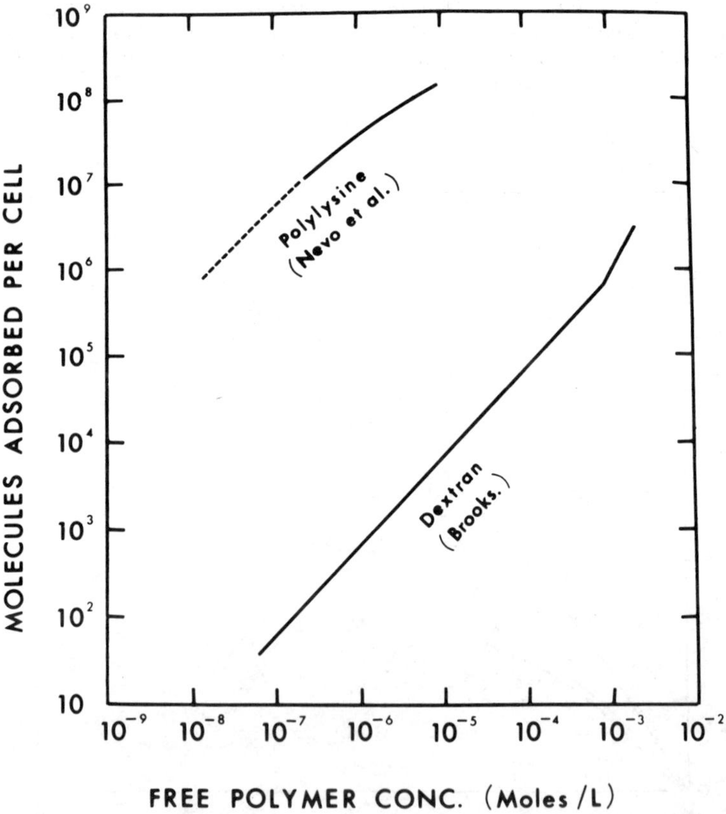

Bibliotheca Anatomica

Figure 4. Adsorption isotherms for polylysine (polymer containing 36 lysine molecules, data from Ref. 15*) and dextran (molecular weight 77,600, data from Ref.* 13*) on human RBC surfaces* (3)

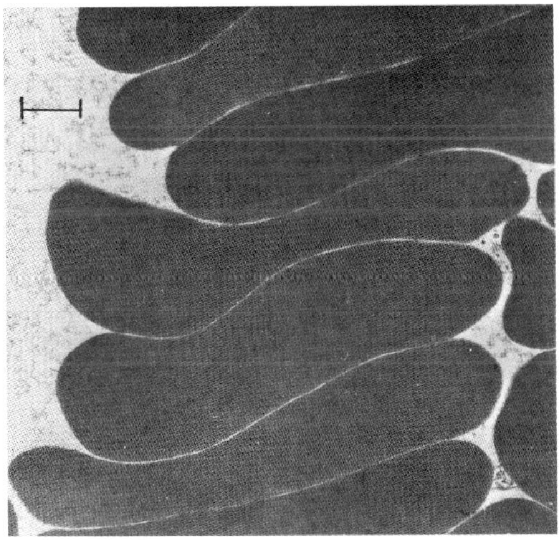

Journal of Supramolecular Structure

Figure 5. Transmission electron micrograph of rouleaux of RBCs in plasma (33). Note the parallel cell surfaces and relatively uniform d. *Length of marker = 1 μm.*

The surface adsorption of all three dextran fractions on normal RBCs rises with the bulk concentration. Over a bulk dextran concentration range of up to 5–6 g/dL, the surface concentrations (in picograms/cell) are similar for Dx 70 and Dx 500, but that for Dx 40 is higher. As the bulk dextran concentration is raised beyond 6 g %, the surface concentration curve for Dx 40 follows an essentially smooth curve, but the curves for Dx 70 and Dx 500 rise sharply to become similar to that for Dx 40.

The adsorption isotherm has been determined also for the positively charged macromolecules, polylysines, on RBC surfaces (*15*). At a given molar concentration in the bulk, the number of polylysine molecules adsorbed per cell is several orders of magnitude higher than that of dextran (*see* Figure 4).

Ultrastructural Studies on Intercellular Distance in Rouleaux Formed in Dextrans with Various Molecular Weights and Concentrations. Electron microscopic studies on RBC rouleaux formed in dextrans with various molecular sizes (*2, 16*) have shown that the surfaces of the adjacent cells in the aggregates are essentially parallel to each other and that the intercellular distance is rather uniform (*see* Figure 5). This intercellular distance, which ranges from 19 to 32 nm, increases with the increasing molecular size of the dextran used to induce RBC aggregation (*see* Figure 6). For a given dextran fraction, this distance does not vary with the dextran concentration (*see* Figure 7) . For each molecular size

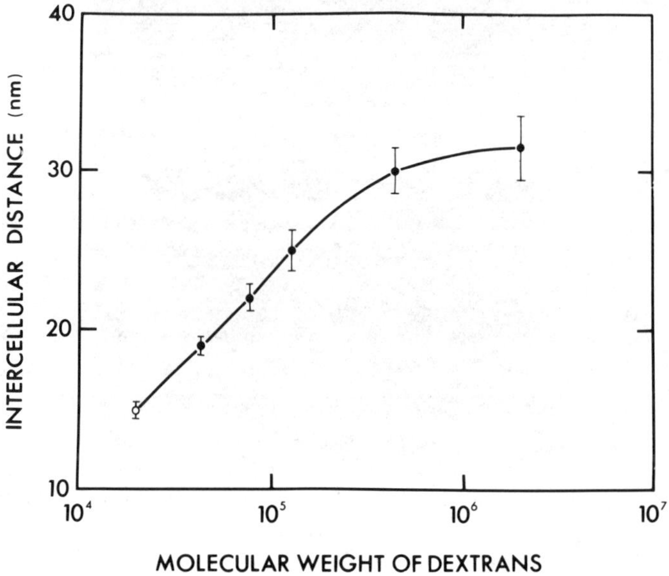

Journal of Supramolecular Structure

Figure 6. Variation of d *with the molecular weight of dextrans (33). The data on Dx 20 (○) were obtained for neuraminidase-treated RBCs, and the results on larger dextrans were the same for normal and neuraminidase-treated cells. Vertical bars denote SEM.*

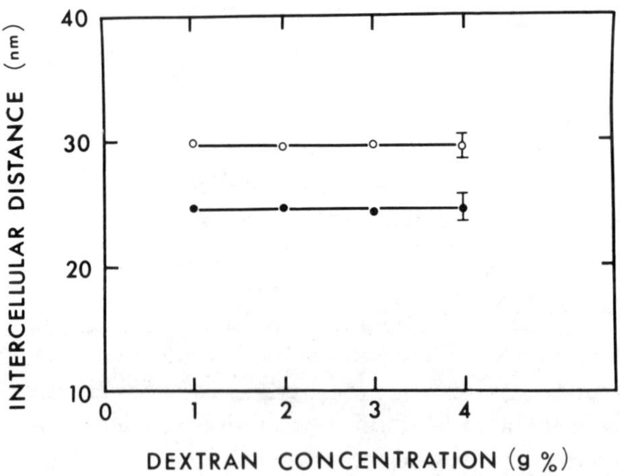

Microvascular Research

Figure 7. The d *of RBC rouleaux formed in Dx 150 and Dx 500 as a function of dextran concentration: (○), Dx 500; (●), Dx 150 (16).*

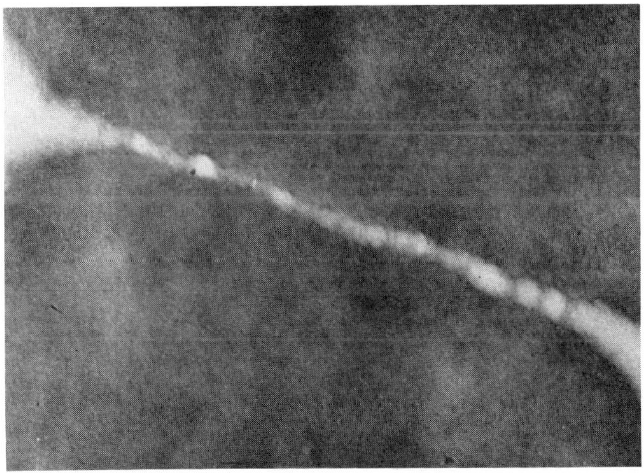

Biochimica et Biophysica Acta

Figure 8. Transmission electron micrograph of aggregates of RBCs in polylysine (19). Note the nonuniform intercellular spacing which contains strands of polylysine molecules (70,000× magnification).

of dextran, the intercellular distance (*see* Figure 6) is considerably shorter than the available information on estimated dextran molecular length (*17, 18*). These findings suggest that significant portions of the dextran molecules are adsorbed on the adjacent cell surfaces, leaving a central portion forming the intercellular distance. It should be pointed out, however, that the absolute value of the intercellular distance may be subject to errors owing to artifacts introduced during preparation for electron microscopic examination and that the estimation of molecular dimensions also involves considerable uncertainties.

Katchalsky et al. (*19*) performed electron microscopic studies on RBC aggregation induced by polylysines with different molecular sizes (*see* Figure 8). The intercellular distance was more irregular than that found in RBC aggregates induced by dextrans or fibrinogen. Nevertheless, the intercellular distance varied directly with, and was shorter than, the molecular length of the polylysines used. Electron-dense strands of polylysines were observed nearly perpendicular to the wavy cell surfaces (Dextrans are not electron dense and hence not visible by electron microscopy.) Studies using purified fibrinogen show that the intercellular distance is approximately 25 nm (*20*) which is again shorter than the molecular length of fibrinogen (*21*).

Electrostatic Repulsive Energy Between Red Cell Surfaces

The surfaces of human RBCs are negatively charged, mainly owing to the carboxyl group of the *N*-acetylneuraminic acid residue of the membrane glycoprotein (*22, 23*). Measurements of electrophoretic mobility show that the normal human RBCs in 0.9% NaCl solution (ionic strength = 150m*M*) have a zeta potential (ζ) averaging − 15 mV (*24, 25*). According to models developed in colloid chemistry (*26*), the electrostatic repulsive pressure between adjacent RBC surfaces can be calculated from the surface potential (estimated by ζ in millivolts), cationic valency (z), ionic strength (I, in moles per liter) and the intercellular distance (d). Since electron microscopic studies have shown that the adjacent RBCs in the rouleaux are parallel to each other (*16*), the geometry of the system is essentially that of parallel plates with uniform separation. Hence, the electrostatic repulsive pressure (σ_e, in dynes per centimeters squared) can be calculated as

$$\sigma_e = 0.064\, IRT \tanh^2 (za\zeta/4kT)\, e^{-\kappa d} \qquad (1)$$

where R is the gas constant , T is the absolute temperature, the function $\tanh(x)$ is equal to $(e^x - e^{-x})/(e^x + e^{-x})$, where e is the base of the natural logarithm, a is the electronic charge, k is the Boltzmann constant, and κ is the Debye–Hückel function or the slope of the potential drop in the double layer (i.e. the reciprocal of the double-layer thickness)

$$\kappa = (8\Pi a^2 N I z^2/1000\epsilon kT)^{1/2} \qquad (2)$$

where N is Avogadro's number and ϵ is the dielectric constant.

Effects of Alterations in Ionic Composition on Electrostatic Repulsion and Cell Aggregation. According to Equation 2, κ varies directly with the square root of I. Therefore a reduced I causes κ to decrease and ζ to increase, leading to an increase in σ_e and a reduction in RBC aggregation (*see* Figure 9A) (*27*).

Partial replacement of the monovalent Na^+ with the divalent cations Ca^{++}, Mg^{++}, or Ba^{++} causes κ to increase and ζ to decrease, leading to a reduction in σ_e and an enhancement of RBC aggregation (*see* Figure 10A) (*27*).

Effects of Reduction of Surface Charge Density on Cell Aggregation. The *N*-acetylneuraminic acid in the RBC membrane can be removed by incubating RBCs with neuraminidase. Incubation of a dilute human RBC suspension (1% by volume) at 37°C for 1 hr with 15 μg/mL of neuraminidase causes the complete removal of *N*-acetylneuraminic acid and a reduction of surface potential by 90% (*24*). The residual negative surface charges can be attributed to the phosphate groups of the membrane phospholipids or the carboxyl groups of the membrane proteins (*28*).

AGGREGATION OF NEURAMINIDASE-TREATED RED CELLS BY DEXTRANS. Following the depletion of RBC surface charge by neuraminidase treatment, the cells can be aggregated by Dx 20 with concentrations above

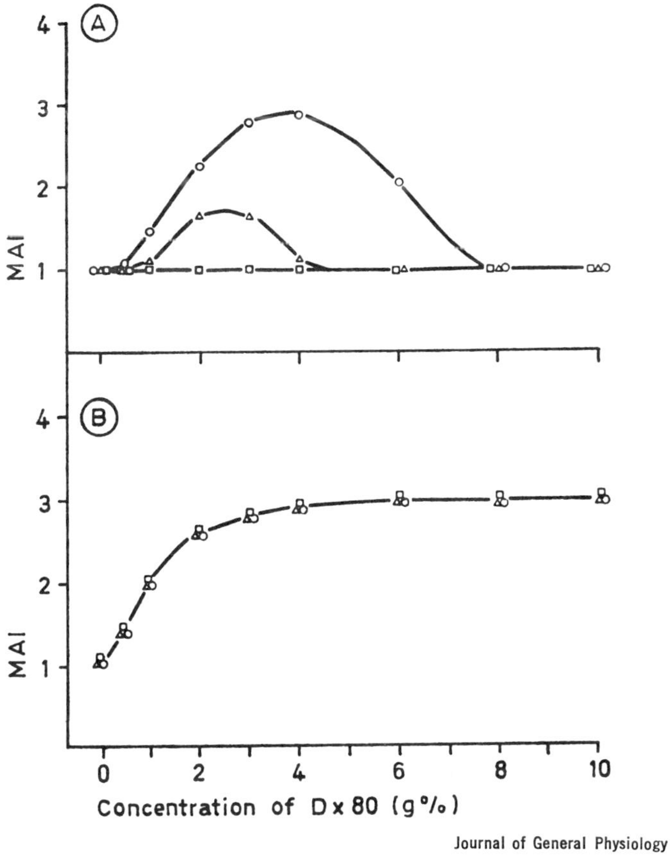

Figure 9. Effects of I *on the MAI of normal RBCs (A) and neuraminidase-treated RBCs (B) in NaCl solutions plus Dx 80 (27).* I *was varied while maintaining the total osmolality (= 300* mOsm*) by using sucrose: (○),* I = *150m*M*; (△),* I = *100m*M*; (□),* I = *100m*M.

3 g/dL (*see* Figure 11A) (*24*). The aggregation phase is enhanced greatly for Dx 40 and to a lesser degree with the larger-molecular-weight dextran fractions. Furthermore, there is no disaggregation phase for the neuraminidase-treated RBCs in all dextran fractions. These results indicate that electrostatic repulsion owing to the presence of normal cell surface charges plays a significant role in preventing RBC aggregation by low-molecular-weight dextrans and in causing RBC disaggregation by high concentrations of dextrans.

SURFACE ADSORPTION OF DEXTRANS ON NEURAMINIDASE-TREATED RBCS. The surface adsorption curves for Dx 40, Dx 70, and Dx 500 on neuraminidase-treated RBCs are closely similar (*see* Figure 12). With bulk dextran cancentrations below 5–6 g/dL, these curves on neuraminidase-

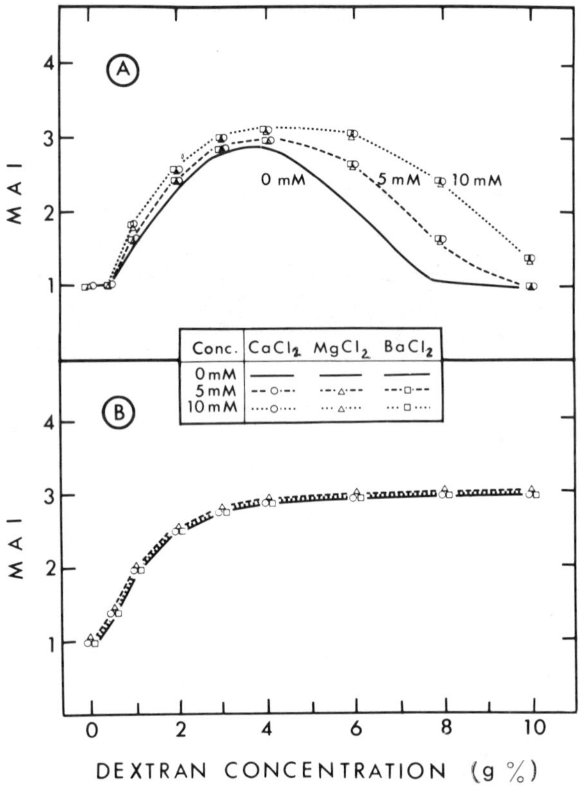

Journal of General Physiology

Figure 10. Effects of replacement of Na^+ with various divalent cations on the MAI of normal RBCs (A) and neuraminidase-treated RBCs (B) in Dx 80 (27)

treated RBCs are similar to those attained for Dx 70 and Dx 500 on normal RBCs. At high-bulk concentrations, the adsorption curves for all dextran fractions on neuraminidase-treated RBCs attain a lower plateau (approximately 0.10×10^{-12} g/cell) than those on normal RBCs (approximately 0.15×10^{-12} g/cell).

ULTRASTRUCTURAL STUDIES ON INTERCELLULAR DISTANCE IN ROULEAUX OF NEURAMINIDASE-TREATED RBCS. Electron microscopic studies on rouleaux of neuraminidase-treated RBCs indicate that their intercellular distance in each dextran fraction was not significantly different from that of normal RBCs as shown in Figure 7.

Effects of Dextrans on Electrostatic Repulsion. In the absence of externally applied forces, the net aggregation energy (E_a) in a rouleau is equal to E_b minus electrostatic repulsive energy (E_e). The aggregation behavior of each cell–dextran system reflects the E_a. Since electrostatic repulsion is negligible for neuraminidase-treated RBCs, their aggregation behavior reflects E_b. It follows that the difference in aggregation behaviors between neuraminidase-treated RBCs and normal RBCs (*see* Figure 11) gives an indication of the E_e in the normal RBC system. As shown in Figure 13, such an index of E_e calculated from aggregation measurements shows that it rises steeply when Dx 70 concentration is raised beyond 4–5 g/dL.

The ζ of normal RBCs suspended in dextran solutions can be determined from measurements of electrophoretic mobility (u), fluid viscosity (η), and the dielectric constant (ϵ) (*29*).

$$\zeta = 360\ \pi\eta u/\epsilon \tag{3}$$

In the absence of dextran, the ζ of normal RBCs is -15 mV, which corresponds to a charge density of 4×10^3 esu/cm^2. This is approximately one-half of the charge density calculated from the determination of sialic acid on normal RBC surfaces (*30*). The addition of dextrans to distilled water causes no significant change in ϵ (*13, 24*) as determined by an ac current bridge method (*31*). The viscosity of the saline solution increases progressively with a rise in dextran concentration. The electrophoretic mobility of normal RBCs decreases with rising dextran concentration, but proportionately less than the viscosity effect. Therefore, the calculated ζ (*see* Equation 3) increases progressively with rising dextran concentration (*see* Figure 14A). This increase in ζ is pronounced more as the molecular weight of dextrans is increased (*see* Figure 14B).

Brooks (*32*) has explained the increase in ζ of normal RBCs in the presence of the neutral polymer dextran on the basis of a reduction in counterion screening. That is, counterions are excluded from the cell surface where a significant fluid volume is occupied by the dextran, thus leading to a reduction in the effective I and a decrease in κ. Thus, Equation 1 can be written as (*33, 34*)

$$\sigma_e = 0.064\ I_\beta RT \tanh^2\ (za\zeta_\beta/4kT)\ e^{-\kappa_\beta d} \tag{4}$$

where the subscript β indicates the presence of dextran and the subscript o indicates the absence of dextran.

$$\kappa_\beta = \kappa_o\zeta_o/\zeta_\beta \tag{5}$$

$$I_\beta = I_o(\zeta_o/\zeta_\beta)^2 \tag{6}$$

By determining ζ_o and ζ_β, calculating κ_o from Equation 2, and using the known value of I_o, one can obtain κ_β and I_β from Equations 5 and 6,

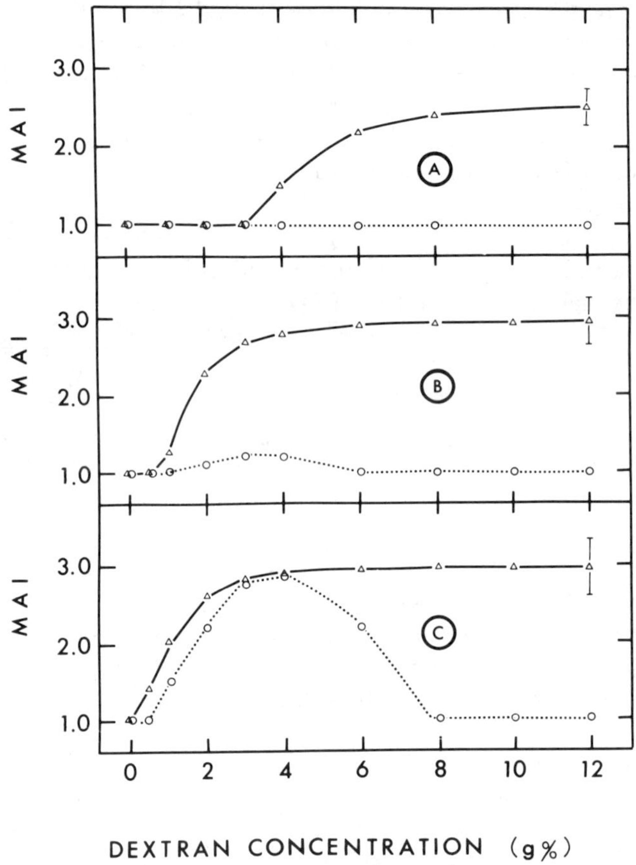

Journal of General Physiology

Figure 11. Effects of dextrans on aggregation of normal RBCs (○) and neuraminidase-treated RBCs (△) (24). A, B, C show results in Dextrans 20, 40, and 80, respectively. Note that neuraminidase-treated cells exhibit stronger aggregation than normal cells and do not show the characteristic disaggregation phase. The vertical bars represent SEM. I = *150m*M.

respectively. Substitution of these values into Equation 4 allows the computation of the electrostatic repulsive pressure as a function of the distance of separation between the surfaces of RBCs aggregated by dextran bridging. Figure 15A shows such curves for normal RBCs in 150m*M* NaCl solution containing different concentrations of Dx 70. An increase in Dx 70 concentration causes an upward shift of the curve. Similar calculations for a reduced ionic strength (I_o = 50m*M* NaCl) are plotted in Figure 15B. At each Dx 70 concentration, the σ_e curve for I_o = 50m*M* is higher than that for I_o = 150m*M*, showing the synergistic effects of dextran and low I in raising σ_e. Since electron microscopic studies have shown that the intercellular distance of rouleaux of RBCs in Dx 70 averages 22 nm (*16*), the σ_e values at $d = 22$ nm (vertical dotted lines in Figures 15A and 15B) are taken as the electrostatic repulsive pressure opposing the aggregation of normal RBCs in the presence of various concentrations of Dx 70. The relationship between σ_e (at $d = 22$ nm) and Dx 70 is plotted in Figure 16. The electrostatic repulsive energy per unit area of interacting cell surface (E_e/A_i) is equal to the integral of σ_e with respect to d as the cell surfaces move from an infinite separation to $d = 22$ nm. Since σ_e is an exponential function of d, the relation between E_e and d would follow the same shape as the σ_e–d curve shown in Figure 16 and agree with the E_e–d relationship deduced from aggregation measurements (*see* Figure 13).

Another possible explanation of the increase in electrostatic repulsive pressure with the addition of dextran is that the adsorption of dextran on the RBC surface may cause a displacement of the sialic acid groups away from the cell surface. Thus, it is possible that the extramembrane portion of glycoproteins with the attached sialic acid groups normally droops over the cell surface and that it becomes progressively straightened

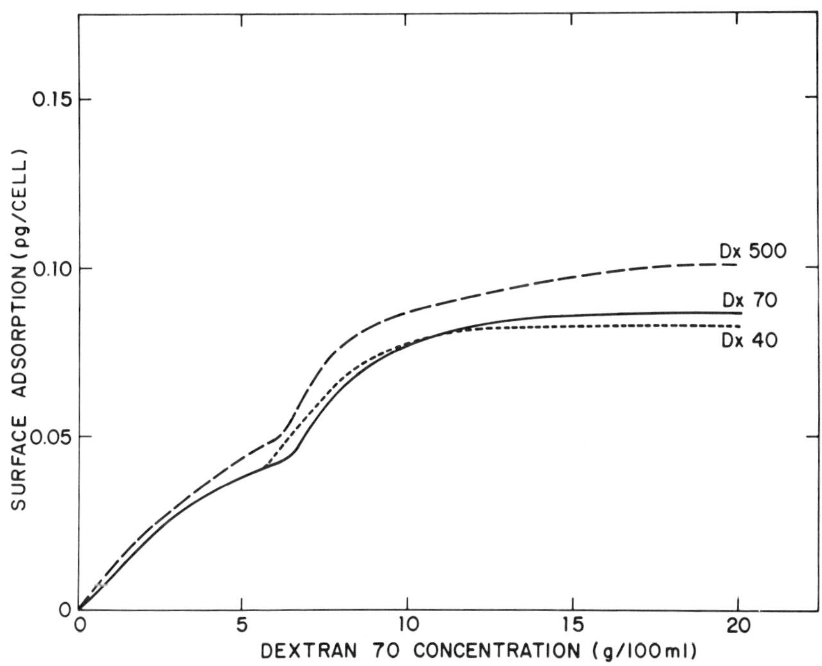

Figure 12. Adsorption isotherms of Dx 40, Dx 70, and Dx 500 on the surfaces of neuraminidase-treated RBCs. Abscissa is bulk concentration.

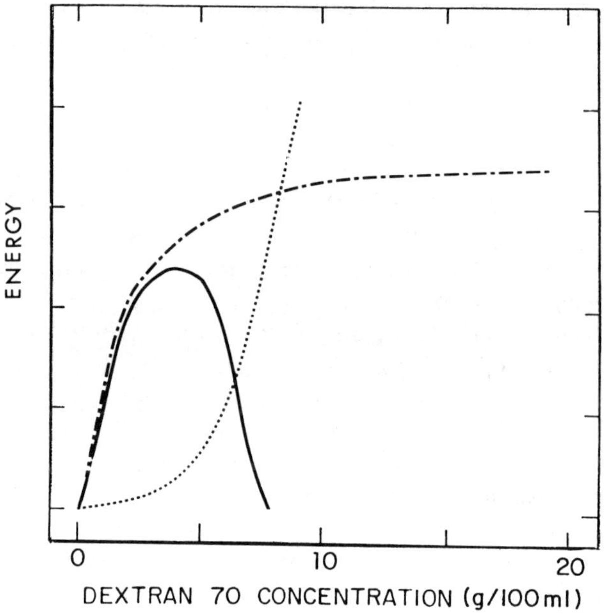

Figure 13. E_a *for normal RBCs* ($E_a = E_b - E_r$) *and for neuraminidase-treated RBCs* ($E'_A = E_B$) *as a function of Dx 70 concentration.* $E'_A - E_A$ *gives an estimation of* E_e *of normal RBCs.* (· · ·), $E_e = E'_a - E_a$; (– · –), E'_a (N-*treated RBC*); (——), E_a (*normal RBC*).

and extended as increasing areas of the cell surface are occupied by the adsorbed dextran. As a result, although *d* remains unchanged with increasing dextran concentration, the distance between the planes of negative surface charges (d_β) may be shortened to cause an enhancement of E_e (*see* Figure 17). Using the σ_e profile derived from Equation 1, one can calculate the narrowing of d_β at various Dx 70 concentrations in order to achieve the same rise in σ_e as shown in Figure 16. Such estimation of d_β indicates that it would decrease from a value of 22 nm toward a low plateau of 6–8 nm at high Dx 70 concentrations. This maximum decrease in d_β by 14–16 nm would correspond to a maximum extension of the glycoprotein molecules from each cell surface by approximately 7–8 nm.

The Aggregating Energy

Evidence for Macromolecular Bridging. Our current concept of the mechanism of RBC aggregation is that it results from macromolecular bridging (2) (*see* Figure 18). In this model the macromolecules in the suspending medium are adsorbed to the RBC membrane surface. When RBCs are brought into close proximity of each other by thermal motion, mechanical agitation, or gravitational sedimentation, the same macromolecule then can be absorbed simultaneously to two adjacent cell surfaces, resulting in the formation of rouleaux. This model is based on several lines of evidence discussed above. First, RBC aggregation does not occur in the absence of the appropriate macromolecules (*1*). Second, electron microscopic studies have shown an intercellular distance between adjacent RBC surfaces in rouleaux, and this distance varies directly with the macromolecular size used to induce rouleau formation (*16, 19*). Third, dextran (*13, 14*) and other macromolecules which can cause RBC aggregation, e.g. fibrinogen (*35, 36*), are adsorbed to the surface of RBCs. Finally, the relationship between the adsorption isotherms of various

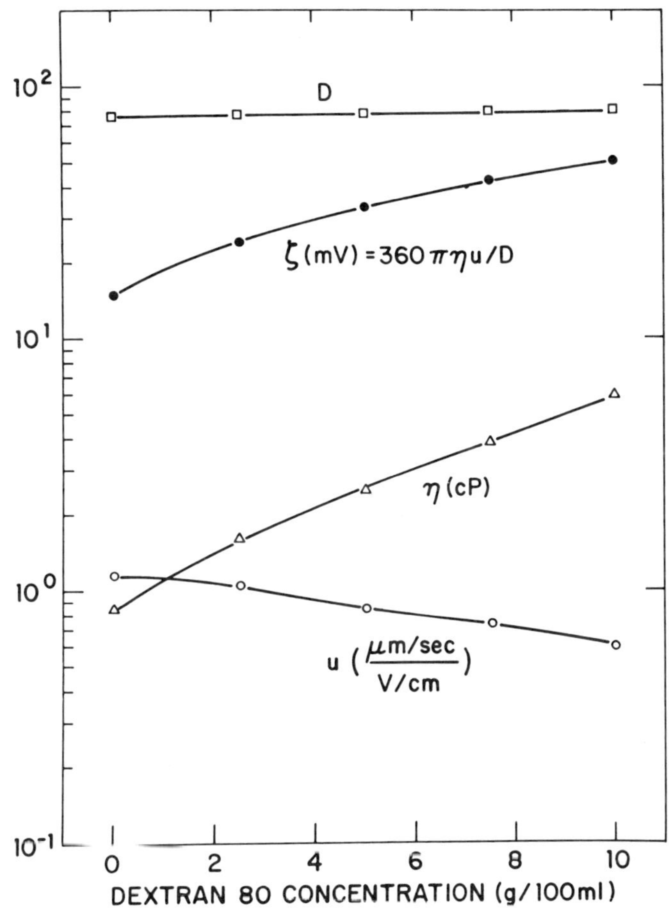

Thrombosis Research Supplement II

Figure 14A. Semilogarithmic plots of D *(same as* ε*),* ζ*,* η*, and* μ *of normal RBCs as a function of Dx 80 concentration* (34). I = *150* mM.

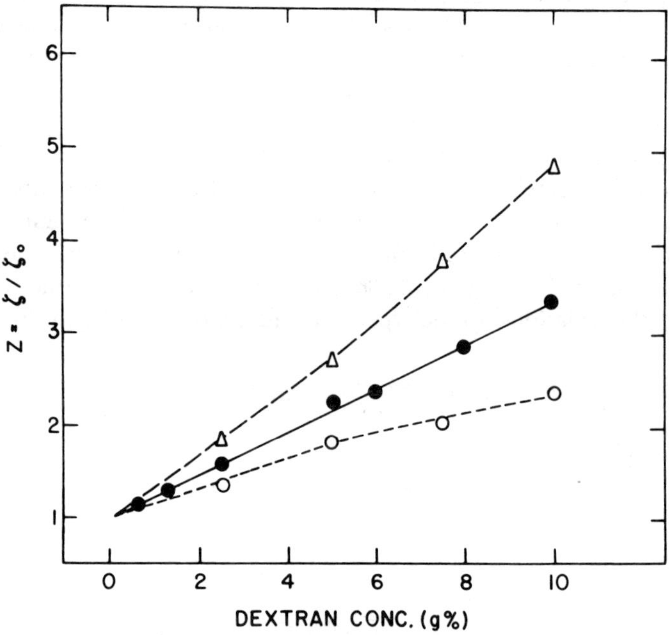

Figure 14B. Effect of dextrans on ζ of normal RBCs in 150mM NaCl solution. The term $\zeta_o = \zeta$ in the dextran-free NaCl solution. Z $(= \zeta/\zeta_o)$ is plotted against the concentration of dextrans with three different molecular weight. (○), Dx 40; (●), Dx 80; (△), Dx 150.

dextran fractions (*see* Figures 3 and 12) and the aggregation behavior of normal and neuraminidase-treated RBCs in these dextran fractions (*see* Figure 11) can be explained best on the basis of macromolecular bridging, as detailed below.

Adsorption isotherms have been obtained for three different fractions of dextrans (Dx 500, Dx 70, and Dx 40) on two types of cells (normal and neuraminidase treated). Over a bulk dextran concentration range up to 5–6 g/dL, the surface adsorption is similar among five of the six systems studied, but the curve is higher for the adsorption of Dx 40 on normal RBCs, which is the only system showing very little aggregation in this range. As the bulk dextran concentration is raised above 5–6 g/dL, normal RBCs become progressively disaggregated in all dextran fractions, whereas the aggregation of neuraminidase-treated cells continues (*see* Figure 11). One possible explanation for the findings of greater adsorption in systems showing reduced aggregation is that disaggregation makes a great surface area for dextran adsorption available. In other words, aggregation of RBCs by macromolecular bridging results in the sharing of the same dextran molecule by two adjacent cells, thus reducing the number of dextran molecules that are adsorbed. This possibility has been tested by geometric considerations (*14*) that are illustrated in Figure 19. When n RBCs are stacked to form an aggregate, there are $(n - 1)$ of such intercellular double surfaces where dextran molecules are shared between adjacent cells, thus reducing cell surface area available for dextran adsorption. The fraction of cell surface area that is available for dextran adsorption (A_a/A_c) in an n-cell aggregate is given as

$$\frac{A_a}{A_c} = 1 - \frac{A_i}{A_c}\frac{n-1}{n} \tag{7}$$

where A_i is the intercellular surface area located inside the aggregate between two adjacent cells and A_c is the total surface area of each cell. Using the values of $A_i = 50\ \mu m^2$ and $A_c = 140\ \mu m^2$, one finds that when the aggregates consist of eight cells or more (up to $n = \infty$), the fraction of A_a would be reduced to approximately two-thirds of that in the dispersed, nonaggregated state. This result obtained from cell geometric

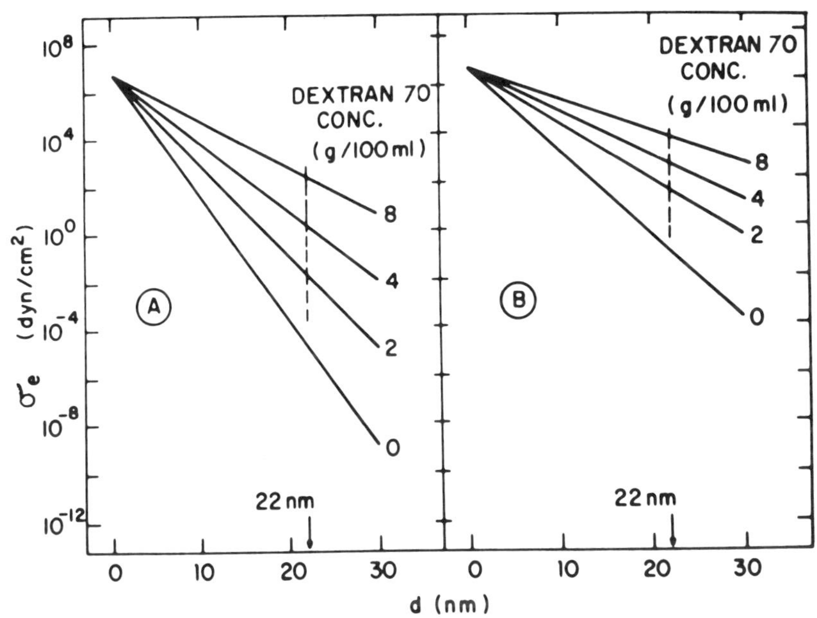

Thrombosis Research Supplement II

Figure 15. Semilogarithmic plot of σ_e from Equation 4 as a function of d *between parallel surfaces of normal RBCs (34). (A),* $I_o = 150$*mM; (B),* $I_o = 50$*mM. For each I, the results for four suspending media are shown with Dx 70 concentrations equal to 0, 2, 4, and 2 g %. The vertical dotted lines are drawn at* d = 22 *nm, which corresponds to the distance of separation between adjacent cells in rouleaux formed in Dx 70.*

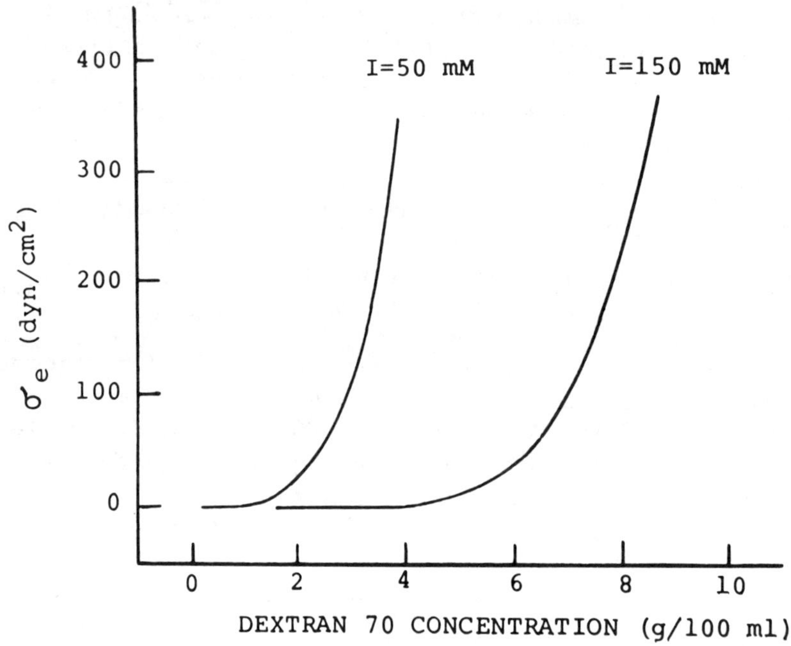

Figure 16. The term σ_e as a function of Dx 70 (same as Dx 80) concentration for $I_o = 50$ and 100mM.

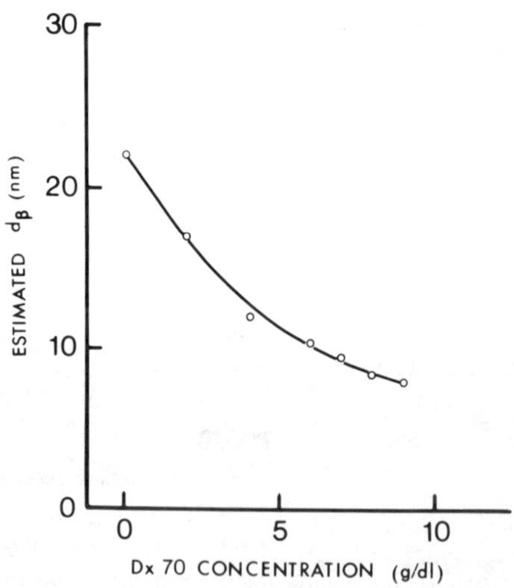

Figure 17. Estimated d_β between charged planes of adjacent RBCs aggregated by Dx 70 at various concentrations.

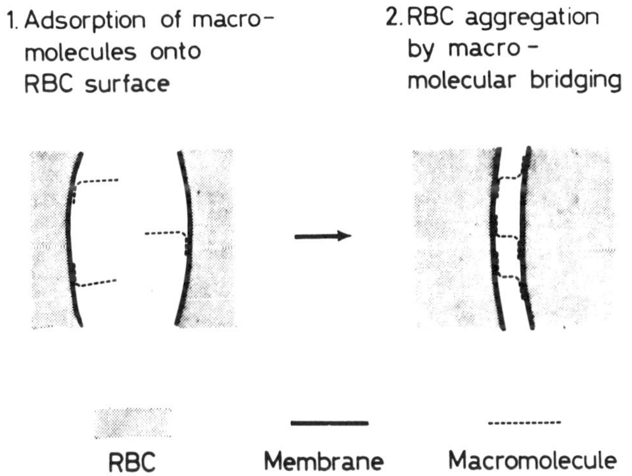

6th European Conference on Microcirculation

Figure 18. Diagram showing the mechanism of RBC aggregation by macromolecular bridging (2)

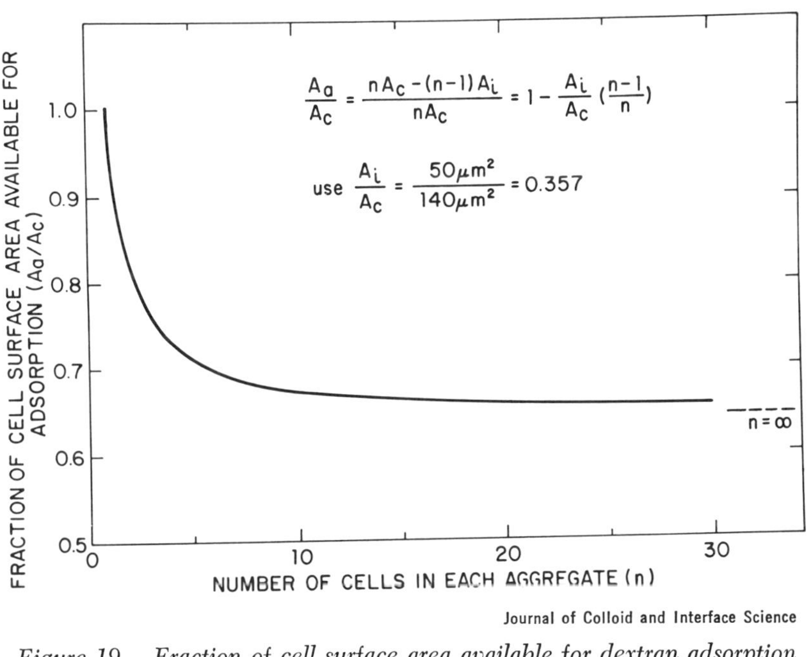

Journal of Colloid and Interface Science

Figure 19. Fraction of cell surface area available for dextran adsorption plotted as a function of the number of cells in each aggregate (14). Note that A_c/A_c is approximately equal to two-thirds for n = 8 to ∞.

considerations is in excellent agreement with the experimental data that show that the plateau values for the adsorption of all fractions of dextrans on the aggregated, neuraminidase-treated RBCs (*see* Figure 12) are approximately two-thirds of those on the disaggregated, normal RBCs (*see* Figure 3). Therefore, the reduction of surface adsorption in an aggregated system can be explained quantitatively on the basis of a decrease in available adsorption sites due to the sharing of dextran molecules by adjacent cell surfaces. The above findings on the adsorption of three types of dextrans on normal and neuraminidase-treated RBCs (*14*) provide quantitative evidence in upport of the concept of macromolecular bridging.

The Macromolecular Bridging Energy. Based on the model of cell aggregation by macromolecular bridging, E_b for a cell pair is a function of the number of bridging macromolecules between these cells (m), the number of bonds with cell surface per bridging molecule (b_m), and the interaction energy per bond (e_b)

$$E_b = (m)\ (b_m)\ (e_b) \tag{8}$$

When the bulk concentration of the bridging macromolecule (e.g. dextran) is raised, the increase in surface adsorption leads to an increase in m. By using higher-molecular-weight dextrans, a comparison of the molecular size and the intercellular distance suggests that a larger fraction of the dextran molecule is attached to the cell surface, thus increasing b_m. Therefore, by increasing m and b_m, respectively, increases in macromolecular concentration and molecular size can raise E_b.

The e_b depends on the physicochemical properties of the bridging macromolecules and the cell surface. The adsorption of macromolecules on solid surfaces can occur by several types of attractive forces (*37*) including covalent bonding, electrostatic attraction, hydrogen bonding, and van der Waals (VDW) attraction. Hydrogen or VDW bonding energy is of the order of 10 kcal/mol, and that of a covalent or electrostatic nature is much higher, being of the order of 100 kcal/mol or more (*38*).

Agglutination of RBCs by polycations (e.g. polylysines) involves electrostatic attraction between the positively charged lysine groups of the polylysine molecule and the negatively charged *N*-acetylneuraminic acid on the RBC surface. Reduction of RBC surface charge by neuraminidase treatment markedly decreases the effectiveness of RBC agglutination by polylysine.

The nature of the binding between dextran and the RBC surface has not been established, but it probably involves the weaker energy of hydrogen bonding or VDW attraction. Since the dextran-induced RBC aggregation can be inhibited by using urea (*39*), hydrogen bonding probably plays a role. Liposomes prepared from lecithin and other lipid components of the RBC membrane are aggregated by dextrans, and the liposome aggregation shows the same variations with dextran molecular weight and concentration as RBC aggregation (*40*). Therefore, dextrans probably interact with the lipid components of the RBC membrane. The difference in the adsorption isotherms of polylysines (*15*) and dextrans (*14*) on RBC surfaces probably reflects the much stronger electrostatic attraction energy between polylysine and the surface negative charge in comparison with the hydrogen bonding of dextrans to membrane components. Plasma proteins (e.g. albumin and fibrinogen) are related more closely to dextran in their interaction with RBC membrane surface. Therefore, studies on RBC aggregation by dextrans with various molecular sizes serve to elucidate the mechanism of fibrinogen or globulin-induced rouleau formation of RBCs, which is observed in the living circulation under low-flow conditions. The agglutination of RBCs by blood group antibodies, which occurs in transfusion reactions, and by lectins involves bonding energy with a magnitude between that for dextrans and for the polycations (unpublished observations). The binding of cells in most tissues probably involves bonding energy even greater than that found in the erythrocyte–polycation system. In the desmosomes between cell

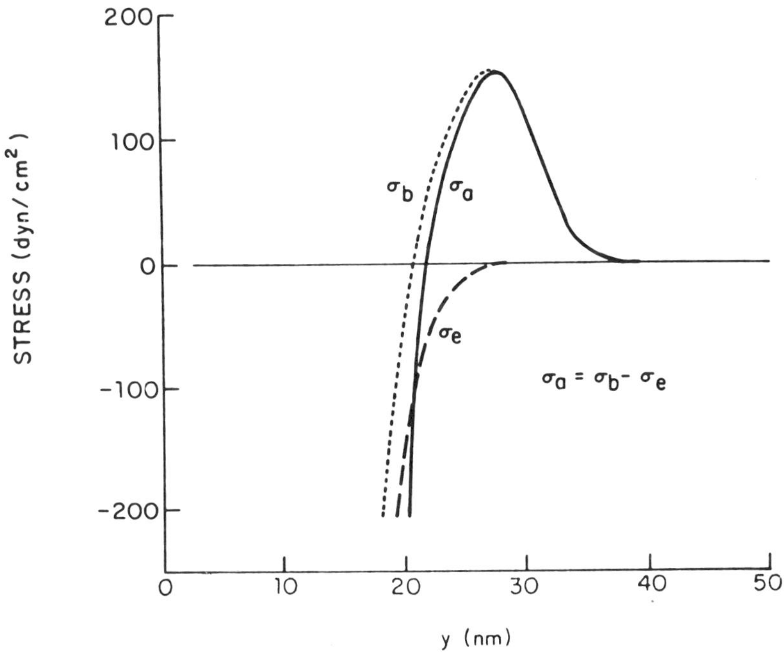

Figure 20A. The σ_b and σ_e between two RBC surfaces as a function of the surface separation in the y direction. The system is normal RBCs suspended in 6 g/dL Dx 70. The σ_a is equal to $\sigma_b - \sigma_e$.

surfaces, the macromolecular strands that are visible under electron microscopy may have strong interaction forces with cell membrane surfaces and provide the energy for stable cell adhesion.

Balance Between Macromolecular Bridging Energy and Electrostatic Repulsive Energy. In the absence of externally applied forces, the net aggregation force per unit cell surface area (σ_a) is equal to the difference between the macromolecular bridging force per unit area (σ_b) and the electrostatic repulsive force per unit area (σ_e)

$$\sigma_a = \sigma_b - \sigma_e \tag{9}$$

In Figure 20A these forces per unit area are plotted as functions of the distance of separation between adjacent cells in the direction (y) perpendicular to the cell surfaces. As mentioned above, σ_e has an exponential relationship with y (*see* Figure 20A). On the other hand, σ_b, is assumed to have a maximum value at some y value. This occurs because the macromolecule would be ineffective in bridging at very large y values and because the macromolecule may impart a repulsive force owing to either molecular rigidity or intermolecular repulsion when y becomes too small. The potentials (or energies) per unit cell surface area owing to macromolecular bridging (ϕ_b) and electrostatic repulsion (ϕ_e) can be obtained by integrating these corresponding forces from $y = \infty$ to $y = d$

$$\phi_b = \int_{\infty}^{d} \sigma_b dy \tag{10}$$

$$\phi_e = \int_{\infty}^{d} \sigma_e dy \tag{11}$$

and the net aggregation potential per unit area (ϕ_a) is given as

$$\phi_a = \phi_b - \phi_e \tag{12}$$

The variations in these potentials along the y axis are shown in Figure 20B for the aggregation of normal RBCs in 6 g/dL Dx 70. The σ_b and ϕ_b curves are hypothetical, whereas the σ_e and ϕ_e curves are computed as described earlier. The ϕ_a curve goes through a minimum value of approximately 10^{-4} dyn/cm at $d = 22$ nm. This potential minimum gives an estimation of the E_a per unit interacting area, and the corresponding y value ($y = d$) gives the equilibrium d.

Figure 20B indicates that the equilibrium d at which a potential minimum for ϕ_a is attained is dominated by the shape of the ϕ_b curves.

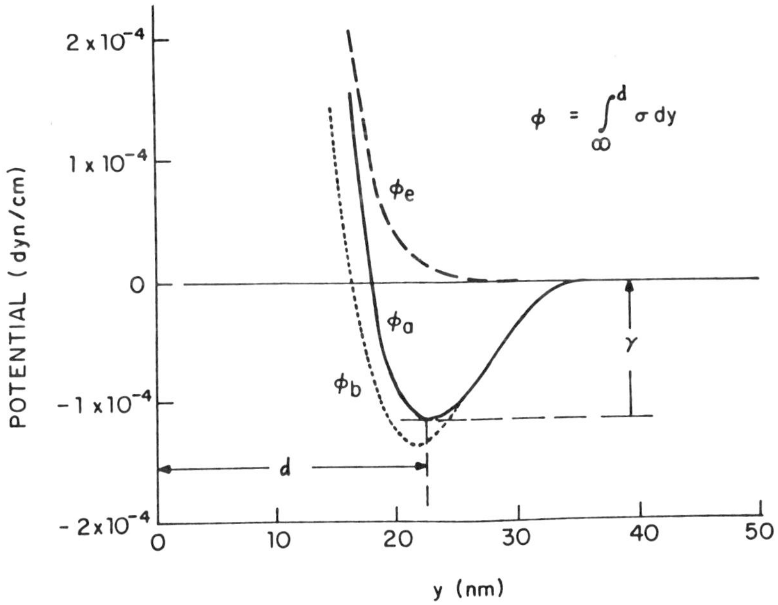

Figure 20B. The ϕ_b, ϕ_e, and $\phi_a = \phi_b - \phi_e$ between RBC surfaces as a function of y. *The potentials are calculated by integrating the corresponding forces from ∞ to* y *and expressed in terms of per unit cell surface area. At* y = d, *Φ_a is at a minimum and $\sigma_a = 0$.*

Thus, even in the absence of ϕ_e (i.e. $\phi_a = \phi_b$), this equilibrium *d* would show very little change although the magnitude of ϕ_a would increase. These are in agreement with the experimental findings that depletion of cell surface charge does not significantly alter *d* as determined by electron microscopy (as discussed earlier) and that it enhances RBC aggregation (*see* Figure 11). When Dx 70 concentration is altered, the magnitude of both ϕ_b and ϕ_e would change accordingly. The ϕ_b curve would shift upward or downward, but without a change in its shape characteristic. Therefore, the equilibrium *d* for the ϕ_a minimum would remain essentially the same. This is in agreement with the electron microscopic finding that *d* for a given dextran fraction does not vary significantly with the dextran concentration (*see* Figure 7). Since ϕ_b and ϕ_e do not vary to the same extent with dextran concentration, the magnitude of ϕ_a varies with dextran concentrations, increasing with concentrations up to approximately 4 g/dL but decreasing at higher concentrations (*see* Figures 2 and 11). An increase in the molecular size of dextran would cause a right shift of the ϕ_b curve. This would explain the variation of the equilibrium *d* with the molecular weight of dextrans as found by electron microscopy (*see* Figure 6).

Possible Role of Long-Range VDW Forces Between Cell Membrane Surfaces. The above discussion has been based on the model of macromolecular bridging, and the E_b is caused by the interaction between the bridging macromolecule and the cell surface. Recent theoretical and experimental studies (*41, 42*) have shown that VDW attractive forces can be exerted between cell surfaces over distances in the range found in RBC rouleaux.

It appears rather unlikely, however, that such long-range VDW forces between cell surfaces can be the primary source of E_b for RBC aggregation. First, RBCs, even after charge depletion by neuraminidase treatment to greatly reduce electrostatic repulsion, do not aggregate unless appropriate macromolecules are present (*see* Figure 11). Second, an increase in dextran concentration up to 4 g/dL causes an increase in E_a. If this energy were VDW in nature, its increase should have caused a reduction in d. Electron microscopic studies, however, indicate that d does not vary with dextran concentration (*see* Figure 7). Third, by the same consideration, an increase in E_a owing to a reduction in E_e should have caused a reduction in d; but d in a given type of dextran is not affected by neuraminidase treatment. Fourth, an increase in dextran molecular weight increases d (*see* Figure 6), and this should reduce the long-range VDW forces between adjacent cell surfaces. Therefore, if VDW forces were the important source of E_b, one would predict a decrease in RBC aggregation with increasing dextran molecular weight, but the opposite is found in experimental observations (*see* Figures 1 and 2).

Although long-range VDW forces do not appear to provide the primary source of E_b, there is the possibility that these forces may contribute additional energy for aggregation. Thus, the bridging of adjacent cell surfaces by macromolecules may provide a sufficiently short and stable d over which the electrodynamic forces can exert a significant attraction, which then would act synergistically with E_b. This possible role of the long-range VDW force existing between adjacent cell surfaces in the rouleaux requires further investigations.

Changes in Strain Energy in RBC Membrane

In order to achieve a stable aggregation in the absence of external forces, E_b must exceed E_e. This E_a is dissipated partially in the process of aggregation and is stored partially as a change in strain energy of the RBC membrane (E_m). As the adjacent cell surfaces become essentially parallel to each other by macromolecular bridging, changes in E_m are induced also in the remainder of the cell, which undergoes deformation at a constant total cell surface area. With the progression of the

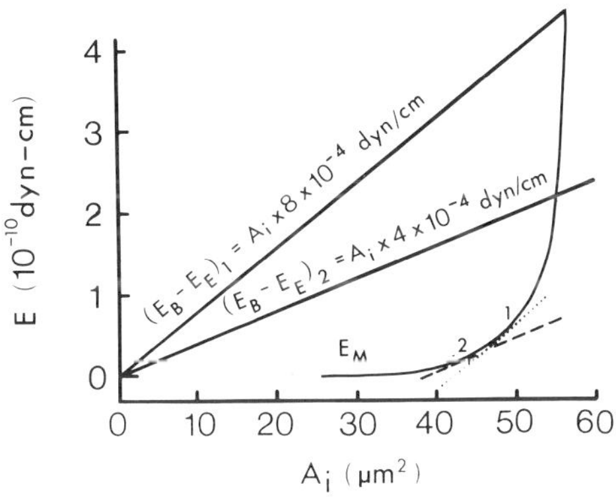

Figure 21A. Variations of E_a $(= E_b - E_e)$ *and* E_m *with* A_i *between two cells in a rouleaux.* $E_b - E_e$ *varies linearly with* A_i*, and two levels are shown* $(= A_i \times 8 \times 10^{-4}$ *and* $A_i \times 4 \times 10^{-4}$ *dyn/cm).* E_m *varies nonlinearly with* A_i*. At equilibrium,* $(E_b - E_e)/A_i = dE_m/dA_i$*. Thus the equilibrium* A_i *values are found from the* E_m *curve at locations where the tangent of the curve is parallel to the* $E_b - E_e$ *line.*

aggregation process, as A_i increases, there is a progressive increase in E_m, which rises steeply as a nonlinear function of A_i (*see* Figure 21A). An equilibrium A_i is attained when the rate of change of E_a is balanced by that of E_m, both of which are with respect to changes in A_i. That is, at equilibrium

$$\frac{dE_m}{dA_i} = \frac{d(E_b - E_e)}{dA_i} \qquad (13)$$

Since both E_b and E_e vary linearly with A_i, Equation 13 can be written also as

$$\frac{dE_m}{dA_i} = \frac{E_b - E_e}{A_i} \qquad (14)$$

Since $(E_b - E_e)/A_i$ is equal to E_a per unit area, dE_m/dA_i also gives a measure of this parameter, which usually is referred to as the fracture energy or adhesive energy (γ in Figure 20B) in the fracture literature (*43, 44*). Figure 21A illustrates two different cases with $(E_b - E_e)/A_i = 8 \times 10^{-4}$ and 4×10^{-4} dyn/cm. These variations may be achieved by having changes in E_b (e.g. by varying the bridging macromolecule) and/or E_e (e.g. by neuraminidase treatment of RBCs or by varying I). The equilibrium A_i can be identified by finding the portion of the E_m curve where the slope (tangent of the curve) is equal to that of the $(E_b - E_e)/A_i$ given. The equilibrium A_i at which the tangent of the

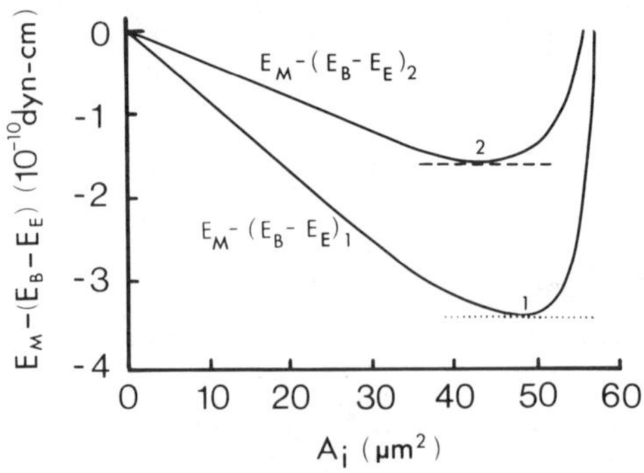

Figure 21B. $E_m - (E_b - E_e)$ plotted against A_i. At equilibrium, $d[E_m - (E_b - E_e)]/dA_i = 0$. Note the difference in the height of these curves at equilibrium and the relatively small difference in equilibrium A_i.

E_m curve equals 8×10^{-4} dyn/cm (Case 1, dotted line) is higher than that at which the tangent equals 4×10^{-4} dyn/cm (Case 2, broken line). These equilibrium A_i values can be found also by plotting $E_m - (E_b - E_e)$ against A_i and locating the minimum, i.e. when $d[E_m - (E_b - E_e)]/dA_i = 0$ (*see* Figure 21B). The results indicate that, owing to the nonlinearity of the change in E_m with A_i, a relatively large change in $(E_b - E_e)/A_i$ causes only a small variation in A_i.

The discussions in this and the preceding sections indicate that the potential minimum concept can be used to estimate the geometric features of cell aggregates, both in terms of the d and the A_i. Consideration of energy balance in a direction perpendicular to the membrane surfaces allows estimations of the E_a between adjacent cells and the equilibrium d of cell separation (*see* Figure 20). Analysis of energy balance in a direction parallel to the membrane surfaces permits an appraisal of equilibrium A_i (*see* Figure 21). Although both d and A_i change with alterations in energy balance, the magnitude of these changes are usually small.

E_m in Equation 13 includes all forms of recoverable elastic energy which may be stored in the system. In the case of RBCs, the elasticity can be taken as being associated entirely with the cell membrane, and E_m is regarded conveniently as being caused by membrane stresses (tensions in the plane of the membrane) and bending stresses owing to changes of curvature. Using the values of 5×10^{-3} dyn/cm as the elastic shear modulus (*45*), and 10^{-12} dyn–cm as the bending stiffness (*46*) of the normal human RBC membrane, and using the unstressed shape of the cell (*47*), the changes in RBC shape resulting from rouleau formation can be computed (*48*). The RBCs are axisymmetric and have a constant surface area. The equilibrium position can be achieved numerically by a finite element method and an iteration procedure (*46*). The value of A_i is specified in advance as an indication of the degree of A_i. In this way, the equilibrium shape of the cell may be found, and the adhesion may be deduced for any specified A_i. Figure 22 shows the results of computing RBC shape in rouleaux composed of two cells. Since axisymmetry is assumed, the intercellular area in each case is a plane disk. The internal pressure of the cell is assumed to be zero in the unstressed position, and it is treated as an unknown with the condition that its adjustment preserves a constant cell volume. With a rise in E_a, which is represented by an increase in A_i in the computation, the increase in E_m alters the shape of the free cell surfaces. When A_i is small, the curvature of the

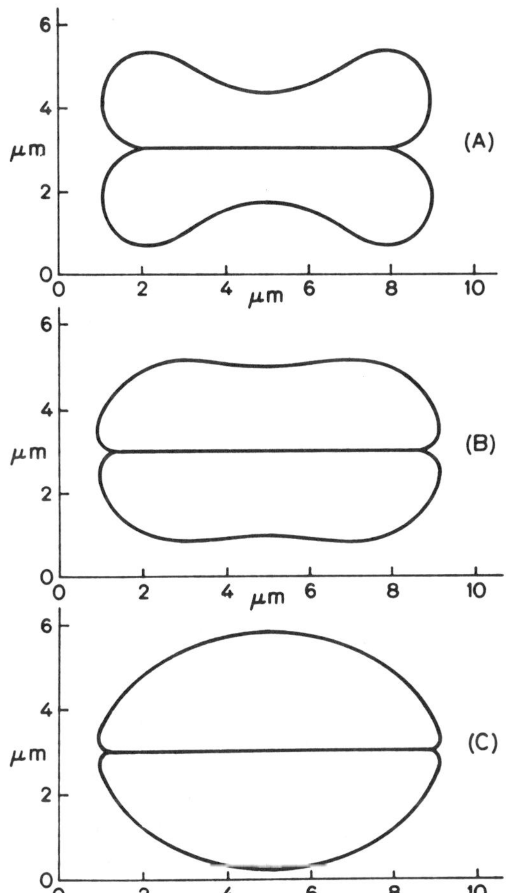

Figure 22. Theoretical shapes of rouleaux composed of two cells, each with increasing adhesive energy in order A, B, and C (48)

free cell surfaces remains concave as in the unstressed cell (*see* Figure 22A). As A_i increases, the concavity decreases to become flat (*see* Figure 22B) and the free cell surfaces eventually become convex with even greater A_i (*see* Figure 22C). These computed changes in cell shape with alterations in E_a are in agreement with experimental observations. Thus, the free cell surfaces of normal RBC aggregates are concaved in 1 g/dL Dx 70, become essentially flat in 2 g/dL Dx 70 and are convexed in 4 g/dL Dx 70. At 6 g/dL Dx 70, as E_e between cell surfaces increases (*see* earlier discussion), the reduction in E_a leads to a flattening of the free cell surfaces again. In neuraminidase-treated RBCs, where $E_e \rightarrow 0$, the free cell surfaces retain their convexity as the Dx 70 concentration is raised to as high as 20 g/dL.

With the use of Equation 14 the adhesive energy (ϕ_a) required to hold each value of A_i can be computed. In the cases of Figures 22B and 22C, it is estimated to be on the order of 10^{-4} dyn/cm, which is the same order of magnitude as that deduced from considering the balance between aggregating potential and electrostatic repulsive potential (*see* earlier discussion).

Disaggregation of RBC Rouleaux by Shear Stress

The discussions in the preceding sections apply primarily to situations without externally applied forces. The E_a present in the rouleaux can be probed by the work needed by externally applied shear stress to cause disaggregation. The mechanical shearing force required to disperse red cell aggregates has been estimated from light reflection (*33*), light transmission (*49*), or viscometric measurements (*50*) made on concentrated cell suspensions. In these studies, however, the geometric relation between the applied shear stress and the RBC rouleaux is not defined clearly. Furthermore, optical and viscometric estimations of aggregation–disaggregation are indirect, and it is difficult to establish a quantitative relationship. A fluid mechanical method has been used to quantify the shear stress required to disaggregate individual rouleaux in a flow channel under microscopic observation (*51*). In this method the shear stress is applied in a direction parallel to the aggregating surfaces and both the applied stress and the resulting disaggregation can be quantified directly. Serial photomicrographs of a two-cell rouleau subject to progressively increasing levels of shear stress are shown in Figure 23. The degree of disaggregation at each shear stress level was determined on the photomicrographs from the percentage of the area of bottom cell exposed by shearing. A plot of this percentage of separation of cells against the shear stress allows the determination of the shear stress required to cause 50% separation (τ_{50}). The τ_{50} value varies with the nature and concen-

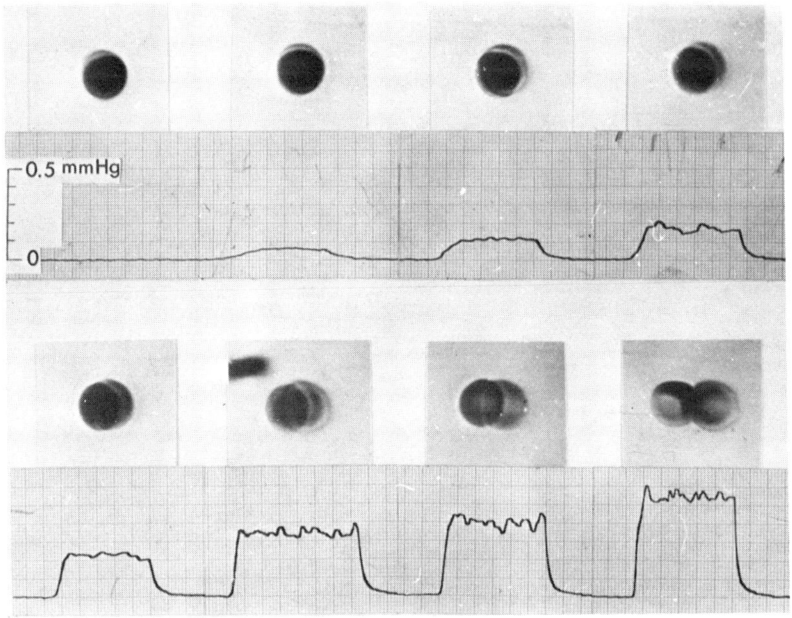

Microvascular Research

Figure 23. Serial photomicrographs of a two-cell rouleau under stationary condition (A) and following increasing shear stresses (B–H) (51). The pressure tracing coresoponding to each stress level (in dynes/centimeters squared) is shown beneath the photomicrograph.

tration of the macromolecule used to induce cell aggregation. For normal RBCs suspended in 4 g/dL Dx 70, τ_{50} is 0.25 ± 0.01 dyn/cm^2 (mean ± SEM). At 50% separation, the work done is approximately $\tau_{50} \times 50$ μm^2 × 4 μm = 5 × 10^{-11} dyn–cm, where 50 μm^2 is the cell surface area sheared and 4 μm is the displacement of the top cell in the direction of shear. Since the change in interaction area at τ_{50} is approximately 25 μm^2, this corresponds to a work done per unit interaction area of approximately 2 × 10^{-4} dyn/cm, which agrees well with ϕ_a deduced from considerations of E_e or membrane strain energy (*see* earlier discussions). The τ_{50} for red cell rouleaux is plasma is slightly lower than that in 4g/dL Dx 70, with ϕ_{a_c} equaling approximately 10^{-4} dyn/cm.

Scanning electron microscopic pictures obtained on RBC rouleaux under shear indicate that the cell shape (e.g. end-cell curvature) is unchanged essentially during partial disaggregation, suggesting that shearing may have caused the same E_m to occur at a lower A_i than in the stationary state. Therefore, it appears that the work done by the shear stress may lead to a left shift of the curve relating E_m and A_i along the abscissa A_i (*see* Figure 24). For a given adhesive energy, i.e. (E_b −

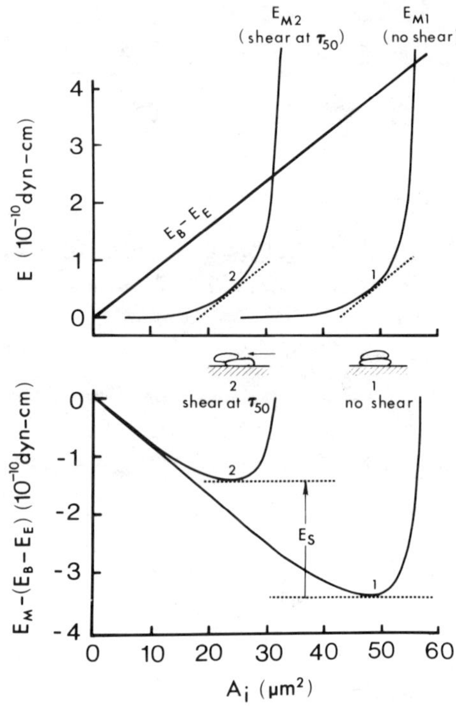

Figure 24. (Top) Variations in E_a ($= E_b - E_e$) and E_m with A_i between two cells in a rouleaux. Two curves are shown for E_m: one with no shear (E_{m1}) and the other under shear stress τ_{59} to 50% cell separation (E_{m2}). (Bottom) $E_m - (E_b - E_e)$ as a function of A_i for the two-cell rouleau under no shear and under shear stress to 50% separation. Note the decreases in equilibrium A_i and equilibrium energy under shear. The difference in equilibrium energy between the two cases reflects the work done by E_s.

$E_e)/A_i$, the mechanical shearing then would lead to a shift of the energy minimum to a smaller A_i with a shallower energy well (*see* Figure 24). This explanation of energy balance in shear disaggregation as outlined in Figure 24 requires further theoretical and experimental testing.

Summary: Energy Balance at Cell Surfaces in RBC Aggregation

The above discussions indicate that RBC aggregation is a phenomenon involving energy balance at the cell surfaces. The aggregating energy supplied by E_b is counteracted by the disaggregating energies owing to E_e and mechanical shearing (E_s). A portion of the E_a is stored in the RBC membrane as E_m, which interacts with the mechanical properties of the cell membrane and any mechanical shear stress in determining the area of interaction and the cell shape. Thus E_b, in an aggregation system can be estimated from considerations of E_e existing between cell surfaces, computation of adhesive energy based on cell shape and membrane properties, and determination of the work needed to cause disaggregation by E_s. Such estimations have been performed for the aggregation of normal RBCs in Dx 70, and all three methods yield a a value on the order of 10^{-4} dyn/cm. The concepts of energy (or potential) minimum can be used to understand the geometric features of the RBC rouleaux at equilibrium, as well as the balance among various types of energies. The equilibrium d perpendicular to the cell surface is primarily a function of the balance between E_b and E_e, with the former usually being the dominant factor. The equilibrium area of interacting cell surfaces is determined by the balance between $(E_b - E_e)$ and E_m, as these two terms vary in different manners with changes in A_i. The disaggregation resulting from work done by the shear stress can be explained by a shift in the E_m vs. A_i relationship. Considerably more theoretical and experimental work is needed, however, to improve our understanding of the interaction of energies involved between cell surfaces.

This research has benefited greatly from knowledge available in electrochemistry and colloid chemistry, and the results obtained may have implications in these basic chemical disciplines, e.g. the energy balance in flocculation and stability of colloid systems. It is hoped that this chapter will stimulate further interactions between electrochemistry–colloid chemistry and biology–medicine. Advances in such interdisciplinary research will not only elucidate the energy balance in RBC aggregation by macromolecules, but also will shed light on interactions of other biological systems, e.g. hemagglutination by antibodies (*52, 53*), platelet aggregation and thrombosis (*54*), and interactions of cancer cells (*55*).

Acknowledgments

This chapter was supported in part by the National Heart, Lung, and Blood Institute, Research Grant HL 16851. The author also wishes to thank R. Skalak for his valuable discussions and comments.

Glossary of Symbols

a = electronic charge (4.803×10^{-10} esu)
A_a = cell surface area available for dextran adsorption
A_c = total surface area of a cell
A_i = area of interacting cell surface
b_m = number of bonds made by each bridging molecule with a cell surface

d = intercellular distance
d_β = distance between the planes of surface charge of adjacent cells in the presence of dextran
Dx = dextran (the number that follows Dx indicates the approximate molecular weight in thousands.)
E_a = net aggregation energy
E_b = aggregating energy
e_b = interaction energy for each macromolecular bond with cell surface
E_e = electrostatic repulsive energy
E_m = change in strain energy of cell membrane
E_s = disaggregating energy owing to mechanical shearing
I = ionic strength
I_o = ionic strength in the absence of dextran
I_β = ionic strength in the presence of dextran
k = Boltzmann constant (1.380×10^{-16} erg/deg)
m = number of bridging macromolecules between cells
N = Avogadro's number
n = number of RBCs in an aggregate
T = absolute temperature
u = electrophoretic mobility
y = axis perpendicular to cell surface (axis for measurement of cell separation)
z = cationic valency
ESR = erythrocyte sedimentation rate
MAI = microscopic aggregation index
RAI = reflectometric aggregation index

Greek Letters

γ = fracture energy (or energy needed for disaggregation), with a unit of energy per unit interacting cell surface area
ϵ = dielectric constant
ζ = zeta potential
ζ_o = zeta potential in the absence of dextran
ζ_β = zeta potential in the presence of dextran
η = viscosity
κ = Debye–Hückel function or the reciprocal of double-layer thickness
κ_o = Debye–Hückel function in the absence of dextran
κ_β = Debye–Hückel function in the presence of dextran
σ_a = net aggregation force per unit interacting cell surface area
σ_b = macromolecular bridging force per unit interacting cell surface area

σ_e = electrostatic repulsive force per unit interacting cell surface area
τ_{50} = shear stress required for 50% cell separation
ϕ_a = net aggregation potential (or energy) per unit area
ϕ_b = potential (or energy) per unit area owing to macromolecular bridging
ϕ_e = electrostatic repulsive potential (or energy) per unit area

Literature Cited

1. Fåhraeus, R. *Physiol. Rev.* **1929,** *9,* 241.
2. Chien, S.; Luse, S. A.; Jan. K.–M.; Usami, S.; Miller, L. H.; Fremount, H. "Proc. 6th Eur. Conf. Microcirculation"; Ditzel, J., Lewis, D. H., Eds.; Karger: Basel, Switzerland, 1971; p. 29.
3. Chien, S. *Bibl. Anat.* **1973,** *11,* 244.
4. Ingelman, B. *Acta Acad. Regiae Sci. Ups.* **1969,** *12,* 9.
5. Chien, S.; Usami, S.; Dellenback, R. J.; Gregersen, M. I. *Science* **1967,** *157,* 827.
6. Fåhraeus, R. *Acta Med. Scand.* **1921,** *55,* 1.
7. Dintenfass, L. "Blood Microrheology: Viscosity Factors in Blood Flow, Ischaemia and Thrombosis"; Appleton: New York, 1971.
8. Chien, S. In "The Red Blood Cell," 2nd ed.; Surgenor, D. M., Ed.; Academic: New York, 1975; Vol. II, p. 1031.
9. Brooks, D. E.; Goodwin, J. W.; Seaman, G. V. F. *Biorheology* **1974,** *11,* 69.
10. Klose, H. J.; Volger, E.; Brechtelsbauer, L. H.; Schmid–Schönbein, H. *Pflügers Arch.* **1972,** *333,* 126.
11. Berman, H. J.; Fuhro, R. L. *Bibl. Anat.* **1973,** *11,* 117.
12. Usami, S.; Chien, S. *Bibl. Anat.* **1973,** *11,* 91.
13. Brooks, D. E. *J. Colloid Interface Sci.* **1973,** *43,* 700.
14. Chien, S.; Simchon, S.; Abbott, R. E.; Jan, K.–M. *J. Colloid Interface Sci.* **1977,** *62,* 461.
15. Nevo, A.; deVries, A.; Katchalsky, A. *Biochim. Biophys. Acta* **1955,** *17,* 536.
16. Chien, S.; Jan, K.–M. *Microvas. Res.* **1973,** *5,* 155.
17. Ingelman, B.; Halling, M. *Ark. Kemi.* **1949.**
18. Thorsén, G.; Hint, H. *Acta Chir. Scand. Suppl.* **1950,** *154,* 1.
19. Katchalsky, A.; Danon, D.; Nevo, A.; deVries, A. *Biochim. Biophys. Acta* **1959,** *33,* 120.
20. Jan, K.–M.; Chien, S. *Bibl. Anat.* **1973,** *11,* 281.
21. Edsall, J. T. In "The Proteins, Chemistry, Biological Activity and Methods"; Neurath, Bailey, Eds.; Academic: New York, 1953; Vol. I.
22. Klenk, E.; Lempfrid, H. *Hoppe - Seyl. Z. Physiol. Chem.* **1957,** *307,* 278.
23. Banham, A. D.; Flemans, R.; Heard, D. H.; Seaman, G. V. F. *Nature* **1958,** *182,* 642.
24. Jan, K.–M.; Chien, S. *J. Gen. Physiol.* **1973,** *61,* 638.
25. Seaman, G. V. F. In "The Red Blood Cell," 2nd ed.; Surgenor, D. M., Ed.; Academic: New York, 1975; Vol. II, p. 1135.
26. Overbeek, J. Th. G. In "Colloid Science"; Kruyt, H. B., Ed.; 1952; Vol. I, Chapters 4–8.
27. Jan, K.–M.; Chien, S. *J. Gen. Physiol.* **1973,** *61,* 655.
28. Edsall, J. T.; Wyman, J. "Biophys. Chem."; Academic: New York, 1958; Vol. I, Chapters 8-9.
29. Shaw, D. J. "Electrophoresis"; Academic: New York, 1969; Chapter 2.
30. Jan, K.–M. "Role of the Surface Charge in Red Blood Cell Interactions"; Ph.D. Dissertation, Columbia University, NY, 1971.

31. Powles, J. G.; Smyth, C. P. In "Physical Methods of Organic Chemistry"; Weissberger, A., Ed.; Interscience: New York, 1960; Vol. I, part 3, Chapter 38.
32. Brooks, D. E. *J. Colloid Interface Sci.* **1973**, *43*, 689.
33. Chien, S.; Jan, K.–M. *J. Supramol. Struct.* **1973**, *1*, 385.
34. Chien, S. *Thromb. Res. Suppl. II* **1976**, *8*, 189.
35. Traber, D. L.; Kolmen, S. N. *Tex. Rep. Biol. Med.* **1963**, *23*, 782.
36. Müller, H. E.; Grämlich, F. *Acta Haematol.* **1965**, *134*, 239.
37. Kipling, J. J. "Adsorption from Solutions of Non-electrolytes"; Academic: New York, 1965.
38. Houwink, R. In "Colloid Science"; Kruyt, H. K., Ed.; Elsevier: Amsterdam, 1952; Vol. II, Chapters 2 and 7.
39. Jan, G.–M. *J. Cell. Physiol.* **1979**, *101*, 49.
40. Schachter, D. *Biochem. Biophys. Res. Comm.* **1978**, *84*, 840.
41. Parsegian, V. A. In "Enriching Topics from Colloid and Surface Science"; van Olphen, H., Mysels, K. J., Eds.; Thronex: La Jolla, CA, 1975; p. 27.
42. Gingell, D.; Fornés, J. A. *Biophys. J.* **1976**, *16*, 1131.
43. Griffith, A. A. *Philos. Trans. R. Soc. London Ser. A* **1921**, *221*, 163.
44. Williams, M. L. "Applications of Continuum Mechanics in Adhesive Fracture"; American Chemical Society, Organic Coatings Div.: Washington, D.C., 1971.
45. Skalak, R.; Tözeren, A.; Zarda, P. R.; Chien, S. *Biophys. J.* **1973**, *13*, 245.
46. Zarda, P. R.; Chien, S.; Skalak, R. *J. Biomech.* **1977**, *10*, 211.
47. Evans, E. A.; Fung, Y. C. *Microvas. Res.* **1972**, *4*, 335.
48. Skalak, R.; Zarda, P. R.; Jan, K.–M.; Chien, S. In "Cardiovascular and Pulmonary Dynamics"; Jaffrin, M.–Y., Ed.; INSERM-Euromech 92, 1977; p. 299.
49. Schmid–Schönbein, H.; Volger, E.; Klose, H. J. *Pflügers Arch.* **1972**, *333*, 140.
50. Brooks, D. E. In "Microcirculation"; Grayson, J., Zingg, W., Eds.; Plenum: New York, 1976; Vol. I, p. 33.
51. Chien, S.; Sung, L. A.; Kim, S.; Burke, A. M.; Usami, S. *Microvas. Res.* **1977**, *13*, 327.
52. Pollack, W.; Reckel, R. P. *Int. Arch. Allergy Appl. Immunol.* **1977**, *54*, 29.
53. van Oss, C. J.; Mohn, J. F.; Cunningham, R. K. *Vox Sang.* **1978**, *34*, 351.
54. Weiss, H. J. *New England J. Med.* **1975**, *293*, 531.
55. Curtis, A. S. G. "The Cell Surface"; Academic: New York, 1967.

RECEIVED October 17, 1978.

SURFACE INTERACTIONS

2

Measurement of the Adsorption of Ca^{2+} and Mg^{2+} to Phosphatidyl Choline Bilayers

L. J. LIS and R. P. RAND

Department of Biological Sciences, Brock University,
St. Catharines, Ontario L2S 3A1 Canada

V. A. PARSEGIAN

National Institutes of Health, Bethesda, MD 20014

We have used osmotic stress (4) to examine the effect of Mg^{2+} and Ca^{2+} on the interactions between dioleoylphosphatidylcholine or dipalmitoyl phosphatidyl choline bilayers. From the net repulsive forces between bilayers we are able to infer electrostatic potentials and charge densities at the site of ion binding; these quantities are sensitive to bilayer separation. We find that at any particular bilayer separation dioleoyl phosphatidyl choline bilayers (melted hydrocarbon chains) adsorb less charge than dipalmitoyl phosphatidyl choline bilayers (frozen hydrocarbon chains) and that the binding of Ca^{2+} is greater than that of Mg^{2+} for both kinds of bilayers.

We have previously measured the net repulsive force between dipalmitoyl phosphatidyl choline (DPPC) bilayers in a multilayer lattice charged by the adsorption of Ca^{2+} (*1*). From these forces we were able to infer electrostatic potentials and charge densities at the site of ion binding by integrating a nonlinear Poisson–Boltzmann equation. This initial study on DPPC showed that as bilayers are pushed together, bound Ca^{2+} desorbs. When NaCl is added to the $CaCl_2$ solution, membrane repulsion decreases even though more cationic charge appears to adsorb to the bilayer (*1*). In this chapter we report the net repulsive force and estimate the associated electrostatic potentials and surface charge densities as a function of bilayer separation between bilayers of DPPC or of

0-8412-0473-X/80/33-188-041$05.00/1

dioleoyl phosphatidyl choline (DOPC) in 30m*M* $CaCl_2$ or $MgCl_2$ solutions. The calculated surface potentials and charge densities, at least for DPPC bilayers in 30m*M* $CaCl_2$ and 100m*M* NaCl, are higher than those inferred by Lau et al. (*2*) from electrophoresis and NMR spectroscopy on multilamellar vesicles. Our present force measurements as well as the electrophoretic studies (*2*) show that DOPC bilayers bind less divalent charge than do those of DPPC. We find a preference for Ca^{2+} over Mg^{2+} for both phospholipids while electrophoresis (*2*) suggests ion preference by saturated chain PCs only.

Experimental

The phosphatidyl cholines PCs used in this study were obtained from the Sigma Chemical Co. The DPPC used here showed swelling behavior in $CaCl_2$ solutions similar to that of DPPC previously obtained from other sources (*1*). All lipids showed less than 1% impurity by thin-layer chromatography (TLC). Water was doubly distilled and salts were of reagent grade. Dextran was obtained from Pharmacia Chemicals and mixed in known proportions with salt solutions prior to contact with lipid.

Samples were made either by adding lipid to known amounts of salt solution or by adding lipid to dextran-plus-salt solutions. Mixtures were allowed to equilibrate for 48 hr and then were mounted between mica windows. Mixing and waiting at 50°C for up to 48 hr provided no advantage in reaching equilibria nor did the addition of a Ca^{2+} ionophore, A23187. All measurements were at room temperature (25°C).

The x-ray diffraction and force-measuring techniques have been described elsewhere (*3, 4*). Lamellar phases produce a series of x-ray reflections whose spacings are integral multiples of the bilayer repeat spacing, d. The lipid bilayer thickness, d_l, and bilayer separation, d_w, can be determined for each value of d. When bilayers are forced apart by electrostatic repulsion (*1, 5*), they maintain the same thickness they have in excess pure water (44 Å for DPPC and 30.5 Å for DOPC at 25°C). The bilayer separation differs from the multilayer repeat spacing, d, by this constant lipid bilayer thickness. The net repulsive force between bilayers is given by the experimentally measured osmotic pressure of the dextran solutions (*6*). The difference between the osmotic pressure exerted by dextran in salt solutions and dextran in pure water (*6*) is no greater than the error in measuring the dextran concentration (*7*).

Simultaneous measurements of d and osmotic pressure provide a relation between the separation of bilayers and their mutual repulsive pressure. Measurement of the electrostatic repulsion is, in fact, a determination of the electrostatic potential midway between bilayers relative to the zero of potential in the dextran reservoir. The full nonlinear Poisson–Boltzmann differential equation governing this potential has been integrated (*1*) from the midpoint to the bilayer surface to let us infer the surface potential. The slope of this potential at the surface gives a measure of the charge bound.

Results

DPPC. Dry DPPC was mixed with solutions of $CaCl_2$ or $MgCl_2$ to make samples of approximately 30 wt % lipid. The lamellar repeat distances *d* for these mixtures are shown in Figure 1 as a function of salt concentration. In 0.7m*M* $CaCl_2$ or 9m*M* $MgCl_2$ solutions, DPPC multilayers suddenly swell to the maximum allowed by the amount of water in the mixture. In $CaCl_2$ solutions, they remain maximally swollen at concentrations up to 30m*M* $CaCl_2$; but in $MgCl_2$ the swelling decreases steadily with increasing salt concentration.

Figure 2 shows the relationship between net repulsive force between bilayers and the bilayer separation for DPPC bilayers (frozen chains) either in 30m*M* $CaCl_2$ or in 30m*M* $MgCl_2$. In all cases where the DPPC bilayers are separated by more than 20 Å of salt solution, the force to maintain a particular bilayer separation is greater in $CaCl_2$ than in $MgCl_2$.

As described above, we calculate surface potential, Ψ_s, and area per Me^{2+}, S, from the data in Figure 2. In Figures 3 and 4, the variations in Ψ_s and S respectively are shown for both Ca^{2+} and Mg^{2+} data sets, respectively. More Ca^{2+} is bound than Mg^{2+}.

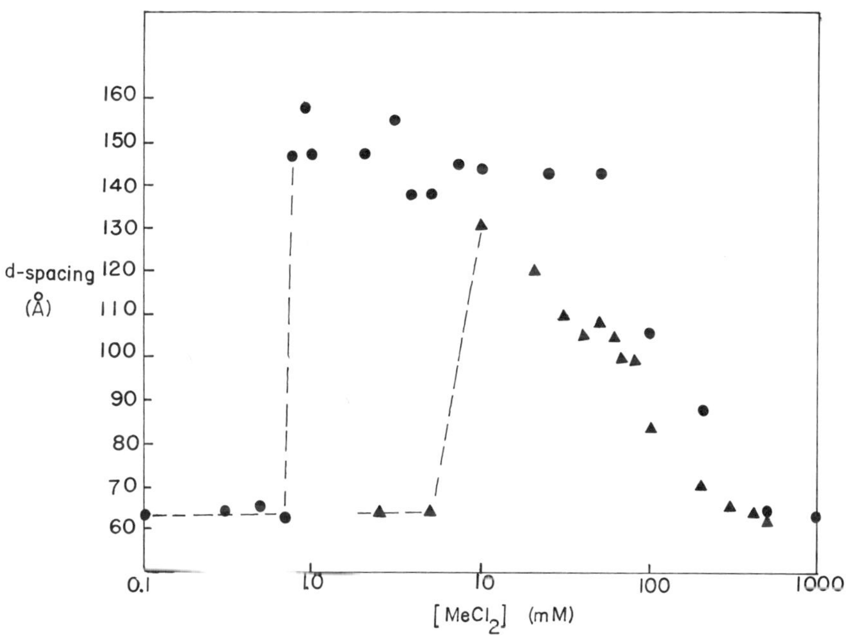

Figure 1. Repeat distance, d, *of the lamellar phase formed by 30 wt % DPPC in various concentrations of $CaCl_2$ or $MgCl_2$.* T = *25°C: (●), $CaCl_2$ and (▲), $MgCl_2$.*

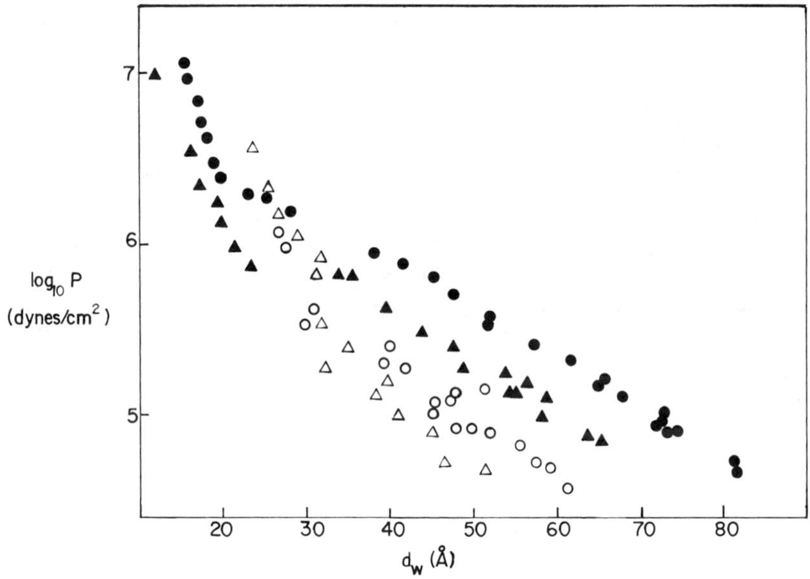

Figure 2. Interbilayer pressure, P, *as a function of DPPC or DOPC bilayer separation,* d_w, *in 30mM $CaCl_2$ (●, DPPC; ○, DOPC) or 30 mM $MgCl_2$ (▲, DPPC; △, DOPC) solutions.* T = 25°C.

Both ions appear to desorb from the DPPC bilayer surfaces as they come closer. We cannot infer an association constant for the binding of Mg^{2+} or Ca^{2+} to DPPC if we use a definition based on mass action. The apparent binding coefficient as well as surface potential and charge density vary with bilayer separation (*1*).

DOPC. The force between DOPC bilayers is less than between DPPC bilayers at the same d_w and ionic solution. The difference in forces conferred by Ca^{2+} and Mg^{2+} is less pronounced than for DPPC; repulsion at a particular DOPC bilayer separation in only slightly greater in $CaCl_2$ than in MgCl. These differences are reflected in the Ψ_s and S for DOPC bilayers shown in Figures 3 and 4.

Discussion

The swelling of DPPC in $CaCl_2$ and $MgCl_2$ solutions (Figure 1) supports the suggestion (*6, 8*) that these solutions confer a repulsion between bilayers by divalent cation adsorption to the bilayer surface. Ca^{2+} is more effective than Mg^{2+}. The force curves in Figure 2 resemble those obtained for charged phospholipid bilayers (*5*). At any particular bilayer separation, DPPC (frozen chains) bilayers absorb Ca^{2+} or Mg^{2+} to a greater extent than do DOPC (melted chains) bilayers, and Ca^{2+}

adsorbs to a greater extent than Mg^{2+} with a greater difference in ion binding to DPPC.

To go beyond these qualitative statements requires further assumptions. We have converted measured forces and distances into bound charge potentials and densities by assuming that the electrostatic potential between bilayers is governed by a nonlinear Poisson–Boltzmann equation (*1*). As expected, for a given d_w, inferred surface potentials and charge densities increase with the measured force. Contrary to expectation, the absolute values of these quantities are larger than in other studies and there is a rapid change in apparent potential and charge density with bilayer separation.

Lau et al. (*2*) have used electrophoresis and ^{31}P NMR to estimate divalent cation adsorption potentials and the fraction of phosphorous nuclei whose magnetic resonance is affected by the presence of divalent cobalt. Although they see the same preference for solid vs. melted chains and the preference of (saturated chain) DPPC and DMPC for Ca^{2+} over Mg^{2+}, they note no similar ion preference by egg lecithin. They used

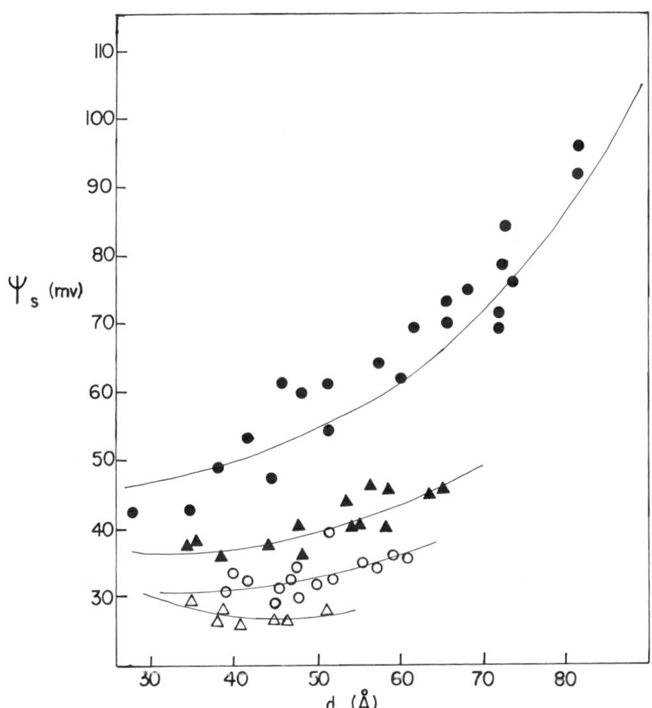

Figure 3. Surface potential, Ψ_s, for DPPC or DOPC bilayers in 30mM $CaCl_2$ (●, DPPC; ○, DOPC) or 30mM $MgCl_2$ (▲, DPPC; △, DOPC) solutions as a function of the bilayer separation, d_w. Curves through points are derived from best-fit lines for the data in Figure 2.

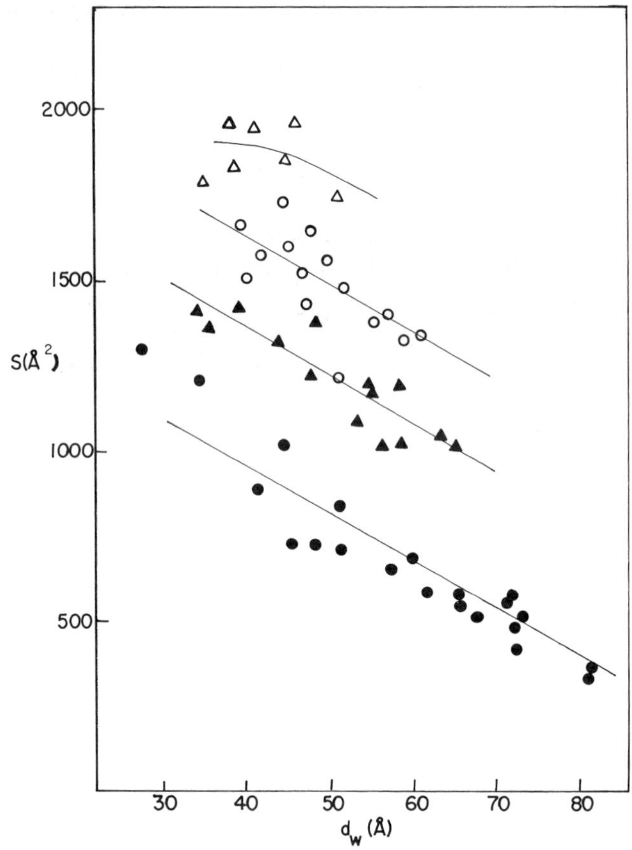

Figure 4. Surface area per bound divalent charge, S, *for DPPC or DOPC bilayers in 30mM $CaCl_2$ (●, DPPC; ○, DOPC) or 30mM $MgCl_2$ (▲, DPPC; △, DOPC) solutions as a function of the bilayer separation,* d_w. *Curves through points are derived from best-fit lines for the data in Figure 2.*

some of our DOPC and found results virtually identical to those on their egg lecithin. We find (unpublished observation) that chromatographically pure egg lecithin immersed into $CaCl_2$ or $MgCl_2$ separates to form two distinct lamellar phases. Egg lecithin is not completely equivalent to DOPC.

We are unable so far to explain the systematic difference in surface potential and density estimates from the two methods. Our numbers decrease if we assume that charge binds some distance into the water away from the bilayer surface (*10*), but this distance must be an improbable 16 Å to achieve reconciliation with the electrophoretic measurements. Conversely, the electrophoretic slip plane for the zeta potential would have to be an unlikely 16 Å away from the actual locus of charge in order to match our numbers. Using NMR measurements (*9, 11*) requires several assumptions, but the range of likely error in these assumptions is not wide enough to encompass all other estimates. Recognition of differences in NaCl concentrations in the various systems only aggravates the disagreement between these indirect methods for determining potential.

The influence of divalent cations on membrane–membrane forces is of great interest in biological systems such as presynaptic or secretory cells. As long as the qualitative parallels and the quantitative disagreements persist, we suggest that it is best to follow ion binding by all available methods. When phase separation occurs, as with egg lecithin (and should occur if, in fact, charge binding depends on hydrocarbon chain composition), x-ray diffraction is able to show this better than macroscopic observations. Force measurements with x-ray diffraction offer the advantages of concomitant bilayer structure determination and ion-binding information at bilayer separations. These direct force measurements rather than a desnsity chain of interference are the best way to learn how divalent chains might be acting to modify forces between natural membranes.

Acknowledgments

We wish to thank S. G. A. McLaughlin and J. B. Stamatoff for providing us with helpful information and preprints of their papers. This work was supported by a National Research Council of Canada grant (A4920) to R. P. Rand and a Muscular Dystrophy Association of Canada post-doctoral fellowship to L. J. Lis.

Literature Cited

1. Lis, L. J.; Rand, R. P.; Parsegian, V. A. *Biochemistry,* in press.
2. Lau, A. L. Y.; McLaughlin, A. C.; MacDonald, R. C.; McLaughlin, S. G. A., chapter 00 in this book.
3. Luzzati, V.; Rand, R. P. *Biophys. J.* **1968,** *8,* 125–139.
4. LeNeveu, D. M.; Rand, R. P.; Parsegian, V. A.; Gingell, D. *Biophys. J.* **1977,** *18,* 209–230.
5. Cowley, A. C.; Fuller, N. F.; Rand, R. P.; Parsegian, V. A. *Biochemistry* **1978,** *17,* 3163–3168.
6. LeNeveu, D. M.; Rand, R. P.; Parsegian, V. A. *Nature* **1976,** *259,* 601–603.
7. Millman, B. M., personal communication.
8. Inoko, Y.; Yamaguchi, T.; Furuya, F.; Mitsui, T. *Biochim. Biophys. Acta* **1975,** *413,* 24–32.
9. McLaughlin, A.; Grathwohl, C.; McLaughlin, S. G. A. *Biochim. Biophys. Acta,* in press.
10. Stamatoff, J. B., personal communication.
11. McLaughlin, A. C.; Grathwohl, C.; Richards, R. E. *J. Magn. Reson.,* in press.

RECEIVED November 20, 1978.

3

The Adsorption of Alkaline Earth Cations to Phosphatidyl Choline Bilayer Membranes: A Unique Effect of Calcium

ARTHUR L. Y. LAU and ALAN C. McLAUGHLIN

Department of Biology, Brookhaven National Laboratory, Upton, NY 11739

ROBERT C. MacDONALD

Department of Biological Sciences, Northwestern University, Evanston, IL 60201

STUART G. A. McLAUGHLIN

Department of Physiology & Biophysics, Health Science Center, SUNY, Stony Brook, NY 11794

Electrophoretic mobility and ^{31}P-NMR measurements were made to investigate the binding of the alkaline earth cations to membranes formed from phosphatidyl choline molecules with either saturated or unsaturated hydrocarbon chains. Calcium and magnesium bind to the same degree to membranes formed from lipids with unsaturated chains. These phosphatidylcholine molecules are present in the liquid crystalline state at all temperatures under consideration. Calcium binds more strongly than magnesium to membranes formed from lipids with saturated chains, even when the lipids are in the liquid crystalline state. The selectivity is enhanced when the temperature is lowered and the saturated chain lipids are in the gel state.

We previously have used electrophoretic mobility and ^{31}P-NMR measurements to study the adsorption of alkaline earth cations to bilayer membranes formed from the zwitterionic lipid phosphatidyl choline (PC) (*1*). The lipids used in this previous study had unsaturated or

0-8412-0473-X/80/33-188-049$05.00/1

branched chains and were present in the membrane in a fluid or liquid crystalline state. In all cases the adsorption could be well described by a combination of the Henry adsorption isotherm, the Boltzmann relation, and the Grahame equation from the theory of the diffuse double layer, a combination referred to as a Stern equation (*2*). The binding deduced from these measurements decreased in the sequence Mn^{++}, Mg^{++}, Ca^{++}, Co^{++}, Ni^{++}, Sr^{++}, Ba^{++}. If we assume that a lipid molecule occupies an area of 60 $Å^2$ in the bilayer and that there is a 1:1 stoichiometry for the binding of the divalent cation to PC, the intrinsic dissociation constants are, respectively, 0.3, 1.0, 1.0, 1.2, 1.2, 2.8, and 3.6*M*. Specifically, there was no significant difference in the binding of calcium and magnesium to membranes formed from these PC molecules.

We report here the results of a study of the adsorption of the alkaline earth cations to bilayer membranes formed from phosphatidylcholines with saturated chains: dipalmitoyl phosphatidyl choline (DPPC) and dimyristoyl phosphatidyl choline (DMPC). Our salient result is that the adsorption of calcium is distinct from the other alkaline earth cations in two respects. First, only calcium adsorbs significantly more strongly to PCs with saturated chains than to phosphatidyl cholines with unsaturated chains, even when all lipids are present in the liquid crystalline state. Second, when the membranes are present in the frozen or gel state, the binding of calcium is significantly enhanced. We used two independent techniques to demonstrate this unique behavior of calcium.

Zeta Potential Measurements

As illustrated schematically in Figure 1, the adsorption of a divalent cation (e.g. calcium) to the surface of a bilayer membrane formed from a zwitterionic lipid (e.g. PC) will produce a positive charge at the surface of the membrane and induce the formation of a Gouy–Chapman diffuse double layer in the aqueous solution. McLaughlin (*3*) may be consulted for a recent review of surface potentials. According to the Gouy–Chapman theory, the potential will fall off in an approximately exponential manner with distance from the membrane, as illustrated in Figure 1. The Debye length, the distance at which the potential falls to $1/e$ of its value at the surface of the membrane, is about 10 Å for the solutions used in this study. A measurement of the electrophoretic mobility of PC vesicles allows us to calculate the zeta potential, the potential at the hydrodynamic plane of shear (*see* Figure 1).

The multilamellar dispersions or vesicles were formed in the conventional manner (*4*). Drying down the lipid and suspending it in the desired aqueous solution (0.1*M* NaCl, 0.01*M* tris, pH 7.5) yielded, on gentle agitation, vesicles of the appropriate size for microelectrophoresis

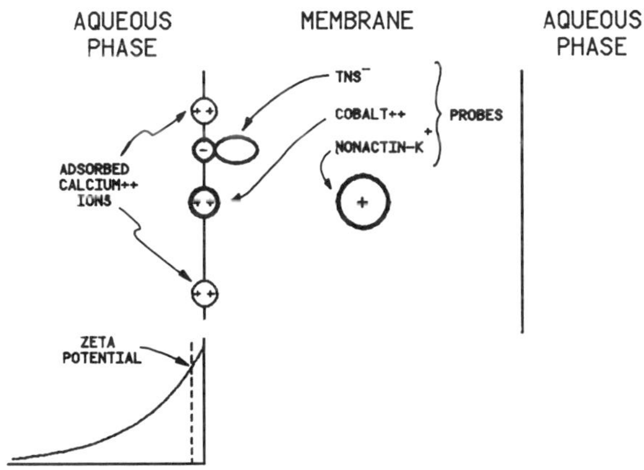

Figure 1. (Bottom) *Diagram of the electrostatic potential adjacent to a membrane bearing a positive charge. The zeta potential is the potential at the hydrodynamic plane of shear, which should be about 2 Å from the surface of the membrane.* (Top) *Schematic of the location of the probe molecules used to detect the potential produced by the adsorption of calcium and other alkaline earth cations to membranes formed from PC. The divalent cation cobalt and the amphipathic, anionic, fluorescent probe TNS will sense the potential at the interface. The non-actin–K^+ complex will sense the potential in the center of the membrane.*

measurements (1–20 μ in diameter). These vesicles were placed in an electrophoresis apparatus (Rank Bros.), a known electric field was applied, and their velocity was measured with a calibrated reticule in a microscope. Care was taken to focus at the stationary layer, the layer at which the fluid in the electrophoresis tube does not move (*5*). The zeta potentials, ζ, were calculated from the electrophoretic mobility measurements by means of the Helmholtz–Smoluchowski equation

$$\zeta = \mu\eta/\epsilon_r\epsilon_0$$

where μ is the electrophoretic mobility, η is the viscosity, ϵ_0 is the permittivity of free space, and ϵ_r is the dielectric constant of the aqueous solution. The assumptions inherent in the derivative of this equation have been discussed elsewhere (*6*). Measurements were made on 10 different vesicles in three different experiments; results are presented in Table I as the average $\pm$ standard deviation of the 30 measurements.

The available evidence suggests that the plane of shear lies 2 Å from the surface of a phospholipid bilayer membrane (*7*) so the zeta potential should be a good approximation to the electrostatic potential at the

Table I. Zeta Potentials of PC Vesicles[a]

Lipid	*Temperature (°C)*	*Divalent Cation*	*Zeta Potential (mV)*
Egg PC	25	Mg^{++}	11.0 ± 1
Egg PC	25	Ca^{++}	10.0 ± 1
DMPC	35	Mg^{++}	13.5 ± 1
DMPC	35	Ca^{++}	16.0 ± 1
DMPC	15	Mg^{++}	13.0 ± .5
DMPC	15	Ca^{++}	20.5 ± 1

[a] All solutions contain 0.1*M* NaCl, 0.01*M* tris, pH 7.5 in addition to either 0.05*M* $MgCl_2$ or 0.05*M* $CaCl_2$.

surface of these membranes. In our previous study (*1*) of membranes formed from PCs with unsaturated chains we used the nonactin–K^+ complex as a probe of the electrostatic potential at the center of the membrane (*3*). These probe and zeta potential measurements gave similar results, which indicates that the adsorption of calcium does not produce any significant change in the dipole potential (*1*). (As discussed elsewhere (*3*), the zeta potential will respond only to a change in the diffuse double layer potential but the nonactin–K^+ complex will respond to changes in the double layer and in dipole potential.) As illustrated in Figure 1, it is also possible to detect the adsorption of divalent cations to PC membranes by means of the anionic fluorescent probe, TNS (*8*). This technique, however, is not accurate enough to detect a difference in the adsorption of calcium and magnesium and we will not discuss these results here.

For brevity we will consider only the results we have obtained when either 50m*M* $CaCl_2$ or 50m*M* $MgCl_2$ is added to the solution containing the DMPC vesicles. In NaCl solutions the zeta potentials of phosphatidyl choline vesicles are zero within experimental error (*1,9*). It is apparent from Table I that the addition of either 50m*M* $CaCl_2$ or 50m*M* $MgCl_2$ causes an increase in the zeta potential of egg PC vesicles to about 10–11 mV. Lowering the temperature to 15°C or raising it to 35°C produced, respectively, a 1-mV decrease and a 1.5-mV increase in the zeta potentials. There was no significant Ca^{++}/Mg^{++} selectivity at any of these temperatures. As shown in Table I the addition of 50m*M* $MgCl_2$ to DMPC vesicles causesd a similar increase in the zeta potential at 15° and 35°C. Since the transition temperature, T_c, of DMPC multilamellar dispersions is 24°C (*10*), we conclude that magnesium binds equally well to DMPC in either the gel or the liquid crystalline state and that the adsorption is not substantially different from the adsorption to egg vesicles. The same conclusion is reached from measurements with lower concentrations of Mg^{++}. The cations Sr^{++}, Ba^{++}, and Co^{++} also bind to approximately the same degree to egg PC membranes and to DMPC membranes, at temperatures above and below the T_c of the latter lipid.

In contrast, when 50m*M* $CaCl_2$ is added to multilamellar dispersions formed from DMPC, the zeta potential increased to 16 mV at 35°C, where the membranes were in the liquid crystalline state (*see* Table I). This is a significantly higher zeta potential than observed for egg PC vesicles (*see* Table I). When the temperature was 15°C, the addition of 50m*M* $CaCl_2$ increased the zeta potentital of DMPC vesicles to 20.5 mV (*see* Table I). Thus, DMPC membranes bind calcium more strongly when the lipids are in the gel rather than the liquid crystalline state. The effect does not depend on the presence of NaCl or tris. We reached the same conclusion from zeta potential measurements when $CaCl_2$ was the only electrolyte present in the aqueous solutions. Similar results were obtained on DPPC membranes above and below the transition temperature.

^{31}P-NMR Measurements

We arrived at the same conclusions reached above by using cobalt as a probe for the binding of calcium and magnesium to sonicated phosphatidylcholine vescicles. Since cobalt is paramagnetic, the binding of cobalt to the surface of the membrane causes a large increase in the linewidth of the NMR (^{31}P-NMR) signal from the phosphate group. Under our experimental conditions, this increase in linewidth is directly proportional to the amount of cobalt bound to the membrane. A complete theoretical analysis allows one to calibrate the ^{31}P-NMR linewidth in terms of the fraction of phosphate groups bound to cobalt ions (*11*). In the presence of 5m*M* $CoCl_2$ (0.1M NaCl, pH 7.5), about 0.5% of the phosphate groups are bound directly to cobalt ions (*1*).

In 0.1M NaCl, 5m*M* $CoCl_2$ does not produce a significant surface potential ($\zeta \cong 2$ mV); it does, however, produce a significant increase in the ^{31}P-NMR linewidth (*1*). Calcium or magnesium ions, at relatively high concentrations (e.g. 50m*M*), bind to the membrane and produce a large electrostatic potential that decreases the concentration of cobalt ions near the surface of the membrane. This reduces the binding of cobalt to the membrane and produces a concomitant decrease in the ^{31}P-NMR linewidth. A comparison of the ^{31}P-NMR linewidths in the presence of cobalt plus calcium or magnesium thus reveals the relative binding strengths of these two alkaline earth metal ions.

The vesicles were prepared as previously described (*1*). All divalent cations were added to the solution prior to sonication. In one experiment 5m*M* praseodymium chloride was added to the vesicle suspension after

Table II. ^{31}P-NMR Linewidth Ratios for Phosphatidylcholine Vesicles[a]

Lipid	*Temperature (°C)*	*Linewidth Ratio, Ca^{++}/Mg^{++}*
Egg PC vesicles	10	1.02 ± 0.04
Egg vesicles	20	1.01 ± 0.02
Egg vesicles	35	0.96 ± 0.004
DMPC vesicles	10	0.75 ± 0.02 (overall)
DMPC vesicles	10	0.78 ± 0.03 (inside)
DMPC vesicles	35	0.82 ± 0.01

[a] Ratio of the ^{31}P-NMR linewidths of lipid vesicles in the presence of 5m*M* $CoCl_2$ and 50m*M* $CaCl_2$ to those in the presence of 5m*M* $CoCl_2$ and 50m*M* $MgCl_2$. All solutions contained 0.1*M* NaCl, buffered by 0.01*M* tris to pH 7.5. The values are the averages of at least four experiments ± standard deviation.

sonication to selectively shift the ^{31}P-NMR signal from the phosphate groups on the outer surface of the vesicles.

Table II gives the ratios of the ^{31}P-NMR linewidths in the presence of calcium or magnesium for egg PC and DMPC vesicles. For egg PC this ratio is 1.0 at 10° and 20°C, indicating that calcium and magnesium have identical dissociation constants in their binding to egg PC at these temperatures. At 35°C the ratio is slightly less than 1.0. This difference, however, is small compared with the effects observed with saturated phosphatidylcholines.

For DMPC vesicles at 35°C, the ratio of ^{31}P-NMR linewidths in the presence of calcium or magnesium is substantially less than 1 (0.82). This indicates that calcium binds more tightly than magnesium to DMPC. At 10°C, the linewidth ratio is further reduced (0.75), indicating that the selectivity of DMPC for calcium over magnesium is enhanced below the phase transition ($T_c \cong 22$°C for sonicated vesicles (*12*)). Very similar results were obtained with DPPC vesicles.

The NMR measurements were made on sonicated phospholipid vesicles to obtain relatively narrow ^{31}P-NMR signals while the electrophoretic mobility measurements were made on unsonicated vesicles. Since there are differences between the two systems (e.g., area per molecule (*13, 14*)), we do not attempt to quantitatively compare the ^{31}P-NMR data with the electrophoretic data, but rather use the ^{31}P-NMR data as an independent demonstration of the difference between the binding of calcium and magnesium. We note, however, that the linewidth ratio for the outer monolayer of the sonicated vesicles were identical within experimental error (*see* Table II). This implies that the Ca^{++}/Mg^{++} selectivity of the two monolayers is identical. We had expected the selectivity to be greater for the inner monolayer because the polar head groups of the lipids in this monolayer occupy a smaller area (*13*).

We assume that the concentration of bound cobalt is proportional to the Boltzmann factor, $\exp(-F\psi/RT)$, where ψ is the potential at the cobalt binding site, F is the Faraday constant, R is the gas constant, and T is the temperature. The ratio of the linewidths, r, could be interpreted in terms of the difference in the surface potential, $\Delta\psi$, in the presence of calcium or magnesium by means of the equation:

$$r = \exp(-2F\Delta\psi/RT)$$

The ^{31}P-NMR technique, using cobalt as a probe, thus provides information about the electrostatic potential at the binding site of the divalent cations (presumably the phosphate group). In contrast, the electrophoretic mobility measurements give information about the potential at the plane of shear.

Discussion

If calcium is unique among the alkaline earth cations in binding more strongly to membranes formed from PCs in the gel rather than in the liquid crystalline state it also should be unique in increasing the transition temperature. Simon et al. (*15*) have noted that calcium is indeed more effective than magnesium, strontium, or barium in increasing the transition temperature of both a DMPC monolayer and a DPPC bilayer. Our results also agree well with x-ray diffraction measurements (*16*), which demonstrate that the adsorption of divalent cations to DPPC bilayers produces a repulsive force between membranes and that calcium produces a larger force than magnesium.

It is known (*17*) that "the coordination geometry of calcium is irregular in both bond angle and bond length" and that "calcium is quite different from magnesium, which maintains six-coordination in a closely regular octahedron." Williams (*17*) also notes that "magnesium, as it strongly demands a certain geometry, is weakened in its ability to bind to the irregular geometries of coordination sites of biological molecules." As the area per lipid molecule decreases in the gel state (*18*) one can speculate that the variables coordination geometry of calcium is in some way responsible for its unique ability to bind more strongly to PC bilayers at temperatures below the transition temperature.

Glossary of Symbols

ζ = zeta potential
ψ = surface potential
μ = electrophorctic mobility
η = viscosity
ϵ_r = dielectric constant of the aqueous solution
ϵ_0 = permittivity of free space
T_c = phospholipid phase transition temperature

Acknowledgments

This research was supported by NIH Grant GM 24971 to A. McLaughlin and S. McLaughlin and NSF Grant PCM 76-04363 to S. McLaughlin. The portion of this work done at Brookhaven National Laboratory was supported in part by the USDOE.

Literature Cited

1. McLaughlin, A. C.; Grathwohl, C.; McLaughlin, S. *Biochim. Biophys. Acta* **1978**, *513*, 338–357.
2. Aveyard, R.; Haydon, D. A. "An Introduction to the Principles of Surface Chemistry;" Cambridge University Press: London, 1974.
3. McLaughlin, S. G. A. In "Current Topics in Membrane Transport," Bronner, F., Kleinzeller, A., Eds.; Academic: New York, 1977; Vol. 9, pp. 71–144.
4. Bangham, A. D.; Hill, M. W.; Miller, N. G. A. *Methods Membr. Biol.* **1974**, *1*, 1–68.
5. Henry, D. C. *J. Chem. Soc.* **1938**, 997–999.
6. Overbeek, J. Th. G.; Wiersema, P. A. In "Electrophoresis," Bier, M., Ed.; Academic: New York, 1967; Vol. 2, pp. 1–52.
7. McLaughlin, S., unpublished data.
8. Eisenberg, M., personal communication.
9. Hanai, T.; Haydon, D. M.; Taylor, J. *J. Theor. Biol.* **1965**, *9*, 278–296.
10. Phillips, M. C.; Ladbrooke, B. D.; Chapman, D. *Biochim. Biophys. Acta* **1970**, *196*, 35–44.
11. McLaughlin, A. C.; Grathwohl, C.; Richards, R. E. *J. Magn. Reson.* **1978**, *31*, 283–294.
12. Van Dijck, P. W. M.; DeKruijff, B.; Aarts, P. A. M. M.; Verkleij, A. J.; DeGier, J. *Biochim. Biophys. Acta* **1978**, *506*, 183–191.
13. Huang, C.; Mason, J. T. *Proc. Nat. Acad. Sci. USA* **1978**, *75*, 308–310.
14. Cowley, A. C.; Fuller, N. L.; Rand, R. P.; Parsegian, V. A. *Biochemistry* **1978**, *17*, 3163–3168.
15. Simon, S. A.; Lis, L. J.; Kauffman, J. W.; MacDonald, R. C. *Biochim. Biophys. Acta* **1975**, 317–326.
16. Lis, L. J.; Rand, R. A.; Parsegian, V. A. Chapter in this book.
17. Williams, R. J. P. In "Calcium in Biological Systems," Symposia of the Society for Experimental Biology, Duncan, C. J., Ed.; Cambridge University Press: Cambridge, 1976, Number 30, pp. 1–17.
18. Träuble, H.; Sackmann, E. *J. Am. Chem. Soc.* **1972**, *94*, 4499–4510.

RECEIVED October 17, 1978.

Electrical Properties of Membrane Systems: On Mechanisms of Interaction of Ca^{++} with Dipalmitoyl Lecithin at the Air–Water Interface

GIUSEPPE COLACICCO[1], MUKUL K. BASU, and FRANCINE A. TANSEY

Albert Einstein College of Medicine, 1300 Morris Park Avenue, Bronx, NY 10461

Penetration of electrolytes into both the air–water interface and films of dipalmitoyl lecithin is accompanied by a relatively small surface potential increase, whereas hydrolysis of $CaCl_2$ produces accumulation of $Ca(OH)_2$ and related species at the interface (1). Although in the absence of ionic lipids a correlation between interfacial ionic populations of the electrolyte and the surface potential changes is not yet possible, the marked surface potential effects of $CaCl_2$ accompanying the presence of small quantities of acidic phospholipids in dipalmitoyl lecithin films suggest that the acidic lipid contaminants are still the only certifiable species whose interaction with $CaCl_2$ produces an appreciable surface potential increase. Surface radioactivity and IR absorption spectra of dipalmitoyl lecithin in the presence of $CaCl_2$ produced no evidence of Ca^{++}–dipalmitoyl lecithin interaction.

In previous reports (*1, 2, 3*) we described the surface potential (ΔV) response of films of isoelectric phospholipids in the presence of anionic and cationic contaminants, either when the lipids were spread on electrolyte solution or when the latter was injected under the lipid film that had been spread on water. In the presence of 10 wt % acidic contaminant (dipalmityl phosphate, dicetyl phosphate, and dicapryl phthalate (DCP)

[1] Current address: Dr. G. Colacicco, Bioelectrochemistry Laboratory, Columbia University, BB Room 1411, 630 West 168th Street, New York, NY 10032.

0-8412-0473-X/80/33-188-057$05.00/1

DCP) the ΔV of mixed dipalmitoyl lecithin (DPL)–DCP film spread on $CaCl_2$ was appreciably higher than those on either NaCl or H_2O. However, when the electrolyte was injected under DPL films preformed on water, the ΔV remained unchanged and was thus the same on H_2O, NaCl, and $CaCl_2$; but it increased swiftly and markedly in the order $CaCl_2 >$ NaCl $> H_2O$ when the preformed DPL film contained 10 wt % or more acidic phospholipid (*1*). We interpreted this to mean that the larger ΔV differences on $CaCl_2$ over NaCl could be attributed to the interaction of the electrolyte cation with the negative fixed charge of the diffuse double layer of the acidic contaminants. The monolayer data were consistent with the observed effects of ionic lipids onto the electrical properties of neutral phospholipids at an oil–water interface (*4*). We also suggested that the ΔV increase of DPL films spread on $CaCl_2$ as compared with films spread on H_2O could be caused by interfacial penetration of the electrolyte—penetration which would not take place when the electrolyte was injected under a preformed DPL film (*1*). Although a quantitative correlation was established between the ΔV effect of Ca^{++} and the concentration of the acidic phospholipid contaminant (*1–4*), no such correlation was possible between ΔV effect and the large quantities of electrolyte ($CaCl_2$) penetrated into the air–lipid–H_2O interface (*1*). At that time IR spectroscopy revealed a novel phenomenon of adsorption of Ca^{++} in the form of $Ca(OH)_2$—adsorption which was largest at highest pH in the absence of lipid and was still appreciable when DPL was spread on $CaCl_2$ solution at pH 5.6 (*1*).

These studies bear on the much debated question whether the ΔV effect of $CaCl_2$ is caused by interaction of Ca^{++} ions with the phosphate group of the neutral phospholipid. We have maintained that such an interaction is absurd (*1–6*), whereas the opinion is divided among various laboratories, and strangely in the same laboratory after using two different techniques (*7, 8*). This indicates that there is much to be clarified regarding architecture and molecular and electrostatic properties of the DPL–H_2O–electrolyte interface.

In this chapter we report some new data regarding: (a) the ΔV effect of small concentrations of acidic phospholipid; (b) the pH dependence of surface radioactivity in the adsorption of $CaCl_2$; and (c) the IR spectra of DPL films on NaCl and $CaCl_2$ solutions. The purpose of this study is to obtain information and formulate new arguments toward the elucidation of molecular and electrostatic models of DPL membranes and the latter's interaction with electrolytes, with particular regard to the DPL–Ca^{++} system.

Materials and Methods

Dipalmitoyl phosphatidyl choline (L-α), alias DPL, and the sodium salt of dicetyl (hexadecyl or palmityl) phosphate were purchased from

Sigma Chemical Co., St. Louis, MO. Dimyristoyl and distearoyl lecithins (DML and DSL) and synthetic dipalmitoyl phosphatidic acid (DPPA) were products of Applied Science, State College, Pa.

The radiotracers $^{45}Ca^{++}$ and $^{36}Cl^{-}$ were obtained in the form of $^{45}CaCl_2$ and $H^{36}Cl$ solutions from New England Nuclear, Boston, MA. The methods for preparing $Ca^{36}Cl_2$ and $Na^{36}Cl$ solutions, distilled H_2O, and the criteria for homogeneity of lipids have been described; so were the monolayer techniques for surface pressure (π), ΔV, and surface radioactivity determinations as well as the procedure for the measurement of IR absorption of films and solutions by attenuated total reflectance (ATR) on germanium plates (*1, 9*). When necessary, details of procedures will be presented with the data under the specific experiment.

Results and Analysis

Numerical values are averages of data from at least three experiments; the deviations are presented under each figure. IR absorption curves are direct tracings from assays which were repeated at least twice.

Surface Potential Studies

Two types of experiments were carried out. In one, the lipid was spread from the organic solvent onto the electrolyte solution; in the other, the lipid film was first spread on distilled H_2O and small volumes of concentrated electrolyte solution (to make the final concentration) then were injected into the aqueous phase with continuous magnetic stirring. The desired pH was obtained by adding calculated volumes of either concentrated HCl or NaOH solutions; the solutions were unbuffered in order to avoid complications from the buffer's ions.

Films of Dipalmitoyl Lecithin and Related Lecithins Spread onto Electrolyte Solutions. In Figure 1, the ΔV of spread DPL films is presented for various values of π; DML and DSL are included for comparison. In line with previous data (*1, 2, 3, 10, 11*) for the three lecithins, the ΔV on $CaCl_2$ was higher than that on NaCl solutions, and the ΔV increments, $\Delta(\Delta V)$, became larger as the film pressure increased. In the absence of molecular correlates of surface tension and ΔV (*1, 2, 5, 6*), we shall refrain from speculations in trying to account for the influences of the lipid structure on the ΔV effect of $CaCl_2$. However, since the three lecithins have the glycerylphosphoryl choline moiety in common, the marked differences in behavior among them (*see* Figure 1) must be ascribed to specific effects of the different hydrophobic chains onto several unknown lipid functions at the air–lipid–H_2O interface; among such functions we shall mention lipid–H_2O and lipid–lipid interactions, dielectric constant effects, the influence of the lipid on the penetration, the organization and structure of both H_2O and electrolytes, and so on.

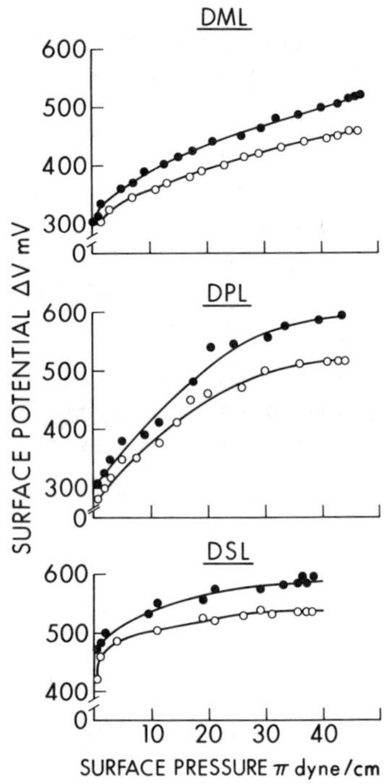

Figure 1. Lipid spread on the electrolyte solution. Change of ΔV with the surface pressure of lecithin films spread on 150 mequiv NaCl (○) and $CaCl_2$ (●), pH 5.6, 25°C: DML, DPL, and DSL. Deviations: ±5 mV; ±0.2 dyn/cm.

Influence of the Acidic Lipid. To a rapid inspection (*see* Figure 2), the presence of the acidic DPPA (in mixed DPPA–DPL films spread onto the electrolyte solution) produced $\Delta V - \pi$ curves which are similar to those in the absence of the acidic lipid. Two features are striking. First, in the $\Delta V - \pi$ curves, the ΔV difference between $CaCl_2$ (filled symbols) and NaCl (empty symbols) is nearly the same whether the concentration of acidic lipid was 10 or 50%; differential concentration effects, however, show some specificities in the low film pressure regions (*see* Figure 2) and in the ΔV — area curves at given film pressure values (*2, 3, 4*). Secondly, inasmuch as the appearance of the negative fixed

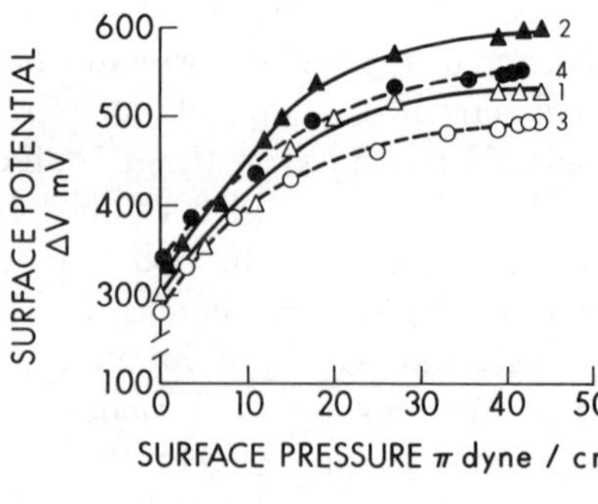

Figure 2. Lipid spread on the electrolyte solution. ΔV–π curves of mixed films of DPPA and DPL spread on 150 mequiv aqueous $CaCl_2$ (Curves 2 and 4) or NaCl (Curves 1 and 3), pH 5.6, 25°C. The concentrations of acidic lipid are expressed in weight percent. Deviations: ±10 mV; ±0.2 dyn/cm. DPPA: (——), 10%; (– – –), 50%.

charge of the anionic lipid at the membrane–H_2O interface is known to produce a negative change in ΔV, i.e. a negative zeta potentional (*2, 3, 4, 6, 12*), the greater the concentration of the acidic lipid, the lower the ΔV was and, in spite of the ΔV increase caused by Ca^{++} over Na^{+}, the more difficult it was for the acidic mixed film to obtain the higher ΔV of the isoelectric lecithin. This means that at a given $CaCl_2$ concentration the greater the acidic lipid:DPL ratio, the more the fixed negative charge remained unneutralized by Ca^{++} in the electrical diffuse double layer. Further detailed analyses must be postponed past these simple interpretations until we gain a better understanding of the molecular correlates of the surface parameters.

Electrolyte Injected under the Lipid Film. There is a marked difference in the ΔV response of phosphatidyl choline films to the electrolyte, depending on whether the lipid film is spread onto the electrolyte solution or the electrolyte is injected under the lecithin film that was spread first onto distilled H_2O (*1*).

Dipalmitoyl Lecithin. In contrast to the ΔV–π curves presented in Figure 1, no difference between NaCl (empty circles) and $CaCl_2$ (filled circles) is seen in the $\Delta V - \pi$ curves at 4 min after the electrolyte was injected under the DPL films that had been spread on distilled H_2O at a given initial pressure (π_i) value (*see* Figure 3); a modest increment in ΔV up to 25 mV occurred between 4 and 10 min. Electrolyte injection was accompanied by small surface pressure changes about whose significance we wish not to speculate. The rise of ΔV with the increase of film pressure (*see* Figure 3) is part of the general behavior of lipid films, irrespective of the composition of the aqueous phase.

Influence of the Acidic Lipid. Unlike DPL films, mixed films containing 10 and 50% DPPA showed a prompt response to the introduction

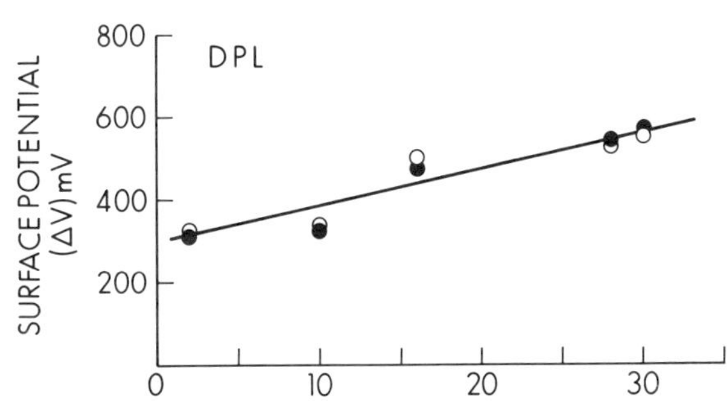

Figure 3. Electrolyte injected under DPL films. ΔV–π values of DPL films at 4 min after injections of NaCl (○) or $CaCl_2$ (●) in the aqueous phase (distilled H_2O) to a final 150-mequiv concentration. Error as in Figure 1.

of $CaCl_2$ into the aqueous phase. Most of the rise in ΔV occurred within the first 5 min (*see* Figures 4 and 5). Interestingly, the magnitude of the effect is not too different in the cases in which the concentration of the acidic lipid was 10 or 50%, suggesting that saturation or most of the effect occurs at concentrations less than 10 mol % DPPA. In a rapid inspection of the curves in Figure 4, two points of interest are noticeable. First, the initial ΔV values at time zero were not the same for the two concentrations of the acidic lipid, although no direct correlation is obvious. Secondly, certain features may represent specific effects to be related to structural characteristics of the acidic component, the latter's interaction with DPL in the film, the film pressure, etc.; e.g., notice the marked difference in the ΔV–time curves between 10 and 50% DPPA in the experiment in which the initial film pressure was 10 dyn/cm. Differences were not so conspicuous for other values of the film pressure. Any attempt at interpretation is unwarranted.

A more refined study of the effect of the acidic contaminants is described in Figure 5. In order to assess the significance of small quantities of anionic film constituents, it was important to establish the smallest concentration (of such contaminants) at which the ΔV effect can distinguish between Ca^{++} and Na^{+} or H_2O. For an electrolyte concentration of 150 mequiv and various concentrations of DCP the ΔV increment of $CaCl_2$ over water was related to the time which elapsed after electrolyte injection. In complete absence of exogenous acidic lipid, the kinetic curves for $CaCl_2$, NaCl, and H_2O were the same ($\Delta(\Delta V)$ = zero), during

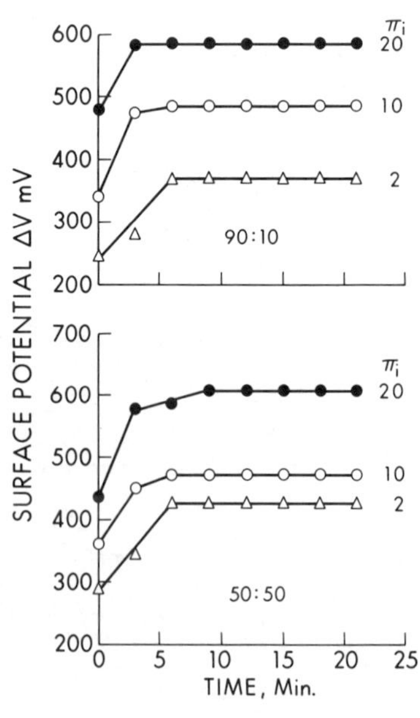

Figure 4. Electrolyte (150 mequiv $CaCl_2$) injected under DPL–DPPA mixed films. Kinetic curves of ΔV of films containing 10 wt % (upper panel) *and 50 wt %* (lower panel) *DPPA at three different values of π (2, 10, and 20 dyn/cm). Aqueous hypophase, pH 5.6, 25°C. The mixed-lipid film at the indicated pressure was spread first on distilled H_2O; the electrolyte then was injected beneath at time zero. Error as in Figure 2.*

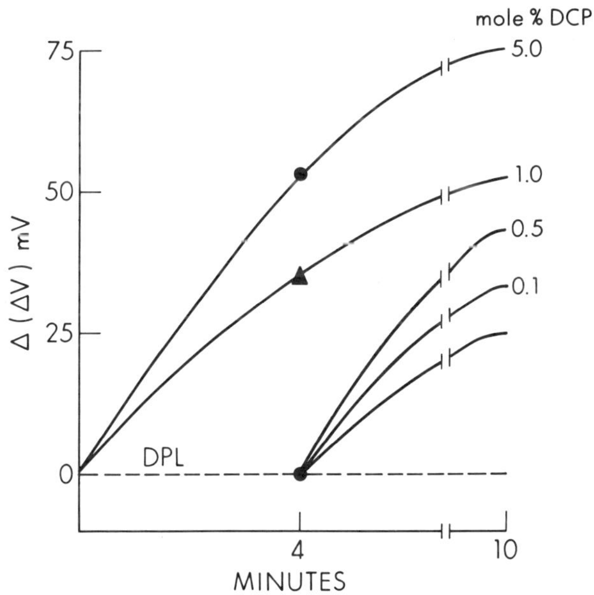

Figure 5. Electrolyte (150 meq $CaCl_2$) injected under DPL–DPC mixed films. Kinetic curves of ΔV increase Δ(ΔV) of mixed films of DPL containing the indicated mole percent concentration of acidic phospholipid, Na–DCP. Aqueous hypophase, pH 5.6, 25°C. The mixed films at 30 dyn/cm pressure were spread on distilled H_2O; the electrolyte then was injected beneath at time zero. The Δ(ΔV) values signifies the increase in ΔV of the DPL–DCP film on $CaCl_2$ over the ΔV on distilled H_2O. By the variance analysis, the rise of the first three curves (0.1, 0.5, and 1.0 mol % DCP) over that for DPL alone was highly significant. The accuracy of the electrometer readings and the π determinations were ±5 mV and 0.1 dyn/cm, respectively.

the first 4 min; but then with $CaCl_2$, the Δ(ΔV) value rose continuously until it reached a saturation value of 25 mV (lowest curve) at 10 min. The same lag period of 4 min was observed when 0.1 and 0.5 mol % DCP was introduced with DPL in the mixed film; however, past that time the Ca^{++} effect increased slightly though significantly above that for DPL alone.

The lag time disappeared when the DCP concentrations were 1 mol % and greater; we did not study 0.75 mol % DCP. Inasmuch as the lag period was the same for the control and for 0.1 and 0.5 mol % acidic lipid, the question arises as to whether the 25-mV effect of Ca^{++} with DPL is merely caused by penetration of the electrolyte into the DPL film, or is actually the effect of concentrations of acidic lipid contaminants below 0.1 mol % (contaminants which at such low levels probably cannot be removed from DPL by any known isolation procedure). A resolution of the two effects must await further experimentation. A definite message, however, is that 1 mol % acidic phospholipid is sufficient to produce a marked differentiation between Ca^{++} and H_2O (*see* Figure 5) and Na^{+} (not shown) by the ΔV technique.

Surface Radioactivity

In various attempts at establishing the stoichiometry of an eventual interaction of Ca^{++} with lecithin films, the increases in surface radioactivity observed in the presence of $^{45}Ca^{++}$ in the aqueous phase were ascribed to adsorption of Ca^{++}. Apart from the interpretations, the data themselves in this field are interestingly discordant (*13, 14, 15*). To appreciate the significance of the statement, we first present our own data from experiments in which the excess surface radioactivity in counts per minute (Δ cpm) in the presence of DPL film (cpm with DPL − cpm without DPL) are related to the hypophase pH; the aqueous phase contained 150m*M* NaCl and 93μ*M* $CaCl_2$ carrier (*see* Figure 6).

Except for a minimum at pH 3, for which we have no explanation, adsorption of Ca^{++} decreased from 90 cpm at pH 1 to zero at pH 7.5; above this pH value, there was a desorption of Ca^{++}, which became dramatic (200 cpm) between pH 9 and pH 10. The desorption of Ca^{++} under lecithin films at high pH was observed also by Rojas and Tobias (*13*), whereas two other groups of investigators observed a marked rise in surface radioactivity in going from pH 4 to pH 10 (*14*) and pH 11 (*15*). Although the discrepancies among several reports are serious, they could be explained in terms of variations in experimental conditions, to which we referred specifically on a previous occasion (*1*). With regard to the Ca^{++} sorption in the high pH region, we note that desorption of $^{45}Ca^{++}$ occurred from lecithin films spread onto solutions of relatively high ionic strength (150m*M* NaCl and 93μ*M* $CaCl_2$ in the present experiments,

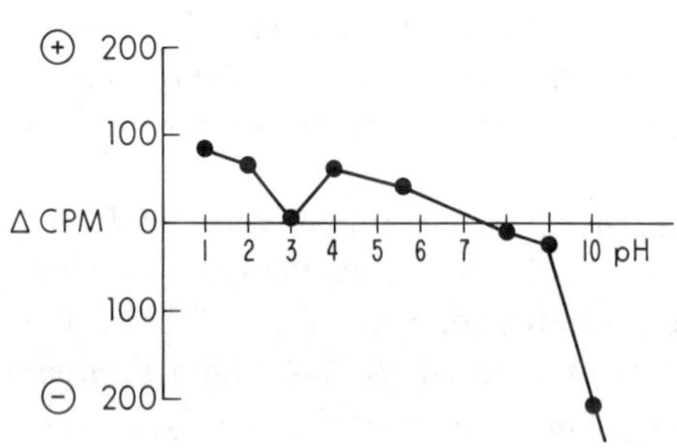

Figure 6. Effect of subphase pH on adsorption and desorption of $^{45}Ca^{++}$. Films of DPL were spread on 0.15M NaCl at 25°C, containing 93μM $CaCl_2$ carrier and brought to the desired pH by addition of either HCl or NaOH solutions. The Δ cpm = cpm with DPL − cpm without DPL.

and 50mM NaCl and 100μM $CaCl_2$ in the report of Rojas and Tobias (*see* Ref. *13*, pp. 400, 401).) Where, in contrast, the $^{45}Ca^{++}$ adsorption increased with the pH, the ionic strength was low—namely 10mM NaCl and 0.13μM $CaCl_2$ in one case (*15*) and only 100μM $CaCl_2$ in the other case (*14*).

This account is consistent with two of our own observations. At a given $CaCl_2$ carrier concentration at pH 5.6, the Ca^{++} sorption rose as the electrolyte concentration in the aqueous phase increased and it plateaued at about 150 mequiv of either NaCl or $CaCl_2$ (*1*). In the absence of lipid, the adsorption of $^{45}Ca^{++}$ was greater as the pH of the suphase increased; this was markedly so between pH 9 and pH 10 (*see* Ref. *1,* Table I). In light of the observation that, in the absence of lipid, the surge of $^{45}Ca^{++}$ at the air–water interface coincided with the accumulation of $Ca(OH)_2$ and/or other possibly related species (*1*), it is conceivable that with DPL films at pH's as high as 9 or 10, greater concentrations of electrolyte promote collapse of the interfacial $Ca(OH)_2$ layer and loss of ^{45}Ca to the aqueous phase. Such a barrier layer, however, would be stable and account for the surge of $^{45}Ca^{++}$ sorption by the lecithin films at high pH (*14, 15*) and low ionic strength, since in the absence of lipid, Ca^{++} sorption was greatest at pH 10 (*1*).

Spectroscopic Studies

In spite of the great confusion in data and interpretations of whether and how Ca^{++} interacts with lecithins, no attempt was made to obtain information from IR spectroscopy. Using this technique, we reported that for a given concentration of $CaCl_2$ in the aqueous phase, the IR absorption by the air–water interface increased with the pH of the aqueous phase, presumably from $Ca(OH)_2$ (*1*). Furthermore, at pH 5.6 addition of DPL-spread films produced a significant increase in the intensities of the IR absorption peaks of interfacial $Ca(OH)_2$ at 2.90, 4.65, and 6.15 μm, and conferred to the 6.15 μm peak the doublet structure typical of $Ca(OH)_2$ at pH 10. In the present study we expand on the significance of the IR absorption of interfacial $Ca(OH)_2$ and we approach the question of the Ca^{++}–lipid phosphate interaction by probing the IR absorption of DPL films on $CaCl_2$ solutions at the frequencies attributed to the P = O and P–O–C bonds of the lipid phosphate at wavelengths between 8 and 10 μm.

Interfacial $Ca(OH)_2$. After either the germanium plate was dipped through the interface of the $CaCl_2$ solution, or the $CaCl_2$ solution was deposited onto the germanium plate, the ATR IR spectra showed three peaks at wavelengths of 2.90, 4.65, and 6.15-μm. The peak at 4.65 μm, which was of uncertain origin, had the smallest intensity. Since in the absence of $CaCl_2$, there were no peaks (in either the absence or presence of DPL film), and since the peaks at 2.90 μm and 6.15 μm are attributed to O–H stretching frequencies such as those of H_2O and presumably $Ca(OH)_2$, this IR absorption must be related either to the O–H bond of $Ca(OH)_2$ or to H_2O structures that are modified by Ca^{++} ions and related species.

The data in Figure 7 feature two experiments—one with and the other without DPL film. The upper panel describes the dependence of those IR absorption peaks on the pH of the aqueous phase containing

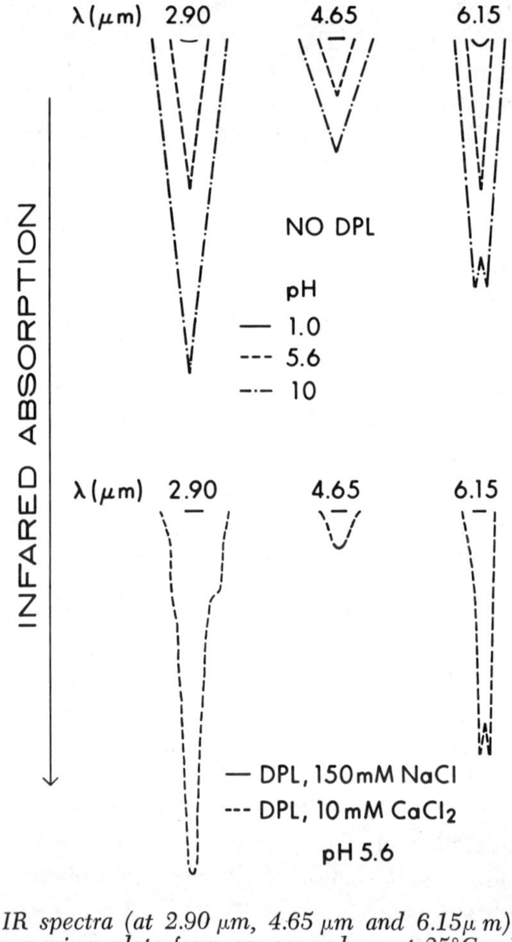

Figure 7. IR spectra (at 2.90 μm, 4.65 μm and 6.15μ m) of films collected on germanium plate from aqueous phase at 25°C, pH 5.6. Upper panel: *0.075M $CaCl_2$ at different pH values without DPL films.* Lower panel: *DPL films spread on either 0.15M NaCl or 0.010M $CaCl_2$. The solution, with or without the lipid film, was left to stand at room temperature for a given time. The germanium plate then was dipped through the interface four times, dried under a stream of N_2 gas, and examined in the ATR accessory (for details,* see *Refs.* 3 *and* 9*).*

75m*M* $CaCl_2$ in the absence of lipid film. In the lower panel are two curves describing the influence on the same peaks by 150m*M* NaCl and 10m*M* $CaCl_2$ at pH 5.6 in the presence of DPL films.

For a given $CaCl_2$ concentration, in the absence of DPL, the intensity of the IR absorption at all three frequencies increased with the pH (*shown*) and with the $CaCl_2$ concentration at a given pH value above 4 (*not shown*). A feature of interest in the upper panel is the splitting of the 6.15-μm peak at pH 10.

In the presence of DPL (*lower panel*), the IR absorption was nil on 150m*M* NaCl and very large on 10m*M* $CaCl_2$, suggesting a specific effect of $CaCl_2$. Moreover, DPL on 10m*M* $CaCl_2$ (*see* Figure 7) and on 75m*M* $CaCl_2$ (*1*), pH 5.6, produced the frequency-splitting effect which was typical of $Ca(OH)_2$ at pH 10 (*upper panel*); the doublet did not appear on $CaCl_2$, pH 5.6, in the absence of DPL (*1*). This suggests that at pH 5.6, DPL confers upon the interfacial H_2O or $Ca(OH)_2$ structures certain characteristics of the $Ca(OH)_2$ at pH 10. Such a spectroscopic feature might be brought about by the modifications that the lecithin operates on the interfacial lattices of the aqueous solvent. Another modification is evident also in the peak at 2.90 μm in the form of a partial shift to lower frequencies. This can be interpreted to mean that, in the presence of DPL (and not in its absence, *upper panel*), certain populations of O–H of either H_2O or $Ca(OH)_2$ are engaged in hydrogen bonding (*18*). Irrespective of whether this bonding is to other H_2O or $Ca(OH)_2$ molecules or to the polar groups of DPL, the central thesis is that it is Ca^{++} that brings about such modifications in the molecular organization of the air–H_2O and DPL–H_2O interfaces. Whenever Ca^{++} produces effects that are demonstrable in terms of modifications of the spectroscopic characteristics of the phosphate group of the isoelectric lecithin by either NMR (*15, 17*) or IR (*see* Figure 8, Curve 3), the effect does not necessarily represent direct Ca^{++}–phosphate binding, but it could mean some kinds of water binding promoted by Ca^{++}. The detailed mechanisms of such phenomena remain to be established.

Probing Ca^{++}–Phosphate Binding. Although this question has been at the center of the current investigations, little experimentation was reported on direct evidence which could be furnished by IR spectroscopy. In a brief mention of IR absorption of DPL–uranyl nitrate in nujol paste, it was concluded that indeed Ca^{++} interacts with the lipid phosphate group (*7*). Subsequently, after thermodynamic analysis of calorimetric studies with DPL dispersions in aqueous electrolyte, the same laboratory suggested the absence of direct Ca^{++}–phosphate interaction (*8*). Strangely, the authors of the second work (*8*) failed to provide explanations for the discrepancy between this and a previous report (*7*). The particular predicament implies that either one of the two experiments had to be

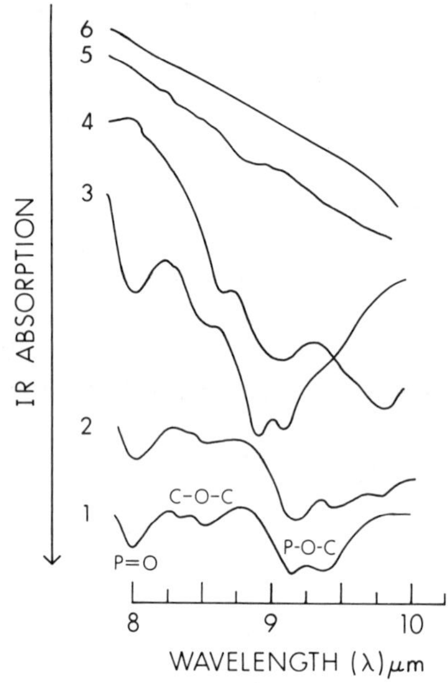

*Figure 8. IR spectra (at 8 to 10 μm) of DPL films either deposited directly (Curve 1) or collected from the interface onto germanium plate (other curves). Curve 2, DPL on 0.15*M *NaCl; Curve 3, DPL on 0.001*M *$CaCl_2$; Curve 4, DPL on 0.010*M *$CaCl_2$; Curve 5, DPL on 0.075*M *$CaCl_2$; and Curve 6, interface of 0.075*M *without DPL film. Spectroscopy technique as in Figure 7.*

reinterpreted. In the IR study (*7*), we see a deficiency, which will be apparent during the illustration of our own work.

In Figure 8 are the tracings of the IR absorption in the 8 to 10-μm region of the spectrum, namely, the P = O, C–O–C, and P–O–C bonds of DPL, in films collected from the interface of aqueous solutions with different concentrations of NaCl and $CaCl_2$. For comparison the spectrum of DPL deposited on the germanium plate from organic solvent is presented in Curve 1. The spectrum of the DPL film collected on the same plate from the surface of 150 mequiv NaCl (Curve 2) shows little modification in the absorption of the phosphate region. With 1m*M* $CaCl_2$ (Curve 3), the P=O bond was very slightly affected (red shift from 8.00 to 8.05 μm), whereas the peak of the P–O–C stretching was shifted to a higher frequency and the slope of the absorption–wavelength curve between the P=O and the P–O–C frequencies was increased markedly. Both effects suggest strengthening of the P–O–C bond, probably owing to conformational and environmental changes (*18*) induced on the DPL molecules by the solvent (the $CaCl_2$ solution). A shift of the C–O–C frequency to greater wavelengths is also apparent, but we have no explanation for it. The shift of 5/100 μm for the P=O bond may be too small to be significant, but it should not be ignored, irrespective of whether the partner in the interaction is Ca^{++} itself or related $Ca(OH)_2$ and H_2O structures. More dramatic effects were observed with DPL films collected from the surface of 10 and 75m*M* $CaCl_2$; the P=O absorption peak at 8 μm vanished, and the P–O–C peak was distorted in 10m*M* $CaCl_2$ and absent in 75m*M* $CaCl_2$. We should hasten to say that disappearance of the P=O and P–O–C peaks cannot be interpreted to mean interaction of Ca^{++} with those bonds because the IR absorption of the solvent's interface without DPL was so great (Curve 6) as to mask any specific effect of Ca^{++}. The absorption blank of 10m*M* $CaCl_2$ (not shown) bespeaks the same argument. The huge IR absorption of the $CaCl_2$ blank solutions in the 8 to 10-μm region could be caused by the particular interfacial water structures brought about by the accumulation of $Ca(OH)_2$ at the interface. This is consistent with the increase of the 2.9 and 6.15-μm peaks (which can be ascribed to the O–H stretching frequencies of water and possibly also of $Ca(OH)_2$) with increasing $CaCl_2$ concentration. Masking of the 8 to 10-μm peaks of DPL by the solvent is a difficulty which was ignored by other investigators. From a marked red shift of the P=O peak of a DPL–uranyl nitrate–nujol paste, Chapman et al. concluded that the uranyl ion interacted with the lecithin's phosphate group (*7*). Unfortunately, separate spectra of lecithin and uranyl nitrate in nujol were not shown; indeed, we could not interpret the IR spectra of DPL films collected from the surface of 10m*M* uranyl nitrate because a huge absorption peak of uranyl nitrate masked the P=O peak of lecithin

Discussion and Conclusions

We have used a multiple approach in trying to establish interfacial mechanisms and structures pertaining to the interaction of Ca^{++} with films of DPL.

Using electrolyte injection under DPL films and ΔV measurements under conditions in which $CaCl_2$ has no effect (*see* Figure 5), we conclude beyond a doubt that acidic lipid contaminants are the only certifiable species whose interaction with Ca^{++} can be identified with a prompt increase in ΔV. Similarly the ΔV effects that generally are observed when lecithin films are spread onto $CaCl_2$ solutions could be produced either by the presence of undetectable quantities of anionic contaminant lipids or by the penetration and orientation of the ions of the electrolyte and water (*1, 5*). In the light of previous experience (*see* Ref. *16,* Figure 8), we used the injection techniques in order to gain appreciation of phenomena that could transpire from the effects of variations in the order of addition of the interaction participants. This criterion, so well known to biochemists, should be exploited in the study of interfacial phenomena.

By measuring surface radioactivity of DPL films spread onto aqueous solutions containing the radiotracers $^{45}Ca^{++}$ and $^{36}Cl^{-}$, we have established that in the presence of relatively high electrolyte concentration (150 mequiv NaCl (*see* Figure 6)), the adsorption of Ca^{++} decreased from a highest value at pH 1 to zero at pH 7.5, and desorption to very large values occurred in going from pH 7.5 to pH 10; the very large desorption of Ca^{++} from DPL films at high pH is conceivable in light of the very large adsorption of Ca^{++} as $Ca(OH)_2$ at pH 10 in the absence of DPL (*see* Ref. *1*), as if DPL breaks a preformed $Ca(OH)_2$ lattice, which at this pH is otherwise stable.

The IR studies greatly facilitated our interpretation of the data from the other studies. For instance, in the absence of DPL film, accumulation of Ca^{++} at the interface was accompanied (a) by formation of interfacial $Ca(OH)_2$ as suggested by IR spectroscopy data and (b) by desorption of Cl^-, as suggested by the measurements of surface radioactivity on $CaCl_2$ solutions containing the $^{36}Cl^-$ nuclide (*1*). These data are consistent with a mechanism of $CaCl_2$ hydrolysis; the reaction is driven to the right and thus to more $Ca(OH)_2$ accumulation in base and to the left and thus to depletion of interfacial $Ca(OH)_2$ in acid (*see* Figure 7). A concept of some biological consequence in this respect is that although $Ca(OH)_2$ and HCl are products of the same hydrolysis reaction, in the absence of DPL, $Ca(OH)_2$ accumulates at the interface whereas the constituents of HCl seek the solution. This implies that the interfacial $Ca(OH)_2$ structures have some mechanism of defense from or inpenetrability to HCl. This barrier to the transport of HCl into the lipid-free interfacial lattice could serve as an elementary model in the approach to mechanisms of membrane transport.

Relation to Other Studies. Interaction of Ca^{++} with phosphatidyl choline (PC) mono- and bimolecular membranes has been investigated carefully by Hauser and Phillips (*15*) and McLaughlin (*17*). Using a variety of approaches such as ΔV, surface radioactivity, zeta potential measurements, and NMR studies, both groups of investigators concluded that Ca^{++} interacts with the phosphate residues of lecithin; the extent of the interaction, however, is very small since it involves less than 5% of the phosphate residues (*17*).

The paucity of the interaction raises new questions such as whether the interacting phosphate groups belong to the phosphatidyl choline or to some acidic lipid–phosphate contaminant, which could be silent to ΔV and zeta potential measurements. Such a possibility finds support in the observation (*see* Figure 5) that up to 0.5 mol %, exogenous dipalmityl phosphate was not detected promptly by ΔV, but it was detected after several minutes. Certain techniques involving equilibration and saturation might reveal such a reservoir of acidic contaminants that are not apparent in the early phase of kinetic studies. Another possibility is the partial interaction by the hydrogen bond of the O–H groups of interfacial $Ca(OH)_2$ and specially structured H_2O with the phosphate group of either lecithin or acidic phospholipid contaminants. This possibility finds support in the observation of a marked though partial red shift of the 2.9 μm IR peak of either water or $Ca(OH)_2$ in the presence of DPL films (*see* Figure 7, *lower panel*). The special water structures inevitably are brought about by interfacial Ca^{++} and may very well be so by other polyvalent cations such as Co^{++} (*17*) which mimic or resemble Ca^{++} in their interaction with acidic lipids and interfacial water. Inasmuch as even 1m*M* $CaCl_2$ caused a marked deformation of the IR absorption of the P–O–C bond, whereas this and the P═O absorptions were masked totally in 10 and 75m*M* $CaCl_2$, the question of a precise mechanism for the interaction of the lecithin–phosphate with Ca^{++} must be extricated from the complexity of the viscous interfacial lattices that are formed at the $CaCl_2$ concentrations used in our experiments (*19*). Much more work is required to resolve the various molecular and ionic contributions of these membranous structures to the surface parameters such as electrical potential, surface tension, surface viscosity, and spectroscopic absorption.

Mechanisms and Interfacial Structures. Meanwhile, we may provide some schematic models (*see* Figure 9) and a critical assay of the mechanisms that reflect the views either expressed in the literature or presented in this communication.

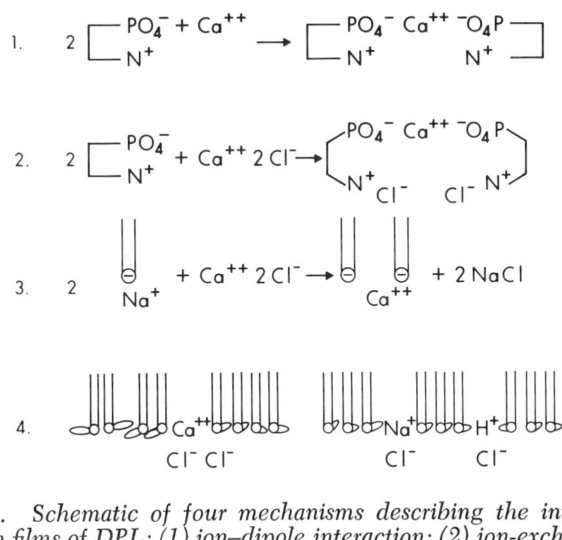

Figure 9. Schematic of four mechanisms describing the interaction of Ca^{++} with films of DPL: (1) ion–dipole interaction; (2) ion-exchange mechanism; (3) ion–ion interaction with ionized anionic lipid contaminant; and (4) penetration of electrolyte, H_2O, and derived ions into the air–water or the lipid–water interface. A highlight of Mechanism 4 (consistent with the surface radioactivity data, Ref. 3) is the adsorption of the ions of HCl resulting from th ehydrolysis of $CaCl_2$. The coexistence of $Ca(OH)_2$ and aqueous HCl at the interface requires the formation of compartments or pools that permit the separation of the acid from the base. Such a coexistence of acidic and basic pools is conceivable in the light of the $Ca(OH)_2$ film on the HCl solution following the hydrolysis of $CaCl_2$ in the absence of DLP films and is probably a characteristic of DPL films, since the adsorption of Cl^- was nil without DPL.

According to Mechanism 1, the ion–dipole interaction proposed by Shah and Schulman (*11*), Ca^{++} binds to the oxygen of a polarized P–O bond in the structure of the phosphorylcholine zwitterion. In spite of the speculative claims of Shah and Schulman (*11*) and the theoretical affirmations of Gillespie (*20*), this mechanism must be rejected on the basis of the theoretical arguments presented here and elsewhere (*2, 5, 6*). In brief, if such a bond existed (and the evidence (IR or NMR) is not available), it should not bear any direct relation to ΔV and surface dipole moments, which are generated only by ionized species and not by silent ion pairs nor by partial charges of polarized covalent bonds (*2, 5*).

In Mechanism 2, which was mentioned by myself (*2*) and Shah and Schulman (*11*) and is consistent with the tenets of ion-exchange theory, Ca^{++} and Cl^- bind respectively to the ionic phosphate and quaternary nitrogen of the unfolded salt linkage. Although consistent with the observed increase in ΔV (*2, 6, 12, 21*), this mechanism is contradicted by two observations. First, in the range between pH 4 and pH 9, the zwitterion is not titratable with either H^+ or ^-OH (*22, 23*). Since in the Hofmeister series H^+ is adsorbed much more effectively than Ca^{++} or Al^{+++} onto acidic surfaces, it is inconceivable that Ca^{++} would displace the proton at the low Ca^{++} concentrations used by those who claim Ca^{++}–DPL phosphate binding, unless both a special mechanism and the related energetics are accounted for fully. Secondly, the binding constant of Ca^{++} with an acidic phospholipid monolayer varied with the Ca^{++} concentration (*15*); the anomaly could be explained by large excesses of Ca^{++} ions that penetrate the interface outside the binding or screening stoichiometry that is calculated from ΔV data, and such excesses were very large in the case of DPL (*see* Figure 5, *lowest curve,* and Ref. *1*). Finally, for the small percentage of binding (*see* Figure 5 and Refs. *15* and *17*), one cannot discount the possibility that the binding lipid species be acidic contaminants and not lecithin itself.

According to Mechanism 3, Ca^{++} binds to the negative fixed charge in the electrical double layer of an acidic lipid. Although this mechanism, proposed by myself (*1, 2, 3, 6*), accounts for the observed increase in ΔV, it does not account for the large quantities of Ca^{++} that other approaches have shown to accumulate at the interface (*see* Figures 6 and 7 and Ref. *1*).

Mechanism 4 also was born in this laboratory (*1–6, 24*). A rise in ΔV could be determined by the penetration and $\pm$ orientation of the ions of electrolyte, H_2O, and their products of interaction. Such ions could be part of either the hydration shell of the lipid's molecular organization or the thick interfacial viscosity layers that are located above and/or below the mathematical line (*25*) of surface tension. The measurements of IR absorption and surface radioactivity (*see* Figures 6, 7, 8 and Ref. *1*) provided the evidence for the complexity of the molecular composition of the interface. Detailed knowledge of the latter's topography will be indispensable for some understanding of the structure–function relationship of artificial and natural membranes and will be possible only by applying direct spectroscopic techniques (*26*) to the air–water and lipid–water interfaces in situ, in vivo.

Acknowledgments

This work was supported by Grants from the NIH Heart, Lung, and Blood Institute.

Literature Cited

1. Colacicco, G.; Basu, M. K. *Biochim. Biophys. Acta* **1978,** *509,* 230–238.
2. Colacicco, G. *Biochim. Biophys. Acta* **1972,** *266,* 313–319.
3. Colacicco, G. *Chem. Phys. Lipids* **1973,** *10,* 66–72.
4. Colacicco, G. *Nature* **1965,** *207,* 936–938.
5. Colacicco, G. *Biophys. J.* **1978,** *21,* 48a.
6. Colacicco, G. In "Biological Horizons in Surface Science," Prince,M. L., Sears, D. F., Eds.; Academic: New York, 1973; pp. 247–284.
7. Chapman, D.; Urbina, J.; Keough, K. M. *J. Biol. Chem.* **1974,** *249,* 2512–2521.
8. Chapman, D.; Peel, W. E.; Kingston, B.; Lilley, T. H. *Biochim. Biophys. Acta* **1977,** *464,* 260–275.
9. Colacicco, G.; Basu, M. K.; Buckelew, A. R., Jr.; Bernheimer, A. W. *Biochim. Biophys. Acta* **1977,** *465,* 378–390.
10. Vilallonga, F.; Fernandez, M.; Rotunno, C.; Cereijido, M. *Biochim. Biophys. Acta* **1969,** *183,* 98–109.
11. Shah, D. O.; Schulman, J. H. *J. Lipid Res.* **1965,** *6,* 341–349.
12. Davies, J. T.; Rideal, E. K. "Interfacial Phenomena"; Academic: New York, 1961.
13. Rojas, E.; Tobas, J. M. *Biochim. Biophys. Acta* **1965,** *94,* 394–404.
14. Kimizuka, H.; Kabahara, T.; Nejo, H.; Yamauchi, A. *Biochim. Biophys. Acta* **1967,** *137,* 549–556.
15. Hauser, H.; Philips, M. C. *Proc. Intl. Congr. Surf. Act. Subst., 6th* **1973,** *2,* 371–380.
16. Colacicco, G. *J. Colloid Interface Sci.* **1969,** *29,* 345–364.
17. Lau, A. L.; McLaughlin, A.; MacDonald, R. C.; McLaughlin, S. G. Chapter 3 in this book.
18. Bellamy, L. J. "Advances in Infrared Group Frequencies"; Methuen and Co. Ltd.; Bungay, Suffolk, England, 1968.
19. Colacicco, G.; Buckelew, A. R., Jr.; Scarpelli, E. M. *J. Colloid Interface Sci.* **1974,** *46,* 147–151.
20. Gillespie, C. J. *Biochim. Biophys. Acta* **1970,** *203,* 47–61.
21. Davies, J. T. *Proc. R. Soc.* **1951,** *A208,* 224–247.
22. Bangham, A. D.; Papahadjopoulos, D. *Biochim. Biophys. Acta* **1966,** *126,* 181–184.
23. Abramson, M. B. *J. Colloid Interface Sci.* **1970,** *34,* 571–579.
24. Colacicco, G.; Basu, M. K. "Abstracts of Papers, Colloid Surf. Sci. Symp.," 51st National Meeting, ACS, 1977.
25. Adam, N. K. "The Physics and Chemistry of Surfaces"; Oxford University Press: Oxford, 1971.
26. Ockman, N. *Biochim. Biophys. Acta* **1971,** *345,* 263–282.

RECEIVED October 17, 1978.

5

Effect of Cation Concentrations and Temperature on the Rates of Aggregation of Acidic Phospholipid Vesicles: Application to Fusion

SHLOMO NIR, JOE BENTZ, and ARCHIE R. PORTIS, JR.[1]

Department of Experimental Pathology, Roswell Park Memorial Institute, 666 Elm Street, Buffalo, NY 14263

Rates of aggregation of acidic phospholipid vesicles are calculated from potential curves obtained as a sum of van der Waals and electrostatic interactions. The effect of temperature on rates of aggregation and on the equilibrium distribution is analyzed. The applicability of activation energies is discussed critically. The effects of cation binding (Ca^{2+}, Mg^{2+}, and Na^{+}) to phosphatidylserine vesicles are analyzed, and the rates of vesicle aggregation are shown to be very sensitive to cation concentration. Experimental results from light scattering and release of trapped fluorescent molecules are presented. The possibility is raised that Ca^{2+} or Mg^{2+} can produce destabilization of PS vesicles followed by aggregation and fusion. High concentrations of Na^{+} inhibit the fusion process although Na^{+} alone ($>$ .4M) produces fast aggregation.

The extension of the Smoluchowski (*1*) theory for rates of aggregation of colloidal particles by Fuchs (*2*) requires a detailed knowledge of the free energy potential curves (*see* review in Verwey and Overbeek (*3*)). In the case of charged particles, these potentials have been taken

[1] Current address: U.S. Department of Agriculture, Urbana, IL 61801.

0-8412-0473-X/80/33-188-075$08.00/1

as a sum of van der Waals (VDW) and electrostatic interactions. As reviewed (*3*), this approach gives qualitative predictions regarding the stability of many colloidal systems, and it has been expected that the prediction of rates of aggregation should improve with corresponding progress in the estimate of the potential curves. Overbeek (*4*) pointed out that in spite of improvements in the estimate of potentials and the introduction of various hydrodynamic corrections, the theory could not explain the approximate independence of size in the measured rates of aggregation of colloidal particles. According to the theory, the rate of aggregation of colloidal particles should be reduced significantly with an increase in colloidal particle size. At the same time, the direct-force measurements in vacuum or in air (*5–12*) showed that the long-range interactions between particles are explained well.

In a previous work (*13*) on phospholipid vesicles, we proposed that large particles, i.e., of radii larger than 1,000 Å, can form stable aggregates in which the distances of separation between particles correspond to secondary minima, i.e., a few nanometers. The aggregation in secondary minima is a rapid process since it does not involve a high potential barrier. We also proved that for sonicated vesicles of radii between 150 and 500 Å, the first step in the aggregation process, i.e., the formation of dimers, must occur at the primary minimum. Thus, our study of sonicated phosphatidyl serine (PS) vesicles considered the process of aggregation to occur at the primary minimum. It qualitatively explained the existing data on vesicle aggregation for given concentrations of Ca^{2+}, Mg^{2+}, and Na^{+} in solution, and provided some more predictions which were confirmed. One of the main elements of this treatment has been the detailed consideration of the neutralization of the negative charges of PS by Ca^{2+}, Mg^{2+}, and Na^{+}. In the present study, we elaborate on some aspects of the computations of rates of aggregation of phospholipid vesicles. We further examine the sensitivity of rates of aggregation to cation concentrations and study the dependence of rates of aggregation on temperature. Our analysis will show that in many cases the rates of aggregation should vary appreciably with temperature. Thus the experimental examination of the variation of the rate of aggregation with temperature can provide another test for the theory.

We have another purpose in studying the temperature dependence of the rate of aggregation. It may be expected that the close approach of vesicles is the first stage in their fusion. The process of membrane fusion is known to be sensitive to temperature (*14, 15, 16*). In order to elucidate the temperature dependence of the rate of fusion per se, we have first to determine the temperature dependence of the process of close approach. We provide here light-scattering measurements which were designed to test our predictions for the aggregation of sonicated

PS vesicles. In addition, we present the results of fluorescent probe leakage experiments which together with the aggregation experiments enable us to draw some conclusions on the mechanism of fusion of PS vesicles.

Expressions for the Rate of Aggregation of Vesicles

We assume that the initial phase of the aggregation process consists of forming pairs of vesicles or dimers and will focus on the dimerization reaction. Light-scattering experiments (*17*) can elucidate the kinetics of dimerization when carried over periods of time short enough to preclude a significant accumulation of larger aggregates. The dimerization reaction is given stoichiometrically as,

$$2\mathrm{C}_1 \underset{k_2}{\overset{k_1}{\rightleftharpoons}} \mathrm{C}_2$$

with reaction rate equation

$$-\frac{d\{C_1\}}{dt} = 2k_1\{C_1\}^2 - 2k_2\{C_2\} \quad (1)$$

where C_1 and C_2 are the monomeric and dimeric species, $\{C_1\}$ and $\{C_2\}$ are their respective molar concentrations, and k_1 and k_2 are the reaction rate constants.

According to Smoluchowski (*1*), k_1 is given by

$$k_1 = 4kT\mathrm{N_A}/(3\eta) \quad (2)$$

in which k is Boltzmann's constant, T is the absolute temperature, $\mathrm{N_A}$ is Avogadro's number, and η is the viscosity of the medium. For sufficiently short times, the second term in Equation 1 can be neglected and an integration yields a half time, $t_{1/2}$, for dimerization given by

$$t_{1/2} = 1/2k_1C_1(0), \quad (3)$$

where $C_1(0)$ is the initial molar concentration of vesicles. Thus a measurement of these half times should show that they are inversely proportional to $C_1(0)$. If such a relation is not obeyed, it means that effects of higher stages in the aggregation process are superimposed. The neglect of the second term in Equation 1 can yield other inverse linear relations such as $t_{3/4} = 1/6k_1C_1(0)$ where we refer to a shorter time for which $C_1(t) = 3/4C_1(0)$. For any number q between zero and one, we could define the time t^* by $C_1(t^*) = qC_1(0)$ and the integration would yield

$t^* = (1/q - 1)/2k_1C_1(0)$. For a given value of q the inverse linear relation between t^* and the initial monomer concentration $C_1(0)$ still holds. The neglect of higher-order aggregates becomes more justified as t^* gets smaller or as q approaches 1. Clearly the best estimates of the rate constant k_1 will be obtained by taking measurements (e.g., light scattering) for a value of q near 1, but this requires greater machine sensitivity than would be the case for measuring $t_{1/2}$ — when $q = 1/2$.

Overbeek (*18*) has shown that taking higher-order aggregates into account, with certain approximations, also will give an inverse linear relation between the initial number of particles (monomers) and the time over which the number of particles is halved. Therefore, from an inverse linear plot between initial concentrations and the time it takes to achieve a certain increase in light scattering, we cannot necessarily deduce the half times for dimerization. Increases in light scattering of less than 50% would be necessary to minimize the contribution of higher-order aggregates.

According to the treatment of Fuchs (*2*), k_1 is given by

$$k_1 = 4kT\mathrm{N_A}/(3\eta W) \tag{4}$$

where

$$W = 2a \int_{2a}^{\infty} \frac{dR}{R^2} \exp\left(\frac{V_T(R)}{kT}\right) \tag{5}$$

The term $2a$ is the center-to-center distance between the vesicles in the dimeric state and $V_T(R)$ is the total free energy caused by intermolecular interactions—as a function of the center-to-center distance R. A convenient simplification is used frequently (*3*),

$$W \lesssim \exp\,(V_T{}^*/kT) \tag{6}$$

in which $V_T{}^*$ is the height of the maximum of the potential barrier. The introduction of hydrodynamic corrections may result in an order-of-magnitude reduction of the rate of aggregation (*19–25*). Following Spielman (*22*), the expression for W in Equation 5 becomes

$$W = 2a \int_{2a}^{\infty} \frac{1}{R^2} \frac{D_{12}(\infty)}{D_{12}(R)} \exp\left(\frac{V_T(R)}{kT}\right) \mathrm{d}R \tag{7}$$

in which $D_{12}(R)$ and $D_{12}(\infty)$ are the diffusion coefficients of the relative motions of spherical particles at a distance R and at infinity. For small separations, i.e., when $d = R - 2a$ is small when compared with R, there is a simplification (*21, 22*),

$$\frac{D_{12}(\infty)}{D_{12}(R)} \cong \frac{a}{2d} \tag{8}$$

so that

$$W = a \int_{2a}^{\infty} \frac{1}{R^2} \left(\frac{a}{R - 2a} \right) \exp \left(\frac{V_T(R)}{kT} \right) dR \tag{9}$$

For spherical particles, our calculations of the total free energy V_T are in most cases the same as in Ref. *13*. Here we outline the approach and present a minimal number of equations needed for the analysis in the subsequent section. The potential V_T is given as

$$V_T = V_e + V_w \tag{10}$$

in which V_e is the electrostatic free energy, and V_w is the VDW free energy. In cases of slow aggregation, the most important term for determining the rate constant k_1 is V_e, which is positive and forms the repulsive barrier in a system of negatively charged PS vesicles. We use the equation of Wiese and Healy (*26*), with the boundary conditions of fixed surface charges.

$$V_e = \frac{4\pi^2}{\epsilon\kappa^2} \left(\frac{a_1 a_2}{a_1 + a_2} \right) \left\{ (\sigma_1 + \sigma_2)^2 \ln \left(\frac{1}{1 - \exp(-\kappa d)} \right) + (\sigma_1 - \sigma_2)^2 \ln \left(\frac{1}{1 + \exp(-\kappa d)} \right) \right\} \tag{11}$$

in which σ_i is the surface charge density of vesicle i of radius a_i, ϵ is the dielectric constant of the medium, and $1/\kappa$ is the Debye–Hückel length. The Greek letter κ is given by

$$\kappa^2 = \frac{4\pi e^2}{\epsilon kT} \sum n_j z_j \tag{12}$$

where n_j and z_j are the density numbers and valences of ions of type j. Note that the surface charge densities in Equation 11 are functions of cation concentrations and are taken from recent binding studies (*27, 28*). In these studies, a modified Gouy Chapman equation (*28, 29*) was used instead of the linear approximation. Having the surface charge densities for given cation concentrations is perhaps the most crucial element in this treatment, since the sensitivity of the rate constants to small differences in cation concentrations is mainly caused by variations in the surface charge densities.

It should be realized that Equation 11 was chosen because of the unavailability of a solution for the electrostatic potential $\psi(x)$ for a system of two interacting bodies in contact with solutions containing both monovalent and divalent cations. At the same time, we have solutions for the surface charge and surface potential for isolated plates, in contact with both monovalent and divalent ions, which bind to the surface to some degree. Our solution for the isolated plates also gives the distance dependence of $\psi(x)$ (*30*). The potential $\psi(x)$ falls off with x, the distance from the surface, more steeply than according to the linear approximation. Therefore, the linear approximation in Equation 11 is regarded as an overestimate of V_e.

The applicability of Equation 11 improves when the rate of aggregation is not too slow, e.g., when the surface potentials (calculated from the modified Gouy Chapman equation (*see* Refs. *28* and *29*)) are not too high. These are the only cases of interest for quantitative comparison with experiment, since detection becomes impossible when the rate of aggregation is too slow. We point out below how we estimate rates for other cases.

Since the linear approximation does not take a full account of the screening effect of divalent cations, it is clear that the linear approximation (*see* Equation 11) could yield serious overestimates of V_e when the electrolyte solutions contain 10mM or less of Na^+ and several millimoles of Ca^{2+} or Mg^{2+}. For such cases we used the approach of Bell, Levine, and McCartney (*31*), which gives for cases of symmetric ions (of valency 1–1 or 2–2)

$$V_e = \frac{128\pi nkT}{\kappa^2} \frac{\gamma_1\gamma_2 a_1 a_2}{R} \exp(-kd) \tag{11'}$$

in which n is the number of cations per unit volume and $\gamma_i = \tanh(ze\psi i/4kT)$, where $\psi_i(i = 1, 2)$ is the surface potential of an isolated surface.

This expression, which is very close to that of Gregory (*32*), was shown to give an underestimate to V_e. Thus fast aggregation can be ruled out unambiguously when the rate as computed with Equation 11′ turns out to be slow. For a few limiting cases, the predictions based on Equations 11 and 11′ give conflicting answers, and reference to a related experiment is necessary—as explained in the section dealing with the sensitivity of rates of aggregation to cation concentrations.

The expressions and calculated values for VDW interactions are given in Refs. *13* and *33*. The calculations are based on a pair-summation procedure which includes retardation corrections and many-body orientation effects, and as shown Ref. *34* gives approximately the same results as the macroscopic Lifshitz approach (*35, 36*) or other macroscopic treatments (*37, 38, 39, 40*). The VDW parameters of water are taken from Ref. *41* and those of the phospholipids from Ref. *42*. In the case of unilamellar vesicles of external radii of 150 Å the values of V_w are -1.44, -0.51, and -0.09 kT for separation distances of 10, 20, and 50 Å, respectively.

Experimental

The experiments described here were designed to test predictions regarding the effect of variations in cation concentrations on aggregation and to further elucidate the effect of Ca^{2+} and Mg^{2+} on the process of fusion of sonicated PS vesicles. We provide results obtained by two techniques: light scattering and release of fluorescent molecules. Since the intensity of light scattered by a particle is approximately proportional to the square of its volume, the light-scattering experiments provide information about the size of the vesicle aggregates as a function of cation concentrations and temperature. We expect that on a time scale of several minutes, any increase in the size of sonicated vesicles will be the result of aggregation and/or fusion. In the second set of experiments, carboxyfluorescein (CF) was encapsulated initially in the vesicles at high concentrations, i.e., under conditions of self-quenching (*43, 44*). Its release was followed by an increase in its fluorescence that occurs upon dilution in the external medium. Thus the CF release experiments provide information regarding the leakage from vesicles, which might occur at a level of isolated vesicles or during the process of aggregation and fusion.

In the following, we provide a brief description of the procedures while the experimental results are presented in following three sections.

Materials and Methods. PS was purified from bovine brain in this laboratory as previously described (*45*). Dispersions of multilamellar vesicles and sonicated unilamellar vesicles were prepared as described earlier (*15, 45*) in a standard buffer containing 100m*m* NaCl, 7m*m* L-histidine *N-tris*-(hydroxymethyl) methyl 2-amino ethane-sulfonic acid (TES), and approximately 0.1m*m* EDTA, adjusted to pH 7.4. CF containing vesicles were prepared by hydration and sonication in a solution of 100m*m* CF, 0.1m*m* EDTA, and 1/10 (*v/v*) of the standard buffer, adjusted to pH 7.4, separated by passage through a Sephadex G-75 (1.0 × 20 cm) column equilibrated with the standard (0.1*M NaCl*) buffer and stored on ice.

Fluorescence of the CF-containing vesicles was measured with an Amino–Bowan spectrofluorometer (excitation 490, emission 550) using a Corning cut-off filter (#3–68, ~ 520 nm). Complete release was obtained by adding Triton X-100 (0.05% *v/v*). Light scattering was measured with the same instrument with excitation and emission both set to 400 nm. PS concentrations were assayed by phosphate determination (*46*).

Sensitivity of Rates of Aggregation to Cation Concentrations

As shown in the section dealing with the expressions for the rate of aggregation of vesicles, the sensitivity of rates of aggregation to cation concentrations enters solely through the electrostatic free energy term. Furthermore, since the aggregation rates depend exponentially upon the total free energy (*see* Equations 5 and 6) these rates are extremely sensitive to changes in the electrostatic free energy, and so to cation concentrations. From Equations 11 and 11′, it is seen that the ionic environment affects the electrostatic free energy in two distinct ways. An increase in the ionic concentrations will reduce the surface charge density and will increase the magnitude of κ and both of these changes reduce the magnitude of V_e.

For a moment, let us consider just the effect of κ on V_e. If d is chosen to be near the value of $1/\kappa$, then doubling κ would diminish V_e by an order of magnitude, with either Equations 11 and 11′. The effect of doubling κ would be most pronounced in systems showing slow aggregation. For example, if V_T was initially about $20kT$, then this doubling of κ would increase the rate of aggregation by nearly eight orders of magnitude. Whereas if V_T was $5kT$, then the rate of aggregation would increase by about two orders of magnitude.

To estimate the kinetic parameters for sonicated vesicles, we have chosen to evaluate the function W by Equation 6, i.e. choosing a value which overestimates W. The rational behind this was discussed in Ref. *13;* however, one point must be emphasized. For many cases, the V_T curves show monotonic increase as the distance of separation between the vesicle surfaces decreases to 10 Å, below which the theory cannot be applied. The absence of a maximum in the V_T curve occurs with either the linear approximation, Equation 11 which overestimates V_e or Equation 11′ which underestimates V_e (*32*). Estimating the W factor, using Equation 6 with hydrodynamic corrections rather than integrating the curve is at present not an important point since integrating the curve does not obviate the problem of the behavior of the free energy curve. An arbitrary cut-off separation would have to be chosen still, since the contribution to W from the integral in the region of 10–20 Å is significant. However, the macroscopic theories used are not valid at distances of separation below 10 Å, and their validity in the regime of 10–20 Å of separation is open to question. Hence, neither the estimation of W nor its integrated values gives a fully satisfactory use of the Smoluchowski–Fuchs aggregation theory (*1, 2*) for these cases.

In general, whenever the value of $V_T{}^*$ at 20 Å, as computed from Equation 11′, is large enough to preclude aggregation, we would predict no aggregation. Whenever the value of $V_T{}^*$ at 10 Å, as computed from Equation 11, is small enough we would predict rapid aggregation, except in the case where κ is very large, such as a $1M$ solution of a monovalent cation. For large values of κ the electrostatic potential barrier is effective only at short distances of separation and a cut-off at 10 Å might be misleading. Table I provides calculated values that demonstrate the dependence of rates of aggregation on cation concentrations.

High Ca^{2+} Concentrations. In these cases, the calculations unambiguously predict fast aggregation because even the overestimate of the free energies at 10 Å (Equation 11) gives fast aggregation. The results indicate only little variation in the half-life times for changes of several millimoles in Ca^{2+} concentrations above 20–30m*M*. The magnitudes of the half-life times are in agreement (within an order of magnitude) with the experimental values of Lansman and Haynes (*17*). However, the use of light scattering to follow Ca^{2+}- or Mg^{2+}-induced aggregation is complicated by the fusion of the vesicles that also occurs under most conditions. As shown by freeze–fracture electron microscopy (*47*), NaCl-induced aggregation does not cause fusion of PS vesicles. (Of course, we cannot rule out the possibility that a few percent of the vesicles do fuse.) Therefore, measurements of rates of aggregation are best done without divalent cations in the system.

Sodium Alone. We have investigated NaCl-induced aggregation by using light scattering to follow aggregation.

As noted before (*13, 48*), there is fast aggregation with 1*M* of Na^+ and none is visible with 1*M* tetramethyl ammonium (TMA^+), which,

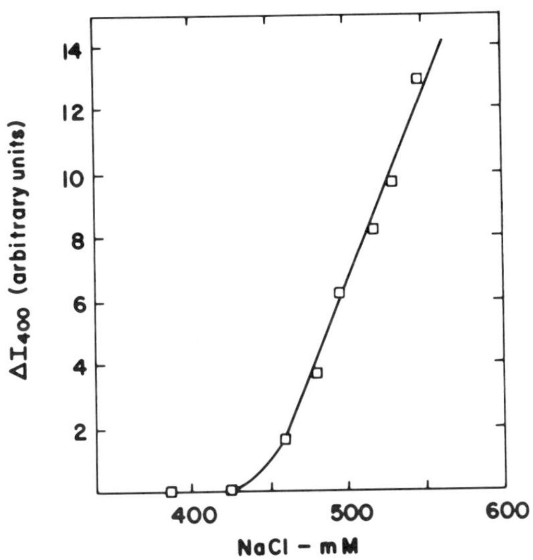

Figure 1. Increase in light scattering at 400 nm of PS vesicles (50μM PS or 4 × 10^{-2} mg/cm³) in the presence of various concentrations of NaCl at 22°C. The increase in light scattering was determined following the addition of appropriate aliquots of buffer containing 2M NaCl to the vesicles. Light scattering owing to the vesicles (sonicated but uncentrifuged) was 7.5 units. The increase in the presence of high NaCl was complete in 2–3 min.

Table I. Effect of Cation Concentrations on the

Cation Concentrations in Solution	$V_e(10)$ (kT_o)	$k_1(10)$ (M^{-1} sec^{-1})
.1M Na[c]		6.3×10^{-2}
.35M Na^+	8.4	4.6×10^{5}
.5M Na^+	3.6	4.9×10^{7}
.86M Na^+	.85	8.0×10^{8}
1M Na^+	.5	1.1×10^{9}
.1M Na^+ + .1mM Ca^{2+}[c]		1.8
.1M Na^+ + .1mM Ca^{2+}	68.0	1.2×10^{-20}
.1M Na^+ + .5mM Ca^{2+}	44.0	2.4×10^{-10}
.1M Na^+ + .5mM Ca^{2+}[c]		48.
.1M Na^+ + 1mM Ca^{2+}	24.4	3.8×10^{-2}
.1M Na^+ + 2mM Ca^{2+}		34.
.1M Na^+ + 10mM Ca^{2+}		3.2×10^{6}
.1M Na^+ + 20mM Ca^{2+}		7.5×10^{7}
.1M Na^+ + 30mM Ca^{2+}		2.7×10^{8}
.1M Na^+ + 40mM Ca^{2+}		5.4×10^{8}
.1M Na^+ + 50mM Ca^{2+}		8.0×10^{8}
.1M Na^+ + 1mM Mg^{2+}	53.8	1.5×10^{-14}
.1M Na^+ + 2mM Mg^{2+}	43.4	4.3×10^{-10}
.1M Na^+ + 5mM Mg^{2+}	30.1	2.1×10^{-4}
.1M Na^+ + 10mM Mg^{2+}	20.7	2.5
.1M Na^+ + 50mM Mg^{2+}	4.3	2.7×10^{7}
.2M Na^+ + .5mM Ca^{2+}	17.7	4.6×10^{1}
.2M Na^+ + 3mM Mg^{2+}	19.4	8.4
.25M Na^+ + 4mM Mg^{2+}	12.5	7.7×10^{3}

[a] Results are for sonicated PS vesicles of having a radius of 150° Å at 20°C. V_e (Equation 11) is given in units of 4×10^{-14} erg or kT_o ($T_o = 16.6°C$). The values of V_w are -1.44 and 0.51 kT_o for $d = 10$ and 20 Å, respectively (*13*). The binding constants used were $0.8M^{-1}$ for Na^+, $10M^{-1}$ for Mg^{2+}, $75M^{-1}$ for Ca^{2+} when its concentration is 1mM or higher, and $35M^{-1}$ for Ca^{2+} when its concentration is below 1mM. These values give better adherence to the experimental binding results (*27, 28*) in this range of concentrations than the corresponding values 35 and $4M^{-1}$ for Ca^{2+} and Mg^{2+}, respectively.

contrary to Na^+, does not cause a significant charge neutralization of PS (*49*). The results in Table I indicate predictions for relatively fast aggregation in the presence of smaller concentrations of Na^+, e.g. 0.5M. These predictions are confirmed by the experimental results presented in Figures 1 and 2.

The results in Figure 2 indicate that the degree of aggregation increases with Na^+ concentration. However, at present, the measurements may not yield half-life times (or shorter times for the dimerization reaction) mainly because initially we had a heterogeneous distribution of vesicle sizes. What we actually see might reflect an advanced stage in

Rate of Aggregation of PS-Sonicated Vesicles[a]

$k_1(20)$ $(M^{-1}\ sec^{-1})$	$t_{1/2}(10)$ (sec)[b]	$t_{1/2}(20)$ (sec)[b]
3.7×10^5	1.9×10^8	32
	2.5×10^1	
	2.4×10^{-1}	
	1.5×10^{-2}	
	1.1×10^{-2}	
1.2×10^6	6.6×10^6	10
1.6	9.8×10^{26}	7.3×10^6
2.5×10^3	2.9×10^{16}	2.8×10^3
3.6×10^6	2.5×10^5	3.2
8.5×10^5	3.1×10^8	14
7.2×10^6	3.5×10^5	1.6
2.7×10^8	3.7	4.4×10^{-2}
6.8×10^8	1.6×10^{-1}	1.7×10^{-2}
9.8×10^8	4.3×10^{-2}	1.2×10^{-2}
1.2×10^9	2.2×10^{-2}	1.0×10^{-2}
1.3×10^9	1.5×10^{-2}	9.2×10^{-3}
1.5×10^2	7.7×10^{20}	7.9×10^4
4.0×10^3	2.7×10^{16}	3.0×10^3
2.8×10^5	5.4×10^{10}	4.1×10^1
6.0×10^6	4.7×10^6	2.0
	.43	
3.8×10^7	2.6×10^5	3.1×10^{-1}
2.9×10^7	1.4×10^6	4.0×10^{-1}
	1.5×10^3	

[b] The term $t_{1/2}$ is computed for $2.5 \times 10^{-4}M$ of PS which corresponds to $4.3 \times 10^{-8}M$ of PS vesicles using a 70-Å^2 surface area per PS head group and 50 Å for the bilayer thickness (2.6×10^{13} vesicle/cm^3). This concentration of PS would correspond to 0.20 mg/mL of PS using 790g/mol for PS based upon dioleyl chains. In Refs. *13* and *17* the value 710 g/mol for PS was used, based upon dipalmitoyl chains.

[c] In these cases, Equation 11′ was used instead of Equation 11.

the process of aggregation. If we somewhat arbitrarily compare the times it takes to reach a given amount of scattered light, say, corresponding to an ordinate of 5 in Figure 2, which is certainly longer than $t_{1/2}$ for dimerization, we find times of 22 and 12 sec for solution concentrations of 860 and 1050m*M* Na^+. These times give the ratio of 1.8 whereas the calculations yield a ratio of the half-times of 1.5. Thus, this crude comparison shows a qualitative agreement for the effect of Na^+ concentrations.

Ca^{2+} or Mg^{2+} and Na^+. The results in Table I indicate that when Equation 11 is used, $t_{1/2}(10)$ is extremely large in many cases of interest (for which aggregation and fusion occur), such as .1*M* Na^+ + 2m*M* Ca^{2+}.

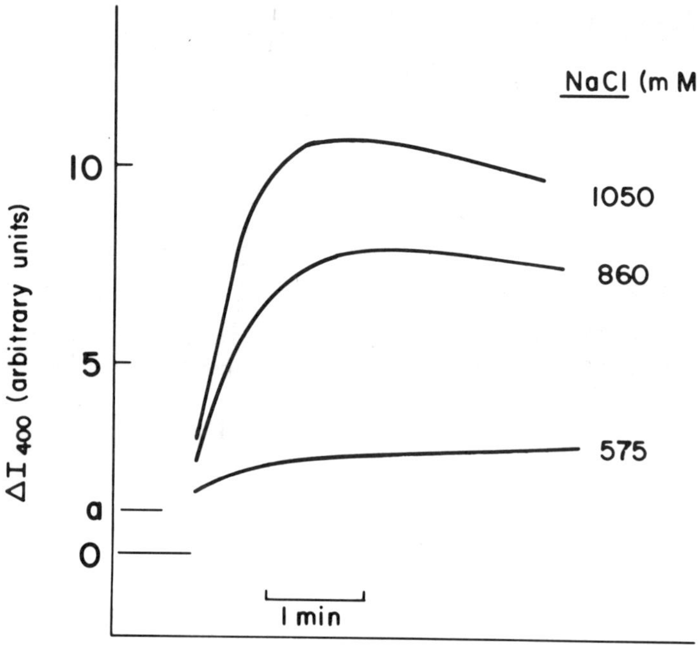

Figure 2. Time course of light scattering increase from addition of NaCl to 5μM PS vesicles at 20°C: (a) scattering from 0.1M NaCl buffer and PS vesicles (5μM) is 1.1 units; background scattering from 0.1M and 1M NaCl buffer was 0.4 and 0.7 units, respectively.

The results for this case show that $t_{1/2}(20)$ is small; therefore slow aggregation cannot be claimed. Guided by the results of this case, we see that the rate constant is significantly larger than that obtained by using $V_e(10)$. Therefore we conjecture that whenever κ is near 10^7 cm^{-1}, the rate of aggregation would be better estimated by taking $V_e(20)$. When more aggregation data become available, it will be possible to better define the cut-off distance. Using this criterion, we would predict rapid aggregation for other cases such as .1*M* Na^+ + 1m*M* Ca^{2+} or .1*M* Na^+ + 5m*M* Mg^{2+}. For cases such as .1*M* Na^+ with .5m*M* Ca^{2+} and with .1m*M* Ca^{2+}, we would predict slow and very slow rates, respectively. Similar statements can be made for cases of low Mg^{2+} concentrations.

In cases such as .1*M* Na^+ + .5m*M* Ca^{2+} (or less Ca^{2+}) the applicability of the linear approximation (Equation 11) may be questionable because the value of the surface potential for the isolated surfaces is $\psi_0 = -60$mV. In order to get estimates for the lower bounds of $t_{1/2}(10)$ and $t_{1/2}(20)$, we took the values of κ and ψ_0 for this case and used Equation 11′. This is justified because our studies (*30*) indicate that the distance dependence of ψ (for the isolated surfaces using the modified Gouy Chapman equation) is very close to the case of the Na^+ alone. If these values of $V_e(10)$ are used, then the rates of aggregation would be predicted to be very slow. If values of $V_e(20)$ are used, then this method would predict fast aggregation. However, a comparison with the case of .1*M* Na^+ indicates that this method would predict a fast rate for the .1*M* Na^+ alone if $V_e(20)$ is used. This is contrary to the experimental results. Estimating W using $V_e(10)$ from Equation 11′ would predict slow aggregation, in agreement with these experiments. Thus, we tend to believe that the estimates provided by using the linear approximation (Equation 11) and taking the cut-off at 20 Å still will provide qualitatively correct predictions for values of κ near 10^7 cm^{-1}.

The results of Table I show a high sensitivity in the rate of aggregation to variations in Ca^{2+} concentrations over a range of a few millimoles. For instance, when using either Equation 11 or Equation 11′, an increase in the Ca^{2+} concentration from 0.1m*M* to 2m*M* (with 0.1*M* Na^+ in the medium) results in an increase in the rate of aggregation of over five orders of magnitude. A similar increase in the rate with Mg^{2+} is obtained by increasing its concentrations. This sensitivity of the rate of aggregation to the divalent cation concentration is to a large extent caused by charge neutralization. Clearly the differences in effects of Ca^{2+} and Mg^{2+} are the result of differences in their interactions with PS vesicles (in aqueous solutions), which is seen by the significant difference between their intrinsic binding constants (*27, 28, 50*)

Effect of Temperature on the Aggregation Process

Variations in temperature are expected to affect both the size of aggregates and the rates of aggregation. Owing to a lack of data, we will avoid the question of the dependence of the shape of an aggregate on temperature. When shape changes with temperature are ignored, an increase of temperature will lead to two different opposing effects. In the following, we will show that with our present knowledge of the parameters of the system, the equilibrium distribution should favor a more dispersed state of vesicles (i.e., a smaller average aggregate size) with an increase in temperature, whereas the initial rate of aggregation will increase generally with temperature.

Effect of Temperature on the Equilibrium Distribution. It is not yet possible to show explicitly the temperature dependence of the equilibrium distribution for large aggregates. However, the results of Figure 3 (showing a decrease in light scattering with temperature) are anticipated in the framework of an analysis dealing only with monomers and dimers.

For this system, the ratio between the number of dimers (N_2) to the number of monomers (N_1) is given approximately by the following

$$\frac{N_2}{N_1} = 8N_o v \exp\{-V_{min}/kT\} \tag{13}$$

expression (*13, 51*) in which N_o is the initial number of monomers per unit volume, v is the volume of a vesicle, and V_{min} is the value of the potential at the minimum, corresponding to the dimer state. We already have pointed out (*13*) that, in order to have a significant degree of aggregation, the depth of the potential well must be at least several kT (for lipid concentrations of $\sim$ 1 mg/cm^3). For a potential independent of temperature and of a depth of $-20\ kT_o$ (where T_o corresponds to 25°C), Equation 13 will give an increase in N_2/N_1 by exp(2.6) when the temperature is lowered from 50°C to 10°C. For a shallower potential, the corresponding increase will be smaller. We will continue in the analysis as if the minimum of the potential, V_{min}, is obtained by a sum of V_e and V_w, and that the temperature dependence of these functions is not significant. An analysis of the temperature dependence of V_T is given in the next subsection.

The experimental results presented in Figure 3 illustrate the effect of temperature on the equilibrium distribution. As we have indicated in previous sections, the growth in vesicle size over this time span cannot

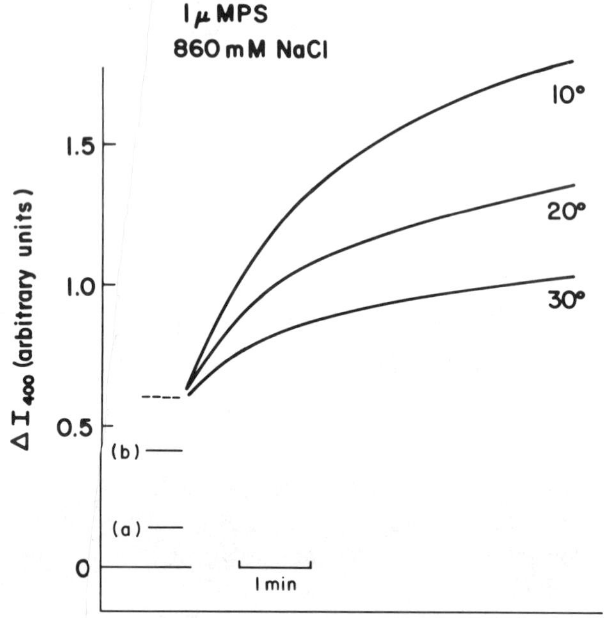

Figure 3. Temperature dependence of the increase in light scattering upon addition of 860mM NaCl to PS vesicles (1μM PS or 8 × 10⁻⁴ mg/mL): (a) scattering from vesicles in the presence of 100mM NaCl buffer; (b) scattering from 860mM NaCl buffer alone.

provide accurate information regarding the rates of aggregation which, as we show in the next subsection, would increase with temperature. This figure illustrates that sonicated PS vesicles in the presence of relatively high Na^+ concentrations favor a more aggregated state at lower temperatures. Such behavior also occurs in cellular systems under certain conditions, where intramembrane particles aggregate in the cold (52).

Temperature Dependence of Rates of Aggregation: General Analysis. In this section we present a general analysis, although the numbers appearing pertain mostly to PS vesicles. To simplify the analysis, we use W as given by Equation 6. From Equations 4 and 6 we have

$$k_1 = \frac{4kT}{3\eta_0} \exp\left(- \frac{(V_T{}^* + E_a)}{kT} \right) \quad (14)$$

in which E_a is the activation energy for the viscosity (53). If the quantities E_a and $V_T{}^*$ are assumed to be independent of temperature, then Equation 14 predicts an increase of the rate with temperature. An approximate value of E_a for water at room temperature is 4,000 cal/mol, which corresponds to 6.7 kT_0 in the exponential (in which the denominator is kT rather than RT, where R is the gas constant). For sonicated unilamellar vesicles, values of $V_T{}^*$ in Table I range between 0 and 65 kT_0 (depending on the concentrations of cations), whereas for multilamellar or big unilamellar vesicles of radii of Iμm, $V_T{}^*$ corresponds to $\geqslant 100kT_0$ (*13, 33*).

In the case of relatively fast aggregation of small vesicles, we take $E_A + V_T{}^* = 12\ kT_0$. Raising the temperature from 10° to 50°C would increase the exponential term in Equation 14 by $\approx \exp\{1.6\}$. If the aggregation is slow, say with $E_a + V_T{}^* = 20\ kT_0$, then the corresponding increase would be $\exp\{2.6\}$.

The pre-exponential term would have a relatively small effect on the temperature dependence of the rate. According to Equation 14, the pre-exponential term would cause a slight increase of the rate with temperature.

In the case of big vesicles, $V_T{}^* + E_a \gtrsim 100\ kT_0$, the rates would be enhanced by a factor of exp(13) over this temperature range. It should be cautioned that the large increase in the rate of aggregation of large vesicles with temperature may not be realistic because such vesicles may undergo fast aggregation at the secondary minimum (*13*) and the degree of observed aggregation will be determined by the equilibrium distribution.

The above analysis indicates that it is possible to choose systems of sonicated vesicles for which an appreciable increase of the rate of aggregation with temperature is predicted, provided that V_T does not increase significantly with temperature.

TEMPERATURE DEPENDENCE OF V_T. The magnitude of the attractive VDW interactions at a distance of separation of 10 Å between sonicated phospholipid vesicles is approximately $1\,kT_o$. Hence, for cases of slow aggregation where $V_T^* \gtrsim 15kT_o$, the VDW interaction would have little effect on the rate of aggregation or its temperature dependence. Our calculations indicate (*see* expressions in Ref. *33*) that the change of V_w with temperature may be neglected given the available accuracy of the VDW theory.

In regard to the temperature dependence of V_e, it can be seen from Equation 11 that it is sensitive to both σ, the surface charge density, and to κ, the Debye–Hückel parameter. V_e decreases with κ, which according to Equation 12 only slightly increases with temperature because ϵ decreases somewhat faster with temperature than $1/T$ (*see* Table II). In Table II, we illustrate the dependence of the surface charge density on temperature. The method of calculation of σ is described in Ref. *28*. When no temperature dependence of the binding constants is considered, the calculated values of σ decrease by about 2% over the temperature range 10°–50°C. For this case, Table II indicates an approximate increase of less than 10% in V_e over the above temperature range.

As we noted before, the rates of aggregation are expected to increase with temperature provided that V_T^* remains constant or does not increase significantly with temperature. This behavior is illustrated in Table III.

A full set of data regarding the temperature dependence of the binding constants has not been available. Since the binding constant of Na^+ to PS is relatively small, $.8M^{-1}$ (*28*), its temperature dependence over this temperature range is not expected to result in a significant increase in

Table II. Temperature Dependence of Quantities Affecting Rates of Aggregation of Vesicles[a, b]

		Temperature (°C)				
		10	*20*	*30*	*40*	*50*
ϵ		84.1	80.4	76.7	73.3	69.9
$k(\times 10^7\ cm^{-1})$		1.05	1.05	1.06	1.07	1.07
$\sigma(e/70\ Å^2)$		.227	.226	.225	.224	.223
V_e	$d = 10$ Å	33.7	34.4	34.9	35.4	35.8
	$d = 20$ Å	10.2	10.4	10.5	10.6	10.6
$\eta(10^{-2}\ cP)$		1.307	1.002	0.798	0.653	0.547

[a] V_e Equation 11 is given in units of 4×10^{-14} erg or kT_o ($T_o = 16.6$°C). Results are for sonicated PS vesicles with a 150-Å radius in contact with a solution of $.1M$ Na^+ and $1mM$ Ca^{2+}. The values of V_w are -1.44 and -0.51 kT_o for $d = 10$ and 20 Å, respectively (*13*). The results for V_e are overestimates because they are calculated with the value of $35M^{-1}$ for the binding constant of Ca^{2+} to PS.

[b] The values of the viscosity, η, and the values of the dielectric constant, ϵ, are from Ref. *64*

V_e with temperature. Regarding the temperature dependence of the binding of Ca^{2+} or Mg^{2+} to PS, preliminary results (*54*) indicate that the measured ratios Ca^{2+}/PS and Mg^{2+}/PS do not vary siginificantly or even slightly increase with temperature over the range of 0°–37°C. Therefore, the order of magnitude increase in the rates of aggregation which is seen in Cases 2 and 3 in Table III is predicted by the present theory to be observed for PS vesicles.

ON THE ENERGY OF ACTIVATION. The temperature dependence of many processes, such as diffusion, permeability, and partition coefficients, has been represented often in terms of the Arrhenius plots (*52*). It would appear that k_1 in Equation 14 also could be treated in this manner by defining

$$E^* = E_a + V_T^* \tag{15}$$

and writing

$$k_1(T) = A \exp\{-E^*/RT\} \tag{16}$$

where $A = (4kT/3\eta_0)$ is essentially temperature independent.

The graph of ln k_1 vs. $1/T$ then would yield E^* as its slope and A as the intercept. Since E_a is known ($\sim$ 4kcal/mol), it then would be possible to obtain an estimate of V_T^* which could be compared with the value computed from the theory. Likewise, A could be compared with the theory's value.

Such a comparison between experiment and theory would be interesting insofar as V_T^* is computed at distances of ca. 10 Å, where the validity of the macroscopic theories used may be questionable. However, using the Arrhenius equation (*see* Equation 16) is not necessarily valid as it is written. It is assumed implicitly in writing Equation 16 that E^* is temperature independent. If rewriting the functional form of k_1 as given in Equation 14 into the form given in Equation 16 embeds a temperature dependence into E^*, then the Arrhenius plots will yield anomalous results.

In particular, the linearization of the Arrhenius equation will yield erroneously large values for the activation energies when E^* decreases with temperature, and, conversely, erroneously small values when E^* increases with temperature (*see also* Ref. *55*). This may be understood easily since from Equation 16

$$\frac{d \ln k_1}{d(1/T)} = -\frac{E^*}{R} + \frac{T}{R}\frac{dE^*}{dT} = -\frac{1}{R}\left(E^* - T\frac{dE^*}{dT}\right) \tag{17}$$

Let E_p denote the apparent activation energy obtained from the slope of the graph of ln k_1 vs. $1/T$. If E^* is not temperature independent, then

Table III. Temperature Dependence of the

System	*Temperature (°C)*	V_e*(10)*
Case 1:	10	2.6
Solution contains .2*M* Na^+	20	2.7
	30	2.7
	40	2.7
	50	2.8
Case 2:	10	33.7
Solution contains .1*M* Na^+ and 1m*M* Ca^{2+}	20	34.4
	30	34.9
	40	35.4
	50	35.8
Case 3:	10	43.5
Solution contains .1*M* Na^+ and 5m*M* Mg^{2+}	20	44.4
	30	45.1
	40	45.8
	50	46.3

[a] The results for $V_e(10)$, $t_{1/2}(10)$, and $t_{1/2}(20)$ in Cases 2 and 3 overestimates the values for PS vesicles because they are calculated with binding constants to PS of $35M^{-1}$ (Ca^{2+}) and $4M^{-1}$ (Mg^{2+}). Equation 11 is used to calculated V_e. The term

E_p will differ from E^* by $\pm\ |T(dE^*/dT)|$, since $E_p = E^* - T(dE^*/dT)$. If E^* increases with temperature $[(dE^*/dT) > 0]$, then E_p is smaller than E^*. For example, suppose that E^* changes by 1 kcal/mol as the temperature goes from 10°–50°C; then $|(dE^*/dT)| \simeq 0.025$ kcal/mol deg. For $T = 300$ K, the term $|T(dE^*/dT)| = 7.5$ kcal/mol.

Inspection of the results in Table IV illustrates our point. In the first case, the apparent activation energy E_p (from the Arrhenius plot) is much larger than the value of E*. Here the magnitude of the potential barrier is relatively small, and although it increases from .7 to .8 kcal/mol over this temperature range, the decrease with temperature of the activation energy for viscosity is greater. Thus E^* decreases with temperature. In Cases 2 and 3, $V_T(10)$ is large and its increase with temperature by about 1 kcal/mol results in an apparent reduced value of E^*. Note that merely from the estimate of errors in E^* or from the RMSE obtained from the regression line, it is difficult to suspect what are the real (theoretical) values of E^*. Furthermore, the linear fit is not worse than that obtained in typical cases in the literature (*55*).

Hence, using the Arrhenius plots may lead to serious overestimates or underestimates of the average activation energies. In order to avoid anomalous results, it is necessary to know either that the activation energy is temperature independent or the basic form of that temperature dependence, i.e., whether dE^*/dT is positive or negative.

Kinetic Parameters of Aggregation of Vesicles[a]

$k_1(10)$ (M^{-1} sec^{-1})	$k(20)$ (M^{-1} sec^{-1})	$t_{1/2}(10)$ (sec)	$t_{1/2}(20)$ (sec)
3.5×10^{-3}		2×10^{9}	
6.5×10^{-3}		1.1×10^{9}	
13×10^{-3}		5.3×10^{8}	
25×10^{-3}		2.8×10^{8}	
51×10^{-3}		1.4×10^{8}	
1.6×10^{-6}	3.0×10^{4}	4.4×10^{12}	2.3×10^{3}
3.2×10^{-6}	4.9×10^{4}	2.2×10^{12}	1.4×10^{3}
7.3×10^{-6}	8×10^{4}	9.5×10^{12}	9×10^{2}
1.6×10^{-5}	1.3×10^{5}	4.4×10^{11}	6×10^{2}
3.7×10^{-5}	2×10^{5}	1.9×10^{11}	3×10^{2}
7×10^{-11}	2.9×10^{3}	1×10^{17}	2.4×10^{3}
2×10^{-10}	4.8×10^{3}	4×10^{17}	1.4×10^{3}
4×10^{-10}	8.3×10^{3}	2×10^{16}	8×10^{2}
1×10^{-9}	1.4×10^{4}	7×10^{15}	5×10^{2}
3×10^{-9}	2.3×10^{4}	2×10^{15}	3×10^{2}

$t_{1/2}$ is computed here for $4.2 \times 10^{-4} M$ PS which corresponds to $7.2 \times 10^{-8} M$ of PS vesicles. *See* Footnote *b* to Table I.) With 790 g/mol for PS, this implies 0.33 mg/mL PS.

Table IV. Apparent Activation Energies Obtained from a Linearization of Log k_1 vs. $1/T$

System[a]	E^* (30°C)[a] (kcal/mol)	E_p[b,c] (kcal/mol)	*RMSE*[c]
Case 1[d]	4.7	$12.2 \pm .4$	.07
Case 2[d]	24	$14.4 \pm .6$	.10
Case 3[d]	30.2	$16.7 \pm .9$	.16

[a] From Table III.

[b] The error in the value E_p from the Arrhenius plot is from the error in the slope, which is given by $(\text{RMSE}/\{\sum_{i=1}^{n}(x_i - \bar{x})^2\}^{1/2})$, in which x_i stands for $1/T_i$ and $\bar{x}$ is the average of $1/T$ in the given range.

[c] The statistical measures are given by Mood and Graybill (*56*). The RMSE is given by $\{\sum_{i=1}^{n}(Y_i - Y_{ic})^2/(n-2)\}^{1/2}$, in which $n = 5$ is the number of data points, Y_i is the value of $k_1(10)$ at the particular temperature, and Y_{ic} indicates the ordinate on the regression line.

[d] The values of E^* at 30°C are obtained from $E^* = V_T^* + E_a$ in Equation 1. The quantity V_T^* is obtained from the sum of $V_e(10\ \text{Å})$ in Table I and $V_w(10\ \text{Å})$, which is $-1.44kT_o$.

Vesicle Aggregation and Fusion

The results of Papahadjopoulos et al. (*15*) on sonicated PS vesicles indicate that at Ca^{2+} concentrations of 1–2m*M* (and .1*M* Na^{+}), several dramatic changes occur including aggregation, fusion, and leakage. With these cation concentrations, the hydrocarbon chains of the lipid bilayers undergo a phase change from a fluid to a crystalline state. Similar effects occur in the presence of higher concentrations of Mg^{2+} (5m*M* or above). For instance, Mg^{2+} in the range of 5m*M* induces aggregation of PS vesicles, but without significant fusion and without a phase change. (This was done at 25°C. At lower temperatures, the degree of fusion increases (*50*)). In addition to the qualitative differences, in terms of binding (*27, 28*) and promotion of fusion (*15*), these two divalent cations produced structurally different complexes with PS in respect to lamellar spacings and hydrocarbon packing (*27*). As shown in studies on several systems (*15, 16*), temperature has an important effect on membrane fusion. In the case of PS vesicles (with Ca^{2+} or Mg^{2+}), the addition of EDTA can reverse some of the size increase—but only that part which is the result of aggregation alone (*17, 57*).

Clearly, close approach of vesicles is a prerequisite for their fusion and, as was shown (*15*), no fusion of PS vesicles was observed under conditions which are unfavorable for aggregation, such as .5m*M* Ca^{2+} (+.1*M* Na^{+}). We already have pointed out that the results in Table I for this case show a relatively slow rate of aggregation. However, an increase in permeability has been observed with Ca^{2+} concentrations at or slightly above .5m*M* (*15*).

The last observation could suggest that the vesicles lose some of their structural integrity with a certain amount of Ca^{2+} in the medium, under such conditions that aggregation is still quite slow because of the electrostatic potential barriers. Another possibility is that some small degree of fusion occurs with .1*M* Na^{+} + .5m*M* Ca^{2+}, but that it is more difficult to detect such fusion as opposed to the increase in permeability. The combined experiments of light scattering and leakage of CF presented here were designed to shed more light on this question.

As a related question, we want to examine whether a certain amount of Ca^{2+}/PS or Mg^{2+}/PS is required for obtaining a significant degree of fusion, apart from the requirement for lowering the potential barrier. The binding studies (*27, 28*) indicate that with .1*M* Na^{+} and Ca^{2+} or Mg^{2+}, a significant degree of fusion occurs when the binding ratios for both Ca^{2+}/PS and Mg^{2+}/PS are near or exceed 0.35. The results of Papahadjopoulos et al. (*15*) show that in .1*M* Na^{+} a significant degree of fusion is seen with Ca^{2+} > 1.5m*M*. An increase in the amount of Na^{+} above .1*M* (with 1.5m*M* Ca^{2+}) would result in a reduction in the ratio Ca^{2+}/PS below ·35. On the other hand, the potential barrier for aggregation is lowered by an addition of Na^+, as is illustrated in Tables I and V. In contrast, when the amount of Na^+ in the medium is reduced to 10m*M* or less, the ratio of Ca^{2+}/PS will exceed .35 even with .1m*M* Ca^{2+} (*27, 28*). However, with such low Na^+ concentrations, the calculated rates of aggregation are extremely slow even in the presence of 2m*M* Ca^{2+} (*see* Tables I and V). The experiments presented in Figures 4 through 7 were designed to show the effect which the amount of divalent cation bound (to the vescile) has on the processes of vesicle aggregation, leakage, and fusion.

The results in Figure 4A indicate a significant amount of release of CF from PS vesicles in a solution of .1*M* Na^+ + 1.5m*M* Ca^{2+}. The amount, or at least the rate, of CF released increases with 5m*M* Ca^{2+}. The light-scattering results in Figure 4B indicate a significant increase in particle

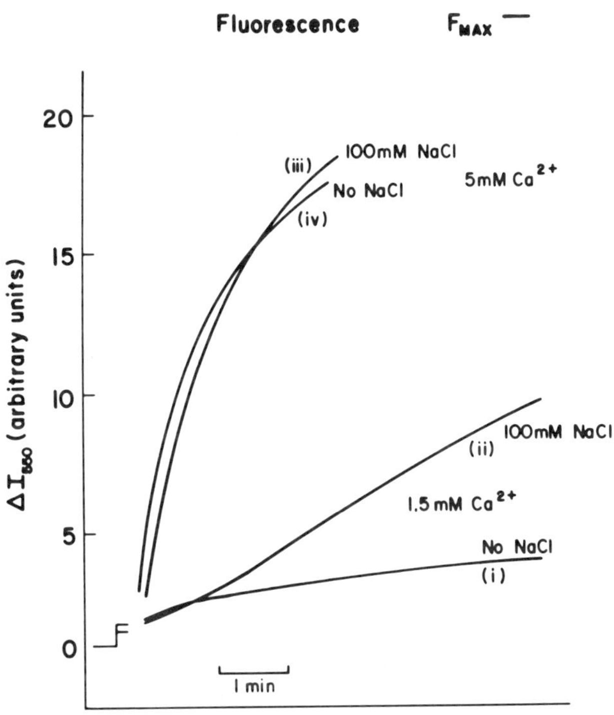

Figure 4A. Time course of CF release from PS vesicles: (i), 1.5mM Ca^{2+} added to medium containing less than 2mM Na^+; (ii), 1.5mM Ca^{2+} + .1M Na^+; (iii), 5mM Ca^{2+} + 1M Na^+; and (iv), 5mM Ca^{2+} + < 2mMNa^+. Without Ca^{2+}, the scattering from vesicles in .1M Na^+ and no Na^+ was 1.8 and 2.2 units, respectively, and the fluorescence of vesicles in .1M Na^+ and no Na^+ was .6 and .8 units, respectively. Fluorescence of the vesicles following addition of 0.05% Triton X-100, corresponding to complete release, was 22.2 units.

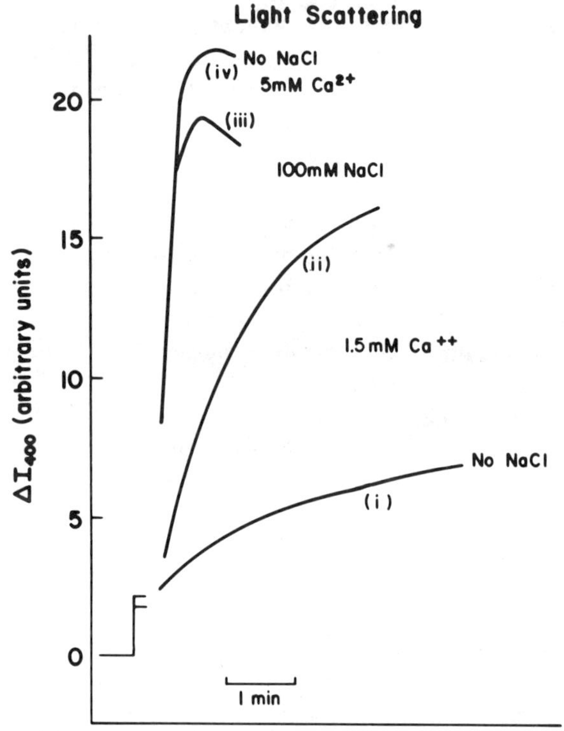

Figure 4B. Time course of aggregation of PS vesicles following addition of Ca^{2+} in buffer containing 100mM NaCl and no NaCl (< 2mMNa^{+}) at 22°C: (i), 1.5mM Ca^{2+} added to medium containing less than 2mM Na^{+}; (ii), 1.5mM Ca^{2+} + .1M Na^{+}; (iii), 5mM Ca^{2+} + .1M Na^{+}; and (iv), 5mM Ca^{2+} + < 2mM Na^{+}. Without Ca^{2+}, the scattering from vesicles in .1M Na^{+} and no Na^{+} was 1.8 and 2.2 units, respectively.

sizes which is caused by aggregation and fusion (*15, 57*). The other curves in Figures 4A and 4B (for 1.5m*M* Ca^{2+}) show dramatic inhibitions of Ca^{2+}-induced aggregation and CF release in the presence of low Na^{+} concentration as compared with 100m*M* NaCl. With low Na^{+}, the addition of higher Ca^{2+} (5m*M*) concentrations caused a rapid aggregation of the vesicles. These results are explained by the calculated rates of aggregation. In the presence of 1.5m*M* Ca^{2+} + 2m*M* Na^{+}, the rate of aggregation is slow (even when overestimating it by using Equation 11′ at 20 Å); therefore, although the calculated ratio of Ca^{2+}/PS is .45, the fusion and accompanying leakage are reduced significantly. In the presence of 5m*M* Ca^{2+} + 2m*M* Na^{+}, our calculations indicate that fast aggregation should occur; therefore both the requirements for aggregation and fusion are satisfied in this case. The results in Figure 4A (with 2m*M* Na^{+} + 1.5m*M* Ca^{2+}) indicate a certain degree of CF release and that could mean a certain degree of destabilization of the vesicles under conditions which are unfavorable for aggregation. However, the corresponding results of Figure 4B also indicate a certain degree of vesicle aggregation. Thus, the small degree of CF release and vesicle aggregation (and probably fusion) seen in the case of 1.5mM Ca^{2+} + 2m*M* Na^{+} could result from some uneven distribution in Ca^{2+} binding, i.e., some vesicles (or some domains on the vesicles) bind more Ca^{2+} than others, and therefore have a lowered potential barrier for aggregation. Another possibility is that a few vesicles become destabilized by the large amount of Ca^{2+} bound to them, and that these vesicles tend to break at one or more regions. These (partially broken) vesicles are more susceptible to aggregation (and fusion (*33*)) since the potential barriers are reduced for fragments of vesicles.

In marked contrast to the effects of low Na^{+} on the Ca^{2+}-induced aggregation and release of CF, Mg^{2+}-induced aggregation and release were enhanced in the presence of low Na^{+}. Our calculations for the cases presented in Figure 5 indicate that if the vesicles preserve their integrity, then their close approach would be very slow, as seen in Table V. Thus, one possible conclusion would be that under the conditions of the experiment, where the ratio Mg^{2+}/PS exceeds .35, the vesicles break up and thus are susceptible for aggregation and fusion.

In 100m*M* NaCl, the amount of CF released is dependent on the concentration of Mg^{2+} present (*see* Figure 5A). The release is correlated with incomplete fusion of the vesicles (*50*) where the extent of release is related to the relative increase in size. Ca^{2+}-induced CF release, however, proceeds until complete release occurs, which is expected since Ca^{2+}-induced fusion eventually results in the formation of cochleate lipid cylinders with no internal aqueous space (*15, 27*)

The light-scattering results in Figure 6 show that with a given amount of Ca^{2+} in solution, the amount of light scattered (after 1 or 2 min) is a monotonically decreasing function of the Na^{+} concentrations in the range .1–.75*M*. This may seem paradoxical when these results are compared with those of Figures 1 through 3, where the amount of scattered light increased with Na^{+} concentration in accord with the calculated reduction in the electrostatic potential barriers with Na^{+} concentrations because of screening and binding. Thus, if the vesicles retain their integrity(as they do with Na^{+} and no Ca^{2+}), the electrostatic potential barriers should be reduced whenever the concentration of Na^{+} is increased. We must conclude that the enhancement in light scattering seen in Figure 6 is not only an expression of simple aggregation, but also of fusion. For the purpose of illustration, let us examine the situation by using the approximate relation wherein the amount of scattered light is proportional to the square of the volume of the scattering particle (*58, 59*). Under these conditions, vesicles in a fused unilamellar state would scatter more

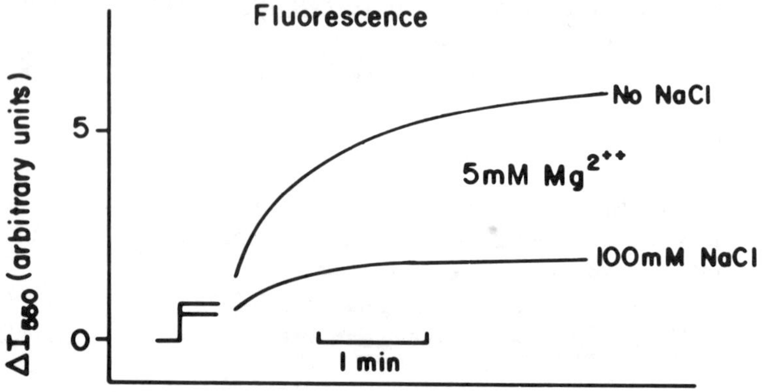

Figure 5A. Time course of CF release from PS vesicles (15μM) and aggregation: (i), 2.5mM Mg^{2+} + 1M Na^+; (ii), 5mM Mg^{2+} + .1M Na^+; (iii) 2.5mM Mg^{2+} added to medium containing less than 2mM Na^+; and (iv), 5mM Mg^{2+} + < 2mM Na^+. The scattering from the vesicles (no Mg^{2+}) was 1.6 units. Without Mg^{2+}, the fluorescence of vesicles in .1M Na^+ and no Na^+ was .7 and .9 units, respectively. Fluorescence of the vesicles following addition of 0.05% Triton X-100, corresponding to complete release, was 20 units.

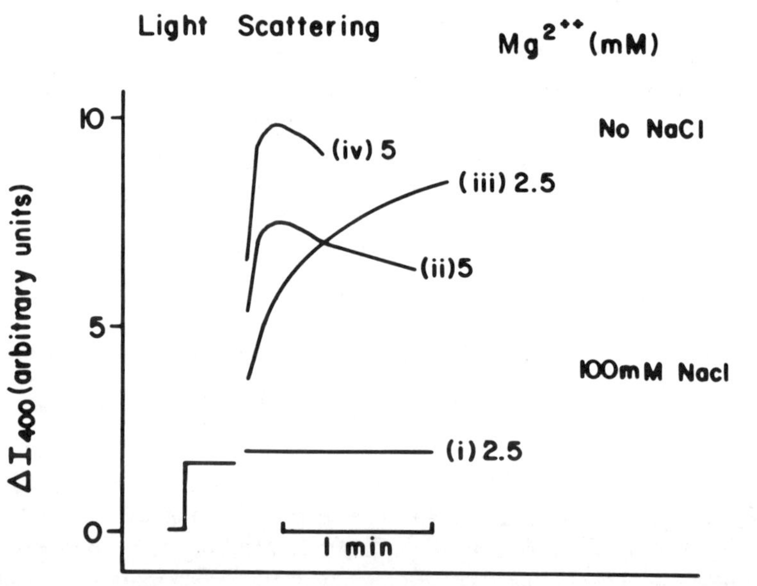

Figure 5B. Time course of aggregation of PS vesicles following addition of Mg^{2+} in buffer containing 100mM NaCl and no NaCl (< 2mM Na^+) at 22°C: (i), 2.5mM Mg^{2+} + .1M Na^+; (ii), 5mM Mg^{2+} + .1M Na^+; (iii), 2.5mM Mg^{2+} added to medium containing less than 2mM Na^+; and (iv), 5mM Mg^{2+} + < 2mM Na^+. The scattering from the vesicles (no Mg^{2+}) was 1.6 units.

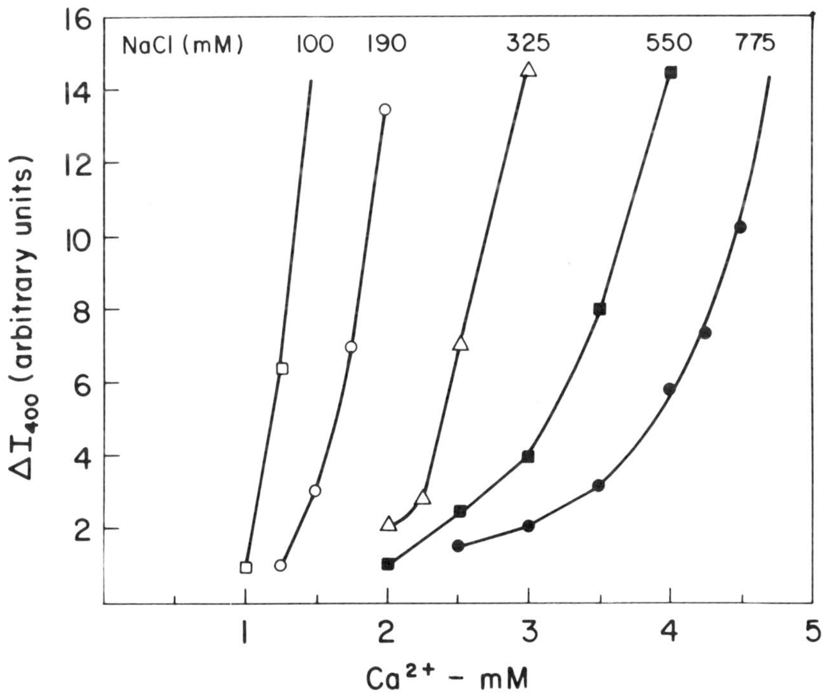

Figure 6. Increase in light scattering after 1 min following addition of Ca^{2+} to PS vesicles (15μM) in the presence of buffer containing the indicated concentrations of NaCl.

light than vesicles in an aggregate. For example, let us compare the scattered intensities, *I*, caused by a group of *n* vesicles (i) in the monomeric state (*I* monomer); (ii) fully aggregated in a spherical geometry (*I* aggregate); and (iii) fused to a unilamellar vesicle (*I* fused). If we assume that *n* is small (~ 10) (so that the diameters of vesicles are sufficiently smaller than the wavelength of light), then the corresponding intensities of scattered light are given by *I* monomer $= nAv^2$, *I* aggregate $= n^2Av^2$, and *I* fused $= n^3Av^2$, in which *v* is the volume of a vesicle in a monomeric state and *A* is assumed to be the same constant in all three cases.

Our explanation of the results of Figure 6 is that apparently the process of vesicle fusion is enhanced with an increase in the amount of Ca^{2+}/PS, and therefore for a given concentration of Na^+ an increase in the concentration of Ca^{2+} not only produces more aggregation, but also produces more fusion. However, an increase in the amount of Na^+ results in a reduction in the ratio Ca^{2+}/PS as shown in Table V. By increasing the concentration of Na^+, we promote the aggregation process, but at

Table V. Effect of Na^+ Concentrations on the Binding of Ca^{2+} or Mg^{2+} to PS Vesicles and on Rates of Aggregation[a]

Cation Concentrations in Solution	*Ca^{2+}/PS or Mg^{2+}/PS*[b] *Bound*	*Ca^{2+}/PS or Mg^{2+}/PS*[c] *Total*	$t_{1/2}$*(10) (sec)*	$t_{1/2}$*(20) (sec)*
2m*M* Na^+ + 1.5m*M* Ca^{2+}	.45	.49	[d]	2.7×10^7
.1*M* Na^+ + 1.5m*M* Ca^{2+}	.37	.38		3.8
2m*M* Na^+ + 5m*M* Ca^{2+}	.45	.49	[d]	370
.1*M* Na^+ + 5m*M* Ca^{2+}	.41	.42	120	.09
2m*M* Na^+ + 2.5m*M* Mg^{2+}	.41	.49	[d]	3.3×10^7
.1*M* Na^+ + 2.5m*M* Mg^{2+}	.28	.30		1000
2m*M* Na^+ + 5m*M* Mg^{2+}	.41	.49	[d]	4.3×10^4
.1*M* Na^+ + 5m*M* Mg^{2+}	.32	.34		41
.1*M* Na^+ + 1m*M* Ca^{2+}	.35	.36		14
.1*M* Na^+ + 1.25m*M* Ca^{2+}	.36	.37		6.7
.1*M* Na^+ + 1.5m*M* Ca^{2+}	.37	.38		3.8
.19*M* Na^+ + 1.25m*M* Ca^{2+}	.30	.31	250	.08
.19*M* Na^+ + 1.5m*M* Ca^{2+}	.31	.32	110	.06
.19*M* Na^+ + 1.75m*M* Ca^{2+}	.32	.33	58	.05
.19*M* Na^+ + 2m*M* Ca^{2+}	.33	.34	33	.05
.325*M* Na^+ + 2m*M* Ca^{2+}	.27	.27	.34	
.325*M* Na^+ + 2.25m*M* Ca^{2+}	.28	.28	.28	
.325*M* Na^+ + 2.5m*M* Ca^{2+}	.29	.29	.24	
.325*M* Na^+ + 3m*M* Ca^{2+}	.30	.30	.17	
.55*M* Na^+ + 2m*M* Ca^{2+}	.20	.20	.03	
.55*M* Na^+ + 2.5m*M* Ca^{2+}	.22	.22	.03	
.55*M* Na^+ + 3m*M* Ca^{2+}	.23	.24	.02	
.55*M* Na^+ + 3.5m*M* Ca^{2+}	.25	.25	.02	
.55*M* Na^+ + 4m*M* Ca^{2+}	.26	.26	.02	
.775*M* Na^+ + 3m*M* Ca^{2+}	.19	.19	.01	
.775*M* Na^+ + 4.25m*M* Ca^{2+}	.22	.22	.01	
.775*M* Na^+ + 4.5m*M* Ca^{2+}	.23	.23	.01	

[a] The binding constants and other parameters used are the same as in Table I. *See* Ref. *28* for explanation of the procedure for calculating the amount of cation tightly bound or trapped in the double layer region. Equation 11 was used unless otherwise indicated.
[b] Tightly bound.
[c] Total including amount trapped in double layer.
[d] In these cases, Equation 11′ was used.

the same time we inhibit the destabilization of vesicles and their fusion. The fact that fusion was inhibited by an increase in Na^+ concentrations in the .1–75*M* range is supported also by the inhibition of CF release in experiments similar to those shown in Figures 4A and 5A (data not shown).

Our current view of the mechanism of membrane fusion in the case of PS vesicles in media containing divalent cations is along the lines of Papahadjopoulos et al. (*15*), where it was proposed that the key event leading to vesicle membrane fusion is the isothermic phase change induced by the divalent metals. In the present study, we have further supplemented the explanations (*15*) by showing that apparently the process of fusion and destabilization of the membranes is critically dependent on the amount of divalent cation that is bound (and not the amount present in solution). An increase in the NaCl concentration causes a lowering of the potential barrier and promotes vesicle aggregation and vesicle fusion—up to the extent that it is not at the expense of reducing the amount of divalent cation bound below the level necessary for the promotion of vesicle destabilization and fusion. An increase in the amount of divalent cation bound, when conditions are unfavorable for the close approach of intact vesicles (up to 20 Å), results in the vesicle's destabilization and (apparently) breakage. This in turn enables surface contact between fragments followed by fusion.

More detailed experiments on Ca^{2+}- and Mg^{2+}-induced aggregation and fusion are planned in order to further test the conclusions we have reached. Of particular interest is the conjugate system of PS mixed with a neutral phospholipid (such as lecithin), which requires much higher concentrations of Ca^{2+} in order to induce fusion.

More experiments still are required to answer the question regarding the changes in the intrinsic binding constants in a range of cation concentrations where fusion occurs. In the present study, the binding constant of Ca^{2+} to PS was set at $75M^{-1}$ whenever the bulk concentration of Ca^{2+} was above 1m*M*, i.e., the range where fusion occurs; otherwise, we used the value of $35M^{-1}$ as in Refs. *27* and *28*. However, this choice was based on data taken at .1*M* Na^{+}, and as we have seen here, the degree of fusion is reduced in the presence of significantly higher Na^{+} concentrations.

Summary and Concluding Remarks

The theory as developed so far has succeeded in qualitatively accounting for the available data on the rates of aggregation of PS vesicles. In particular, the calculations emphasize the sensitivity of these rates to small variations in the cation concentrations within specific (for each cation treated) ranges. The calculations further provide fairly good estimates for the cation concentrations at which the transition from slow to fast aggregation occurs, as defined by the time scale of the light-scattering experiments—wherein aggregation in less than a minute is fast. The theory is most powerful in predicting those cases for which the aggregation would be slow. Our criterion for excluding fast aggregation has been the existence of a high potential barrier at a distance of a 20-Å separation. The conclusion that the free energy potential between vesicles is represented well as a sum of the VDW and electrostatic interactions in the range of a 20-Å or greater separation appears justified.

The VDW interactions seem to have little effect on the rate of aggregation of small vesicles in a primary minimum. However, this statement may be made only because the magnitudes of Hamaker coefficients are less than 10^{-13} erg (5×10^{-14} erg), in contrast to much higher values frequently used in treatments in colloid science (*3*). Our estimates of VDW parameters for phospholipid vesicles are based on the analysis of a significant amount of recent data (*33, 42*).

The most crucial element in the treatment has been a detailed knowledge of the binding of cations to PS. The approximate values of the intrinsic binding constants for these cations were confirmed initially by several independent techniques: direct binding measurements (*27, 28*); ζ potentials in the presence of Na^+ and Ca^{2+}; and by predictions on aggregation (*13*) such as the statement that PS vesicles should aggregate in high Na^+ ($\simeq 0.5M$) concentration, but not in the presence of higher concentrations of cations which have a low affinity for binding to PS. Recently, another confirmation for the degree of binding of Na^+ to PS has been obtained by NMR studies of ^{23}Na (*49*) and by ζ potentials of PS vesicles in solutions of monovalent cations (*61*). Because of (and in spite of) the success of the modified Gouy Chapman theory to account for the binding of cations, it would be interesting to see the effects of various refinements (*62, 63*) such as a distinction between more than the two regions of the solution and the surface. (Such a distinction is included in the VDW calculations.)

We elucidated a difficulty encountered due to the fact that in most cases, the potential curves still were rising for distances of separation below 10 Å. This question is worthy of further investigation, i.e., whether treatments based on more regions (*62, 63*) or on consideration of discrete charges (*62*) would avoid the need for a cut-off at 10 Å—which is certainly unjustified when the solutions have high ionic strengths.

Several predictions have been made here with regard to the temperature dependence of the aggregation process. So far our experimental results have confirmed the prediction regarding the increase in the degree of aggregation upon cooling. However, we do not have data to test the stronger second prediction that in certain cases an increase in the rate of aggregation is found upon heating.

As discussed in the expressions for the rate of aggregation of vesicles, the testing of this prediction may require an observation of the progress of the aggregation reaction at its very early stages. Preliminary results from an analysis which takes into account both back reactions and higher-order aggregates indicate that when the aggregation stage which corresponds to a 50% increase in scattered light, then the presence of trimers and tetramers should not be ignored.

Finally, a combination of calculations and experiments enabled us to propose that the process of vesicle destabilization, leakage, and fusion shows a critical dependence on the ratios Ca^{2+}/PS or Mg^{2+}/PS. Furthermore, an increase in the Na^{+} concentration, while clearly promoting the close approach of vesicles which is necessary for fusion, may inhibit this fusion owing to the simultaneous reduction of the amount of Ca^{2+} and/or Mg^{2+} bound to the vesicle. This reduction is not just the result of the competition between Na^{+} and the divalent cations for binding, but also of the reduction in the magnitude of the surface potential, which results in a smaller enhancement in the concentration of divalent cations near the surface. The results demonstrated that destabilization of the vesicles (by the large amounts of divalent cation bound to them) can promote their aggregation under conditions which would produce only very slow aggregation of intact vesicles.

Acknowledgments

This work was supported by NIH Grant GM23850-01 and partially supported by Grant CA17609. The allotment of time by SUNYAB Computer Center is acknowledged. Some of the ideas in the section dealing with vesicle aggregation and fusion developed during a stay by S. Nir in the laboratory of A. Loyter, Jerusalem, on an ICRETT Program. Thus the project was partially supported by ICRETT and ICRDB under Contract No. NO1-CO-65341 with the International Union Against Cancer. The experimental work was performed in the laboratory of D. Papahadjopoulous under Grant GM18527. Grant CA05467 also is acknowledged.

Addendum

The studies of sodium-induced aggregation of sonicated PS vesicles have been extended by using stopped-flow rapid mixing techniques in the laboratory of Drs. E. P. Day and J. T. Ho (*65*). These experiments showed that even on a time scale of 1 sec, the net rate of aggregation is diminished with increasing temperature in the range of 15°C to 35°C. These observations stimulated the extension of the treatment to account for both the reverse reaction of deaggregation and for higher-order aggregates, as has been reported (*61, 66*). The experiment and analysis (*65, 66*) demonstrated that higher-order aggregates contribute significantly to the scattered light even at times less than 1 sec. In addition it has been demonstrated in the laboratory of Drs. Day and Ho (*65*) that the sodium-induced aggregation is completely reversible to both sodium ion dilution and to temperature changes.

Glossary of Symbols

N_A = Avogadro's number
M, mM, μM = molar, millimolar, and micromolar, respectively
T = absolute temperature
k, R = Boltzmann's constant and the gas constant
E_a = activation energy
D = diffusion coefficient
$D_{12}(R)$, $D_{12}(\infty)$ = diffusion coefficients of the relative motions of Particles 1 and 2 at center-to-center distance, R, and infinity, respectively
V_e = electrostatic free energy of interaction
V_w = VDW free energy of interaction
V_T = total or sum of V_e and V_w free energies
v = vesicle volume
a = vesicle radius
d = distance between external surfaces
n_i = density number, i.e., number per unit volume for ion i
Z_i = valency of ion i
e = unit of electronic charge
$\{C_1\}$, $\{C_2\}$ = molar concentrations of vesicles in monomeric and dimeric forms, respectively
N_o = initial number density of monomers
N_1, N_2 = number density of vesicles in monomeric and dimeric forms, respectively
k_1, k_2 = forward and backward rate constants of the aggregation reaction $2C_1 \underset{k_2}{\overset{k_1}{\rightleftharpoons}} C_2$
$t_{1/2}$ = half-life time for the dimerization reaction
W = a factor which expresses the decrease in k_1 owing to a potential barrier
I = scattered intensity
RMSE = root mean square error

Greek Letters

μ = micron
η = viscosity
ϵ = dielectric constant
κ = inverse of Debye–Hückel length
σ = surface charge density
$\psi(x)$ = electrostatic potential at distance x from surface
ζ = the electrostatic potential at the slip plane
$\psi(o)$, ψ_0 = surface potential

Literature Cited

1. Smoluchowski, M. Z. *Physik. Chem.* **1917,** *92,* 129.
2. Fuchs, N. Z. *Phys.* **1934,** *89,* 736.
3. Verwey, E. J. W.; Overbeek, J. TH. G. "Theory of Stability of Lyophobic Colloids"; Elsevier: Amsterdam, 1948.
4. Overbeek, J. TH. G. *J. Colloid Interface Sci.* **1977,** *58,* 408.
5. Derjaguin, B. V.; Abrikosova, I. I.; Lifshitz, E. M. *Q. Rev.* **1956,** *10,* 295.
6. Kitchener, J. A.; Prosser, A. P. *Proc. R. Soc. Edinburgh, Sect. A* **1957,** 242, 403.
7. Van Silfhout, A. *Proc. K. Ned. Akad. Wet. Ser. B* **1966,** *69,* 501, 513, 532.
8. Sparnaay, M. J.; Jochems, P. W. J. *Int. Congr. Surface Activity, 3rd, Cologne* **1960,** *2,* Sect. B/III/I, 375.
9. Rouweler, G. C. J.; Overbeek, J. TH. G. *J. Chem. Soc., Faraday Trans. 2* **1971,** *67,* 2117.
10. Hunklinger, S.; Geisselmann, H.; Arnold, W. *Rev. Scient. Instrum.* **1974,** *43,* 584.
11. Tabor, D; Winterton, R. H. S. *Proc. R. Soc. Edinburgh, Sect. A* **1969,** *312,* 435.
12. Israelachvili, J. N.; Tabor, D. *Proc. R. Soc. Edinburgh, Sect. A* **1972,** *331,* 19.
13. Nir, S.; Bentz, J. *J. Colloid Interface Sci. Edinburgh, Sect. A* **1978,** *65,* 399.
14. Papahajhopoulos, D.; Poste, G.; Schaeffer, B. E.; Vail, W. J. *Biochim. Biophys. Acta* 1974, *352,* 10.
15. Papahadjopoulos, D.; Vail, W. J.; Newton, C.; Nir, S.; Jacobson, K.; Poste, G,; Lazo, R. *Biochim. Biophys. Acta* **1977,** *465,* 579.
16. Kantor, H. L.; Prestegard, J. H. *Biochemistry* **1975,** *14,* 1790.
17. Lansman, J.; Haynes, D. H. *Biochim. Biophys. Acta* **1975,** *394,* 335.
18. Overbeek, J. TH. G. "Colloid Science"; Kruyt, H. R., Ed.; Elsevier: Amsterdam, 1952; p 245.
19. Brenner, H. *J. Colloid Sci.* **1965,** *20,* 104.
20. Derjaguin, B. V. *Discuss. Faraday Soc.* **1966,** *42,* 317.
21. Charles, G. E.; Mason, S. G. *J. Colloid Sci.* **1960,** *15,* 236.
22. Spielman, L. *J. Colloid Interface Sci.* **1970,** *33,* 562.
23. Hull, M.; Kitchener, J. A. *Trans. Faraday Soc.* **1969,** *65,* 3093.
24. Honig, E. P.; Roebersen, G. J.; Wiersema, P. H. *J. Colloid Interface Sci.* **1971,** *36,* 97.
25. Ruckenstein, E.; Marmur, A.; Gill, W. N. *J. Theor. Biol.* **1976,** *58,* 439.
26. Wiese, G. R.; Healy, T. W. *Trans. Faraday Soc.* **1970,** *66,* 490.
27. Newton, C.; Pangborn, W.; Nir, S.; Papahadjopoulos, D. *Biochim. Biophys. Acta* **1978,** *506,* 281.
28. Nir, S.; Newton, C.; Papahadjopoulos, D. *Bioelectrochem. Bioenerg.* **1978,** *5,* 116.
29. McLaughlin, S. G. A.; Szabo, G.; Eisenman, G. *J. Gen. Physiol.* **1971,** *58,* 667.
30. Bentz, J.; Nir, S., Bull. Math. Biol. **1980,** in press.
31. Bell, G. M.; Levine, S.; McCartney, L. N. *J. Colloid Interface Sci.* **1970,** *33,* 335.
32. Gregory, J. *J. Colloid Interface Sci.* **1975,** *51,* 44.
33. Nir, S. "Progress in Surface Science"; Davison, S. G., Ed.; Pergamon: Oxford, 1977; Vol. 8, p 1.
34. Nir, S. *J. Theor. Biol.* **1975,** *53,* 83.
35. Lifshitz, E. M. *Zh. Eksp. Teor. Fiz.* **1955,** *29,* 94.
36. Dzyaloshinskii, I. E.; Lifshitz, E. M.; Pitaevskii, L. P. *Adv. Phys.* **1961,** *10,* 165.
37. McLachlan, A. D. *Proc. R. Soc. Edinburgh, Sect. A* **1963,** *274,* 80.

38. Van Kampen, N. G.; Nijboer, B. R. A.; Schram, K. *Phys. Lett. A* **1968,** *26,* 307.
39. Ninham, B. W.; Parsegian, V. A.; Weiss, G. H. *J. Stat. Phys.* **1970,** *2,* 233.
40. Langbein, D. *Phys. Rev. Sect. B.* **1970,** *2,* 3371.
41. Nir, S.; Adams, S.; Rein, R. *J. Chem. Phys.* **1973,** *59,* 3341.
42. Nir, S.; Andersen, M. *J. Membrane Biol.* **1977,** *31,* 1.
43. Weinstein, J. N.; Yoshikami, S.; Henkart, P.; Blumenthal, R.; Hagins, W. A. *Science* **1977,** *195,* 489.
44. Blumenthal, R.; Weinstein, J. N.; Sharrow, S. O.; Henkart, P. *Proc. Natl. Acad. Sci. USA* **1978,** *74,* 5603.
45. Papahadjopoulos, D.; Miller, N. *Biochim. Biophys. Acta* **1967,** *135,* 624.
46. Fiske, C. H.; Subbarow, Y. *J. Biol. Chem.* **1925,** *66,* 2606.
47. Vail, W. J., personal communication.
48. Hauser, H.; Phillips, M. C.; Marchbanks, R. M. *Biochem. Journ.* **1970,** *120,* 329.
49. Kurland, R.; Newton, C.; Nir, S.; Papahadjopoulos, D. *Biochim. Biophys. Acta* **1979,** *551,* 137.
50. Portis, A., Newton, C., Pangborn, W., Papahadjopoulos, D. *Biochemistry* **1979,** *18,* 780.
51. Fuoss, R. M. *J. Am. Chem. Soc.* **1957,** *79,* 3301.
52. Volsky, D.; Loyter, A. *Biochim. Biophys. Acta* **1977,** *471,* 243.
53. Eyring, H. *J. Chem. Phys.* **1936,** *4,* 283.
54. Newton, C., personal communication.
55. Papahadjopoulos, D.; Nir, S.; Ohki, S. *Biochim. Biophys. Acta* **1971,** *266,* 561.
56. Mood, A. M.; Graybill, F. A. "Introduction to the Theory of Statistics"; 2nd ed., McGraw-Hill: New York, 1963.
57. Day, E. P.; Ho, J. R.; Kunze, R. K., Jr.; Sun, S. T. *Biochim. Biophys. Acta* **1977,** *470,* 503.
58. Tanford, C. "Physical Chemistry of Macromolecules;" John Wiley & Sons: New York, 1961.
59. Yi, P. N.; MacDonald, R. C. *Chem. Phys. Lipids* **1973,** *11,* 111.
60. Papahadjopoulos, D. *Biochim. Biophys. Acta* **1968,** *163,* 240.
61. Eisenberg, M.; Gresalfi, T.; Riccio, T.; McLaughlin, S. *Biochemistry* **1979,** *18,* 5213.
62. Levine, S.; Mingins, J.; Bell, J. M. *J. Phys. Chem.* **1963,** *67,* 2095.
63. Suave, R.; Ohki, S. *J. Theor. Biol.* **1979,** *81,* 157.
64. "The Handbook of Chemistry and Physics," 50th ed.; The Chemical Rubber Publishing Co., Cleveland, OH, 1969–70.
65. Bentz, J.; Nir, S.; Vail, W. J.; Kwok, A. Y. W.; Hark, S. K.; Ho, J. T.; Day, E. P., unpublished data.
66. Bentz, J.; Nir, S. 1979, Abstracts, 23rd Annual Meeting, *Biophys. J.* **25,** 261a.
67. Bentz, J. 1979. PhD Thesis, State University of New York at Buffalo.

RECEIVED October 10, 1978.

6

The Influence of Extremely Small Attractive As Well As of Repulsive van der Waals–London Forces on Cell Interactions

C. J. VAN OSS—Department of Microbiology, State University of New York at Buffalo, Buffalo, New York 14214

A. W. NEUMANN—Department of Mechanical Engineering, University of Toronto, Toronto, Ontario M5S 1A4, Canada

R. J. GOOD—Department of Chemical Engineering, State University of New York at Buffalo, Buffalo, New York 14214

D. R. ABSOLOM—Department of Microbiology, State University of New York at Buffalo, Buffalo, New York 14214

The van der Waals interaction between two types of cells in a liquid (neglecting for the moment their zeta potentials) can be approached via the expression for the free energy of interaction of Cells 1 and 2 across Fluid 3: $\Delta F_{132} = -A_{132}/12\pi d^2$ *for Solids 1 and 2 in Fluid Medium 3, separated by a distance* d. A_{132} *is the Hamaker coefficient for the interacting system, which normally is positive; it is always positive if Fluid 3 is a gas. But when Fluid 3 is a liquid, under rather common conditions,* A_{132} *can be negative. Then Cells 1 and 2 will repel one another. Cells of the same kind always will attract one another, i.e.* A_{131} *is always positive. However,* A_{131} *can be very close to zero, and under the influence of even very low zeta potentials the cells also will repel each other. The intrinsic interfacial free energies of biological fluids have values close to those of the cells immersed in them, so that both effects can occur quite commonly.*

When the sign of the net van der Waals (VDW) interaction between two different solid bodies (*1*) or between two different dissolved polymers (*2*) in liquids is negative, their interaction is repulsive. The possibility of the existence of such repulsive VDW interactions in liquids

0-8412-9473-X/80/33-188-107$05.00/1

was already implicit in Hamaker's classical treatment (*3*), and was stated explicitly by Visser (*4*): "when two different materials are immersed in a liquid medium and the interaction of each of these materials with that of the liquid medium is larger than the interaction between these materials themselves, spontaneous separation can occur due to dispersion forces only." Such repulsive action was found for the system poly-(tetrafluorethylene)glycol–iron oxide (*5*).

The VDW interaction between two different substances, 1 and 2, immersed or dissolved in Liquid 3 will be repulsive when the Hamaker coefficients of the two substances A_{11} and A_{22} stand in either of the following relations to that of the liquid, A_{33} (*4*)

$$A_{11} < A_{33} < A_{22} \tag{1}$$

$$A_{11} > A_{33} > A_{22} \tag{2}$$

Assuming for the moment that the interaction between particles and/or macromolecules in a liquid medium may be represented as occurring between semi-infinite, homogeneous slabs, the interaction energy ΔF, at separation distance d may be represented as

$$\Delta F_{132} = -A_{132}/12\pi d^2 \tag{3}$$

if A_{132} is the Hamaker coefficient of the system. Expressions 1 and 2 then can be written also as

$$\Delta F^{c}_{11} > \Delta F^{c}_{33} > \Delta F^{c}_{22} \tag{4}$$

$$\Delta F^{c}_{11} < \Delta F^{c}_{33} < \Delta F^{c}_{22} \tag{5}$$

as conditions for repulsion, where ΔF^{c}_{ii} is the free energy of cohesion of Substance i. This may be restated as follows: in Liquid 3, Materials 1 and 2 will undergo a VDW repulsion when the interfacial free energy of Material 1 is larger and that of Material 2 is smaller than that of the liquid medium or vice versa (*1, 2*). The derivation of Equation 3 assumes that Fluid 3, between Bodies 1 and 2, is chemically homogeneous. That is to say, if it is a solution there is no adsorption or orientation of any component at either the 1, 3 or 2, 3 interface. It also assumes that the intermolecular forces in the system obey an inverse sixth power law of attraction.

Not only can negative VDW forces be correlated with particle engulfment or rejection phenomena at solidification fronts (*1*), and of the phenomenon of phase separation in solutions of pairs of polymers (*2*), but these repulsive forces also can be applied to novel separation methods. Antigen–antibody precipitates can be dissociated completely by lowering the interfacial free energy of the liquid medium to a value intermediate between those of the antigenic determinant and the antibody–active site (*6*). The empirical separation method that is known as "hydrophobic chromatography" depends entirely upon adsorption under conditions where positive VDW attractions prevail between adsorbent and adsorbed compounds. It also depends on VDW repulsion between these compounds and the adsorbent surface, when the interfacial free energy of the liquid medium is lowered to a value intermediate between those of the adsorbent and the adsorbed compounds in the elution step (*7*).

Interactions Between Biological Materials

Practically all biological materials have a lower surface free energy than water, and therefore will always exert a VDW attraction on each other in that medium

$$A_{11} \leqslant A_{22} < A_{33} \tag{6}$$

because Equation 6 disobeys Equations 1 and 2. Also, the same materials

$$A_{11} = A_{22} \tag{7}$$

always will exert a VDW attraction upon one another, in our medium, as Equation 7 disobeys Equations 1 and 2. (These considerations neglect any interaction between such materials that may be caused by their ζ-potentials).

However, when the interfacial free energies of the two materials and that of the liquid medium approximate being equal

$$A_{11} \approx A_{33} \approx A_{22} \tag{8}$$

A_{132}, and thus the value of the VDW interaction, will approach zero.

Interfacial Free Energies of Biological Liquids

The liquid media that are most germane to studies on interactions between cells and/or biopolymers are blood plasma and serum. A closer investigation into the surface tensions of blood plasma and serum thus seems essential. As early as 1913 the surface tension of human blood serum at 37°C was reported as $\approx$ 45.4 dyn/cm (measured by the falling-drop method) (*8*). More recently, Lewin (using platinum ring torsiometry) found values of $\approx$ 47.8 and $\approx$ 50.5 dyn/cm at 37°C and 20°C re-

Table I. Surface Tensions of Normal Human Blood Serum and Plasma and Their Ultrafiltrates at 22°C

	Surface Tension in dyn/cm[a]
Serum	45.7
Plasma	47.8
Serum ultrafiltrates[b]	70.3
Plasma ultrafiltrates[b]	70.5
Saline water (0.15*M* NaCl)	73.8

[a] Measured by the pendant-drop method (*10*).
[b] *See* Refs. *11* and *12*.

spectively for normal human serum (*9*). With the pendant-drop method (*10*) we found, for one sample of human serum, 45.7 dyn/cm and 47.8 dyn/cm for heparinized plasma from the same donor at 22°C. Ultrafiltrates through membranes with a molecular weight cut-off of $\approx$ 20,000 (*11, 12*) of the same serum and plasma yielded values of 70.3 and 70.4 dyn/cm, respectively (*see* Table I).

Now, while it is possible, by lowering the surface tension of the liquid medium to $\approx$ 50 dyn/cm by means of solutes of low molecular weight, to dissociate antigen–antibody precipitates (of the pure VDW type) (*6*), whole mammalian serum (with a surface tension $<$ 50 dyn/cm) definitely has no such dissociating power. Upon some reflection this is not as anomalous as it might seem. Table I shows that it is mainly because of the presence of proteins (of molecular weight $>$ 20,000) that blood (or plasma) has a surface tension $\gamma < 50$ (and not $\gamma \approx 70$), and it cannot be expected that molecules with dimensions of the order of 100 Å will effect a separation by purely physical means between other proteins of the same approximate size that are only about 2 Å apart at their site of interaction (*13*).

For cell–protein interaction this reasoning also holds. However, for cell–cell interaction the dimensions of the proteins that cause the major decrease in surface tensions is not ipso facto a hindrance in contributing to the VDW attraction or repulsion between cells. Nevertheless, serum or plasma proteins (which take up roughly 5–6% of the total volume of the liquid) are not likely to play a significant role in intracellular VDW interactions for reason(s) stated below.

The surface tension of a pure liquid is a direct measure of the free energy or cohesion. For a solid, the surface free energy, which has the same dimensions as surface tension, is the measure of free energy of cohesion. Thus, for Substance *i*

$$\Delta F^{c}{}_{ii} = -2\gamma_{iv} \tag{9}$$

where v stands for vapor. The expression relating free energy of cohesion, and hence surface tension, to the Hamaker coefficient is of a form that is analogous to Equation 3 (*14, 15*)

$$\Delta F^{c}_{ii} = -\frac{A_{ii}}{16\pi d^{2}_{oii}} \tag{10}$$

where d_{oii} is the equilibrium intermolecular distance for Substance *i*. (In Ref. *15*, the symbol A is used to represent a quantity that is proportional to the Hamaker coefficient as used in this chapter, but smaller by the factor, $(n_i/\pi)^2$, where n_i is the concentration of *i* in molecules/cubic centimeters.) For a pure liquid, A_{ii} can be estimated from the molecular properties of Substance *i* (*4, 15*). The restrictions indicated earlier on the derivation of Equation 3 also apply to Equation 10.

The situation is somewhat different for a solution, particularly if a surface-active component is present. The measured surface tension is strongly influenced by adsorption at the liquid–vapor surface (*16*). Yet adsorption does not influence the values of the Hamaker coefficients that must be used in Equations 3 and 10, which are related to those of the pure substances by the volume fraction weighted averages. Thus, for solution *j*

$$A_{jj} = \sum_{i} \varphi_i A_{ii} \tag{11}$$

where φ_i is the volume fraction of the component whose Hamaker coefficient is A_{ii} in Solution *j*.

We now may examine the relevance of surface tension measurements to interactions of cells in contact with serum or plasma. The surface tensions reported in Table I for the unfiltered liquids are much lower than those of the ultrafiltrates, at least in part, because of the adsorption of proteins having molecular weight $> 20{,}000$ at the liquid–air interface. And the measurements made on the ultrafiltrates are likely to be a good approximation to the zero-time surface tensions of the whole serum and plasma. The solutes that remain in the solution after ultrafiltration evidently are of a relatively low level of surface activity, and do not affect the surface tension to any greater extent than would be expected from their volume fractions.

Finally, we can estimate the comparative influence of plasma and serum proteins with molecular weights $> 20{,}000$ on the cell–serum and cell–plasma interfaces, as opposed to surface-active solutes such as ethylene glycol or dimethylsulfoxide (*6*). The latter will adsorb at the interfaces to an extent that is comparable with their adsorption at the liquid–

vapor surface, i.e. they will form an oriented monolayer. On the other hand the cells are bathed normally in serum or plasma and their surfaces already are at a condition of saturation with the serum or plasma proteins. There will be little tendency for additional adsorption at the cell–fluid interfaces of protein components of serum or plasma which are stable solutes in the natural environment of the cells. Hence these proteins will not cause the dissociation of cell–cell aggregates.

Intrinsic Interfacial Free Energies of Biological Liquids

Thus it would seem that the actual intrinsic surface tension of biological liquids, i.e. the one that plays a role in the interactions between cells among one another, and between cells and biopolymers, must be close to that of serum or plasma ultrafiltrates, i.e. $\gamma \approx 70$ dyn/cm or the intrinsic interfacial free energy of the interstitial mammalian liquid, serum or plasma, or $\Delta F_i \approx -140$ ergs/cm^2.

Interfacial Free Energies of Blood Cells

From contact angle measurements (*17, 18*), interfacial free energies for blood cells can be derived easily (*17, 19*) (*see* Table II).

Clearly, the interfacial free energies of blood cells all are exceedingly close to the -140 ergs/cm^2 of the surrounding liquid (disregarding the proteins, *see* Table I). Thus, for all blood cells in their surrounding liquids, the Hamaker coefficient A_{131}, according to Visser (*4*)

$$A_{131} = A_{11} + A_{33} - 2A_{13} \approx A_{11} + A_{33} - 2\sqrt{A_{11}A_{33}} \qquad (12)$$

is extremely close to zero. Also, with the slightest negative zeta potentials, all these cells in their medium should be in very stable suspension, as indeed they are.

Table II. Contact Angles with Sessile Drops of Saline Water (*17, 18*) and Interfacial Free Energies of Various Blood Cells Derived from These (*17*)

Cell Type	*Contact Angle (in degrees)*	*Interfacial Free Energy* ΔF_i *(in ergs/cm²)*
Polymorphonuclear leukocytes	18°	−138.3
Lymphocytes	15°–16°	−140.3– −139.6
Erythrocytes	15°–16°	−140.3– −139.6
Platelets	16.3°	−139.4

Some virus (herpes simplex)-transformed cells become quite hydrophilic (*17, 20*), with contact angles as low as 13° ($\Delta F_i \approx -141.4$ ergs/cm^2), while many tissue surfaces have interfacial free energies of about −136 ergs/cm^2 (*20*). In such cases A_{132} will be negative and the cells will be repelled by the tissues (even without a negative zeta potential). Such cells clearly would be free to circulate and penetrate in most tissues without adhesion and may suggest a model for metastatic cells.

Glossary of Symbols

A = effective Hamaker coefficient
d = separation distance
d_0 = equilibrium separation distance
ΔF = free energy difference
mol wt = molecular weight
n = concentration, in molecules/cm^3
i, j, 1, 2 = different materials
3 = liquid medium
v = vapor
c = cohesion

Greek Letters

γ = surface tension
ζ = electrokinetic potential
φ = volume fraction of a given component

Literature Cited

1. Neumann, A. W.; Omenyi, S. N.; van Oss, C. J. *Colloid Polym. Sci.* **1978,** *257,* 413.
2. van Oss, C. J.; Omenyi, S. N.; Neumann, A. W. *Colloid Polym. Sci.* **1978,** *257,* 737.
3. Hamaker, H. C. *Physica* **1937,** *4,* 1058.
4. Visser, J. *Adv. Colloid Interface Sci.* **1972,** *3,* 331.
5. Fowkes, C. F. M. In "Surfaces and Interfaces"; Burke, J. J., Ed.; Syracuse University Press: New York, 1967; pp. 197–244.
6. van Oss, C. J.; Absolom, D. R.; Grossberg, A. L.; Neumann, A. W. *Immunol. Commun.* **1979,** *8,* 11.
7. van Oss, C. J.; Absolom, D. R.; Neumann, A. W. *Sep. Sci. Technol.* **1979,** *14,* 305.
8. Morgan, J. L. R.; Woodward, H. E.; *J. Am. Chem. Soc.* **1913,** *35,* 1249.
9. Lewin, S. *Br. J. Haematol* **1972,** *22,* 561
10. Padday, J. F. In "Surface and Colloid Science"; Matijević, E., Ed.; Wiley-Interscience: New York, 1969; Vol. 1, pp. 101–149.
11. van Oss, C. J.; Bronson, P. M. *Sep. Sci.* **1970,** *5,* 63.
12. van Oss, C. J.; Bronson, P. M. In "Membrane Science and Technology"; Flinn, J. E., Ed.; Plenum: New York, 1970; pp. 139–149.
13. van Oss, C. J.; Neumann, A. W. *Immunol. Commun.* **1977,** *6,* 341.
14. Girifalco, L. A.; Good, R. J. *J. Phys. Chem.* **1957,** *61,* 904.
15. Good, R. J.; Elbing, E. *Ind. Eng. Chem.* **1970,** *62,* 54.

16. Gibbs, J. W. "The Equilibrium of Heterogeneous Substances," "The Collected Papers of J. W. Gibbs"; Dover Publications: New York, 1961; Vol. 1, pp. 219–328.
17. van Oss, C. J.; Gillman, C. F.; Neumann, A. W. "Phagocytic Engulfment and Cell Adhesiveness"; Marcel Dekker: New York, 1975; pp. 7–19.
18. van Oss, C. J. *Annu. Rev. Microbiol.* **1978**, *32*, 19.
19. Neumann, A. W.; Good, R. J.; Hope, C. J.; Sejpal, M. *J. Colloid Interface Sci.* **1974**, *49*, 291.
20. van Oss, C. J., unpublished data.

RECEIVED October 17, 1978.

MACROMOLECULES

7

Interaction of Prothrombin and Ca^{++} with Phospholipid Monolayers Containing Phosphatidyl Serine

M.-F. LECOMPTE[1] and I. R. MILLER

Department of Membrane Research, The Weizmann Institute of Science, Rehovot, Israel

This work deals with the stoichiometry and the structural aspects of the interaction between prothrombin, negatively charged phospholipids, and Ca^{++}. We studied the adsorption equilibria of Ca^{++} and of prothrombin on monolayers containing phosphatidyl serine. Surface radioactivity of $^{45}Ca^{++}$ and of ^{3}H-prothrombin served as a measure for adsorption of the two compounds to the phospholipid monolayers. Synergetic attachment of Ca^{++} and prothrombin to the phospholipid water interface was observed and the number of Ca^{++} per attached prothrombin was determined. Electrochemical investigations on lipid monolayers interacting with prothrombin and Ca^{++} showed that prothrombin increases the lipid layer capacitance and produces a pseudocapacitance peak at the cystine–cysteine redox potential, indicating penetration of the lipid layer with a preferential orientation.

Blood coagulation is preceded by a sequence of zymogen-to-enzyme conversions brought about by some very specific proteases (*1*). In these processes the interaction between coagulation factors and phospholipids plays a crucial role (*2*) by accelerating some of the reaction steps leading to clot formation.

The key role played by the thrombin in the blood coagulation process has led to its extensive study and mainly to the understanding of its complex formation from its zymogen, prothrombin.

[1] On leave of absence from the Chemistry Department, C.S.P., Université Paris–Nord, Avenue J. B. Clément, 93430 Villetaneuse, France.

0-8412-0473-X/80/33-188-117$05:00/1

The conversion of prothrombin to thrombin, as well as some other conversions of this series, require membrane-bound complexes of protease, substrate, and a conversion accelerating protein—Factor **V** (*3*). Factor **Xa**, obtained from Factor **X**, is instrumental in the proteolytic decomposition of prothrombin to thrombin. Prothrombin and Factor **X**, being Vitamin K-dependent proteins, contain γ-carboxyglutamic acid residues (*4, 5*) and bind to membranes containing acidic phospholipids (e.g., Platelet Factor 3) in the presence of Ca^{++} (*6*). Intact protein structure, along with the γ-carboxyglutamic acid and Ca^{++} or other cations, is essential for the formation of protein membrane complexes (*7*). Yet our understanding on a molecular level is still fragmentary. Recently efforts have been made to obtain a clue to the following fundamental questions concerning the structure of the lipid–prothrombin interaction products on the membrane. Does prothrombin penetrate the phospholipid layer or is it bound only to the membrane surface? What is the ionic–hydrophobic balance in the protein–membrane phospholipid interaction? Can we say something about the orientation of the interacting prothrombin molecule with respect to the lipid layer from platelets? Nelsestuen and his co-workers (*8*) tried to answer some of these questions by quasielastic light scattering. They observed the increase in the radii of the negatively charged phospholipid vesicles upon adding prothrombin and Ca^{++}. They also determined the binding stoichiometry (*9, 10*). Their conclusion was that Fragment I of prothrombin binds by Ca^{++} bridges to the negative phospholipids with the long axis of the prolate elipsoid mole perpendicular to the plane of the lipid layer.

In the present work we studied the interaction of prothrombin with lipid monolayers in the presence of Ca^{++}. The stoichiometric relations of the interacting components were determined by measuring surface radioactivity of ^{3}H-labelled prothrombin and ^{45}Ca. The penetration was inferred from the effect of the interaction on the capacitance of the surface layer.

Materials and Methods

The phospholipid, supplied in chloroform–methanol solution, was purchased from Lipid Products, Nutfield, England. For spreading, samples were evaporated in a stream of nitrogen, their lipid content was determined by weight, and they were dissolved in hexane (Spectra Grade, Fluka).

The lipid monolayers were maintained above the collapse pressure by spreading a two-to-fivefold excess of the lipid over an aqueous subphase. The experiments were performed in a $0.1M$ KNO_3 solution, buffered with $10^{-3}N$ *tris*-HNO_3 to pH 7.6-7.8. The pH was checked at the beginning of the experiments.

Prothrombin ($\sim$ 74,000 mol wt) was a gift of R. Benarous and J. Elion, Institut de Pathologie Molèculaire, 24 rue du Faubourg Saint-Jacques, 75014 Paris (France).

Radioactive ^3H-labelled prothrombin was prepared by a method described previously (*11*); sialic acid was oxidized with sodium meta-periodate. The obtained aldehyde was reduced by sodium (^3H)-borohydride.

Radioactive sodium-(^3H)-borohydride and $^{45}Ca^{++}$ were purchased from the Radiochemical Center, Amersham, England.

Water was doubly distilled over permanganate. The different inorganic salts and acids were of analytical grades. Purified mercury was triple distilled under vacuum before use.

Surface radioactivity was determined with an end-window, gas-flow counter equipped with a 100–200 nm thick Formvar window supported by a thin, perforated aluminum sheet as described earlier (*12*). The counter was calibrated with a spread monolayer of ^3H oleic acid, the radioactivity of which was determined also in bulk.

The surface concentration of the prothrombin Γ_{pt} was given by

$$\Gamma_{pt} = \frac{(cpm)^{\sigma}{}_{pt}}{A} \frac{N_{pt}}{(cpm)^{b}{}_{pt}} \frac{(cpm)^{b}{}_{O.A}}{(cpm)^{\sigma}{}_{O.A}}$$

where $(cpm)^{\sigma}{}_{O.A}$ and $(cpm)^{b}{}_{O.A}$ are the surface and bulk (scintillation) counts of oleic acid, $(cpm)^{\sigma}{}_{pt}$ is the surface count of pt, $(cpm)^{b}{}_{pt}$ is the scintillation count of N_{pt} molecule of prothrombin, and A is the surface area. The efficiency of the $^{45}Ca^{++}$ surface radioactivity determination was obtained by adsorption of Ca^{++} on a phosphatidyl serine (PS) monolayer of known charge in the absence of any other salt so that only Ca^{++} could be the counterion of the negatively charged monolayer. The calibration curve is given in Figure 1.

At very low concentrations of Ca^{++} the main radioactivity comes from the surface. At higher concentration of Ca^{++} the bulk radioactivity from the not-soft-enough β-radiation of $^{45}Ca^{++}$ comes into play and it increases linearly with the concentration of Ca^{++}. Extrapolation of the linear part to zero concentration gives the radiation from the surface (proportional to the surface concentration).

The slope of the linear part gives the concentration-dependent radiation from the bulk of the solution. The radiation from the bulk of the tritium-labelled prothrombin at the experimental concentrations was not significant (for every microgram per milliliter it corresponds approximately to 5.10^{-4} $\mu g/cm^2$ or at maximal experimental concentrations the radiation from the bulk above 50 cpm corresponds to about 2.5×10^{-3} $\mu g/cm^2$). The total surface counts varied at the same time between 100 cpm (in the absence of Ca^{++} with EDTA added) and 6,000 cpm (in the presence of 10m*M* Ca^{++}).

For the surface radioactivity, the lipid monolayers were spread first on a buffered 0.1*M* KNO_3 solution (pH 7.8) in a trough described elsewhere (*13*). The concentrations of Ca^{++} and prothrombin (one of them radioactive) were adjusted by injection underneath the monolayer with subsequent mixing by repeated retrieval and reinjection of about one-third of the subphase with a syringe, followed by gentle mixing with a magnetic stirrer.

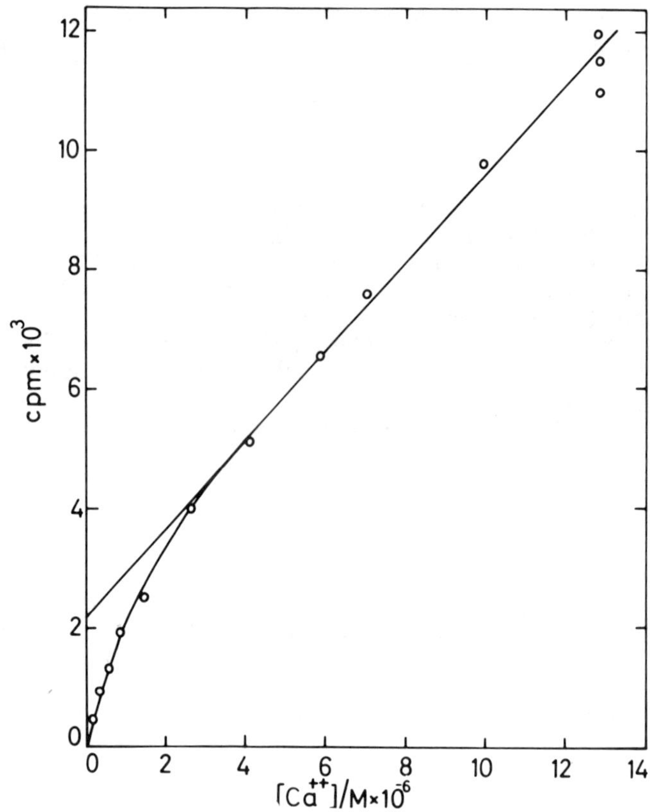

Figure 1. Surface radioactivity from a pure aqueous solution of $^{45}CaCl_2$ covered by a PS monolayer (0.36 μg/cm²) as a function of the concentration of calcium

For the electrodic investigation of the interacting lipid monolayers, we used a hanging-mercury-drop electrode suspended on a thin, drawn capillary. The dropping-mercury electrode described in the previous publications (*14, 15, 16*) gave erratic results since upon interaction the surface layer becomes rigid and does not adjust readily to the expanding mercury surface. The penetration of the prothrombin into the lipid layer was inferred from the increase of capacitance as determined by ac polarography (80 Hz, sweep rate 20 mV/sec) or cyclic voltametry with a P.A.R. Model 170 polarographic instrument.

Results

Surface Composition. The surface composition after spreading a known amount of PS as to obtain a condensed monolayer ($\sim$ 0.36 μg/cm²) was obtained by measuring surface radioactivity of prothrombin or Ca^{++} injected to different concentrations underneath the lipid monolayer. The surface radioactivity of prothrombin was measured in the presence of nonradioactive Ca^{++} and vice versa the surface radioactivity of Ca^{++} was measured in the presence of unlabelled prothrombin.

The surface concentration can be determined safely from the surface radioactivity only when the contribution from the bulk to the surface radiation is less than that of the surface layer. In the case of $^{45}Ca^{++}$ this condition is fulfilled only up to a concentration of 2.10^{-3}m*M*. These concentrations are much lower than the concentrations ($\sim$ 2m*M*) required for a massive adherence of prothrombin, which is a prerequisite for prothrombin–thrombin conversion. As seen from Figure 2 the surface concentrations of Ca^{++} and prothrombin, at constant 1 to 1.5 $\times$ 10^{-3}m*M* of Ca^{++}, increase linearly with the prothrombin concentration in the bulk. This is not surprising since the surface concentrations under these conditions do not exceed 10% of the maximal surface concentration (cf. Figure 3). Every adhering prothrombin molecule is accompanied by about 10 Ca^{++} ions (9.5 $\pm$ 2). This number of Ca^{++} ions per prothrombin molecule is about one and a half of the value found with liposomes at one or two orders of magnitude higher than concentrations of Ca^{++} and prothrombin. A possible explanation of this discrepancy may be based on the different configuration or orientation of the first adsorbed prothrombin molecules as compared with the subsequent ones allowing for a larger number of Ca^{++} binding sites.

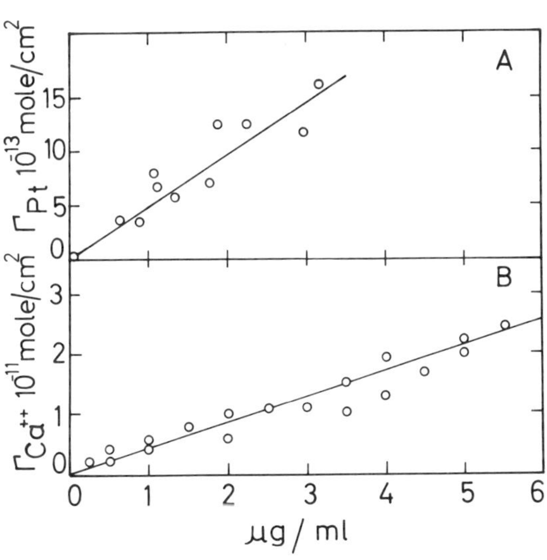

*Figure 2. Adsorption isotherms of (A) (^{3}H)-prothrombin or (B) ^{45}Ca on the PS monolayer (0.3 µg/cm^2) as a function of prothrombin concentration in the bulk; 1 to 1.5 $\times$ 10^{-3}m*M *Ca^{++}.*

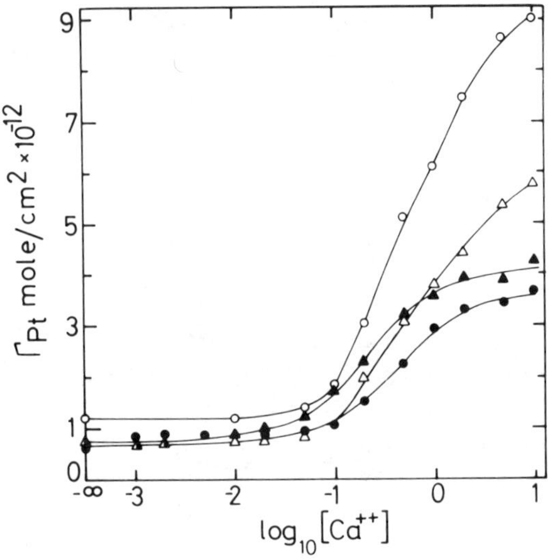

Figure 3. The surface concentration of (^3H)-prothrombin adsorbed onto the PS monolayer as a function of the calcium concentration in millimolar units in the bulk at different concentrations of prothrombin: (●) 1.1 μg/mL; (▲) 1.8 μg/mL; (△) 2.2 μg/mL; (○) 3μg/mL.

By the present method we are not able to determine the surface concentration of calcium adsorbed from more concentrated solutions. This hopefully will be succeeded by the Blodgett method or an equivalent procedure which will remove the surface layer from the solution.

The effect of Ca^{++} on the adsorption of prothrombin can be measured in the calcium concentration region and in a wide region of prothrombin concentration. Nevertheless, we never exceeded the prothrombin concentration of 5–6 μg/mL, which is about 20 times lower than the concentrations used in the prothrombin conversion experiment or in bulk adsorption experiments. However, as seen from Figure 3, the surface concentration of prothrombin at any bulk concentration increases abruptly ca. 0.5m*M* Ca^{++}. The saturation surface concentration is reached only above 10mM Ca^{++}. In the whole region of low concentrations of Ca^{++} between 0 and 0.1m*M*, the surface concentration of prothrombin does not seem to be very dependent on the Ca^{++} concentration even though Ca^{++} evidently is coadsorbed with prothrombin. This behavior is in keeping with the idea that the first molecules of prothrombin which are adsorbed in this region are oriented in the surface so as to bind by a hydrophobic bond rather than by Ca bridges. The adsorption constant of the prothrombin calculated from the slope of the upper line in Figure 2 ($K_{ad} \sim 10^{-2}$ cm) is not very dependent on concentration of calcium. The Ca^{++} ions do not seem to be essential for the protein–lipid-layer interaction. They presumably are attached to the proteins in the bulk and carried along to the surface with the adsorption of the latter.

Lipid Layer Penetration. The penetration of the lipid layers by prothrombin was inferred from the increase in capacitance. This was determined mainly by controlled-phase ac polarography. Similar results, even though less accurate, were obtained by cyclic voltametry. Therefore, only the results obtained by the first method will be presented here. The measurements were carried out with a hanging drop suspended on a thin, drawn-out capillary. The results obtained with a dropping-mercury electrode were erratic. Since the interacting monolayer becomes rigid even above room temperature at 37°C, it does not follow readily the expanding surface of a growing drop. Also, the reformation of the monolayer from the disrupted state at high polarizations is a slow process and many scans were started around the zero charge point where the monolayer is most stable.

In Figure 4 the effect of prothrombin on the differential capacity of a PS monolayer is presented. The overall capacitance, which is proportional to the out-of-phase (quadrature) ac current, increases upon interaction with the prothrombin. Moreover, a pseudocapacitance peak (at −0.7 V relative to the N–AgCl electrode) characteristic of cystine–cysteine redox potential appears. The peak potential moves as expected toward more

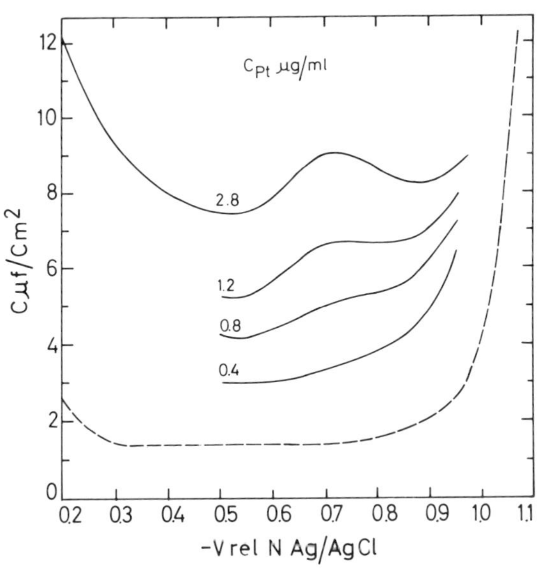

Figure 4. Differential capacitance of condensed PS monolayer with different concentrations of prothrombin injected underneath at 0.5mM Ca^{++} (as a function of the mercury electrode potential): (– – –), lipid monolayer on the presence of calcium.

positive values as the pH decreases. The size of the pseudocapacitance peak depends also on the pH and on the concentration of calcium. The area of the peak represents the number of half-cystines available for the electrode process in the surface complex. It is not understood completely as yet how this number varies under the different conditions, but it is bound to be related to the conformation and the composition of the penetrating protein fractions.

The increase in capacitance with prothrombin concentration is presented in Figure 5. Even in the absence of Ca^{++} there is an appreciable increase in capacitance upon addition of prothrombin. However, the initial increase of capacitance with prothrombin concentration is less steep in the absence or at very low concentrations of calcium ($\sim 10^{-3}$mM) than at the higher concentrations. In the low Ca^{++} concentration region the capacitance increases almost linearly with prothrombin up to concentrations of about 5 μg/mL. In this concentration region the surface concentrations as seen from Figure 2 are also a linear function of the bulk concentration of prothrombin. Every adsorbed prothrombin molecule up to a given surface concentration contributes by the same value to the surface-layer capacitance C^σ

$$C^\sigma = C^\sigma{}_1 + \Theta(C^\sigma{}_{pt} - C^\sigma{}_1)$$

where $C^\sigma{}_1$ and $C^\sigma{}_{pt}$ are the capacitances of the pure lipid and of the prothrombin-penetrated domain, respectively, and Θ is the penetrated

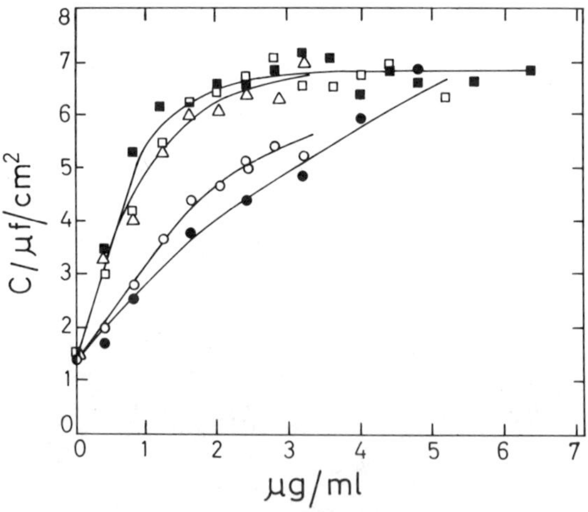

Figure 5. Differential capacitance at −0.5 V relative to the N–Ag/AgCl electrode as a function of the concentration of prothrombin at different concentrations of calcium: (●) 0mM; (○) 10^{-3}mM; (■) 2mM; (□) 0.5mM; (△) 0.1mM.

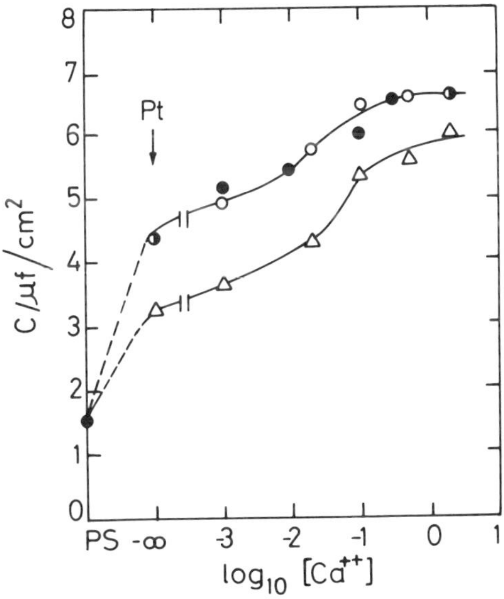

Figure 6. Differential capacitance at −0.5 V relative to the N–Ag/AgCl electrode as a function of the concentration of calcium in millimolar units at different concentrations of prothrombin: (a) (○) 2.4 μg/mL; (b) (△) 1.2 μg/mL ((a) and (b) reconstructed from Figure 5); (c) (●) 2.4 μg/mL (the arrow indicates the injection of prothrombin underneath the monolayer).

monolayer fraction. Above a certain surface concentration, penetration stops. This critical surface concentration is reached at a very low prothrombin concentration in the presence of 0.5m*M* or higher Ca^{++}.

In Figure 6 the dependence of capacitance on calcium concentration for different concentrations of prothrombin is shown. Curves **a** and **b** were reconstructed from the curves obtained by changing the concentration of prothrombin at different constant concentrations of Ca^{++}. Curve **c** was obtained by changing the concentration of Ca^{++} at a prothrombin concentration of 2.4 μg/mL. We obtain a very good agreement by the two different ways at the prothrombin concentration of 2.4 μg/mL. It is evident from the increase in the capacitance that substantial penetration of the lipid layer by the prothrombin is observed even in the absence of Ca^{++}. A further pronounced increase in penetration is observed ca. 0.1m*M* Ca^{++}.

Conclusions

From the previous results one can draw some conclusions with regard to the structure and the arrangements of the prothrombin molecules interacting with a lipid layer consisting of PS. A model based on these conclusions are presented schematically in Figure 7. As seen in this drawing

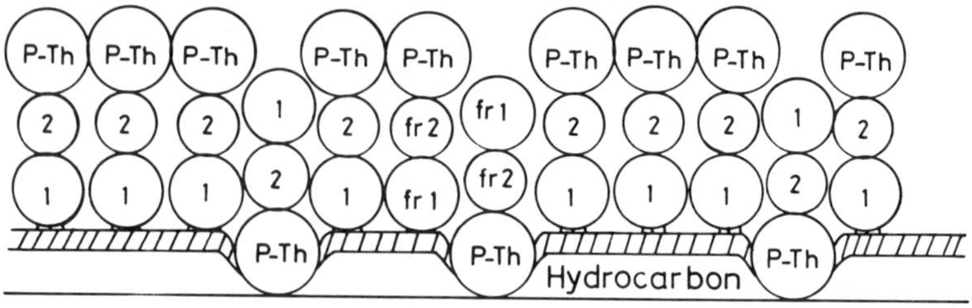

Figure 7. A schematic model of the interaction between prothrombin and the PS monolayer in the presence of calcium: on one side, the hydrocarbon chain is in contact with the air or the mercury drop; on the other side, the layer of prothrombin protrudes in the aqueous solution. The prothrombin schematic configuration is according to Ref. 8.

a fraction of about 10% of a densely packed layer of prothrombin penetrates the lipid layer on which it is adsorbed. Part of the penetration as derived from the increase in capacitance, concurrent with protein adsorption observed by others in suspensions (*17, 18*), occurs already in the absence of Ca^{++}. The degree of penetration increases to some extent with an increasing concentration of Ca^{++}, the largest increase being between 0.03m*M* and 0.3m*M* Ca^{++}. Above this concentration there is an overall increase of adherence of the prothrombin in the lipid layer. Since the prethrombin is composed of more hydrophobic amino acids, it is likely that it provides all or most of the penetrating peptide chains. The major fraction of the adsorbed prothrombin molecules is bound through its Fragment I to the negatively charged lipid layer by calcium bridges, as suggested by Nelsestuen and others (*8, 9*). The penetrating fractions contain some cysteine residues which undergo an electrode redox process giving rise to a pseudocapacitance peak. The peak area represents the number of cysteine residues taking part in the electrode process. This number is not exactly proportional to the overall increase of capacitance of the lipid layer and it seems to depend on the concentration of Ca^{++} and on pH in a way that is not yet clear. However, this dependence is bound to relate to the conformation and the composition of the penetrating polypeptide chain fractions under different conditions. The relation between the configuration of the two populations of prothrombin in the surface and the prothrombin–thrombin transformation is not yet clear. One could, however, speculate that a dynamic equilibrium between the two forms exists. The calcium-bound configuration may assure a high local concentration of prothrombin which may displace the thrombin obtained by proteolytic splitting of the penetrated form.

Acknowledgment

We particularly thank R. Benarous and J. Elion for their purified prothrombin and for the discussions. This work was supported by a grant from the Volkswagen Foundation. Dr. LeCompte was supported by an EMBO Fellowship.

Literature Cited

1. Suttle, J. W.; Jackson, C. M. *Physiol. Reviews* **1977**, *57*, 1–70.
2. Zwaal, R. F. A. *Biochim. Biophys. Acta* **1978**, *515*, 163.
3. Davie, E. W.; Fujikawa, K. *Annu. Rev. Biochem.* **1975**, *44*, 799.
4. Stenflo, J.; Fernlund, P.; Egan, W.; Roepstorff, P. *Proc. Natl. Acad. Sci. USA* **1974**, *71*, 2730.
5. Nelsestuen, G. L.; Zytkovicz, T. H.; Howard, J. B. *J. Biol. Chem.* **1974**, *249*, 6347.
6. Papahadjopoulos, D.; Hanahan, D. J. *Biochim. Biophys. Acta* **1964**, *90*, 436.
7. Nelsestuen, G. L. *J. Biol. Chem.* **1976**, *251*, 5648.
8. Lim, T. K.; Bloomfield, V. A.; Nelsestuen, G. L. *Biochemistry* **1977**, *16*, 4177.
9. Nelsestuen, G. L.; Lim, T. K. *Biochemistry* **1977**, *16*, 4164.
10. Nelsestuen, G. L.; Broderius, M. *Biochemistry* **1977**, *16*, 4172.
11. Butkowski, R. J.; Bajaj, S. P.; Mann, K. G. *J. Biol. Chem.* **1974**, *249*, 6562.
12. Frommer, M. A.; Miller, I. R. *J. Colloid Interface Sci.* **1966**, *21*, 245.
13. Miller, I. R.; Bach, D. *Chem. Phys. Lipids* **1974**, *13*, 453.
14. Rishpon, J.; Miller, I. R. *Bioelectrochem. Bioenerg.* **1975**, *2*, 215.
15. Rishpon, J.; Miller, I. R. *J. Electroanal. Chem.* **1975**, *65*, 453.
16. Miller, I. R.; Rishpon, J. "Electrical Phenomena at the Biological Membrane Level;" Roux, E., Ed.; Elsevier Scientific: Amsterdam, Oxford, New York, 1977; p. 93.
17. Bull, R. K.; Jevons, S.; Barton, P. G. *J. Biol. Chem.* **1972**, *247*, 2747.
18. Nelsestuen, G.; Kisiel, W.; Di Scipio, R. G. *Biochemistry* **1978**, *17*, 2134.

RECEIVED October 17, 1978.

8

Enzymatic Clotting Processes. IV. The Chymosin-Triggered Clotting of *p*-κ-Casein[1]

T. A. PAYENS[2] and PAULA BOTH

Netherlands Institute for Dairy Research, Ede, The Netherlands

The clotting of p-κ-*casein brought about by chymosin has been studied under different conditions of enzyme concentration, ionic strength, and temperature. Flocculation rate constants of* p-κ-*casein have been estimated from the length of the clotting time. The following results have been obtained: (a) the rule of Segelcke and Storch, which states that the product of clotting time and enzyme concentration is constant, is approximately obeyed; (b) the activation energy of the clotting process proper becomes negative beyond 33°C; (c) the flocculation rate of* p-κ-*casein is retarded severely, which is attributed to an orientational constraint in the clotting; and (d) the clotting time shows a pronounced minimum at an ionic strength of about 0.05, which is explained to arise from the opposing effects of an overall electrostatic repulsion and local electrostatic bond formation in the ES complex.*

The kinetics of the protease-triggered clotting of blood and milk has been formulated in a number of recent publications from this laboratory (*1, 2, 3*). In milk clotting, the coagulation is initiated through the limited proteolysis of κ-casein, the milk protein component which normally protects the casein micelles from flocculation by calcium ions (*4*). Kappa-casein is a single polypeptide chain of 169 residues, the sequence

[1] Dedicated to Prof. E. Havinga on the occasion of his retirement from the chair of organic chemistry at the University of Leiden.

[2] Current address: Physico-Chemical Department, Netherlands Institute for Dairy Research, P.O. Box 20, 6710 BA EDE, The Netherlands.

0-8412-0473-X/80/33-188-129$05.00/1

of which imparts the protein with an amphipatic character (*5*). Its specific splitting between residues Phe (105) and Met (106) by chymosin (EC 3.4.23.4) yields insoluble p-κ-casein and starts the coagulation of the casein micelles in milk (*4, 6*). The specificity of this proteolysis has been studied using peptide analogs, which were synthesized in Havinga's laboratory (*7, 8, 9*).

This chapter deals with the clotting of κ-casein which follows its proteolysis. The rate of change of the weight-average molecular weight, M_w, of the solute can be calculated by allowing for the enzymatic production of the clotting species in Von Smoluchowski's rate theory for the coagulation of unstable colloids (*10*). The result is (*1, 2*):

$$\overline{M_w}/M_o = 1 - Mo(1 - f)\sqrt{8\nu/k_s}\,\{f(t/\tau) - (1 - f)(t/\tau)^3/3\}/C_o \quad (1)$$

where $\overline{M_w}$ is the weight-average molecular weight; M_o that of the substrate; f the ratio of the molecular weight of the peptide split off by the enzyme to $\overline{M_o}$; ν the rate of proteolysis; k_s the flocculation rate constant of the product; and C_o the solute concentration in g mL^{-1}. The parameter τ in Equation 1 is defined as

$$\tau = (k_s\nu/2)^{-1/2} \quad (2)$$

and governs the kinetics of the clotting process. Thus it is clear from Equation 1 that the clotting time, t_c, of enzymatic clotting reactions to a fair extent is given by

$$t_c \approx \tau \quad (3)$$

For a proper discussion of some of the results to be described below, it is instructive to investigate the rate of the clotting as predicted by Equation 1 somewhat closer. Therefore in Figure 1 we have plotted the progress curve of the reduced weight-average molecular weight for different values of the parameters ν and k_s. The molecular weight passes through a shallow minimum, which is caused by the dominancy of the proteolysis of the substrate during the lag phase. Indeed, one easily verifies from Equation 1 that the depth of the minimum is given by

$$\overline{(M_w/M_o)}_{min} = 1 - \{M_o f^{3/2}/(1 - f)^{1/2}\}\sqrt{32\nu/9k_s} \quad (4)$$

which becomes more pronounced as the ratio ν/k_s increases. Consequently, the same increase in t_c brought about by either a decrease of k_s or ν works out quite differently. In the first case, the minimum becomes more pronounced, in the latter it tends to disappear.

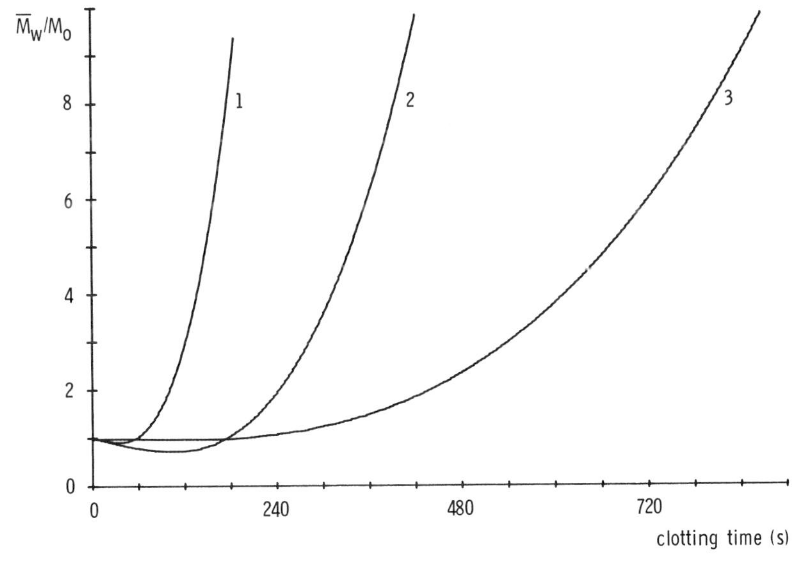

Figure 1. Progress curve of the weight-average molecular weight of clotting p-κ-casein. Model parameters (cf. Equation 1): M_o = 19,000; $f = 0.33$; $C_o = 0.02$ g mL^{-1}. Note that (1) $v = 10^{-8}$ mol mL^{-1} sec^{-1}, $k_8 = 10^5$ mL mol^{-1} sec^{-1}; (2) $v = 10^{-8}$ mol mL^{-1} sec^{-1}; $k_s = 10^4$ mL mol^{-1} sec^{-1}; (3) $v = 10^{-9}$ mol mL^{-1} sec^{-1}; $k_s = 10^5$ mL mol^{-1} sec^{-1}.

Alais and Lagrange (*11*) and Hamdy and Edelstein (*12*) have studied the effect of the ionic strength on the clotting time of milk. Their results are quite remarkable in so far as a pronounced minimum is observed at a total ionic strength of about 0.125. We have confirmed this observation in clotting experiments with pure κ-casein and with micellar casein. As explained above, changes in t_c can be due to either a change in ν or k_s. We shall show, however, that in the present case salt mainly affects the rate of proteolysis ν. A qualitative explanation of the salt effect will be given in terms of Brönsted's theory of the influence of the ionic strength on the rate of ionic reactions (*13*). Besides we have measured the flocculation rate constant of *p*-κ-casein at different temperatures and estimated the activation energies involved.

Experimental Details and Results

Kappa-casein was prepared according to McKenzie and Wake (*14*). The protein was dissolved in 0.001*M* imidazole buffer, pH 6.7, the ionic strength of which was adjusted by adding NaCl.

To study the clotting of micellar casein skim milk prepared from low-heat skim milk powder was diluted 16 times with 0.01*M* $CaCl_2$, to which NaCl was added to adjust the ionic strength.

Clotting was brought about with appropriately diluted rennets (CSKF Leeuwarden, The Netherlands) of an original clotting strength of 10,000 Soxhlet Units (*SU*) (*15*). The clotting was monitored in the Cary 14 spectrophotometer at 500 nm as described previously (*2, 3*). Clotting times were estimated by linear extrapolation of the absorbance–time curves to zero absorbance increase.

Some characteristic clotting curves of micellar casein at 35°C and different ionic strength are shown in Figure 2. It is seen that an increase of the ionic strength brings about an extension of the lag phase with a simultaneous disappearance of the minimum. This indicates, as explained in the introduction, that salt chiefly decreases the rate of proteolysis.

Typical clotting curves for κ-casein at 31.5°C and different enzyme concentrations are presented in Figure 3. Note that the minimum in the absorbance is perceptible only at the highest enzyme concentrations, suggesting that with κ-casein the ratio ν/k_s is smaller than with micellar casein.

The influence of the ionic strength on the clotting time of κ-casein at 35°C is shown in Figure 4. The pronounced minimum is strongly reminiscent of the minimum found with milk (*11, 12*), though it occurs at a considerably lower ionic strength.

In Figure 5 we have given a typical example of the double logarithmic regression of the clotting time on the enzyme concentration. From the slope of such plots we find the exponent γ in the relation

$$t_c\ (\mathrm{E})^{\gamma} = \text{constant} \tag{5}$$

to vary between 0.90 and 1.00 (cf. Table I).

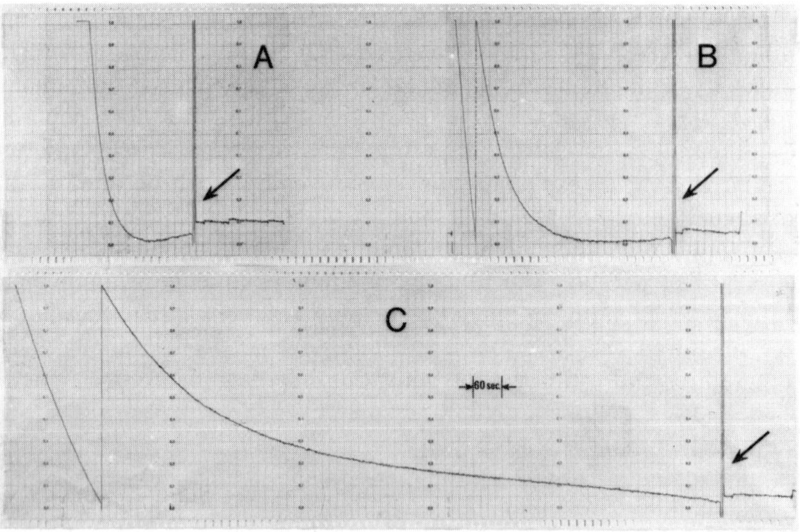

Figure 2. Clotting of micellar casein by rennet at different ionic strengths. Absorbance measurements in the Cary 14 spectrophotometer at 500 nm in 0.5-cf cuvettes; 35°C skim milk diluted 16 times; clotting strength in reaction mixture: 1.43 SU. Time flows to the left. (A) I = *0.04;* (B) I = *0.06;* (C) I = *0.12.*

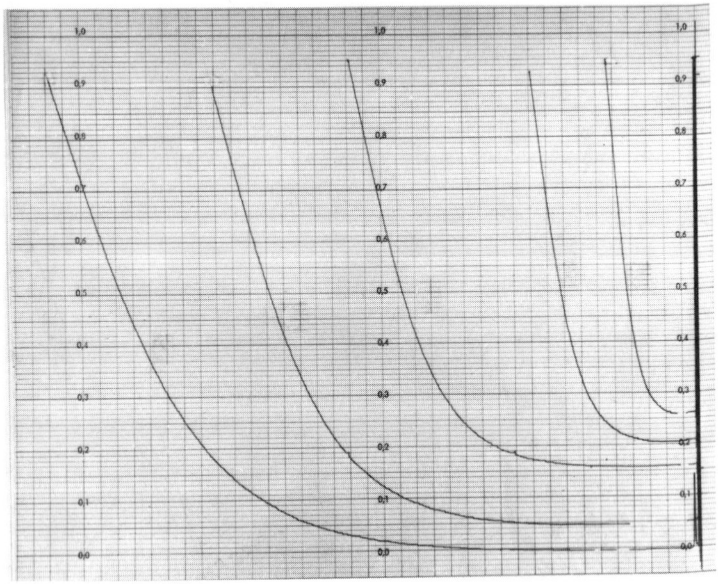

Figure 3. Clotting of κ-casein (0.1 g/100 mL) by rennet at I = 0.2, pH = 6.7, and 31.5°C. Absorbance measurements in the Cary 14 spectrophotometer at 500 nm in 1-cm cuvettes. Time flows to the left. Clotting strength in reaction mixture from left to right: 1.98; 2.65; 3.97; 7.94; and 15.88 SU.

The influence of the temperature on the clotting time of *p*-κ-casein is presented in Figure 6. The least-squares regression curve through the experimental points is

$$\ln t_c = -496.75 + 4.79218 \times 10^5/T - 1.6063 \times 10^8/T^2 + 1.7896 \times 10^{10}/T^3 \quad (6$$

Discussion

The Relationship Between Clotting Time and Enzyme Concentrat Table I shows that with κ-casein the rule of Segelcke and Storch (which states that

$$t_c \cdot (\mathrm{E}) = \text{constant}$$

is approximately obeyed.

It has been explained before (*1*) that this simple result is fo because on the basis of Equations 2 and 3 one would expect

$$t_c \cdot (\mathrm{E})^{1/2} = \text{constant}$$

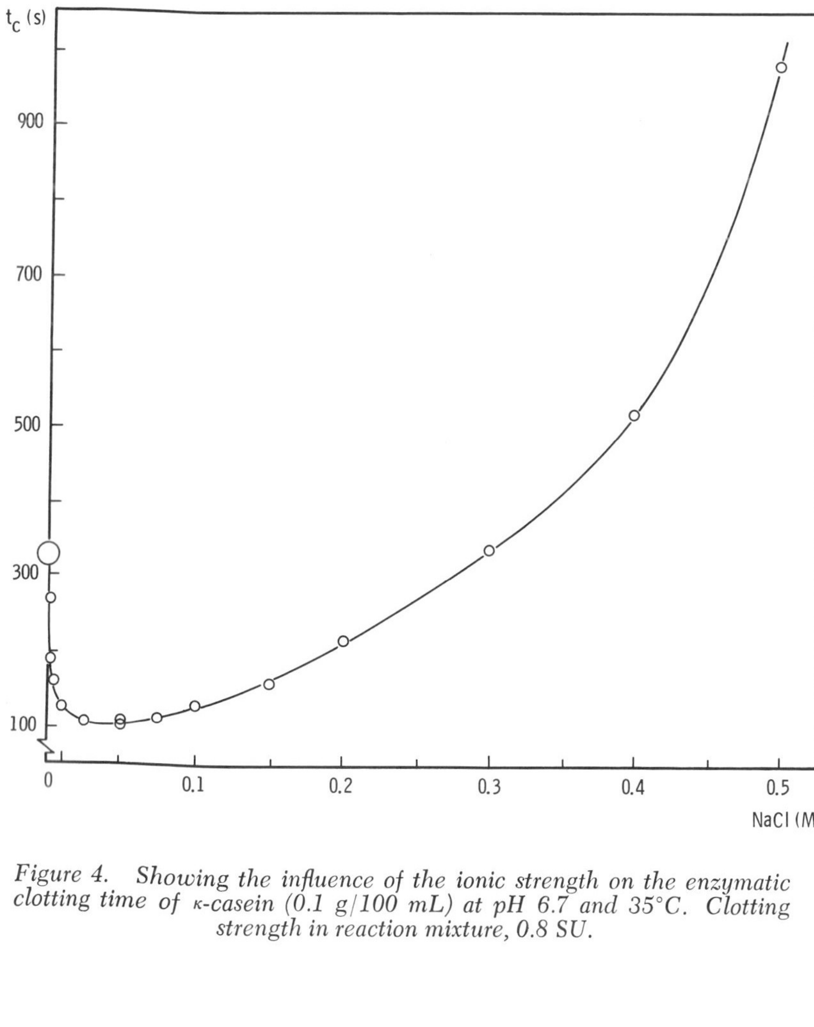

Figure 4. Showing the influence of the ionic strength on the enzymatic clotting time of κ-casein (0.1 g/100 mL) at pH 6.7 and 35°C. Clotting strength in reaction mixture, 0.8 SU.

Table I. Double Logarithmic Regression of Clotting Time on Enzyme Concentration Found with the Clotting of κ-Casein by Rennet. Number of Experimental Points 5 Everywhere

Temperature (°C)	*Slope*	r^2
15.7	0.896	0.9926
21.0	0.982	0.9920
26.4	0.946	0.9981
31.5	1.019	0.9930
35.5	1.055	0.9871
42.5	1.002	0.9997

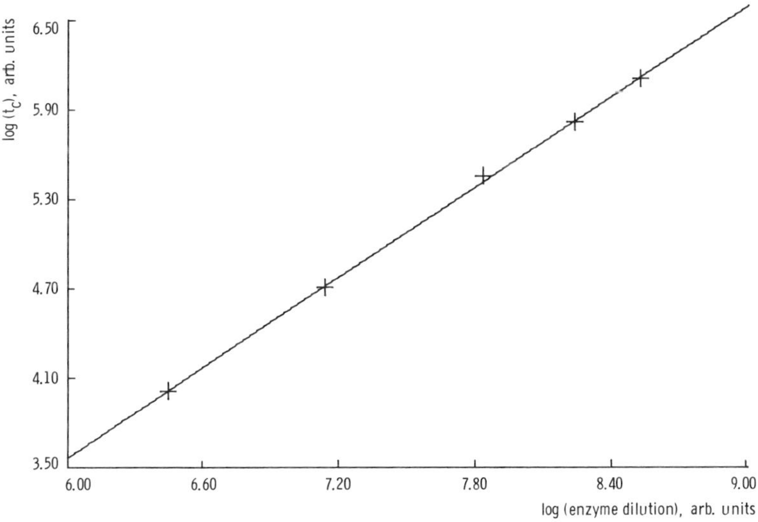

Figure 5. Double logarithmic regression of clotting time on enzyme concentration for κ-casein (0.1 g/100 mL). Experimental conditions: pH = 6.7; I = 0.2; and 31.5°C.

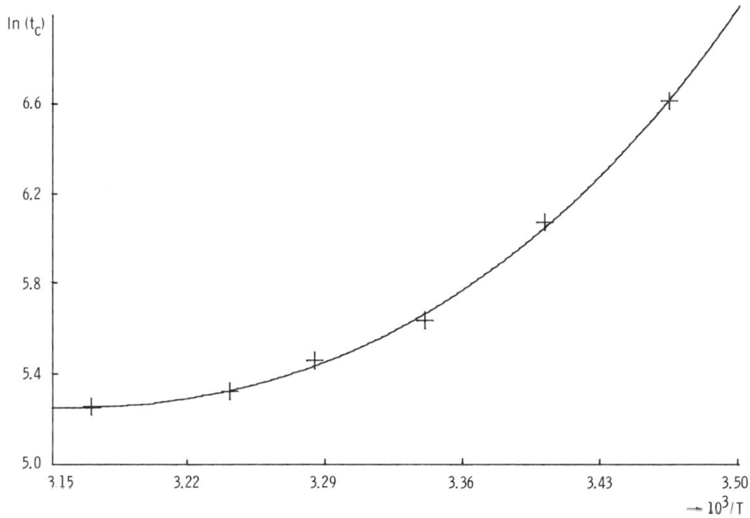

Figure 6. Arrhenius plot for the clotting of κ-casein brought about by rennet. Experimental conditions: pH 6.7; I = 0.2; 31.5°C. Kappa-casein concentration: 0.1 g/100 mL; clotting strength in reaction mixture, 3.97 SU.

The discrepancy between Equations 7 and 8 has been ascribed to a flocculation rate constant depending on enzyme concentration (*1*). Thus in the present case (cf. Equations 2 and 3), one would have approximately

$$k_s \propto (E) \tag{9}$$

This conclusion is in agreement with the recent studies of Vreeman et al. (*16*), which have shown that κ-casein is a strongly associating protein, forming micelles of about 30 subunits. The particle that flocculates is thus polyfunctional and its flocculation rate constant will increase with the number of *p*-κ-casein residues per particle, that is with the concentration of the enzyme.

The Flocculation Rate Constant. In principle Equation 1 contains all of the information to determine both ν and k_s from a single light-scattering experiment as has been demonstrated earlier with the clotting of fibrin(*3*). However, the present turbidity measurements suffer from multiple scattering (*2*), which makes an estimate of ν from the previously proposed plot of $(\overline{M}_w - M_o)/M_o t$ vs. t^2 unreliable. Therefore ν was estimated with the catalytic constants recalculated from Garnier's data (*17*), assuming that the molecular weight of chymosin was 34,000 (*15*) and the clotting strength of the purified enzyme was 10^7 SU (*18*). The flocculation rate constants then could be computed from the clotting times by Equations 2 and 3. Some typical examples have been collected in Table II. One should realize that the rate constants thus obtained are only rough estimates on account of the approximate character of Equation 3, the possible exhaustion of the substrate, and the uncertainties that occur in the computation of ν. Notwithstanding this, there can be little doubt that the flocculation rate constant of *p*-κ-casein, like that of micellar casein and fibrin (*3*), is much smaller than the diffusion-controlled value (*10*).

Table II. Clotting Time (t_c), Flocculation Rate Constants (k_s),[a] and Activation Energies (ϵ_s)[b] Observed with the Clotting of *p*-κ-Casein at pH 6.7, *I* = 0.1, and Different Temperatures

Temperature (°C)	k_{cat} *(s^{-1})*[c]	t_c *(s)*	$10^{-5} \cdot k_s$ *($mL\ mol^{-1}\ sec^{-1}$)*	ϵ_s *($kJ\ mol^{-1}$)*
26.4	21 ± 3	282	1.0	47
35.4	31 ± 3	206	1.3	−13
42.5	39 ± 2	192	1.2	−36

[a] $k_s = 2/(t_c^2\ \nu)$.
[b] $\epsilon_s = 2R\mathrm{d}\ \ln t_c/\mathrm{d}(1/T) - \epsilon_p$ and $\epsilon_p = 41.8$ kJ mol^{-1} (*19*).
[c] Recalculated from Ref. *17* accepting the molecular weight of chymosin to be 34,000 (*15*).

In order to get some insight into the factors reducing the clotting rate, the influence of the temperature on the clotting time was analyzed. By combining Equations 2 and 3 we have

$$\mathrm{d}\ln t_c/d(1/T) = (\epsilon_s + \epsilon_p)/2R \qquad (10)$$

where R is the gas constant, ϵ_s the activation energy of the clotting process proper, and ϵ_p the activation energy of proteolysis. The latter has been determined by Nitschmann and Bohren (*19*) as 41.8 kJ mol^{-1}. With this value one finds from Equations 6 and 10 the activation energies, ϵ_s, given in the last column of Table II. They become negative at about 33°C, suggesting that the retardation of the clotting can be explained only partially by the presence of an energy barrier and must be caused largely by a steric factor. Therefore one concludes that only a small fraction of the surface of the κ-casein micelle consists of sticking residues of *p*-κ-casein and that this constitutes a severe orientational constraint in the clotting. The recent computations of Schurr and Schmitz (*20*) demonstrate also that Brownian movement cannot speed up the clotting rate to the diffusion-controlled limit: rotary diffusion produces only a modest enhancement of the rate by a factor of about 2.

The negative activation energies of Table II point to a dominant attraction between the flocculating particles (*10*). The origin of this attraction is not clear. It probably can be explained by the Coulomb attraction between the positively charged residues of *p*-κ-casein (*5*) and the residual, negatively charged κ-casein on the micelle surface. An alternative though less likely explanation is that long-range van der Waals (VDW) forces become more important at the higher temperatures.

The Influence of the Ionic Strength on the Clotting Time. We have seen that between $I = 0.03$ and 0.5 an increase of the ionic strength chiefly brings about a decrease of the rate of proteolysis. We shall now present a simple argument to explain the effect of salt and to show the importance of the cluster of cationic residues around the labile Phe–Met bond in κ-casein (*5*) in stabilizing the ES-complex. If the latter is identified with Brönsted's activated complex (*13*), one has for the rate of proteolysis

$$\nu = -\mathrm{d(ES)}/\mathrm{d}t = k_{cat}\, K(\gamma^e{}_E \gamma^e{}_S/\gamma^e{}_{ES})\,(\mathrm{E})\,(\mathrm{S}) \qquad (11)$$

where K is the equilibrium constant for the complex formation; $\gamma^e{}_E$, $\gamma^e{}_S$, and $\gamma^e{}_{ES}$ are the electrostatic activity coefficients of the enzyme, the substrate, and the ES-complex; and k_{cat} is the catalytic constant.

By definition we further have

$$\gamma^e_E \gamma^e_S / \gamma^e_{ES} = \exp\,(-\Delta F^e_{ES}/kT), \qquad (12)$$

where ΔF^e_{ES} is the electrical work connected with the formation of the complex.

Let the simple model shown in Figure 7 stand for the ES-complex. The complex, consisting of two globules of the same size and charge, is supposed to be stabilized by a number of electrostatic bonds. It is probable that His 98, 100, and 102 are among the residues in κ-casein that participate in such bonds. Their importance for the proteolysis is indicated by the photooxidation experiments of Hill and Laing (*21*) and Zittle (*22*). On the enzyme, among others, Asp (290) near the active center could be involved, because it is known that Chymosin A, in which this residue is replaced by Gly, is catalytically less active than Chymosin B (*23*). For the above model, according to S. N. Timasheff (*24*), ΔF^e_{ES} is approximately equal to the sum of a Verwey–Overbeek potential to account for the net repulsion plus a sum of screened Coulomb potentials to take care of the electrostatic bonds, viz.

$$\Delta F^e_{ES} = \Psi_o^2\, Dr/2 + (e^2/D) \sum_{i,j} z_i z_j \exp\,(-\kappa r_{ij})/r_{ij} \qquad (13)$$

where Ψ_o is the surface potential of the partners, D the dielectric constant, r the particle radius, r_{ij} the distance between the charges z_ie and z_je, and κ the reciprocal Debye–Hückel length defined as

$$\kappa = 3.33 \cdot 10^7 \sqrt{I} \qquad (14)$$

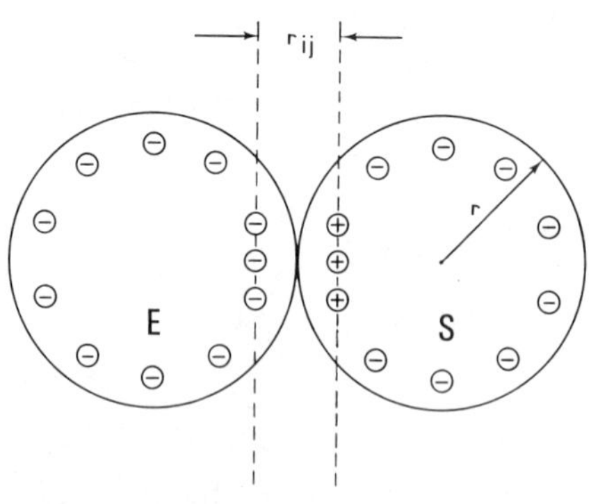

Figure 7. A simple model to account for electrostatic interactions in the ES-complex of chymosin and κ-casein

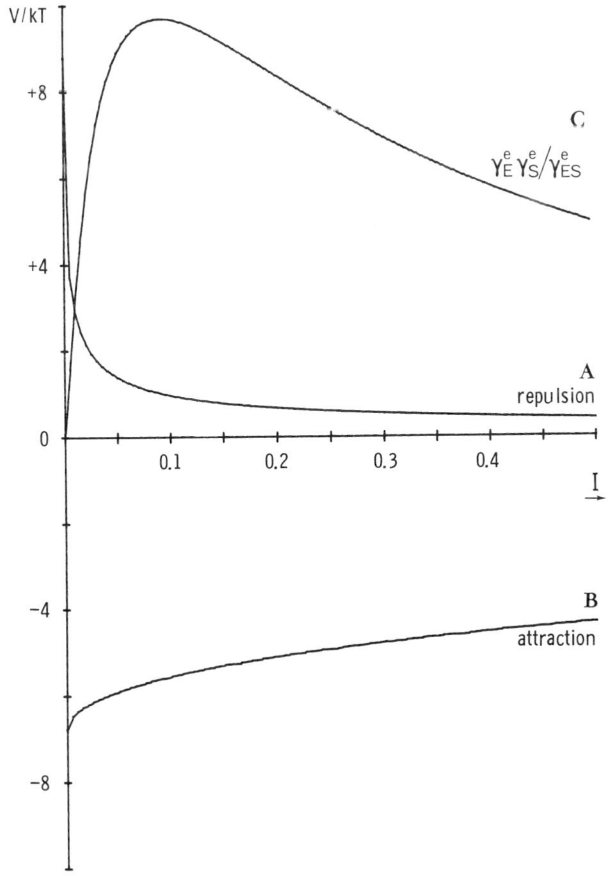

Figure 8. Model calculation of the influence of the ionic strength on the rate of proteolysis of κ-casein by chymosin. Model parameters (cf. Equations 13 and 14): Z = −8; r = 2.5 nm; a = 2.75 nm; 3 pairs of electrostatic bonds formed at r_{ij} = 0.25 nm. (A) Overall electrostatic repulsion (first term on rhs of Equation 13); (B) Local side-chain attractions (second term on rhs of Equation 13); (C) $\gamma^e{}_E\gamma^e{}_S/\gamma^e{}_{ES}$ (Equation 11).

In the Debye–Hückel approximation Ψ_o is calculated from

$$\Psi_o = (Ze/D)\ \{1/r - \kappa/(1 + \kappa a)\} \tag{15}$$

where Z is the valency of the protein and a is the distance of closest approach of the protein and the counterions.

Figure 8 illustrates that on the basis of this simple model and with the parameters adopted there, a maximum in ν appears at $I = 0.125$. The maximum shifts to lower I-values by increasing the number of bonds or decreasing the distances r_{ij}. However, in view of the simplicity of the model chosen such refinements are not considered to be very interesting. The issue of the present discussion is merely to demonstrate how the remarkable minimum in Figure 4 could arise. At low ionic strength, where the thickness of the electrical double layer exceeds the radii of the interacting particles, an increase of the ionic strength chiefly brings about a suppression of the overall repulsion. However, when the thickness of the double layer becomes smaller than the size of the particles, salt mainly affects local sidechain interactions. The argument emphasizes, moreover, the importance of such (probably secondary) interactions in determining the specificity of the enzyme for a particular peptide bond.

Glossary of Symbols

$\overline{M}_w$ = weight-average molecular weight
M_o = molecular weight of substrate
f = ratio of molecular weight of peptide split off and M_o
k_s = flocculation rate constant
k_{cat} = catalytic constant
c_o = solute concentration ($g\ mL^{-1}$)
t_c = clotting time
R = gas constant
T = absolute temperature
(E) = enzyme concentration
(S) = substrate concentration
(ES) = enzyme–substrate complex concentration
ΔF^e_{ES} = electrostatic free energy of formation of ES-complex
D = dielectric constant
r_{ij} = intercharge distance
a = distance of closest approach
$1/\kappa$ = reciprocal Debye–Hückel length
e = elementary charge
Z = valency
I = ionic strength
r = particle radius

Greek Letters

ν = maximum rate of proteolysis
$\tau = (k_s\nu/2)^{-1/2}$
ϵ_s = activation energy of clotting reaction
ϵ_p = activation energy of proteolysis
$\gamma_E{}^e$ = electrostatic activity coefficient of enzyme
$\gamma_S{}^e$ = electrostatic activity coefficient of substrate
$\gamma_{ES}{}^e$ = electrostatic activity coefficient of ES complex
Ψ_o = electrical surface potential

Literature Cited

1. Payens, T. A. J.; Wiersma, A. K.; Brinkhuis, J. *Biophys. Chem.* **1977**, *6*, 253–261.
2. Payens, T. A. J. *Biophys. Chem.* **1977**, *6*, 263–270.
3. Payens, T. A. J. *Discuss. Faraday Soc.* **1978**, *65*, 164.
4. Mackinlay, A. G.; Wake, R. G. In "Milk Proteins: Chemistry and Molecular Biology"; McKenzie, H. A., Ed.; Academic: New York; Vol. 2, p. 175.
5. Mercier, J. C.; Brignon, Gh.; Ribadeau-Dumas, B. *Eur. J. Biochem.* **1973**, *35*, 222–235.
6. Payens, T. A. J. *J. Dairy Res.* **1978**, *46*, 291–306.
7. Schattenkerk, C.; Holtkamp, I.; Hessing, J. G. M.; Kerling, K. E. T.; Havinga, E. *Recl. Trav. Chim. Pays-Bas* **1971**, *90*, 1320–1322.
8. Visser, S.; van Rooyen, P. J.; Schattenkerk, C.; Kerling, K. E. T. *Biochim. Biophys. Acta* **1976**, *438*, 265–272.
9. Ibid., **1977**, *481*, 171–176.
10. Overbeek, J. Th. G. In "Colloid Science"; Kruyt, H. R., Ed.; Elsevier: Amsterdam-New York; Vol. 1, Chapter 7.
11. Alais, C.; Lagrange, A. *Le Lait* **1972**, *52*, 407–427.
12. Handy, A.; Edelstein, D. *Milchwissenschaft* **1970**, *25*, 450–453.
13. Moelwyn-Hughes, E. A. *"Physical Chemistry";* Pergamon: London, New York, Paris, 1957; p. 1225.
14. McKenzie, H. A.; Wake, R. G. *Biochim. Biophys. Acta* **1961**, *47*, 240–242.
15. Foltmann, B. "A Review on Prorennin and Rennin," *C. R. Trav. Lab., Carlsberg* **1966**, *35*, 143–231.
16. Vreeman, H. J.; Both, P.; Brinkhuis, J. A.; van der Spek, C. *Biochim. Biophys. Acta* **1977**, *491*, 93–103.
17. Garnier, J. *Biochim. Biophys. Acta* **1963**, *66*, 366–377.
18. de Koning, P. J. "Studies on Rennin and the Genetic Variants of Casein"; Thesis, University of Amsterdam, 1967; p. 25.
19. Nitschmann, H.; Bohren, H. U. *Helv. Chim. Acta* **1953**, *38*, 1953–1963.
20. Schurr, J. M.; Schmitz, K. S. *J. Phys. Chem.* **1976**, *80*, 1934–1936.
21. Hill, R. D.; Laing, R. R. *J. Dairy Res.* **1965**, *32*, 193–201.
22. Zittle, C. A. *J. Dairy Sci.* **1965**, *48*, 1149–1153.
23. Foltmann, B., personal communication.
24. Timasheff, S. N. In "Biological Polyelectrolytes"; Veis, A., Ed.; Dekker: New York; 1970; Chapter 1.

RECEIVED October 16, 1978.

9

Electrochemical Basis of Heparin-Induced Red Blood Cell Aggregation

K. M. JAN

Department of Physiology, Columbia University, College of Physicians and Surgeons, New York, NY 10032

This chapter was designed to investigate the mechanism of heparin-induced RBC aggregation. RBC aggregation in heparin–saline solutions was quantified by microscopic observation. The effectiveness of purified beef-lung heparin fractions in causing RBC aggregation increases with their molecular weights. Removal of sialic acid from RBC surface by neuraminidase markedly reduced the heparin-induced aggregation. Lowering the ionic strength of the heparin solution enhanced aggregation of normal RBC but not neuraminidase-treated RBC. These results suggest that RBC aggregation by heparin is caused by the binding of the latter to sialic acid residues on adjacent cells. It is postulated that the possible mechanism for heparin to bridge cell surfaces is through a short-range electrostatic attraction between charged molecules of the same sign but of dissimilar charge densities.

The commonly used anticoagulant heparin is composed of glucosamine, uronic acids, *O*-sulfate or sulfate ester groups, and *N*-sulfate or sulfamino groups (*1*). Although some aspects of the structure remain to be defined, the most striking feature of the heparin macromolecule is known to be its large number of negatively charged groups, of which the most prominent are the highly acidic sulfate groups. Heparin in the usual dose used in anticoagulation does not cause RBC aggregation. However it was found that an increase in the concentration of heparin caused not only platelet aggregation (*2*) but also RBC aggregation (*see* Figure 1). The present experiments were designed to investigate the interaction between

0-8412-0473-X/80/33-188-143$05.00/1

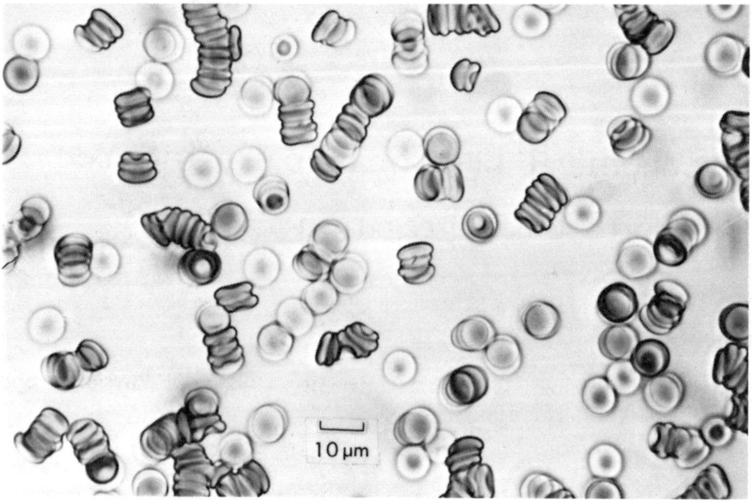

Figure 1. Photomicrograph of heparin-induced RBC aggregation. Human RBCs were washed three times with tris-buffered saline (6) and resuspended in the buffered saline containing 2 g % heparin (Sigma Chemical Co.). RBC's formed typical rouleaux.

the heparin and RBC surface which also is charged negatively primarily because of the presence of sialic acids (*3*). Elucidation of the mechanism of the heparin-induced RBC aggregation helps to define the nature of electrostatic attractions between nonidentical molecules with like charge in a colloid system (*4*) and provides insights into the complex biochemical action of heparin.

Experimental

Commercial heparin was obtained from Sigma Chemical Co. (St. Louis, MO), and purified beef-lung heparin fractions (I–IV) with various mean molecular weights kindly were supplied by J. A. Cifonelli (*5*). Fresh heparinized blood from healthy human subjects was used for the study. After centrifugation and removal of the plasma and buffy coat, the RBCs were washed three times with tri-buffered saline at pH = 7.4. The washed RBCs were suspended in the buffered, saline containing various concentrations of heparin (hematocrit = 1.0%) for the quantification of aggregation by direct microscopic observation in a hemocytometer chamber at 37°C (*6*). The microscopic aggregation index (MAI), which gives the average number of RBCs in each aggregation unit, was calculated as the ratio of the number of RBC's per unit area on the photomicrograph to the number of aggregation units in the same area.

In order to elucidate the role of electrostatic forces in heparin-induced aggregation, the effects of RBC surface charge (*6*) and ionic strength (*7*) were tested systematically in this study.

Results

Normal and neuramiidase-treated (N-treated) RBCs suspended in the saline solution were monodispersed. The addition of Heparin Fraction III (8.1×10^3 mol wt) or Fraction IV (6.0×10^3 mol wt) to RBC suspensions did not cause aggregation of normal or N-treated RBCs (*see* Figure 2). With Fraction II (11.8×10^3 mol wt), normal RBCs showed appreciable aggregation at a heparin concentration of 1 g %. An increase in heparin concentration caused a progressive increase in RBC aggregation. The N-treated RBC's, however, showed essentially no aggregation in Fraction II. With the use of Fraction I (15.5×10^3 mol wt), normal RBCs showed significantly greater aggregation than with Frac-

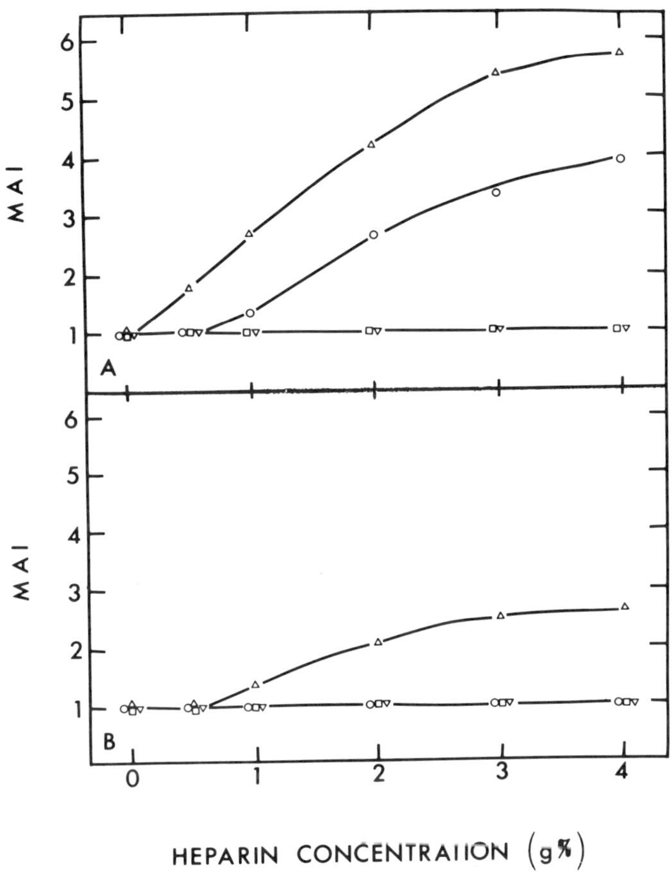

Figure 2. Effects of various fractions of heparin on aggregation of normal (A) and N-treated RBCs (B). Ionic strength = 150mM. Heparin fraction: (△), I; (○), II; (□), III; and (▽), IV.

tion II. N-treated RBCs were aggregated by Fraction I above 1 g % but the degree of aggregation was considerably lower than that of normal RBCs. The commercial heparin preparation (curve marked I = 150m*M* in Figure 3A) is comparable with the purified Heparin Fraction II in their effectiveness in causing RBC aggregation.

A decrease of ionic strength resulted in an increase of aggregation of normal RBCs by heparin (*see* Figure 3A). The N-treated RBCs, however, showed no aggregation despite the reduction in ionic strength (*see* Figure 3B).

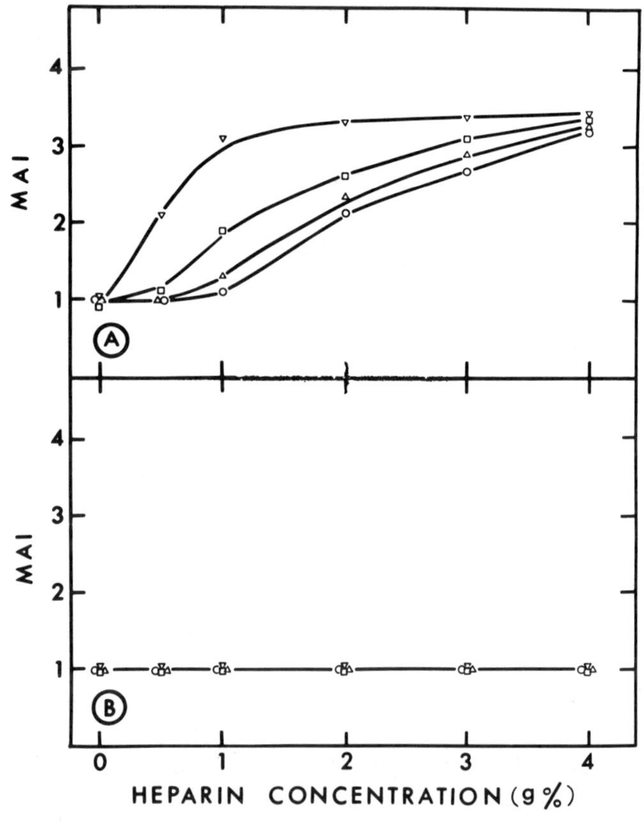

*Figure 3. Effects of ionic strength on RBC aggregation in NaCl solution plus heparin (total osmolality = 300 mOsm, balanced by sucrose). The figure gives the relation between MAI and heparin concentration. A and B show results on normal and N-treated RBCs, respectively. Ionic strength: (▽), 10m*M*; (□), 50m*M*; (△), 100m*M*; and (○), 150m*M*.*

Table I. Effects of RBC Surface Charge and Ionic Strength on RBC Aggregation by Positively Charged and Neutral Polymers

Mechanism of RBC Aggregation	*Role of Electrostatic Force*	*Effects of Experimental Manipulation of Electrostatic Forces on RBC Aggregation*	
		Removal of RBC Sialic Acids with Neuraminidase	*Reduction of Ionic Strength of Fluid Medium*
Electrostatic attraction between RBC sialic acids and polybases (e.g., polylysines)	Attraction between opposite charges	↓ RBC Aggregation	↑ RBC Aggregation
Hydrogen, VDW, and/or covalent bonding between neutral RBC membrane sites and neutral polymers (e.g., dextrans)	Repulsion between like charges of adjacent cell surfaces	↑ RBC Aggregation	↓ RBC Aggregation

Discussion

Studies on RBC aggregation induced by neutral and positively charged macromolecules have led to the concept that RBC aggregation results from macromolecular bridging between adjacent RBC surfaces (*8*, *9*, *10*). As shown in Table I, the nature of RBC aggregation induced by these macromolecules can be classified into two major groups (*10*). RBC aggregation owing to electrostatic attraction between the positively charged polybases and the negatively charged sialic acid on RBC surface is diminished after enzymatic removal of sialic acid; it is enhanced following reduction of ionic strength (decrease of counterion screening of sialic acid). When RBC aggregation results from mechanisms other than electrostatic attraction, the negatively charged sialic acid groups of adjacent RBC's exert an electrostatic repulsive force when cells are brought into close range, and the effects of manipulating the electrostatic force are opposite to the case of polybase-induced RBC aggregation.

The present findings suggest that heparin-induced RBC aggregation requires the presence of the negatively charged sialic acids at RBC surfaces, and is thus similar to the polybase induced RBC aggregation (*see* Table I), where the positively charged macromolecules are attached to the sialic acids on RBC surfaces. Since the polysaccharide unit of heparin is charged negatively, there are two possible mechanisms of electrostatic attractive interaction between heparin molecules and RBC surfaces. First, heparin is a complex macromolecule consisting of mucopolysaccharide and a protein backbone (*11*), and the latter might possess positive charge groups. Chemical analyses, however, have shown only trace amounts of basic amino acids in heparin (12). This mechanism, based on the presence of positive charge groups in heparin molecules is, therefore, not too likely. The second possible mechanism is that a short-range electrostatic attractive force exists between nonidentical molecules of the same sign but dissimilar charge densities (*4, 13*). According to this theory which originated in colloid chemistry, two approaching particles with like charge drive charges off each other's surfaces. If the charge densities are nonidentical, the surface with lower charge density will become neutral and eventually positive when the other is still negative (*4*). As a result, an electrostatic attraction exists between oppositely charged particles. In heparin-induced RBC aggregation, the molecules with strong and weak negative charge densities are heparin and sialic acid, respectively. The present finding that low ionic strength enhances heparin-induced RBC aggregation (*see* Figure 3) is also in agreement with Bierman's theory (*4*) that a repulsive force proportional to ionic concentration counteracts this type of electrostatic attraction. Such a mechanism may lead to a new form of RBC aggregation mediated by electrostatic attraction between RBCs and macromolecules in addition to those given in Table I. The present investigations not only further our understanding of the basic mechanism of cell–cell interaction induced by macromolecules, but also provide experimental evidence in the biological system that an attractive force may exist between nonidentical molcules with like charges. Such interaction forces might form the electrochemical basis of biological functions of heparin and play an important role in molecular interactions in other biological systems (*4, 14*).

The data in the present investigation, however, in no way rule out some possible mechanisms of heparin-induced RBC aggregation without involving charge interactions. Recent reviews by Nir and Andersen (*15*) pointed out that short-range specific molecular interactions are responsible for the specificity of interactions between the RBC surface and macromolecules, especially the blood group substances. Such short-range specific molecular interactions may exist possibly between heparin and the noncarboxyl site of sialic acids of RBCs. An alternation of ionic strength of the fluid medium may have a peculiar effect on the sialic acids and/or heparin molecules such that the degree of heparin-induced RBC aggregation was influenced by variation of the ionic strength (*see* Figure 3A). Another alternative explanation on the mechanisms of heparin-induced RBC aggregation involves the consideration of the long-range van der Waals (VDW) interactions between adjacent cell surfaces. Nir and Andersen (*15*) found that the magnitude of the VDW interactions between molecules follows the sequence: water < phopholipid < cholesterol, protein < sugar. Removal of sialic acids, which comprise a large fraction of the sugar content of RBC surface, may result in a significant reduction in the magnitude of the long-range VDW interactions between cells. According to this theory, as a result of removal of the sialic acids, both the attractive VDW interactions and the repulsive electrostatic interactions will be reduced in magnitude, and their sum can be either reduced or increased (*15*). If the heparin-induced RBC aggregation is caused or initiated by the long-range VDW attraction, and if removal of sialic acids from RBCs results in a reduction of the net attractive force between cells, these changes may provide another reason why the removal of sialic acids from RBCs led to a decrease in the degree of heparin-induced RBC aggregation (*see* Figure 2).

Acknowledgment

The author would like to thank S. Chien for helpful discussions and J. A. Cifonelli for supplying the heparin fractions. This work was supported by National Heart and Lung Institute Grant HL-16851. K. M. Jan is supported by a Research Career Development Award (HL-00036).

Literature Cited

1. Danishefsky, I. *Adv. Exp. Med. Biol.* **1975,** *52,* 105–118.
2. Zucker, M. B. *Federation Proc., Fed. Am. Soc. Exp. Biol.* **1977,** *36,* 47–49.
3. Seaman, G. V. F. In "The Red Blood Cell," 2nd ed.; Surgenor, D. MacN., Ed.; Academic: New York, 1975; Vol. II, pp. 1135–1229.
4. Bierman, A. *J. Coll. Sci.* **1955,** *10,* 231–245.
5. Cifonelli, J. A. *Adv. Exp. Med. Biol.* **1975,** *52,* 95–103.
6. Jan, K. M.; Chien, S. *J. Gen. Physiol.* **1973,** *61,* 638–654.
7. Jan, K. M.; Chien, S. *J. Gen. Physiol.* **1973,** *61,* 655–668.
8. Katchalsky, A.; Danon, D.; Nevo, A.; deVries, A. *Biochim. Biophys. Acta* **1959,** *33,* 120–138.
9. Brooks, D. E. *J. Colloid Interface Sci.* **1973,** *43,* 687–699.
10. Chien, S. *Bibl. Anat.* **1973,** *11,* 244–250.
11. Jeanloz, R. W. *Adv. Exp. Med. Biol.* **1975,** *52,* 3–17.
12. Wang, C. C.; Jaques, L. B. *Arzneim.-Forsh.* **1974,** *24,* 1945–1949.
13. Maroudas, N. G. *J. Theor. Biol.* **1975,** *49,* 417–424.
14. Pitts, E. P.; Jones, R.; Goldberg, L.; Coulston, F. *Biorheology* **1977,** *14,* 33–42.
15. Nir, S.; Andersen, M. *J. Membr. Biol.* **1977,** *31,* 1–18.

RECEIVED October 17, 1978.

10

Ion-Binding Effects on Diffusion of Reversibly Dissociating Lobster Hemocyanin

SUNG HYE YIM[1] and GERSON KEGELES[2]

Section of Biochemistry and Biophysics, University of Connecticut, Storrs, CT 06268

JOHN R. CANN

Department of Biophysics and Genetics, University of Colorado Medical Center, Denver, CO 80220

Diffusion measurements of 0.1% precision were made by the Gouy interference method on hemocyanin from New England lobster, Homarus americanus, *in pH 9.6 glycine buffers containing various levels of calcium ion, such that the degree of dissociation of the protein into half molecules varied from 0 to 0.915 by weight. Nonideality of the diffusion process was examined by comparing experimental Gouy fringe positions with those calculated for ideal diffusion. For dissociating systems, computer simulations of diffusion coupled with binding of protons in nonbuffered solutions predicted diffusion more ideal than observed, yet computations ignoring known ligand coupling agreed more closely with experiment. The pH gradient generated in unbuffered systems exerts a negative feedback on dissociation, causing more ideal diffusion behavior than in an actual buffered experiment.*

The Gouy interference method for studying diffusion in liquids is based on the original observation of G. L. Gouy in 1880 (*1*) that collimated monochromatic light from an illuminated horizontal slit, when passed through a diffusing boundary between two liquids and brought to focus

[1] Experimental work in this chapter was described in the Ph.D. thesis of Sung Hye Yim, University of Connecticut, 1974.
[2] Author to whom correspondence should be directed.

0-8412-0473-X/80/33-188-151$05.00/1

in the eyepiece of a telescope, gives rise to a system of horizontal interference fringes below a bright undeviated slit image. The fringes are spaced very closely just below the slit image and become more widely separated as their deflection below the slit image increases, ending finally in a broad, diffuse bright fringe at the bottom. While Gouy described qualitatively the first-order effect giving rise to interference, he was unable to develop a quantitative theory for the phenomenon, in part, perhaps, because the distribution of concentration in an ideal diffusion process was not developed until 1879 by Stefan (*2*), but more certainly because the theory for the bending of light in refractometrically inhomogeneous media was not developed until 1893 by Wiener (*3*).

The first photograph of Gouy interference fringes was published by Longsworth (*4*) in 1945 in a symposium on diffusion, and this led to the development of the American version of the Gouy interference method, announced in 1947 (*5, 6*). An independently developed British version, interrupted by World War II, was announced in 1948 (*7*).

The present study is based on experimental arrangements (*8*) and improvements in theory (*9, 10*) which utilize the entire fringe system according to the original American version (*5, 6*). An excellent review is available (*11*) which covers the principles and practice of the Gouy interference method. Since this method has fallen into disuse since the time of that review, some background description pertinent to the present study will be presented.

The experiment is arranged by layering a mathematically semi-infinite column of buffer containing the protein sample under a mathematically semi-infinite layer of buffer. The boundary layer between the liquids is, for a convergent light system, carefully located on the optic axis. Monochromatic light from an illuminated horizontal slit is projected through the diffusing boundary and brought to focus, usually at about unit magnification of the source slit. The interference fringes in the focal plane are photographed periodically over several hours, and their deflections below the undeviated slit image at the optic axis are measured with a microcomparator. These deflections at known times after the start of diffusion constitute primary data for the evaluation of diffusion coefficients.

Lobster hemocyanin in glycine–sodium hydroxide buffer at pH 9.6 undergoes a reversible whole molecule–half molecule dissociation which is very sensitive to the level of free calcium ion present (*12*). This dissociation process is also very sensitive to pH (*12*). Cann and colleagues have developed, by simulation techniques (*13, 14*), a general picture of the kinds of partial or complete boundary resolution that may be expected for such coupled systems in various types of transport experiments. Kinetic investigations of the lobster hemocyanin system under the conditions of the present study were developed by using stopped-flow light-scattering experiments (*15, 16*) and it was established that about four calcium ions must bind to two half molecules in order to make it possible for them to reunite to a whole molecule. Protons also enhance reassociation (*12*), but prior ligand binding is not a prerequisite to reassociation (*15, 16*). With the aid of this kinetic information (*15, 16*), a theoretical investigation was made (*17*) which served to predict the degree of boundary resolution to be expected in such systems during analytical ultracentrifugation. The purpose of this chapter is to extend both experimental observation and theoretical predictions to the case of free boundary diffusion.

Equipment and Experimental

The Gouy diffusiometer used in this study was built at Clark University, and its construction has been described in detail by Donoian (*18*). It is arranged with a single-lens, convergent-light system at unity magnification, the 11-cm diameter lens being an air-spaced achromat, color corrected at 5893 Å and 4341 Å, with a focal length of 170 cm. With the aid of a Mann microcomparator, the optical level b from the middle of the cell to the photographic plate was measured by a series of internal calipers (*18*) to be 314.419 cm. The light source was an H-100A4 mercury vapor lamp, the green line $\lambda = 5461$ Å being isolated by a Kodak Wratten 77A filter. The diffusion cell is a modified Tiselius-type cell (*19*) with thick optical flats as clamped-on windows, as described by Gutter and Kegeles (*20*). The solution optical thickness a in the direction of passage of light was measured to be 6.0036 cm. Several previous publications (*21, 22, 23*) have described the specific methodology and application of the diffusiometer used in this study.

Each hemocyanin sample studied was prepared from the blood of a single lobster. To remove nonreacting impurities, the hemocyanin sample, after removal of the clot, was centrifuged, dialyzed against pH 9.6 glycine–sodium hydroxide buffer, and purified by passage through Sephadex G-150 in a two-stage process (*12*). In the presence of 0.01*M* free [Ca^{++}], all reacting hemocyanin should be in the form of whole molecules of molecular weight 940,000. In the presence of 0.001*M* free [Ca^{++}], all reacting hemocyanin should be dissociated in the form of half molecules with a small amount of one-twelfth size (5S) subunits, which do not wholly reaggregate. The first filtration was performed at 0.001*M* free [Ca^{++}] and rapidly emerging fractions which represented aggregates or nondissociating material were discarded. The later emerging fractions were pooled, equilibrated to 0.01*M* free [Ca^{++}], and filtered through Sephadex G-150 under these conditions. The reacting, reaggregated hemocyanin which emerged first was kept for diffusion studies, while any unreactive low-molecular-weight impurities and 5S subunits were retained by the column and discarded. All samples were examined by sedimentation velocity analysis at 42,000 rpm before they were used for diffusion experiments. Samples finally were dialyzed in narrow tubing tied to the end of a glass tube whose other end was previously placed through a rubber stopper fitted to a wide-mouth bottle. After introduction of the protein sample, the upper end of the glass tube was plugged with a small rubber stopper. The sample was removed from the dialysis bag by insertion of a long steel needle, without removal of the dialysis bag from the surrounding buffer in the wide-mouth bottle, an important procedure which was found to be necessary in early diffusion experiments on ovalbumin (*24*) in order to obviate the concentration by surface evaporation of the buffer components inside the dialysis bag. Hemocyanin solutions prepared by dialysis were adjusted slightly, when required, to about 0.3 wt % and placed in the bottom section, on one side, and into a specially elongated top section acting as a solution reservoir (*20*).. The central reference channel was filled with buffer solution (*20*) before clamping together the diffusion cell, and this channel was used to measure the "fractional fringe" portion of the total number of fringes, j_m (*8*).

After a thorough rinsing, the remaining side of the cell and the top section reservoir were filled with buffer solution. A stainless steel needle with capillary tips was lowered to the optic axis by a rack-and-pinion assembly (*18*) and used to siphon the buffer out after the sheared boundary was formed at the interface between bottom and center sections. The boundary was swept up to the capillary tips and collapsed into a very sharp interface as the diffuse regions on either side were washed away by fresh buffer and protein solutions, according to the boundary-sharpening technique of Kahn and Polson (*25*). Fractional fringe photographs were taken with a Rayleigh double-slit system through the reference channel and the diffusion channel, before the boundary was in place and after it had been sharpened at the optic axis (*8*). The sharpening was interrupted several times to allow diffusion to clear unwanted solution from the interior cell surfaces (*5*). When siphoning finally was stopped, timing was started with a stopwatch calibrated against Bureau of Standards signals from station WWV. Deflections Y_j below the slit image (j corresponding to fringe number and the lowest minimum of intensity being numbered zero) were measured in a David Mann microcomparator. According to the theory of the Gouy interference phenomenon (*6*) the quantity Y_j/e^{-z^2j} is a constant for all fringes in a given photograph for ideal diffusion. Tere z_j is the reduced-height $x_j/2\sqrt{Dt}$ at time t for a diffusion coefficient D, where x_j represents the position (distance from the boundary center) from which geometrical optics predicts light will be deflected to the level given by Y_j (*6*).

In addition to the deflection Y_j, primary observable quantities are the fringe number, j, and the total number of fringes, j_m. The total path difference corresponding to the fringe numbered j is given for ideal diffusion by the path difference function (*6*).

$$f(z_j) = \frac{2}{\sqrt{\pi}} \int_0^{z_j} e^{-\beta^2} d\beta - \frac{2}{\sqrt{\pi}} z_j e^{-z_j^2} \quad (1)$$

It is noted that as $z_j \rightarrow 0$ and $x_j \rightarrow 0$ (the center of the boundary), $f(z_j) \rightarrow 0$, and as $z_j \rightarrow \infty$ and $x_j \rightarrow \infty$ (the edges of the boundary), $f(z_j) \rightarrow 1$. The function $f(z_j)$ has been tabulated (*6*) as a known function of $e^{-z_j^2}$, from which either quantity can be obtained if the other is known. The first theory (*6*) gave first-order values of the path difference function as

$$f(z_j) \simeq (j + 3/4)/j_m \tag{2}$$

from which values of $f(z_j)$ are obtained by simple knowledge of the fringe number and the total number of fringes. More detailed theories (*9, 10*) have supplied more nearly correct evaluations of the diffraction problem and more complicated relationships between $f(z_j)$, j, and j_m. The relationships given by Gosting and Onsager (*10*) were inverted by S. W. Provencher (*26*) and programmed for computer use, and a slightly modified version of this program in Fortran IV has been used in this study.

The quantity $Y_j/e^{-z_j^2}$ was given (*5*) the symbol C_t. From the original theory (*6*), the diffusion coefficient D' is obtained from C_t with the relation

$$D' = \frac{j_m{}^2\lambda^2 b^2}{4\pi C_t{}^2 t'} \tag{3}$$

It is noted that an independent value for D' is obtained from every fringe numbered j whose deflection Y_j is measured. For nonideal diffusion, C_t actually will show a drift with fringe number, and the value C_t corresponding approximately to the middle of the diffusing boundary is obtained by an extrapolation procedure (*27*). Akeley and Gosting (*27*) developed the deviation function defined by

$$\Omega_j = e^{-z_j^2} - Y_j/\bar{C}_t \tag{4}$$

to express deviations of the concentration gradient function in the diffusing boundary from the ideal Gaussian error function $e^{-z_j^2}$ predicted for ideal diffusion. These authors used the functional relationship between fringe deviations Ω_j and the path difference function $f(z_j)$ to study diffusion nonideality resulting from the presence of small amounts of aggregated material in purified proteins. For ideal diffusion, Ω_j, is identically equal to zero for every fringe when measurements are made with infinite accuracy.

The recorded time after initiation of diffusion does not take into account the initial lack of sharpness of the boundary. This can be taken into account with a zero-time correction, Δt, which is the length of time the material would have had to diffuse to provide a boundary as unsharp as that observed when timing begins. The relationship used (*5*) to obtain Δ_t is given by

$$D' = D\ (1 + \Delta t/t') \tag{5}$$

The recorded time is actually t', and the apparent diffusion coefficient D' is calculated at t' from Equation 3, while D in Equation 5 is the true diffusion coefficient.

Experimental Results

In order to test technique and calibration of equipment, three separate experiments were performed on the diffusion of 0.5% sucrose vs. water at 25°C. The mean deviations from the mean diffusion coefficient within these three individual experiments were 0.06%, 0.01%, and 0.08%, with maximum Ω_j values of 7, 2, and 6 μ/cm of fringe deflection below the slit image and average Ω_j values of 4.5, 1.0, and 4.0 μ/cm. The average value of the diffusion coefficient for two separate fresh table sugar samples agreed with the value of 5.2066 $(10)^{-5}$ cm^2/sec interpolated from the data of Gosting and Morris (*9*) to within one part in 5000. Reagent-grade sucrose which has been shelved for prolonged periods undergoes a condensation reaction with release of water, and is not trustworthy for molecular purity.

Data obtained for purified hemocyanin samples in pH 9.6 glycine buffer at various levels of free calcium ion are summarized in Table I. The apparent protein equilibrium (formation) constant K_p (**D** $\equiv$ dodecamer or whole molecules and **H** $\equiv$ hexamer or half molecules) and the degree of dissociation by weight, α, are taken from the ultracentrifuge study of Morimoto and Kegeles (*12*).

Table II serves to illustrate the internal precision attainable for diffusion measurements on solutions containing only whole molecules of purified hemocyanin, and Table III indicates the values of average fringe deviations in the same experiment. While the boundary appears ideal within experimental error for the first 30 of 64.498 fringes, negative deviations occur closer to the slit image which cannot be caused by inclusion of higher-molecular-weight impurities (*27*). Similar results were obtained in studies of the many times recrystallized, nonreacting bovine serum mercaptalbumin under isoelectric conditions (*28*). This is a reflection of boundary skewing at the edges, but the origin of this skewing is not known.

Table IV shows the results of one of several diffusion experiments on hemocyanin which has not been purified by the Sephadex filtration procedure. It is noted that the results are still internally self-consistent

Table I. Purified Hemocyanin (0.295%–0.308%) vs. Glycine–NaOH Buffer, (pH 9.6)

Free Ca^{++}	$K_p = (\mathbf{D})/(\mathbf{H})^2$	α	$D \times 10^7$, *cm²/sec*
0.0028	0.68	0.92	3.271
0.0038	3.14	0.74	3.124
0.0043	5.83	0.64	2.978
0.0053	16.6	0.46	2.994
0.0072	76	0.26	2.847
0.0140		0.0	2.697

Table II. 0.297% Hemocyanin vs. Glycine–NaOH Buffer, pH 9.6 (free Ca^{++} = 0.014*M*)

Time, sec[a] (t′)	D × 10^7 [b]	*Deviation*[c]
23979	2.698	0.001
26057	2.694	−0.003
28800	2.699	0.002
32505	2.698	0.001
37261	2.698	0.001
39674	2.696	−0.001

[a] Zero-time correction = 1379 sec.
[b] Least-squared $D = 2.697 \times 10^{-7}$ cm²/sec.
[c] Percent deviation = 0.047%.

Table III. 0.297% Hemocyanin vs. Glycine–NaOH Buffer, pH 9.6 (free Ca^{++} = 0.014*M*; j_m = 64.498)

j	$\overline{\Omega}_j \times 10^4$
0	0.7
1	2.0
2	−0.5
3	0.4
4	2.6
5	4.0
6	3.5
7	4.2
8	1.7
9	2.7
10	2.1
15	1.7
20	0.3
25	−1.5
30	−3.0
40	−10.7
50	−18.9
60	−31.9

Table IV. 0.201% Unpurified Hemocyanin vs. Glycine–NaOH Buffer, pH 9.6 (free Ca^{++} = 0.014*M*)

Time, sec[a] (t′)	D × 10^7 [b]	*Deviation*[c]
12594	2.637	−0.003
16683	2.644	0.004
21211	2.639	−0.001
25306	2.645	0.005
28909	2.644	0.004
36133	2.639	−0.001
39668	2.635	−0.005

[a] Zero-time correction = 1926 sec.
[b] Least-squared $D = 2.640 \times 10^{-7}$ cm²/sec.
[c] Percent deviation = 0.12%.

Table V. 0.201% Unpurified Hemocyanin vs. Glycine–NaOH Buffer, PH 9.6 (free Ca^{++} = 0.014*M*; j_m = 42.605)

j	$\overline{\Omega_j} \times 10^4$
0	9.4
1	17.1
2	24.5
3	27.9
4	30.7
5	34.0
6	36.6
7	37.7
8	41.0
9	42.0
10	43.1
12	43.8
15	44.9
18	43.7
20	42.1
25	34.9
30	23.0
35	5.1

to 0.1% in the diffusion coefficient, and that the mean value of 2.640 $(10)^{-7}$ cm^2/sec differs from that shown in Table II, 2.697 $(10)^{-7}$ cm^2/sec, by only 2%, which is close to the limit of detection with the more popular quasielastic Doppler light-scattering methods (*29*).

On the other hand, if one examines deviations from the ideal Gaussian concentration gradient distribution for this unpurified sample, as shown in Table V, there is, with the Gouy method, a vast distinction between the very large positive Ω_j values for the impure sample, up to about 45 μ/cm of fringe deflection, and the absence of such large positive Ω_j values in the same region of the fringe pattern for the purified sample, in Table III. Thus the fringe deviation procedure is exquisitely sensitive to small amounts of impurity, which are interpreted here as being unreactive or aggregated material (*27*).

Deviations from Ideal Boundary Shape for Reacting Hemocyanin

The major purpose of this study is to investigate the coupling effects between specific ligand binding and the whole-molecule–half-molecule interaction, particularly as they produce nonideality in diffusion. This is best accomplished (*27*) by examining plots of the fringe deviation function, Ω_j, as defined by Equation 4, as a function of the path difference function $f(z_j)$, as defined by Equation 1.

It should be emphasized here that we are not dealing with a simple two-component system, even when the protein is not under interacting conditions. In the case of a simple three-component system, where solvent is designated as Component 0 and two solutes are designated as Components 1 and 2, the flow equations for ideal diffusion in a solvent-fixed frame of reference are (*30*) given by

$$\begin{aligned} J_0 &= 0 \\ J_1 &= -D_{11}\partial C_1/\partial x - D_{12}\partial C_2/\partial x \\ J_2 &= -D_{21}\partial C_1/\partial x - D_{22}\partial C_2\partial x \end{aligned} \tag{6}$$

The diffusion coefficients D_{11} and D_{22} are the principal or "self" diffusion coefficients and the off-diagonal quantities D_{12} and D_{21} are "mutual" diffusion coefficients. Even when Onsager's reciprocal relations (*31*) are valid for the appropriate flow equations so that $D_{12} = D_{21}$, there are still three diffusion coefficients generally required to describe the diffusion process. It is noted that even if $\partial C_1/\partial x = 0$, the flow of Component 1 is linked to that of Component 2 through the term $-D_{12}\partial C_2/\partial x$, and is not zero.

When specific chemical reactions are involved in the coupling of flows, such as in the ionization of weak acids (*32, 33, 34*), the mutual effects are very large. On the other hand, when such specific chemical reactions as complex formation take place in the presence of superposed simple interionic attraction effects, the coupling caused by the chemical reactions becomes the predominant factor responsible for coupled flow (*35*). Thus it is often a reasonable approximation to treat a protein in a buffer as a pseudo two-component system, in the absence of protein dissociation coupled to specific ion-binding reactions. The nonideality of boundary shape in the diffusion of systems involving protein dissociation and ion binding then is ascribed as a good first approximation entirely to the effects of the coupling of the diffusion process with dissociation, which may or may not be linked to ion-binding reactions.

On this basis, models were constructed for computer simulation in which the species whole molecules, **D**, half molecules, **H**, and ligand, **L**, were allowed to diffuse independently for short increments of time, Δt, and then were coupled via chemical interaction. For example, if the overall reaction is

$$2\mathbf{H} + 4\mathbf{L} \rightleftarrows \mathbf{D} \tag{7}$$

then after the short period of independent diffusion Δt, the species were required to re-equilibrate locally everywhere in the boundary to satisfy the equilibrium relation

$$K = \frac{[\mathbf{D}]}{[\mathbf{H}]^2[\mathbf{L}]^4} \tag{8}$$

The calculation then proceeded through another cycle of independent species diffusion, another cycle of re-equilibration, etc.

For the purpose of this calculation it was necessary to know the diffusion coefficient for ligand and whole molecules of hemocyanin (Table I) and the value of the diffusion coefficient for half molecules of hemocyanin. If the level of calcium ion is reduced in pH 9.6 glycine buffer to the point where no whole molecules remain, the half molecules always are contaminated with one-twelfth size (5S) subunits, so that it is impossible to measure directly the diffusion coefficient of half molecules. For a self-associating system, the Gouy method leads to an average diffusion coefficient which is the Z-average, defined by

$$D_Z = \Sigma C_i M_i D_i / \Sigma C_i M_i \tag{9}$$

where C_i are weight concentrations of species, M_i are their molecular weights, and D_i are their corresponding diffusion coefficients. For a system containing only half and whole molecules, a measurement leads to a Z-average diffusion coefficient D_Z at known concentration values of whole molecules and half molecules at a total equilibrium concentration corresponding to half the concentration of protein in the lower solution. Since the diffusion coefficient of whole molecules has been measured (Tables I and II), the only remaining unknown is the diffusion coefficient of half molecules. In this way, all of the experiments on interacting hemocyanin led to the mean value and standard error for the diffusion coefficient of half molecules, $3.55 \pm 0.17 \times 10^{-7}$ cm²/sec.

With the aid of this value and the value 2.697×10^{-5} cm²/sec for whole molecules (Table I and II), two separate types of simulations were made: (1) simulations in which dissociation proceeded with the apparent formation constants of protein shown in Table I, but without taking into account ligand binding and (2) simulations in which ligand binding also was taken into account by using the same apparent formation constants, together with Reaction Scheme 7 and the known ambient (free) ligand ion concentration in each case. The details of the reasoning used in these simulations follow.

Theoretical concentration profiles and gradient curves were computed for the diffusional transport of three model systems: (a) $2\mathbf{H} \rightleftarrows \mathbf{D}$ which ignored ligand mediation; (b) Ca^{++}-mediated association, $2\mathbf{H} + 4Ca^{+} \rightleftarrows \mathbf{D}$ and (c) unbuffered H^+-mediated association, $2\mathbf{H} + 4H^+ \rightleftarrows \mathbf{D}$. To illustrate the computational procedure let us consider ligand-mediated association. Diffusion of the macromolecular species is from an infinitely sharp initial boundary between the macromolecule–ligand solution and a solution of ligand at the same concentration as its equilibrium value in the solution containing the macromolecule. The rates of reaction are sufficiently fast that, in effect, there is local equilibrium among the three interacting species during their differential diffusion. The concentration profiles were computed as a function of time by numerical solution of the set of diffusional transport equations and mass action expressions for constituent macromolecule and constituent ligand. The salient features of the calculation are as follows: the diffusion column is divided into a number of discrete segments of length Δx. Given the initial equilibrium concentrations of the three species in each segment, we calculate the change in their distributions during a short interval of time, Δt, owing to diffusion. There is no re-equilibration by interaction during Δt, each species diffusing independently from its distribution at the beginning of the interval. The changes in concentration are described by Fick's second law of diffusion

$$\frac{\partial C_i}{\partial t} = D_i \frac{\partial^2 C_i}{\partial x^2} \quad i = 1, 2, 3 \tag{10}$$

where C_i is the concentration; D_i is the diffusion coefficient; t is the time; x is the position in the diffusion column; $i = 1$ and 2 designated **H** and **D**, respectively; and $i = 3$, designated the ligand (**L**). Each of the differential Equations 10 is approximated by a finite difference equation through application of the implicit Crank–Nicholson method (*36*), which is unconditionally stable and converges with discretization error of $0[(\Delta t)^2 + (\Delta x)^2]$. The difference equation, in turn, results in a system of linear equations (*37*) whose solution is accomplished by Gaussian elimination (*38*). Given values of C_i at any time t, we calculate their values at $t + \Delta t$ as changed by diffusion using this procedure subject to the boundary conditions that the concentrations at either end of the column remain undisturbed during the diffusion process.

After the concentrations have been advanced, new values of the constituent concentrations on the moles/liter scale of macromolecule, $\overline{C}_{\mathrm{M}}$, and ligand, $\overline{C}_{\mathrm{L}}$, are computed from the C_is as changed by diffusion

$$\overline{C}_{\mathrm{M}} = C_1 + 2C_2$$
$$\overline{C}_{\mathrm{L}} = C_3 + 4C_2$$

and then chemical equilibration is imposed by simultaneous solution of the mass action expressions

$$\overline{C}_{\mathrm{M}} = C_1 + 2K_m C_1^2 C_3^4 \tag{11}$$

$$\overline{C}_{\mathrm{L}} = C_3 + 4K_m C_1^2 C_3^4 \tag{12}$$

for the equilibrium concentrations of C_1 and C_3 (designated for clarity as $\hat{C}_2 = (1/2)(\overline{C}_{\mathrm{M}} - \hat{C}_1)$). Equations 11 and 12, in which K_m is the moles/

liter scale equilibrium constant for the ligand-mediated association, are solved by the Newton–Raphson method as described previously for the sedimentation of the hemocyanin system (*17*). The new values of the equilibrium concentrations serve as the starting concentrations for the next time interval Δt of diffusion followed by re-equilibration and so on. This recursive calculation allows one to follow the evolution of the concentration profiles (C vs. x) and gradient curves ($\partial C/\partial x$ vs. x) of the constituent macromolecule and ligand and individual species.

The computations were made on the University of Colorado's CDC 6400 electronic computer using values of $\Delta x = 0.01$ cm and $\Delta t = 10$ sec, specified values of K_m and constituent concentrations of macromolecule and ligand, and the following values for the diffusion coefficients: **H**, 3.5501×10^{-7} cm^2 sec^{-1}; **D**, 2.6981×10^{-7} cm^2 sec^{-1}; Ca^{++}, 1.2×10^{-5} cm^2 sec^{-1}; and H^+, 9.3×10^{-5} cm^2 sec^{-1}. Material balance was excellent and comparison of control calculations on a noninteracting macromolecule with analytical prediction confirmed the accuracy of the numerical calculations. The three sets of experimental data for which these simulations were developed are listed in Table I for α values of 0.92, 0.46, and 0.26.

Since the direct product of the simulations was a printout of the concentration gradient function and the concentration function for protein and ligand, it was necessary to convert these into predicted Gouy fringe positions in order to develop simulated plots for fringe deviations. The dissociation of the protein produced skew diffusing boundaries, however, and the schemes by which such boundary patterns were used to approximate Gouy fringe positions is described with the aid of Figure 1.

It was assumed, as in the first-order theory for the Gouy interference phenomenon (*6*), that as a first approximation only rays passing through equal values of the refractive index gradient on either side of the boundary center would arrive together at the slit image plane. Thus pairs of positions with equal dn/dx values (*see* upper part of Figure 1) were

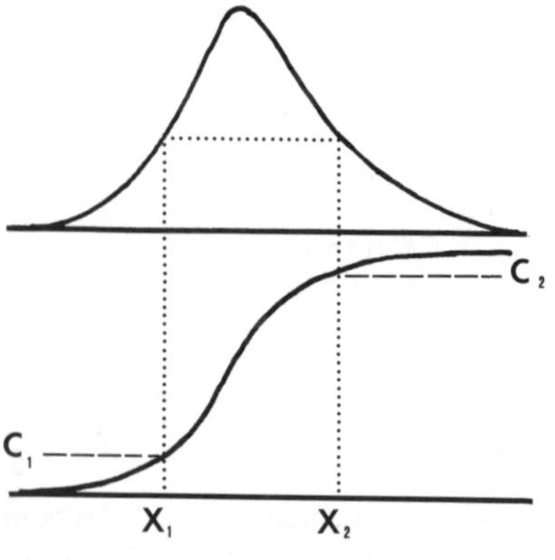

Figure 1. Scheme for interpolation of simulated dc/dx *and* c *values to predict Gouy fringe positions for skew boundaries*

selected, and the corresponding x-values in the boundary were used to locate the effective refractive index values, as shown in the lower part of Figure 1. At this point it was assumed that an adequate wave optical diffraction treatment would result by requiring that

$$Z_j\lambda_o = a(n - n') - (x + x')Y_j/b \tag{13}$$

where Z_j is the Airy integral approximation (*6, 9*) to the requirement for interference. The concentration and concentration gradient functions were converted to the refractive index scale with the factor 0.00197 g^{-1} per 100 mL as determined in this study. The deflection Y_j in Equation 9 was taken as

$$Y_j = ab(dn/dx) \tag{14}$$

Whenever Equations 13 and 14 together predicted the existence of an interference fringe, the position Y_j of the fringe was recorded as a function of the fringe number, thereby constructing one fringe at a time, the simulation of the Gouy interference fringe system. With this in hand, values of the fringe deviation function, Ω_j, (Equation 4) for the simulated model were plotted against the path difference function (Equation 1) and compared directly with experiment.

As a test, this approximation procedure for skew boundaries was used with a nondissociating system as a model, and while the plot for the model showed a scatter of up to 10 μ/cm owing to the graphical interpolation procedures being used, no drift indicative of apparent nonideality caused by any possible deficiency in theory was observed: the Ω_j values scattered about zero over the range of $f(z_j)$ for which they were calculated.

Finally, Figures 2, 3, and 4 show comparisons of experimental data (indicated by circles) with simulated data for the model systems in which ligand binding is ignored (solid curves) and with simulated data in which ligand binding (proton binding at a starting pH of 9.6) is taken into account, assuming that no external buffer system is present (broken curves). (Similar calculations for ligand binding with calcium ion at ambient calcium ion concentrations identical to those shown in Figures 2, 3, and 4 indicated no difference from the case in which no ligand binding was considered.) In Figure 2 are the results corresponding to 0.0028M free [Ca^{++}] and a degree of dissociation $\alpha = 0.92$. Little difference exists between the predictions, both of which suggest a little greater ideality of diffusion than observed experimentally. Results corresponding to 0.0053M free [Ca^{++}] and a degree of dissociation $\alpha = 0.46$ are shown in Figure 3. Surprisingly, the model which ignores ligand binding fits

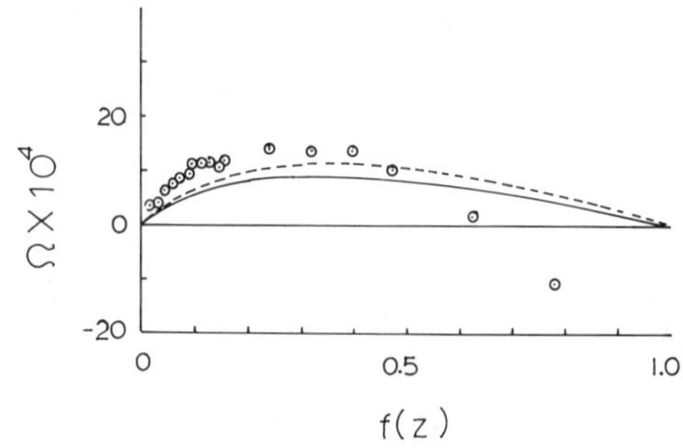

Figure 2. Interference fringe deviation plots for Gouy diffusion of reacting hemocyanin at $\alpha = 0.92$. Points are experimental: (——), simulation for dissociating system, ignoring ligand coupling; (– – –), simulation for dissociating system, assuming proton coupling according to Reaction Scheme 7.

the experimental data quite well, whereas the simulation which includes the coupling shown in Reaction Scheme 7 predicts much greater ideality, which disagrees with experiment.

Figure 4 shows a similar comparison at 0.0072*M* free [Ca^{++}], where $\alpha = 0.26$. Both simulations predict much greater ideality of diffusion than observed. Table I demonstrates that the estimated formation constant under these conditions is increasing very rapidly with relatively small increases in calcium ion, and it is clear that it is becoming increasingly difficult to measure the formation constant with precision under these conditions. Moreover, calculations reported by Yphantis (*39*) on heterogeneous systems containing nonreacting impurities suggest that it is exactly in this region of very small degrees of dissociation where the

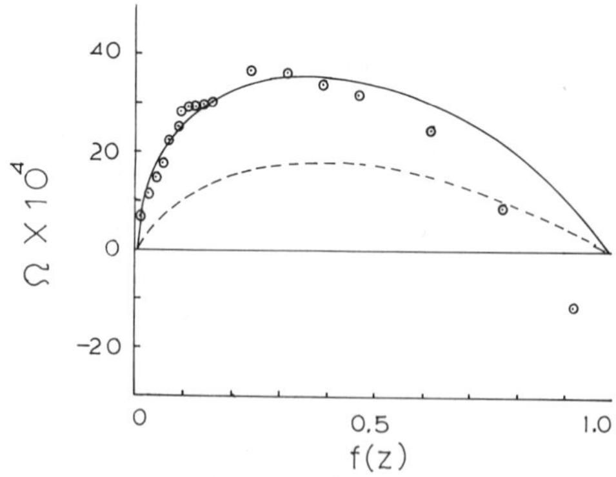

Figure 3. Interference fringe deviation plots for Gouy diffusion of reacting hemocyanin at $\alpha = 0.46$. Remainder of key is as in Figure 2.

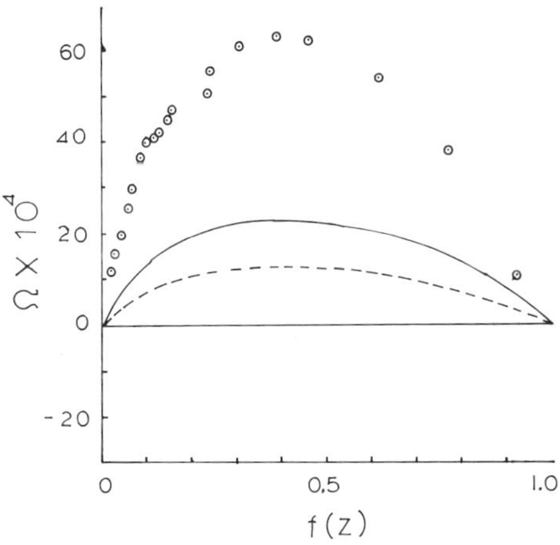

Figure 4. Interference fringe deviation plots for Gouy diffusion of reacting hemocyanin at $\alpha = 0.26$. Remainder of key is as in Figure 2.

demands upon purity become extreme, if one hopes to obtain accurate equilibrium constants by a method such as sedimentation equilibrium. Consequently, it is unclear whether Figure 4 suggests the need for a better model or for purer preparations or better data for the diffusion coefficient of half molecules, or better equilibrium constant data.

The distinction between the two models used in Figure 3 is startling, however, and requires explanation. The dashed curve, including ligand-binding coupling with dissociation during the diffusion process, requires that, by Reaction Scheme 7, excess protons will be released on the solvent side of the diffusing boundary, as dilution of protein takes place. If an external buffer system is not present, as is assumed in this model, then the accumulation of protons exerts a negative feedback effect on dissociation, increasing the relative concentration of whole molecules at the expense of half molecules. This makes the diffusion experiment tend to look more like an experiment in which whole molecules alone are diffusing, so that ideality of the diffusion process is enhanced. On the other hand, in a real experiment in a highly buffered system, the protons which are released as a consequence of dilution at the upper edges of the boundary are absorbed by buffer, and the protein is allowed to undergo dissociation as if it were not coupled with ligand binding. This is why the solid curve in Figure 3, which ignores ligand binding, fits the experimental data obtained in a buffered solvent, generating increased nonideality of diffusion. In living systems, where buffering electrolyte systems are always present, one is led to expect that diffusing and dissociating protein systems will have their diffusion accelerated in regions of low protein concentration, all other factors being equal.

Discussion

While the conditions under which these experiments have been performed (pH 9.6, glycine buffer) do not correspond with in vivo conditions in the lobster, this study does act as a useful model in pointing out the possible in vivo effects of the coupling of intermolecular reactions with ligand binding in other systems.

Figure 3 shows that strong electrolyte buffering in such a system emphasizes the nonideality of the macromolecular diffusion process by destroying the negative feedback accompanying ion release during dissociation. In a strongly buffered system, dissociation of the macromolecules in very dilute solution would appreciably enhance diffusive transport of the macromolecular system, as indicated by the values of the diffusion coefficients given in Table I. Any imposed ligand gradients clearly could exert a strong control over diffusive transport by controlling the degree of dissociation.

Moreover, although no provision was made for such an effect in our model, a certain amount of self-buffering capacity exists, owing to surface binding of the ligand. For example, the number of calcium binding sites is probably about ten times the number (three to five) estimated (*12*) as being coupled directly to the half molecule–whole molecule reaction. In addition, a direct temperature jump experiment (*40*) on weakly buffered ($0.02M$ Veronal–$0.025M$ Na-Veronal–$0.075M$ KNO_3) lobster hemocyanin at pH 7.95 with $5 \times 10^{-5}M$ phenol red as pH indicator, dramatically illustrated the self-buffering of the protein, in the absence of dissociation into half molecules. In this experiment a rapid amplitude shift, proportional to the shift of indicator optical density (OD) occurred during the approximately 100-μsec heating period. The known temperature dependence of the buffer system ionization constant served to calibrate this experiment as an absolute measurement for pH. This shift of OD was followed by a partial reversed relaxation of OD, with a relaxation time of approximately 1 msec. The amplitude of the reversed relaxation corresponded to the uptake of about three protons per whole molecule of hemocyanin. If it is assumed that the ionization reactions involved are driven through about 10% of completion by the 4°C temperature jump, then there may be possibly 30 protons involved per 940,000 mol wt hemocyanin molecule, a number far in excess of that estimated (*12*) as being coupled to the half-molecule–whole molecule interaction. The assignment of these shifts of OD to buffer and protein was verified by subtracting the optical output from a twin cell (*40*) containing only buffer and indicator, the two cells being connected electrically in series. Only the 1-msec relaxation remained in the record, an effect clearly produced by the addition of protein to the system.

On the other hand, when a half molecule–whole molecule reacting system in weak buffer was mixed with buffer containing pH indicator and excess calcium ion in a stopped flow apparatus (*15*), leading to association to whole molecules in the observation cell, an OD change of small amplitude occurred for the pH indicator in the 15-25-sec time range. Clearly the protons involved in this process differ in nature from those just described in the temperature jump experiment, and since the relaxation time in this stopped-flow experiment agrees with those observed in direct 90° light-scattering observation of the macromolecular interaction (*15, 16*), we must assume that we are observing the protons which are coupled directly (*12*) with the half molecule–whole molecule interaction.

The coupling of such an intermolecular reaction with ion binding provides a possible mechanism for the type of slow uptake or release of ions reported by Pilla and Margulies (*41, 42*). These authors reported the disappearance of their slowest relaxation process associated with ion transport in live toad urinary bladder membrane when 1% glutaraldehyde was added to the system. Experience with the reversibly reacting *Escherichia coli* ribosome system when treated with glutaraldehyde (*43, 44*) shows that the cross-linking reactions "freeze" the ribosome–subunit interaction. If the same process occurs in the reacting lobster hemocyanin system when treated with glutaraldehyde, the half molecule–whole molecule reaction will be "frozen," and any previously observable relaxation processes owing to calcium ion or proton binding coupled to this reaction will disappear. This suggests that the disappearance of a relatively slow ionic transport relaxation on treatment of complex structures with cross-linking reagents such as glutaraldehyde may be traced to the "freezing" of an ionically coupled macromolecular interaction at the surface of, or within, such a structure.

These comments attest to the wisdom of the symposium chairman, Dr. Martin Blank, in providing a milieu in which scientists from diverse disciplines may interact.

Acknowledgments

This work was supported in part by NIH Research Grants HL-13166 and HL-13909. G. K. is indebted to D. Miller for historical information revealed in 1979 on the development of the Govy procedure in Great Britain (*7*).

Literature Cited

1. Gouy, G. L. *C. R. Hebd. Seances Acad. Sci.* **1880,** *90,* 307.
2. Stefan, J. *Sitzungsber. Akad. Wiss. Wien, Abt. II* **1879,** *79,* 161.
3. Wiener, O. *Ann. Phys. (Leipzig)* **1893,** *49,* 105.
4. Longworth, L. G. *Ann. N.Y. Acad. Sci.* **1945,** *46,* 211.

5. Longworth, L. G. *J. Am. Chem. Soc.* **1947,** *69,* 2510.
6. Kegeles, G.; Gosting, L. J. *J. Am. Chem. Soc.* **1947,** *69,* 2516.
7. Coulson, C. A.; Cox, J. T.; Ogston, A. G.; Philpot, J. St. L. *Proc. R. Soc. London, Ser. A* **1948,** *192,* 382.
8. Gosting, L. J.; Hanson, E. M.; Kegeles, G.; Morris, M. S. *Rev. Sci. Instrum.* **1949,** *20,* 209.
9. Gosting, L. J.; Morris, M. S. *J. Am. Chem. Soc.* **1949,** *71,* 1998.
10. Gosting, L. J.; Onsager, L. *J. Am. Chem. Soc.* **1952,** *74,* 6066.
11. Gosting, L. J. "Advances in Protein Chemistry"; Academic: New York, 1956; Vol. XI, p. 429.
12. Morimoto, K.; Kegeles, G. *Arch. Biochem. Biophys.* **1971,** *142,* 247.
13. Cann, J. R. "Interacting Macromolecules"; Academic: New York, 1970.
14. Cann, J. R. *Methods Enzymol.* **1978,** *48,* Chap. 11, 242–248.
15. Kegeles, G.; Tai, M. *Biophys. Chem.* **1973,** *1,* 46.
16. Tai, M.; Kegeles, G. *Biophys. Chem.* **1975,** *3,* 307.
17. Cann, J. R.; Kegeles, G. *Biochemistry* **1974,** *13,* 1868.
18. Donoian, H. C. M.A. Thesis, Clark University, Worcester, MA, 1954.
19. Tiselius, A. *Trans. Faraday Soc.* **1937,** *33,* 524.
20. Gutter, F. J.; Kegeles, G. *J. Am. Chem. Soc.* **1953,** *75,* 3893.
21. Biancheria, A.; Kegeles, G. *J. Am. Chem. Soc.* **1957,** *79,* 5908.
22. Donoian, H. C.; Kegeles, G. *J. Am. Chem. Soc.* **1961,** *83,* 255.
23. Christoffers, H. J.; Kegeles, G. *J. Am. Chem. Soc.* **1963,** *85,* 2562.
24. Kegeles, G., 1949, unpublished data.
25. Kahn, D. S.; Polson, A. G. *J. Phys. Colloid Chem.* **1947,** *51,* 516.
26. Provencher, S. W., 1960, private communication.
27. Akeley, D. F.; Gosting, L. J. *J. Am. Chem. Soc.* **1953,** *75,* 5685.
28. James, H. I., Ph.D. Thesis, Clark University, Worcester, MA, 1965.
29. Bloomfield, V. A.; Lim, T. K. *Methods Enzymol.* **1978,** *48,* Chap. 19, 415–494.
30. Kirkwood, J. G.; Baldwin, R. L.; Dunlop, P. J.; Gosting, L. J.; Kegeles, G. *J. Chem. Phys.* **1960,** *33,* 1505.
31. Onsager, L. *Phys. Rev.* **1931,** *37,* 405.
32. Vitagliano, V.; Lyons, P. A. *J. Am. Chem. Soc.* **1956,** *78,* 4538.
33. Müller, G. T. A.; Stokes, R. H. *Trans. Faraday Soc.* **1957,** *53,* 642.
34. Kim, H., Ph.D. Thesis, Clark University, Worcester, MA, 1963.
35. Kim, H. *J. Solution Chem.* **1974,** *3,* 271.
36. Carnahen, B.; Luther, H. A.; Wilkes, H. O. "Applied Numerical Methods"; Wiley: New York, 1969; p. 451.
37. Ibid, pp. 440–441.
38. Ibid, pp. 441–442.
39. Yphantis, D. A.; Correia, J. J.; Johnson, M. L.; Wu, G. M. In "Physical Aspects of Protein Interactions"; Catsimpolas, Ed.; Elsevier North Holland: 1978; 275–303.
40. Tai, M.; Kegeles, G. *Arch. Biochem. Biophys.* **1971,** *142,* 258.
41. Pilla, A. A.; Margulies, G. S. *J. Electrochem. Soc.* **1977,** *124,* 1697.
42. Margulies, G. S.; Doty, S. B.; Pilla, A. A. "Abstracts of Papers," Symposium on Bioelectrochemistry, 1978 Annual Meeting, American Chemical Society, September 1978.
43. Spirin, A. S.; Belitsina, N. V.; Lerman, M. I. *J. Mol. Biol.* **1965,** *14,* 611.
44. Subramanian, A. R. *Biochemistry* **1972,** *11,* 2710.

RECEIVED October 17, 1978.

11

The Effect of Bulk Solvent Structure and Specific Ion Binding on the Temperature Dependence of the Reduction Potential of Horse Heart Cytochrome *c*

GEORGE P. KREISHMAN, CHIH–HO SU, C. WILLIAM ANDERSON, H. BRIAN HALSALL, and WILLIAM R. HEINEMAN

Department of Chemistry, University of Cincinnati, Cincinnati, OH 45221

The temperature dependence of the reduction potential ($E^{\circ\prime}$) *of Cytochrome* c *in NaCl–*H_2O *solutions is biphasic with a transition temperature of 42°C resulting from the previously determined second- or higher-order phase change in the bulk water structure. The phase-transition temperature can be depressed by adding* D_2O *much like a freezing-point depression in binary systems. When comparing* $E^{\circ\prime}$ *and the standard entropy change* (ΔS°) *for the reduction of Cytochrome* c *with those in other halide solutions, the oxidized form of Cytochrome* c *binds the unhydrated chloride ion preferentially. Inclusion of the specific ion binding on the treatment of the data results in a value for* $E^{\circ\prime}$ *of 235 mV at 37°C for Cytochrome* c *in chloride solutions.*

The properties of Cytochrome *c*, a vital component of the electron transport chain, have been studied widely (*1*). The reduction potential ($E^{\circ\prime}$) of Cytochrome *c* has been measured by a variety of techniques and under different conditions, namely in the presence of various salts, as a function of temperature, bound to phospholipid vesicles and in the presence of other components of the electron transport chain. These studies have yielded a wide range of values for $E^{\circ\prime}$ vs. the standard

0-8412-0473-X/80/33-188-169$05.00/1

hydrogen electrode (SHE) from 220 to 275 mV (*2, 3, 4, 5, 6*). It has been speculated that this variation is due partly to the binding of various ions to either the oxidized or reduced forms of Cytochrome *c* (*2*).

In a previous work (*7*), we reported that the temperature dependence of E°′ for Cytochrome *c* was itself dependent on the solvent used (H_2O or D_2O) and the type of alkali halide present (NaCl or NaI). In NaCl–H_2O solutions, the temperature dependence of E°′ exhibited a biphasic character with a change occurring at 42°C. This change was interpreted as a shift in the equilibrium to the larger oxidized form because of an apparent second- or higher-order phase change in the bulk water structure (i.e. water, which is not in the cosphere and strongly interacting with the ions) at 42°C (*8, 9, 10*). The discontinuity in the temperature dependence of the entropy of the bulk water, which is characteristic of higher-order phase changes, was deduced from the temperature dependence of the equilibrium constant for the formation of the trihydrated chloride ion (*8, 9*) and the chemical shift of bulk water in tetramethylammonium chloride solution (*10*). In addition to the dependence of E°′ on the solvent structure, E°′ at a given temperature was different in the presence of NaCl or NaI in H_2O (*7*).

To better understand the reasons for these dependencies, we have measured the temperature dependence of E°′ for Cytochrome *c* as a function of solvent structure (H_2O vs. D_2O and mixtures of the two) and in the presence of a series of alkali halides (NaBr, NaCl, NaF, NaI, and KCl).

Experimental

Materials. H_2O was doubly distilled and deionized. Chemicals and their sources were as follows: Na_2HPO_4, AR grade, Mallinckrodt; $NaH_2PO_4 \cdot H_2O$, certified A.C.S., Fisher Scientific; 2,6-dichlorophenolindophenol (DCIP), 99⁺% pure, Fluka Columbia Organic Chemicals, Columbia, SC; Gold Label grade, Diaprep, Aldrich Chemical Co.; NaBr, NaF, NaI, NaCl, KCl, Suprapur, E. M. Laboratories or Certified A.C.S., Fisher Scientific. Horse heart Cytochrome *c* Type VI, 95–100% pure was purchased from Sigma Chemical Co. and was eluted from a Pharmacia CMC 32 column with 50m*M* phosphate buffer at pH 7.5. The protein was desalted using Sephadex G-25 and lyophilized.

Methods. All solutions were prepared to be 1m*M* Cytochrome *c*, 0.1m*M* DCIP, 0.10*M* alkali halide, and 0.10*M* phosphate buffer at pH 7.0 or pD 7.0. The DCIP served as a mediator–titrant for coupling the Cytochrome *c* with the electrode potential. E°′ values were measured using a previously described spectropotentiostatic technique using an optically transparent thin-layer electrode (OTTLE) (*7, 11, 12*). This method involved incrementally converting the cytochrome from its fully oxidized to fully reduced state by a series of applied potentials. For each potential a spectrum was recorded after equilibrium was attained. The formal redox potential was obtained from a Nernst plot. The *n* value calculated from the slope of all sets of data used was better than 1.00 ± 0.05. The temperature of the solution then was changed and allowed to equilibrate for 30 min at the new temperature before repeating the sequence.

All potentials were measured vs. an SCE reference, the temperature of which was maintained equivalent to that of the OTTLE. The potential of the SCE was measured vs. a commercial SCE after each experiment to verify that no irreversible change in the SCE had been caused by the elevated temperature. The potentials reported here have been corrected by 242 mV for referencing against the SHE at 25°C. The potential change of the SCE over the temperature range of 20°–60°C is less than 4 mV when referenced against an SCE at 25°C (*13*). In the NaF solution, some variation in the potential of the SCE was experienced, and the reported results in NaF solutions should not be relied upon to better than ± 5 mV.

Results and Discussion

Temperature Dependence of E°′ in H_2O and D_2O Solutions. The temperature dependence of E°′ for horse heart Cytochrome *c* in H_2O solutions containing a series of alkali halides at 0.10*M* and 0.10*M* sodium phosphate, pH 7.00, is shown in Figures 1 and 2. For all samples not containing chloride ion, the temperature dependence is linear between

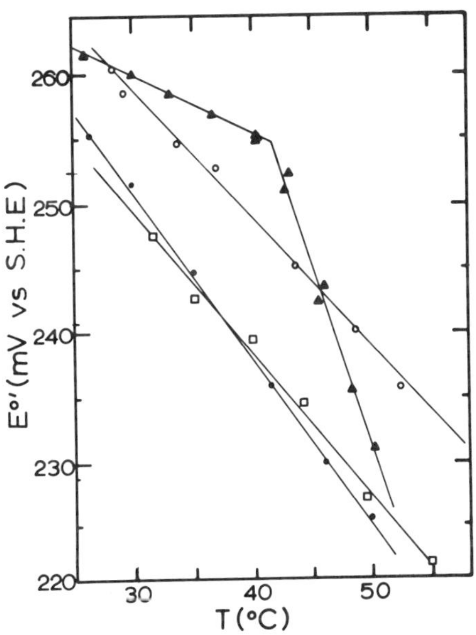

Bioelectrochemistry and Bioenergetics

Figure 1. Temperature dependence of E°′ *for horse heart Cytochrome* c *in 0.10*M *sodium halide solutions in H_2O (12) (○), F^-; (▲), Cl^-; (●), Br^-; and (□), I^-.*

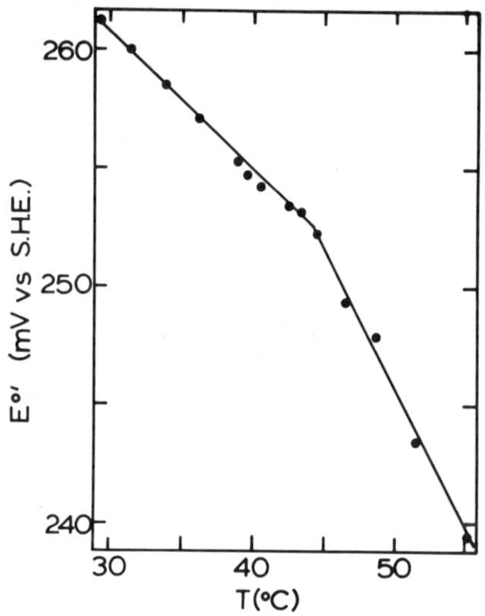

Figure 2. Temperature dependence of E°' *for horse heart Cytochrome* c *in 0.10*M *KCl solutions in H_2O*

25° and 50°C. However, in the presence of either NaCl or KCl, biphasic behavior is observed with a transition occurring at 42°C. In D_2O solutions containing either 0.10*M* NaCl or NaBr in sodium phosphate, pD 7.00, the temperature dependence is linear over the entire range studied (*see* Figure 3).

To eliminate the possibility that the biphasic behavior in chloride-containing solutions may be caused by changes in electrode and mediator properties, the temperature dependence of E°' of the mediator 2,6-

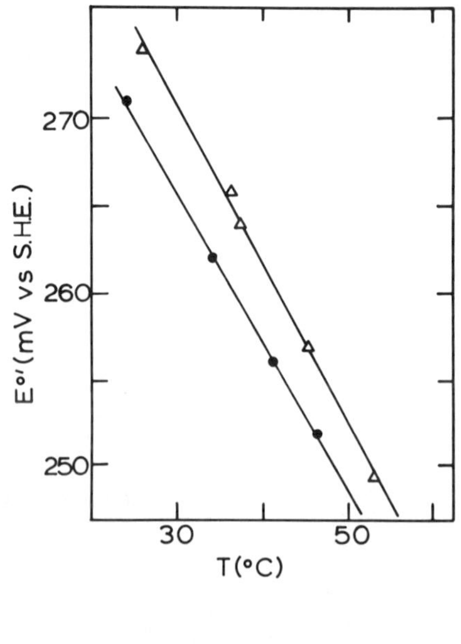

Bioelectrochemistry and Bioenergetics

Figure 3. Temperature dependence of E°' *for horse heart Cytochrome* c *in 0.10*M *NaCl and NaBr solution in D_2O (12): (△), Br^-/D_2O; (●), Cl^-/D_2O.*

dichlorophenolindophenol vs. SCE was determined to be linear from 25° to 55°C.

That the structural integrity of the heme environment of Cytochrome *c* is maintained in the NaCl–H_2O system over the temperature range of 30° to 55°C was confirmed by monitoring the 695-nm band of Cytochrome *c*. The total integrated intensity remained constant over this temperature range indicating that no change in the state of ligation of the iron had occurred (*14*). This is in agreement with the NMR results of McDonald and Phillips who have shown that no variation in the orientation of amino acid residues near the heme group could be observed for either Ferricytochrome *c* or Ferrocytochrome *c* below 60°C in both H_2O and D_2O (*15*). Therefore it must be concluded that the differences in the temperature dependence of E°′ in the various solutions cannot be attributed to changes in the heme environment but must be the result of other causes.

The Variation of the Size Change upon Reduction. The variation of E°′ for horse heart Cytochrome *c* with temperature and solution conditions can be interpreted in terms of a shift in the equilibrium between the oxidized and reduced forms as a result of changes in bulk solvent structure and of different binding affinities of the halide ions. These variables should have an effect on E°′ since it has been suggested that Cytochrome *c* undergoes a size change upon reduction and selectively binds anions to the oxidized form and cations to the reduced form (*16, 17*).

For such a size change, where the oxidized form is slightly larger than the reduced form, the reduction process will be accompanied by water compensation (*12, 18*). This process, as described by Lumry, occurs in two steps: (a) contraction of the protein leaving a hole and (b) water filling that hole. This process allows more hydrogen bonds between water molecules to form, and is accompanied by the expected decrease in both enthalpy and entropy. The magnitude of the enthalpy and entropy changes are directly proportional to the decrease in size of the protein. Using Lumry's value of $\Delta S°/\Delta V = 1.2$ eu/mL and the thermodynamic parameters for the reduction process (*see* Table I), the decrease in volume in going from the oxidized to the reduced form is about 25 to 75 Å^3 for the various solutions (*18*). The size change upon reduction varies depending on the anion present and the nature of the solvent. The largest change occurs for Cytochrome *c* in the presence of NaCl above 42°C; intermediate values are observed for Cytochrome *c* in the presence of NaBr, NaF, NaI, and KCl above 42°C and the smallest values are for NaCl and KCl below 42°C.

These differences in size changes could be the result of two factors: (a) ion binding to the Cytochrome *c* and (b) differences in solvent

Table I. Summary of Thermodynamic Parameters for the Various Solutions of Cytochrome *c* in 0.10*M* Alkali Halide, 0.10*M* Sodium Phosphate, pH or pD 7.00 (*12*)

Alkali Halide	*Solvent*	$\Delta H°$ *kcal/mol*	$\Delta S°$ *cal/mol °C*
NaBr	H_2O	−14.0	−27.1
NaF	H_2O	−12.0	−20.3
NaI	H_2O	−12.7	−23.0
NaCl (< 42°C)	H_2O	−9.1	−10.2
NaCl (> 42°C)	H_2O	−29.4	−75.0
KCl (< 42°C)	H_2O	−9.5	−12.1
KCl (> 42°C)	H_2O	−12.6	−23.7
NaCl	D_2O	−12.0	−19.5
NaBr	D_2O	−13.4	−19.2

Bioelectrochemistry and Bioenergetics

structure. Ion binding is involved in the reduction process given by the following equation

$$\text{Cyt}\ c^{\text{III}} - X^- + M^+ + e^- \rightleftarrows \text{Cyt}\ c^{\text{II}} - M^+ + X^- \quad (1)$$

where the oxidized form binds a halide anion and the reduced form binds an alkali cation (*16*). Binding of a negatively charged ion to the net positively charged oxidized form is expected to reduce charge repulsion in the Cytochrome *c* resulting in a more compact protein. Binding of a cation to the net positively charged Cytochrome c^{II} would increase the charge repulsion in the protein resulting in a slightly larger protein. Thus, in the presence of ion binding the net size change upon reduction would be reduced. These results imply that of the halides, chloride ion binds stronger to the oxidized form than Br^- and I^- ions. For the reduction process in H_2O, the $\Delta S°$ values, which are directly proportional to the size change, are in the expected order with $\Delta S°$ (NaCl) < $\Delta S°$ (NaF) $\simeq \Delta S°$ (NaBr), $\simeq \Delta S°$ (NaI). The larger halide ions (Br^- and I^-) are expected to exhibit weaker binding because of their increased size. Fluoride is not expected to bind strongly because of very strong solvation by H_2O causing the effective ionic size in water to be large (*19*). Very little difference in $\Delta S°$ between cations is expected since Na^+ and K^+ appear to bind to the reduced form with approximately equal affinity (*11*). In H_2O, this is indeed the case below 42°C with $\Delta S°$ (NaCl) $\cong \Delta S°$ (KCl).

Below 42°C, the size change between the oxidized and reduced forms correlates well with the known binding constants of the ions. Above 42°C, however, the $\Delta S°$ for NaCl in H_2O is much larger than for KCl. This difference is related to differences in the bulk water structure. After the higher-order phase change occurs, the larger oxidized form is more favored with decreasing water structure and the $E^{\circ\prime}$ decreases. The large slope of $dE^{\circ\prime}/dT$ for Cytochrome *c* in NaCl–H_2O solution is attributed to this bulk water structural change above 42°C. A similar effect is expected for KCl solutions except that owing to the decreased surface charge density for K^+ as compared with Na^+, one expects that ion–solvent interactions are less marked in KCl and NaCl solutions and, therefore, water as a solvent should behave more uniformly in KCl than in NaCl solutions (*20*). This attenuation of the destructuring is indeed observed.

The effects of solvent structure on the value of $E^{\circ\prime}$ are reflected again by the greater values of $E^{\circ\prime}$ in the more structured D_2O than for all of the solutions in H_2O. ΔS° (NaCl) in D_2O is larger than in H_2O and is approximately equal to ΔS° (NaBr) in D_2O indicating that Cl^- does not bind to the oxidized form as well in D_2O as in H_2O and binds with about the same affinity as Br^-. This may be caused by a more compact form of the oxidized form in the more structured D_2O resulting in a smaller binding constant for the chloride ion.

Possible Specific Anion Binding to the Oxidized Species. When comparing the $E^{\circ\prime}$ values at 37°C for Cytochrome *c* in H_2O and D_2O in the presence of the various halides (*see* Table II), $E^{\circ\prime}$ for Cytochrome *c* in the NaCl–H_2O solution is greater than in NaBr, NaF, or NaI–H_2O solutions, whereas in D_2O solutions, the $E^{\circ\prime}$ in the NaBr solution is greater than in the NaCl solution. From the observed size change and the electrophoretic mobilities, it is implied that the binding constant to the oxidized form for chloride ion is greater than for either the bromide or iodide ion (*5, 17, 21*). From examination of Equation 1, one assumes that as the binding of the anion increases, the oxidized form is more favored, resulting in a lower $E^{\circ\prime}$. This is, however, not the case for the apparent $E^{\circ\prime}$ values of Cytochrome *c* in NaCl–H_2O and NaBr–H_2O solutions. These results are based on the standard manipulation of the Nernst equation

$$E = E^{\circ\prime} - \frac{RT}{nF} \ln Q$$

where Q is the reaction quotient for the assumed reaction

$$\text{Cytochrome } c^{III} + e^- \rightleftarrows \text{Cytochrome } c^{II} \qquad (2)$$

When the concentration of the reduced and oxidized forms are equal, the measured potential is equal to the standard reduction potential. If, however, one includes ion binding to the oxidized and reduced forms (*see* Equation 1), the appropriate Nernst equation is

Table II. $E^{\circ\prime}$ Values at 37°C for Various Solutions of Horse Heart Cytochrome *c*

Alkali Halide	*Solvent*	$E^{\circ\prime}$
KCl	H_2O	257
NaCl	H_2O	255
NaF	H_2O	248
NaBr	H_2O	239
NaI	H_2O	239
NaBr	D_2O	261
NaCl	D_2O	258

$$E_{\text{app}} = E^{\circ\prime} - \frac{RT}{nF} \ln \frac{[\text{Cytochrome } c^{\text{II}} - M^+]\,[X^-]}{[\text{Cytochrome } c^{\text{III}} - X^-]\,[M^+]}$$

We have not included binding of phosphate ions since it appears that they bind to both forms and that under the ionic strength conditions used, both forms would be saturated with phosphate ions (*5, 17*) and these terms in the Nernst equation would cancel. The relative concentrations of X^- and M^+ must be considered for a correct determination of $E^{\circ\prime}$ for this system.

If the binding of the X^- ion is not specific, determination of the correction term would involve utilizing the activities of the ions. The binding of the cation to the reduced form appears to be nonspecific since the electrophoretic mobilities of Cytochrome *c* with added Na^+, K^+, Ca^{+2}, and Mg^{+2} are very similar (*17*). However, the binding constants for the halides vary, with the chloride ion apparently having the highest binding constant (*5*).

Recently studies have shown that in chloride solutions there are predominantly two species—the unhydrated ion (Cl^-) and the trihydrated ion ($Cl(H_2O)_3^-$) (*8, 9*). Theoretical calculations of the interaction of Cl^- with H_2O show that at least one water forms a partial covalent bond with the ion (*22, 23*). The equilibrium constant for the reaction

$$Cl^- + 3H_2O \rightleftharpoons Cl(H_2O)_3^- \tag{3}$$

is 3.83 at 37°C. Thus, the major species of chloride ion present is the trihydrate (*8, 9*). Since the binding of anions to Cytochrome *c* decreases markedly with the increasing size of the anion, it is more probable that the chloride ion binding to the oxidized form is the unhydrated species since the trihydrate has at least one H_2O covalently bonded to it (*22, 23*) and would be larger than the Br^- and I^- ions. Consequently, the concentration of the chloride ion that should be used in the Nernst equation is not that of total chloride, but rather of the unhydrated species. This is related to the total chloride ion concentration $[Cl^-]_{TOT}$ from the equilibrium constant for the hydration reaction

$$K = \frac{[Cl(H_2O)_3^-]}{[Cl^-]}$$

where the activity of H_2O is taken as unity

$$[Cl^-] = \frac{[Cl^-]_{TOT}}{1 + K}$$

and combination with the Nernst equation gives

$$\begin{aligned} E = E^{\circ\prime} - \frac{RT}{nF} \ln \frac{[\text{Cytochrome } c^{II} - M^+]}{[\text{Cytochrome } c^{III} - Cl^-]} \\ \frac{-RT}{nF} \ln \frac{[Cl^-]_{TOT}}{[Na^+]} + RT \ln [1 + K] \end{aligned} \tag{4}$$

Under these conditions, the redox potential is independent of salt concentration in the $0.1M$ range (2).

Thus when [Cytochrome $c^{II} - M^+$] equals [Cytochrome $c^{III} - Cl^-$], one must subtract $RT/nF \ln (1 + K)$ and add the correction for ion concentrations from the midpoint potential to obtain E°′. The value of $RT/nF \ln (1 + K)$ as a function of temperature is shown in Figure 4. Therefore, the corrected redox potential for Cytochrome *c* in the solutions used at 37°C is ~ 235 mV. This value is lower than the other halides

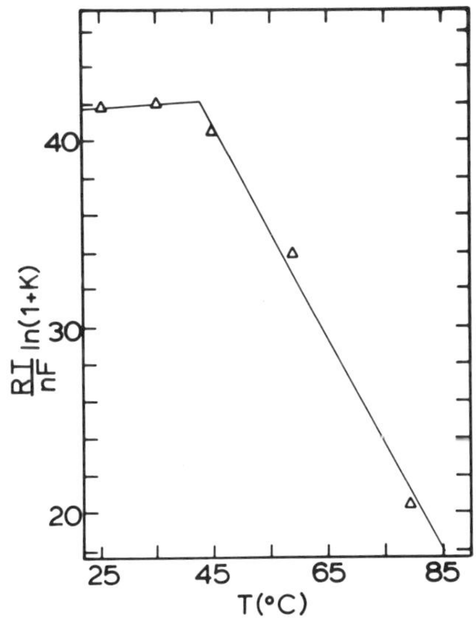

Figure 4. Temperature dependence of the correction term RT/nF *ln (1* + K)

over the entire temperature range studied as is expected for the strongest binding ion. If the cation and anion concentrations are approximately equal, E°′ becomes independent of salt concentrations. The corrected values for these conditions are shown in Figure 5. Also, the fact that the $E^{\circ\prime}$ for Cytochrome *c* is lower in binding medium than in nonbinding medium (*22*) is consistent with the reduction in charge and size of Cytochrome c^{III} – Cl^- as compared with Cytochrome c^{III}, with a resulting decrease of charge repulsion within the protein.

In D_2O solutions, a similar correction term might be applied except that the equilibrium constant for the hydration reaction in D_2O has not been determined. An estimate, however, for *K* can be made from statistical mechanical arguments. *K* is expected to be much lower in D_2O than H_2O. At a given temperature both solutions would have the same total energy, but owing to the greater mass of D_2O, more of the total thermal energy is required to maintain its translational states and hence less energy is available for formation of $Cl(D_2O)_3^-$ (*24*). Thus, the correction term would be smaller and the observed value for E°′ in D_2O is closer to the real value, and for all temperatures, E°′ in NaCl–D_2O solutions is lower than E°′ in NaBr–D_2O solutions (*see* Figure 2).

It must be remembered that this is an approximate value since the coefficients for ion binding have not been established firmly, and this would affect the magnitude of the correction term. Also, no attempt has been made to include the temperature dependence of $\gamma \pm$ for NaCl, which also shows a biphasic behavior (*25*), since in the Nernst equation individual ionic activity coefficients would be required and these have

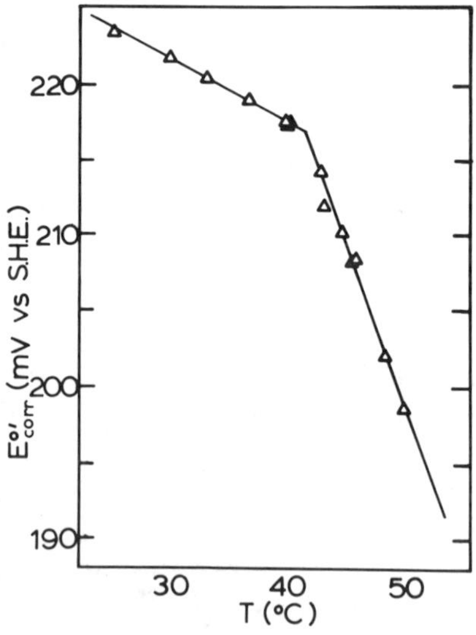

Figure 5. Temperature dependence of the corrected values of E°′ with the inclusion of specific ion binding to horse heart Cytochrome c in 0.10M NaCl in H_2O

yet to be determined. However this correction term is expected to be small. It is interesting that the corrected value for the reduction potential of Cytochrome *c* reported here is now closer to that for Cytochrome *a*, the component that accepts electrons from Cytochrome *c* in the electron transport chain. This treatment of the total reaction for the reduction of Cytochrome *c* removes an ambiguity in the previously reported values of reduction potentials of components of the electron transport chain and may give a more accurate value for the redox potential of Cytochrome *c* under physiological conditions. Specific binding of the unhydrated chloride ion to the oxidized form also could become important as the role of Cytochrome *c* in ion transport across the mitochondrial membrane becomes better understood (*16, 26*). The idea of a protein being able to selectively bind monovalent ions is not new. An example is the well-characterized Na^+–K^+ ATPase system which differentially binds either sodium or potassium ions.

The Temperature Dependence of $E^{\circ\prime}$ in Solutions of Varying H_2O and D_2O Content. The temperature dependence of $E^{\circ\prime}$ for NaCl solutions with the deuterium content varying from 5% to 80% are shown in Figures 6, 7, and 8. In all cases, the temperature dependence is biphasic. The transition temperature, however, is dependent on the amount of deuterium present as shown in Figure 9. With increasing deuterium content, the transition temperature is depressed reaching a minimum of about 10% D and then gradually increasing with increasing deuterium content. This type of behavior is similar to that observed for freezing-point depression phase diagrams for binary systems. Since the

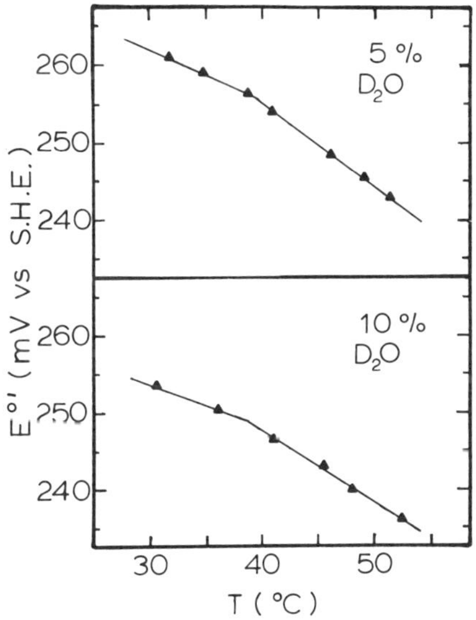

Figure 6. Temperature dependence of E°′ *for horse heart Cytochrome* c *in 0.10*M *NaCl solutions containing 5% and 10%* D_2O

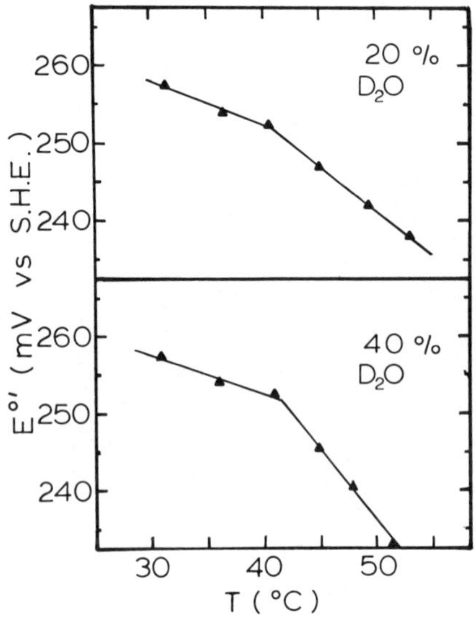

Figure 7. Temperature dependence of E°' *for horse heart Cytochrome* c *in 0.10*M *NaCl solutions containing 20% and 40%* D_2O

only variable in these solutions is the concentration of various solvent species (i.e. H_2O, HOD, and D_2O), the observed behavior clearly demonstrates that the biphasic behavior for the temperature dependence of $E^{\circ\prime}$ is the result of the previously demonstrated higher-order phase change in the bulk solvent structure (*8, 9, 10*).

Upon casual inspection of this phase diagram, it may seem surprising that the maximum depression in the transition temperature occurs at

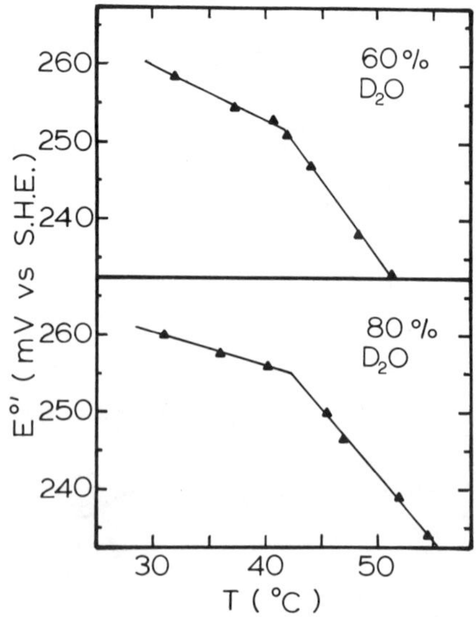

Figure 8. Temperature dependence of E°' *for horse heart Cytochrome* c *in 0.10*M *NaCl solutions containing 60% and 80%* D_2O

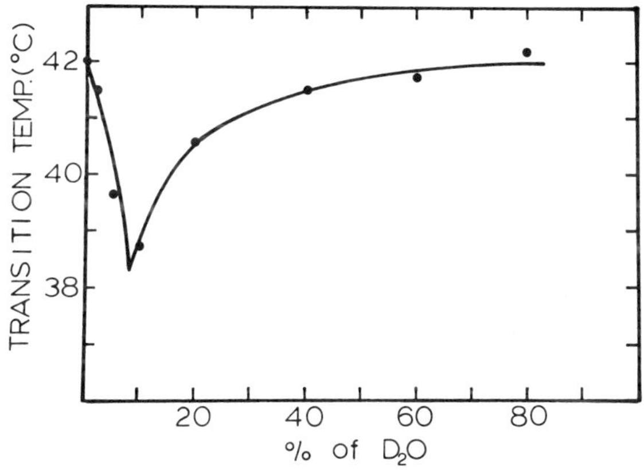

Figure 9. Variation of the transition temperature with D_2O content

~ 10% D and not near 50% D as might be expected. This phase change arises from the perturbation of the solvent structure by the chloride ion. Although the exact nature of this perturbation is not known, it can be speculated that the trihydrated ion is not a random orientation of solvent molecules, as might be expected, but rather is a very specific orientation of solvent molecules. This specific orientation in turn forms a nucleation center for ordering the surrounding solvent molecules into an ordered structure. With increasing temperature, this ordered structure provides an entropic heat sink required in high-order phase transitions. Upon exhaustion of this heat sink at 42°C in pure H_2O, the temperature dependence of the water structure is different. For the higher-order phase change to occur, the nucleating species, $Cl(H_2O)_3^-$, must be present. For deuterated species, the propagation of a new structure in the bulk solvent does not occur, presumably because of the greater stability of the deuterated solvent, and the higher phase change is not observed in D_2O. The maximum effect of the deuterated species on the transition temperature is expected when the active nucleating species is prevalent which is for solutions with a deuterium content of less than 28% (calculated by assuming random distribution of deuterons and protons between the three possible species, H_2O, HOD, and D_2O). Above this concentration, the deuterated species, either HOD or D_2O, predominate. Therefore, the maximum depression might be expected below this deuterium content, as is observed.

The Variation of $E^{\circ\prime}$ and ΔS° of the Reduction Process with Deuterium Percentages. The variation of the values of $E^{\circ\prime}$ and ΔS° with the percent NaCl solutions at 30° and 50°C is shown in Figures 10 and 11. The variation is not linear with the deuterium percentage which is the case for $E^{\circ\prime}$ of Cytochrome *c* in NaBr solutions of varying deuterium

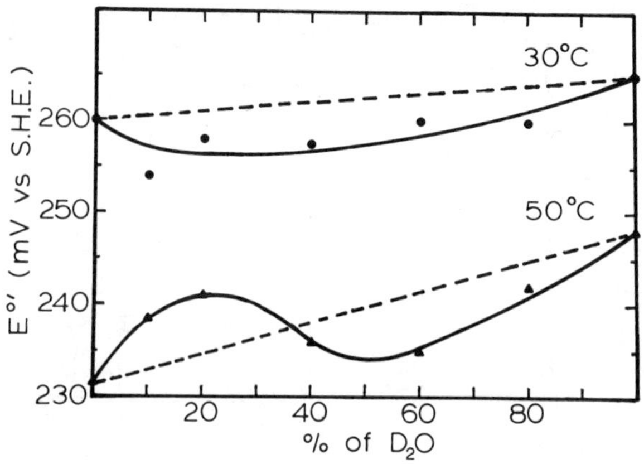

Figure 10. *Variation of* E°' *of horse heart Cytochrome* c *with* D_2O *content at 30° and 50°C*

percentages, indicating that in solutions where the salt does not have a large perturbing effect on the solvent structure, H_2O and D_2O mix in a near ideal manner. This is consistent with near linear behavior of the melting point in solutions of H_2O/D_2O mixtures (*27*). The deviation from linearity, therefore, must be related to the changes with varying deuterium percentages on either the state of hydration of the chloride ion or the nature of the structural perturbation of the chloride ion on the solvent structure.

The equilibrium constant for the hydration reaction (*see* Equation 3) is expected to change and therefore also the correction term in the Nernst equation (*see* Equation 4). Upon mixing H_2O and D_2O, the

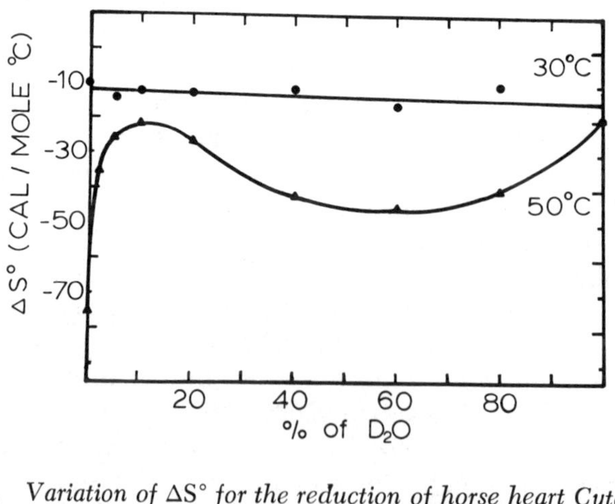

Figure 11. *Variation of* ΔS° *for the reduction of horse heart Cytochrome* c *with* D_2O *content at 30° and 50°C*

entropy of the solvent is greater than the individual components because of the standard entropy of mixing (*28*). This additional entropy is maximum at 50% D. Because of the more entropic solvent, the hydration reaction is expected to be driven to the left resulting in an increase in the concentration of the unhydrated species. Since this is postulated to be the binding species, the oxidized form is more favored with a resultant maximum decrease in $E^{\circ\prime}$. This is apparently the case at 30°C and partly at 50°C (*see* Figure 10).

In addition to change in the equilibrium constant for the hydration reaction and its depression of $E^{\circ\prime}$, for temperatures above the transition temperature, another phenomenon occurs at low D_2O concentrations which results in an increase in $E^{\circ\prime}$ and a decrease in the entropy change for the reduction process. Apparently, the presence of deuterated species aids in stabilizing the bulk water structure above the transition temperature when H_2O is the predominant species (% D < 28%). In the more stable solvent, the smaller reduced form is favored and $E^{\circ\prime}$ is increased.

Although a higher-order phase change occurs in the bulk water structure in chloride solutions (*8, 9, 10*), it must be emphasized that the exact nature of the interaction of the chloride ion with the solvent molecules and its resulting effect on the solvent structure has yet to be determined. For this reason and because the presence of the deuterated species complicate the matter further, the explanation for the obsrved behavior of $E^{\circ\prime}$ and ΔS° with varying % D is still speculative.

The Higher-Order Phase Change and Other Biological Systems. It is interesting to note that the biphasic behavior we observe for $E^{\circ\prime}$ of Cytochrome *c* is similar to the biphasic behavior of other systems. These include the temperature dependence of the mean ionic activity coefficients of NaCl (*25*), the conductance of the chloride ion (*29*), and the $E^{\circ\prime}$ values of the SCE and Ag/AgCl electrodes (*30*) vs. the hydrogen electrode. This biphasic behavior also is observed in both model biological systems like the aggregation properties of purine (*8*), the hydration energy of purine (*31*), and the aggregation properties of membrane vesicles (*32*) and in real systems as indicated by the doubling times of *A. hydrophila* (*33*), a marked increase in the amount of saturated fatty acids in *E. coli* membranes (*34*), and the penicillinase activity in *B. cereus* (*35*). These results are all consistent with the concept that in chloride solutions a higher phase change occurs at 42°C. One would like to speculate as to exactly what this change in structure is and what it is caused by. Even though electrolytic solutions have been studied widely (*36*), a completely rigorous description of their structure has yet to be presented and therefore this type of discussion must await further theoretical as well as experimental development. However, since many advanced forms of life cannot tolerate temperatures much above 42°C, this phenomenon may have important ramifications on the properties of many biological systems in addition to those already studied and may be the underlying cause of hyperthermic death in man.

Glossary of Symbols

$E^{\circ\prime}$ = reduction potential
SHE = standard hydrogen electrode
DCIP = 2,6-dichlorophenolindophenol
n = electron stoichiometry of redox process
SCE = saturated calomel electrode
OTTLE = optically transparent thin layer electrode
mV = millivolts
ΔS° = standard entropy change
eu = entropy units
ΔV = volume change, mL
T = temperature, °C or K
Q = reaction quotient
R = gas constant
F = Faraday constant
E_{app} = potential applied to electrochemical cell
K = equilibrium constant
ΔH° = standard enthalpy change
$\gamma\pm$ = mean activity coefficient

Acknowledgments

The authors wish to thank F. Marchese, L. Leifer, E. Deutsch, and F. Meeks for helpful discussions about the contents of this chapter. This investigation was supported in part by National Science Foundation Grant CHE77-04399 (W.R.H.) and a grant from the University of Cincinnati Research Council (G.P.K.). C. Anderson acknowledges support provided by a Procter and Gamble Fellowship and a University of Cincinnati Summer Research Fellowship. C.–H. Su acknowledges support provided by a University of Cincinnati Summer Research Fellowship. G. Kreishman acknowledges support provided by a University of Cincinnati Summer Faculty Research Fellowship.

Literature Cited

1. Lenberg, R.; Barrett, J. "Cytochromes"; Academic: London, 1973.
2. Margalit, R.; Schejter, A. *Eur. J. Biochem.* **1973,** *32,* 492–499.
3. Vanderkooi, J. M.; Erecinska, M.; Chance, B. *Arch. Biochem. Biophys.* **1973,** *157,* 531–540.
4. Erecinska, M.; Vandekooi, J. M. *Arch. Biochem. Biophys.* **1975,** *166,* 495–500.

5. Nicholls, P. *Biochim. Biophys. Acta* **1974,** *346,* 261–310.
6. Schroedl, N. A., Ph.D. Thesis, Pennsylvania State University, PA, 1976.
7. Anderson, C. W.; Halsall, H. B.; Heineman, W. R.; Kreishman, G. P. *Biochem. Biophys. Res. Commun.* **1977,** *76,* 339–343.
8. Kreishman, G. P.; Foss, D. A.; Inoue, K.; Leifer, L. *Biochemistry* **1976,** *15,* 5431–5435.
9. Inoue, K., Ph.D. Thesis, Michigan Technological University, City, MI 1974.
10. Kreishman, G. P.; Dawson, R. O.; Kuennen, R. W.; Meeks, F. R., submitted for publication in *J. Phys. Chem.*
11. Heineman, W. R.; Norris, B. J.; Goelz, J. F. *Anal. Chem.* **1975,** *47,* 79–84.
12. Kreishman, G. P.; Anderson, C. W.; Su, C.-H.; Halsall, H. B.; Heineman, W. R. *Bioelectrochem. Bioenerg.* **1978,** *5,* 196–203.
13. Covington, A. K. In "Ion Selective Electrodes," Durst, R. A., Ed.; N. B. S. Special Publication 314: Washington, DC, 1969; p. 126.
14. Hines, R. M.; Kreishman, G. P., submitted for publication in *Bioelectrochem. Bioenerg.*
15. McDonald, C. C.; Phillips, W. D. *Biochemistry* **1973,** *12,* 3170–3190.
16. Margalit, R.; Schejter, A. *Eur. J. Biochem.* **1973,** *32,* 500–505.
17. Margoliash, E.; Barlow, G. H.; Byers, V. *Nature* (London) **1970,** *228,* 723–726.
18. Lumry, R. In "Probes of Structure and Function of Macromolecules and Membranes," Chance, B., Yonetani, T., Mildvan, A. S., Eds.; Academic: New York, 1971; Vol. 2, p. 353.
19. Bockris, J. O'M. *Quart. Rev.* **1949,** *3,* 173–180.
20. Leifer, L., private communication.
21. Barlow, G. H.; Margoliash, E. *J. Biol. Chem.* **1966,** *241,* 1473–1477.
22. Kistenmacher, H.; Popkie, H.; Clementi, E. *J. Chem. Phys.* **1973,** *58,* 5627–5638.
23. Clementi, E.; Barsotti, R.; Fromm, J.; Watts, R. O. *Theoret. Chim. Acta (Berlin)* **1976,** *43,* 101–120.
24. Meeks, F., private communication.
25. Harned, H. S.; Owen, B. B. "The Physical Chemistry of Electrolytic Solutions"; Reinhold: New York, 1943; p. 361.
26. Margalit, R.; Schejter, A. *Bioelectrochem. Bioenerg.* **1976,** *3,* 189–202.
27. LaMar, V. K.; Baker, W. N. *J. Am. Chem. Soc.* **1934,** *56,* 2641–2643.
28. Guggenheim, E. A. "Thermodynamics: An Advanced Treatment for Chemists and Physicists"; Interscience: New York, 1949; p. 196.
29. Weast, R. C.; Selby, S. M. "Handbook of Chemistry and Physics"; The Chemical Rubber Co.: Cleveland, 1966; p. D-89.
30. Ives, D. J. G.; Janz, G. J. "Reference Electrodes, Theory and Practice"; Academic: New York, 1961; pp. 161 and 189.
31. Kreishman, G. P.; Gwinner, J. K., submitted for publication in *J. Sol. Chem.*
32. Sheetz, M. P.; Chan, S. I. *Biochemistry* **1972,** *11,* 4573–4581.
33. Kluger, M. J.; Ringler, D. G.; Anver, M. R. *Science* **1975,** *188,* 166–168.
34. Marr, A. G.; Ingraham, J. L. *J. Bacteriol.* **1962,** *84,* 1260–1267.
35. Bernstein, A.; Nickerson, K. W.; Day, R. A. *Arch. Biochem. Biophys.* **1967,** *119,* 50–54.
36. Franks, F. "Water, A Comprehensive Treatise"; Plenum: New York, 1973; Vol. 3.

RECEIVED October 17, 1978.

12

Hemoglobin Oxygenation As a Problem in Surface Electrochemistry

MARTIN BLANK

Department of Physiology, Columbia University, College of Physicians and Surgeons, 630 W. 168 St., New York, NY 10032

It is possible to evaluate the energetics of hemoglobin reactions by considering the hemoglobin molecule as a colloidal droplet and applying the concepts of surface chemistry to the system. The surface free energy of the conformational change of hemoglobin upon oxygenation then can be estimated in terms of the known changes in surface area and charge. The calculated equilibrium constant for the oxygenation reaction varies with pH, as in the acid and alkaline Bohr effects, and also varies with ionic strength. This relatively simple way of estimating the variation of the equilibrium constant may be useful for characterizing conformational changes in globular proteins and also in natural membrane systems.

Before the advent of molecular biology and polymer chemistry, colloid chemistry was the subdivision of science concerned with the study of biopolymers. Although largely overlooked in recent studies of biological problems, the ideas of colloid and surface chemistry appear to provide a suitable level of abstraction for energy calculations. (There is enough detail for physically meaningful calculations, but not so much as to overwhelm.) The application of the ideas of surface electrochemistry in the present calculations on hemoglobin solutions suggests that this relatively simple way of estimating the variation of the equilibrium constant may be useful for characterizing conformational changes in globular proteins and also in natural membrane systems.

0-8412-0473-X/80/33-188-187$05.00/1

Hemoglobin as an Interfacial System

Hemoglobin, a macromolecule containing four polypeptide chains with attached heme groups, can bind four oxygens per molecule. The oxygen binding curve is sigmoid in shape, indicating a change in the affinity with the degree of oxygenation. The curve also shifts with pH (the Bohr effect), indicating a variation of the affinity with pH. These unusual aspects of the reaction of oxygen (or other ligands) with hemoglobin are believed to be caused by the free energy of the conformational change in the hemoglobin molecule upon oxygenation (*1*). The quantitative relation between the known changes in the molecular shape and charge and the variation of the equilibrium constant is not known, in spite of the detailed structural information available. A more classic approach to the study of proteins, i.e., the study of macromolecules as interfacial systems or hydrophilic colloids, appears to be useful for calculating the energy associated with conformational changes (*2, 3, 4*).

The hemoglobin molecule is approximately spherical and has a diameter of about 60 Å, which is in the size range of colloidal systems. If hemoglobin is considered an interfacial system, the conformational changes owing to oxygenation can be estimated in terms of surface free energy. Because of the large surface area of such a system (about 6×10^{11} cm^2/mol), small changes in surface free energy can result in significant changes in the total free energy of the system and in the equilibrium constant.

Energetics of Conformational Changes

The free energy changes in hemoglobin upon oxygenation can be evaluated by smoothing out the structure into a hydrophilic sphere of the same size and surface charge as the macromolecule and assuming that the known changes in the molecular shape and charge upon oxygenation (*5, 6*) can be determined as changes in surface free energy (*see* Figure 1). In practice, we assume a protein monolayer stabilizing a suspended spherical oil droplet, and derive expressions for the changes in the surface free energy (ΔG_s) as a result of changes in the interfacial area (ΔA) and in the surface charge density ($\Delta\sigma$).

$$\Delta G_s = \mathrm{f}(\Delta A, \Delta\sigma) \tag{1}$$

Changes in the surface similar to those that accompany oxygenation (*5, 6*) lead to the variation of $P_{0.5}$, the equilibrium constant (the oxygen pressure for half saturation of hemoglobin).

$$\Delta \ln P_{0.5} = -\frac{\Delta G_s}{2RT} \equiv \frac{\mathrm{k}'}{2} \tag{2}$$

Journal of Theoretical Biology

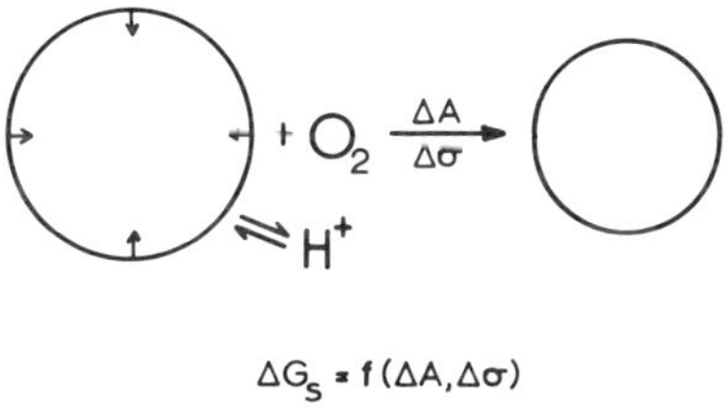

Figure 1. The model used for calculating ΔG_s, the free energy of the conformational change upon oxygenation of hemoglobin. The oxygenation reaction results in a change of area, ΔA, and a change of surface charge density, Δσ (4).

The results of these calculations show that the equilibrium constant varies with pH, as in the acid and alkaline Bohr effects (*see* Figure 2), and also with the ionic strength (*see* Figure 3). In both cases the agreement of the calculations with the observations is quite good, and the deviations can be explained in terms of other properties of hemoglobin (e.g. dissociation) that are not included in the model (*3, 4, 7*). The model also provides a physical meaning for the empirical constant in the oxygen-binding equation (i.e. the Hill coefficient) as well as some insight into the dissociation equilibrium of hemoglobin tetramers(*7*).

The Hill Coefficient as a Surface Excess

The equation used to describe binding equilibria generally considers that a concentration, *C*, of ligand reacts with a fraction, *y*, of available binding sites, and the equilibrium constant or affinity,

$$K = \left[\frac{y}{1-y}\right]\frac{1}{C} \tag{3}$$

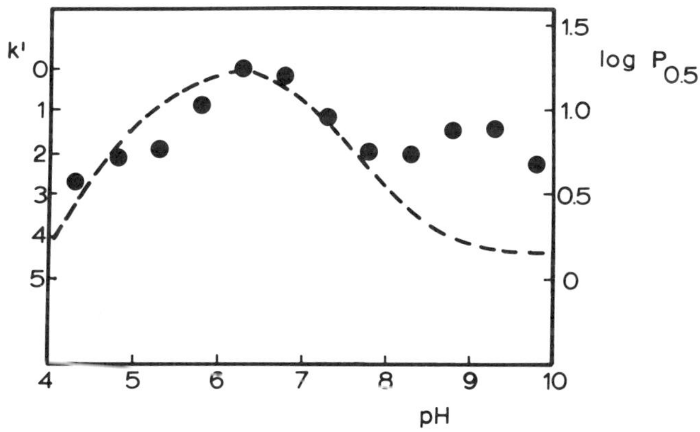

Journal of Theoretical Biology

Figure 2. The calculated values of the equilibrium constant (●) in units of k' superimposed on the dashed curve of the observations in units of log $P_{0.5}$ (4). (The two sets of values are made to coincide at pH 6.3.)

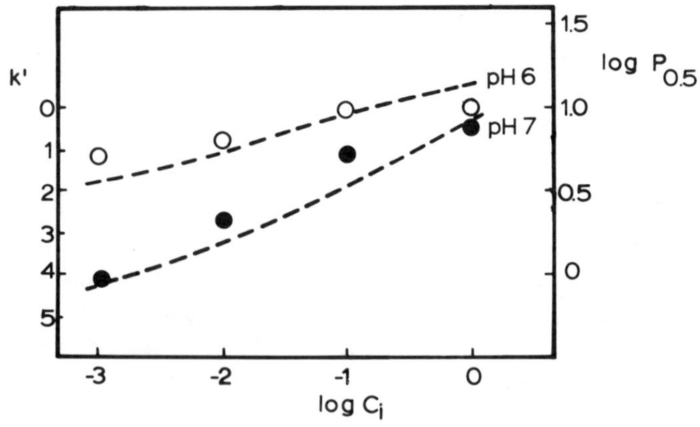

Journal of Theoretical Biology

Figure 3. The calculated values of the equilibrium constant (○) at pH 6.3 and (●) at pH 7.3, in units of k′, as a function of the logarithm of the ionic strength. The measured values of pH 6 and 7 are shown as dashed lines in units of log $P_{0.5}$ (4). (The calculated and observed values are made to coincide at pH 6.3 and log $C_i = -1$.)

The affinity is related to $\Delta G°$, the free energy change per mole when all substances are initially present at unit activity.

$$K = \exp\left[-\frac{\Delta G°}{RT}\right] \tag{4}$$

When additional changes in free energy (ΔG_s) occur as a result of conformational changes during binding, there is a gradual change in the affinity during a reaction (*2, 3*).

$$K'(y) = \mathrm{K} \exp\left[-\frac{\Delta G_s(y)}{RT}\right] \tag{5}$$

In practice, the affinity is assumed to remain constant at a new value, K_e, and the change in binding is given by raising the concentration to a power, n, the Hill coefficient (*1*).

$$K_e = \left[\frac{y}{1-y}\right]\frac{1}{C^n} \tag{6}$$

where $n > 1$ for $\Delta G_s < 0$. (K_e is inversely related to $P_{0.5}$ used to describe oxygenation.)

Using Equations 3, 5, and 6, we can write two equivalent expressions for the effect of an additional free energy term:

$$\Delta G^\circ + RT \ln \left[\frac{y}{1-y} \right] - nRT \ln C = 0 \quad (7)$$

$$\Delta G^\circ + \Delta G_s (y) + RT \ln \left[\frac{y}{1-y} \right] - RT \ln C = 0 \quad (8)$$

If we differentiate the two equations and subtract one from the other

$$n - 1 = - \frac{d\Delta G_s(y)}{RTd \ln C} \quad (9)$$

The term $(n - 1)$ has the form of the surface excess in the Gibbs equation. (It should be noted that n is not constant, but varies with y and is equal to unity at the extremes of the range, $y = 0$ and $y = 1$.) In the case of hemoglobin solutions, the surface excess arises from the free energy of the conformational change during oxygenation, and the Hill coefficient is simply related to the surface excess.

Conclusion

In carrying out these calculations we have in effect evaluated changes in the surface free energy of a single molecule. (Actually, we are dealing with large numbers of molecules and therefore with average values). It may be hard to conceive of a surface free energy for a single molecule, but the idea has proved useful in the past. For example, Langmuir (*8*) described the surface tension of a liquid in terms of the interaction energies of the individual molecules and obtained reasonable estimates of the energetics of simple systems. Frenkel (*9*) also has used this method. From the present work an extension of the approach to a macromolecular system seems to yield valid results as well. The approximation gives the correct magnitudes and also suggests a physical basis for the source of the free energy owing to the conformational change. The surface free energy in this model appears to take into account both the hydrophobicity (*10*) as well as the electrostatic interactions that are known to play a role in determining the shape of proteins in solution (*11*).

Acknowledgment

This study was supported in part by Research Grant PCM 76–11676 from the National Science Foundation.

Literature Cited

1. Antonini, E.; Brunori, M. "Hemoglobin and Myoglobin in their Reactions with Ligands"; North-Holland: Amsterdam, 1971.

2. Blank, M. *J. Colloid Interface Sci.* **1972**, *41*, 97.
3. Blank, M. *J. Colloid Interface Sci.* **1973**, *43*, 557.
4. Blank, M. *J. Theor. Biol.* **1975**, *51*, 127.
5. German, B.; Wyman, J. *J. Biol. Chem.* **1937**, *117*, 533.
6. Muirhead, H.; Cox, J.; Mazzarella, L.; Perutz, M. *J. Mol. Biol.* **1967**, *28*, 117.
7. Blank, M. *J. Electrochem. Soc.* **1976**, *123*, 1654.
8. Langmuir, I. *Chem. Rev.* **1933**, *13*, 147.
9. Frenkel, J. "Kinetic Theory of Liquids"; Dover: New York, 1955.
10. Bigelow, C. C. *J. Theor. Biol.* **1967**, *16*, 187.
11. Tanford, C. "Physical Chemistry of Macromolecules"; Wiley: New York, 1961.

RECEIVED October 17, 1978.

MEMBRANE PROCESSES

13

Transmembrane pH and Electrical Gradients: Evaluation and Possible Role in Oxidative Phosphorylation

CAROL J. DEUTSCH, ANDRIJ HOLIAN, and DAVID F. WILSON

Department of Physiology and Department of Biochemistry and Biophysics, Medical School University of Pennsylvania, Philadelphia, PA 19104

In suspensions of isolated mitochondria under phosphorylating conditions the respiratory chain from the NAD couple to Cytochrome c *is near equilibrium with ATP synthesis. A transmembrane electrical potential of −0.11–−0.17 V (negative inside) and a pH gradient of 0.4–0.8 pH units (alkaline inside) is observed. Approximately 67kJ/mol (16 kcal/mol) is required for ATP synthesis and 3–5 mol of H^+ would have to move across the membrane to provide enough energy to synthesize 1 mol of ATP if these reactions were coupled. The prokaryotic bacterium,* Paracoccus denitrificans, *has a respiratory chain and ATP synthesis efficiency similar to that of mitochondria. At an external pH of 7.6 the transmembrane electrical potential is −0.04 V, the pH gradient is approximately 0, and 50.4 kJ/mol (12 kcal/mol) is required for ATP synthesis. These data suggest that proton transport is not a primary intermediate in the mechanism of oxidative phosphorylation in either mitochondria or* P. denitrificans.

One of the major unsolved problems of modern biochemistry is the mechanism of energy transduction in mitochondrial oxidative phosphorylation. The adenosine triphosphate (ATP) synthesized by the mitochondria serves as the primary source of metabolic energy in higher organisms, being used to drive thermodynamically unfavorable biosynthetic reactions and to do physical work such as ion transport and muscle contraction. In mitochondria the energy for ATP synthesis from adenosine

0-8412-0473-X/80/33-188-195$05.00/1

diphosphate (ADP) and inorganic phosphate (P_i) is provided by the electrochemical oxidation of metabolites by molecular oxygen utilizing a series of highly organized oxidation–reduction reactions. These reactions provide the negative free-energy change required for reversible formation of ATP from ADP and P_i. The energy transduction mechanism is the process by which the electrical (oxidation–reduction) negative free energy is transduced into the group transfer potential of the terminal phosphate bond of ATP. The chemical nature of the transduction mechanism remains a question—the subject of much speculation and debate.

In any sequence of chemical reactions with a single-reaction pathway, net steady-state synthesis of product from reactant occurs only if there is a net negative free-energy change ($-\Delta G$) for the reaction. Moreover if the reaction pathway is such that all of the intermediates are in a steady state (no oscillations), then the ΔG for each reaction of the sequence must be negative. For each reaction, the negative ΔG may be very close to 0 (near equilibrium), but it cannot be positive. Thus there are two criteria which any proposed intermediate must meet in order to be part of the reaction pathway: (a) the ΔG for the formation of the intermediate and its breakdown to product must be either negative or 0; (b) the rate of formation of the intermediate and its breakdown to product must be equal to or greater than the rate of formation of product from reactant.

In this chapter we will summarize the currently available knowledge of the free-energy relationships in the reactions of oxidative phosphorylation in rat mitochondria and a prokaryotic microorganism, *Paracoccus denitrificans.* These data include measurements of the transmembrane electrical potential (E) and the transmembrane pH gradient (ΔpH) values essential to understanding the transport of molecules across the membrane. The free-energy change associated with the movement of protons across the membrane has been calculated in order to evaluate its proposed (*1, 2*) role as an intermediate in oxidative phosphorylation. It is concluded that this free-energy change is too small for proton transport to be an intermediate in oxidative phosphorylation.

The Mitochondrial Respiratory Chain and the Free-Energy Changes of Its Oxidation–Reduction Reactions. Any proposed intermediate must fulfill the thermodynamic condition

$$-a\Delta G_{o-r} \geq -b\Delta G_i \geq -c\Delta G_{ATP} \tag{1}$$

where ΔG_{o-r}, ΔG_i, and ΔG_{ATP} are the molar free energies associated with electron transfer, breakdown of the intermediate, and hydrolysis of ATP, respectively; *a, b,* and *c* are the stoichiometries for the balanced equations. In cases for which *a, b,* or *c* are unknown, the measured ΔG values may be substituted into the inequality and limiting values calculated for the relative stoichiometry. For any oxidation–reduction reaction of the form

$$A_{r} + B_{o} \rightleftarrows A_{o} + B_{r} \tag{2}$$

where the subscripts o and r indicate the oxidized and reduced forms of components A and B. The oxidation–reduction behavior of each component may be treated as one-half of an electrical cell. It is conventional to describe the electrical potential of each half-cell relative to a standard hydrogen electrode. Thus

$$E_{hA} = E_{mA} + \frac{RT}{nF} \ln \frac{A_{o}}{A_{r}} \tag{3}$$

and

$$E_{hB} = E_{mB} + \frac{RT}{nF} \ln \frac{B_{o}}{B_{r}} \tag{4}$$

where E_{hA} and E_{hB} are the reduction potentials of Components A and B relative to a standard hydrogen electrode while E_{mA} and E_{mB} are characteristic half-reduction potentials for the two redox components. The electrical potential (ΔE) generated by transfer of reducing equivalents from A to B is expressed

$$\Delta E = E_{hB} - E_{hA} = E_{mB} - E_{mB} + \frac{RT}{nF} \ln \frac{B_{r}}{B_{o}} \frac{A_{o}}{A_{r}} \tag{5}$$

and the free-energy change ($\Delta G_{o\text{-}r}$) is expressed as

$$\Delta G_{o\text{-}r} = -nF\,\Delta E \tag{6}$$

Experimentally it is essential to know the half-reduction potentials of both A and B and to either measure the oxidized and reduced forms of A and B or to determine by other methods the ratio of the oxidized and reduced forms.

The oxidation–reduction components of the respiratory chain have been studied extensively. It would appear that essentially all of the oxidation–reduction components now are known although some uncertainty remains in the region of Energy-Coupling Site 1 (NADH dehydrogenase). A summary of the components of the respiratory chain is presented in Table I along with their measured half-reduction potentials

Table I. The Half-Reduction Potentials of the Oxidation–Reduction Components of the Mitochondrial Respiratory Chain

Respiratory Chain Fragment	*Component*	n-*Value*	E_m[a]
	Flavin mononucleotide	?	?
NADH dehydrogenase	(Fe–S) N – 1a	1.0	−0.380
	(Fe–S) N – 1b	1.0	−0.250
	(Fe–S) N2	1.0	−0.030
Succinate	Flavin adenine dinucleotide	2.0	−0.040
Dehydrogenase	(Fe–S) S-1	1.0	0.030
	(Fe–S) S-3	1.0	0.060
Ubiquinone 10		2.0	0.045
Cytochrome b–c_1 complex	Cytochrome b_{561}	1.0	0.030
	Cytochrome b_{565}	1.0	−0.030
	Cytochrome c_1	1.0	0.215
	Rieske Fe–S	1.0	0.280
Cytochrome c		1.0	0.235
Cytochrome c oxidase	Cytochrome a	1.0	0.210
	"visible" copper	1.0	0.245
	Cytochrome a_3	1.0	0.385
	"invisible" copper	1.0	0.350

[a] The listed half-reduction potentials (E_m) are for measurements made using intact mitochondria or submitochondrial particles suspended at pH 7.2. (For original references and discussion *see* Wilson et al. (*3, 4*) and Ohnishi (*5, 6*).)

at pH 7.2 and the number of reducing equivalents accepted per mole (n value). Reviews of the data on which the table is based may be found elsewhere (*3, 4, 5, 6*).

The half-reduction potentials of the components fall into four groups. The most negative group is associated with NADH dehydrogenase and is matched with the E_m of the NAD couple at about −0.30 V. The next more positive group of components is associated with Cytochrome b— the ubiquinone region of the respiratory chain and succinate dehydrogenase and is about 0.00 V. The third group includes the Cytochrome c– Cytochrome a region of the respiratory chain (E_m is near 0.24 V) while the fourth group is not really a group but includes the oxygen active site which utilizes Cytochrome a_3 and the invisible copper with E_m values near 0.36 V in the reduction of oxygen to water (*7*). In coupled mitochondria carrying out net ATP synthesis, the components within each of the first three groups behave as equilibrated pools of reducing equivalents in which they have, within experimental error, the same E_h value (*3, 8, 9*). Thus the respiratory chain may be represented as shown in Figure 1 in which the individual components appear only as symbols in the rectangular boxes indicating redox pools; transfer of reducing equivalents between the pools is indicated as being coupled to the phosphorylation of ADP.

The Relationship of the Oxidation–Reduction Reactions and ATP Synthesis in Oxidative Phosphorylation. The free energy of synthesis (or hydrolysis) of ATP can be analyzed readily. Given the reaction

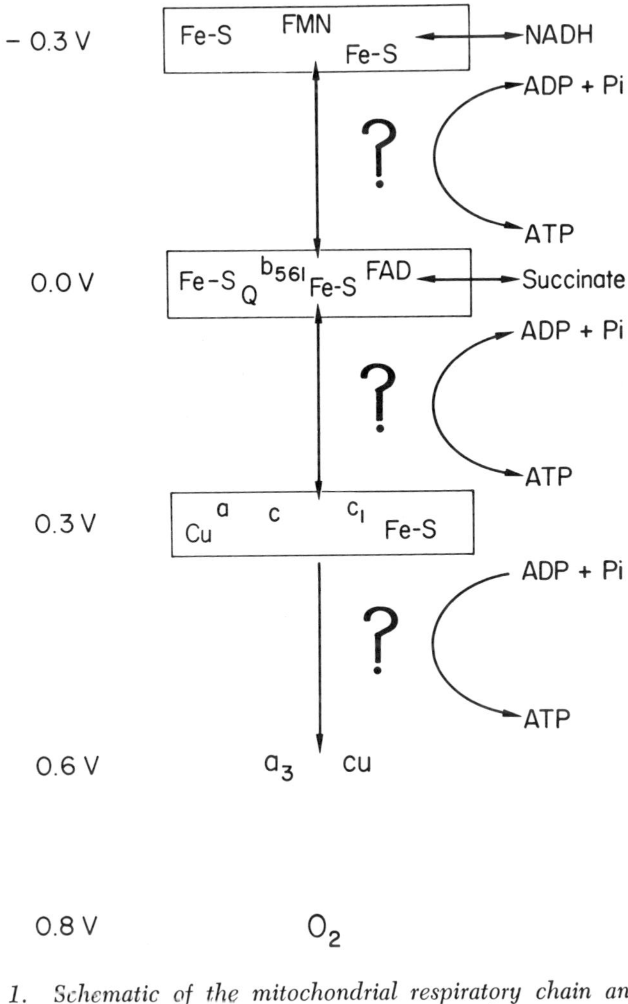

Figure 1. Schematic of the mitochondrial respiratory chain and ATP synthesis. The oxidation–reduction components of the respiratory chain are represented as symbols placed at their approximate E_h *values according to the scale on the left. Each box represents a pool of oxidation–reduction components with the same* E_h *value (3, 8, 9). The transfer of reducing equivalents is coupled to ATP synthesis by an as yet unknown energy transduction mechanism.*

Table II. The Free-Energy Relationships Between the Oxidation–

Material	E_{hNAD}[a] (V)	E_{hC}[a] (V)	ΔG_{o-r} kJ/2 eq	(kcal/2 eq)
Pigeon heart mitochondria	−0.343	0.270	−119	(−28.4)
	−0.308	0.312	−119.7	(−28.6)
Rat liver cells	−0.260	0.269	−102	(−24.4)
Ascites tumor cells	−0.270	0.260	−102	(−24.4)
Culture kidney cells	−0.252	0.271	−100.8	(−24.1)
P. denitrificans	−0.244	0.276	−100.4	(−24.0)

[a] $E_{h\mathrm{NAD}}$ and $E_{h\mathrm{C}}$ are the oxidation–reduction potentials of the NAD couple and the Cytochrome *c* couple, respectively, and ΔG_{o-r} is the free-energy change accompanying transfer of two reducing equivalents from NADH to Cytochrome *c*.

$$\mathrm{ATP} \rightleftarrows \mathrm{ADP} + \mathrm{P_i} \tag{7}$$

the free energy change (ΔG_{ATP}) may be calculated from

$$\Delta G_{\mathrm{ATP}} = \Delta G^{\circ\prime}{}_{\mathrm{ATP}} + RT \ln \frac{[\mathrm{ADP}][\mathrm{P_i}]}{[\mathrm{ATP}]} \tag{8}$$

The $\Delta G_{\mathrm{ATP}}{}^{\circ\prime}$ is −35 kJ/mol (−8.4 kcal/mol) and −31.8 kJ/mol (−7.6 kcal/mol) at zero and 1m*M* [Mg^{2+}], respectively (*10, 11*) and experimental measurement of the concentration of ATP, ADP, and phosphate allows ΔG_{ATP} to be calculated. All three phosphorylation sites are coupled to the same [ATP]/[ADP][P_i] as well as being coupled through a common high-energy intermediate (*12*). These common reactants, combined with the fact that reducing equivalents from NADH must pass sequentially through the phosphorylation sites, assures that all of the transduction sites in well coupled mitochondria have the same ΔE. In order to do precise analyses of the free-energy changes associated with the respiratory chain, it is necessary to accurately estimate the E_h values of redox components on the reducing and oxidizing side of the site(s). We have chosen to utilize the intramitochondrial NAD couple on the reducing side and Cytochrome *c* on the oxidizing side because their E_m values at pH 7.0 are known and their E_h values can be estimated both in vitro and in vivo.

Two phosphorylation sites are associated with the transfer of reducing equivalents from NADH to Cytochrome *c* and the equation for the coupled reaction is:

$$\mathrm{NADH_m} + 2c^{3+} + 2\mathrm{ADP_c} + 2\mathrm{P_{i_c}} \rightleftarrows \mathrm{NAD_m}^+ + 2c^{2+} + 2\mathrm{ATP_c} \tag{9}$$

Experimentally, (*13, 14, 15*) mitochondrial oxidative phosphorylation is regulated by, and equilibrates with, the cytosolic (extramitochondrial)

Reduction Reactions of the Respiratory Chain and ATP Synthesis

ΔG_{ATP}		$\Delta\Delta G^{b}$		
kJ/2 ATP	*(kcal/2 ATP)*	*kJ*	*(kcal)*	*Ref.*
−124.7	(−29.8)	5.7	(1.4)	*9*
−121.3	(−29.0)	1.6	(0.4)	*9*
−101.3	(−24.2)	−0.7	(0.2)	*16*
−98.7	(−23.6)	−3.3	(−0.8)	*16*
−101.3	(−24.2)	0.4	(0.1)	*17*
−100.8	(−24.1)	0.4	(0.1)	*18*

[b] $\Delta\Delta G$ is calculated as $\Delta G_{o-r} - \Delta G_{ATP}$ and under the experimental conditions it must be negative. The accuracy of the measurements is approximately ±4kJ (± 1 kcal) except in the case of isolated mitochondria where the accuracy with which the potential of the NAD couple can be measured is less (*9*).

[ATP], [ADP], and [P_i] and this is indicated in Equation 9 by the subscript c associated with these reactants.

In Table II the available measurements of the thermodynamic parameters associated with the transfer of reducing equivalents from NADH to Cytochrome *c* are presented as the measured E_h values for the NAD couple and Cytochrome *c*, the ΔE for the redox reaction, and the free-energy change for the transfer of two reducing eqiuvalents (*9, 16, 17, 18*). The E_h of the NAD couple ranges from −0.35 V in suspensions of isolated mitochondria to −0.24 V in *P. denitrificans* cells. These ΔE values correspond to ΔG_{o-r} values from −124 kJ/2 eq (−29.8 kcal/2 eq) to −100 kJ/2 eq (−24.0 kcal/2 eq).

The free energy of ATP hydrolysis has been determined in each case by measuring either the extramitochondrial or whole cell [ATP], [ADP], and [P_i] and calculating the free-energy change using the appropriate $\Delta G^{\circ\prime}$ for ATP hydrolysis (*see* Table II). This value then was used to calculate the ΔG for the synthesis of 2 mol of ATP (ΔG_{ATP}) as required for comparing the transfer of reducing equivalents from NADH to Cytochrome *c*. The data show that within experimental error the coupled reactions are near equilibrium (ΔG is near 0). Thus any proposed intermediate in oxidative phosphorylation also must be in near equilibrium with both the oxidation–reduction reactions and the [ATP]/[ADP][P_i].

The Free-Energy Changes Associated with Proton Translocation Across the Mitochondrial Inner Membrane. The inner mitochondrial membrane not only contains the respiratory chain but also determines the osmotic properties of mitochondria because the outer membrane is permeable to molecules with molecular weights less than approximately 2,000 daltons. Thus mitochondrial transport is concerned with the movement of various compounds across the inner membrane that is permeable only to uncharged hydrophilic molecules of less than approximately 80 daltons. The change in free energy which accompanies the movement of a mole of substance (A) from outside to inside a vesicular membrane in aqueous suspension is dependent on the chemical activity on each side of the membrane

$$\Delta G = RT \ln \frac{[a_i]}{[a_e]} \tag{10}$$

where $[a_i]$ and $[a_e]$ represent the activities (concentrations) of the compound in the internal and external aqueous environment. If the compound is electrically charged, its movement across the membrane is also dependent on the transmembrane electrical potential (E) such that

$$\Delta G = nFE \tag{11}$$

where n is the charge on the transported compound. Both equations are written with the correct sign for transport from outside the vesicle to inside the vesicle. Proton movement then would be accompanied by a free-energy change (ΔG_{H^+}) which can be calculated from Equation 12 or 13

$$\Delta G_{H^+} = RT \ln \frac{[H_i^+]}{[H_e^+]} + nFE \tag{12}$$

as

$$\Delta G_{H^+} = RT\ (pH_e\text{-}pH_i) + nFE \tag{13}$$

In order to evaluate ΔG_{H^+} it is necessary to measure both the ΔpH $(pH_e - pH_i)$ and E.

Measurements of the Transmembrane pH Difference. The difference in pH between the internal and external spaces may be determined from the distribution of weak acids or bases (*19*). If the membrane is permeable only to the unionized form of the weak acid, then in the steady state $HA_i = HA_e$, and intracellular pH is given by

$$pH_i = pK_a + \log\,[T_i/T_e\ (10^{(pH_e - pK_a)} + 1) - 1] \tag{14}$$

where T_i and T_e are the intracellular and extracellular concentrations of weak acid, respectively. Similarly, for the case of a weak base, assuming permeability to only the uncharged form of the weak base

$$pH_i = pK_a - \log\,[T_i/T_e\ (10^{(pK_a - pH_e)} + 1) - 1] \tag{15}$$

There is no provision for evaluating the amount of intravesicular binding of the weak acid or base and we (*20, 21, 22*) have adopted the procedure of using both a weak acid and a weak base. For any given pH gradient one of these probes should be concentrated inside the vesicle while the other is excluded. Thus any binding would affect the two probes in opposite ways, making the pH gradient larger for the included species and smaller for the excluded species. If the two probes agree, the internal pH is measured accurately while if they disagree one may only consider that the pH gradient must lie between the two calculated values.

Measurements of the Transmembrane Electrical Potential. The electrical potential across a vesicular membrane can be determined by using electrodes in macroscopic systems or microelectrodes in large cells. In microscopic systems molecules must be used as indicators. At equilibrium an ion (C) which is permeable to the membrane only in its charged form will distribute across the membrane according to the Nernst equation

$$E = \frac{-RT}{zF} \ln \frac{[C_i]}{[C_e]} \tag{16}$$

where z is the electrical charge on the molecule and $[C_i]$ and $[C_e]$ are the internal and external concentrations. The ion must be passively permeable since any active transport of the ion and/or coupling of its transport to other ions which are not at equilibrium would invalidate application of the Nernst equation. Most cell systems actively transport some inorganic ions (Na^+, K^+, Rb^+, Ca^{2+}, Cl^-, etc.) and therefore another class of ions is necessary to satisfy the criteria of passive permeability. Skulachev and co-workers (*23, 24*) first introduced the use of organic lipid soluble ions for this purpose. Considerable evidence has accumulated that, with proper selection of the organic ions and experimental conditions, the membrane potential can be measured accurately (*20–30*). Wherever possible both positively and negatively charged ions should be used because, as was the case with weak acids and bases, binding has opposite effects on the transmembrane electrical potential calculated from the distribution of positively and negatively charged ions.

The Transmembrane pH and Electrical Gradients in Suspensions of Rat Liver Mitochondria. The distributions of weak acids such as acetic acid and 5,5-dimethyloxazolidine-2,4-dione (DMO) give calculated pH gradients which are in good agreement with the distribution of the weak base, methylamine, in mitochondria (*22, 31*). This provides evidence that the pH gradients across the mitochondrial inner membrane can be measured accurately. The transmembrane electrical gradient measurements may be assessed by using the organic cation triphenylmethyl phosphonium ($TPMP^+$) or K^+ in the presence of valinomycin, but the indicated values are too large and negative to be measured using a negatively charged ion. This ion is excluded from the internal space by a negative transmembrane electrical potential gradient. As the value of this negative potential increases to approximately −60 mV the ion concentration inside is only approximately one-tenth that of the concentration outside. Corrections which must be made for advantitious water (trapped space) become larger than the internal ion content and the calculated internal ion concentration falls within the experimental error. In addition, even very slight binding of the probe to the external membrane surface introduces major errors in the calculations. In the absence of accurate values for the distribution of a negative ion, the contribution of binding to the total measured mitochondrial $TPMP^+$ or K^+ accumulation is more difficult to estimate.

As shown in Table III, the pH gradient across the mitochondrial membrane decreases with the increasing pH of the suspending medium, ranging from ca. 0.8 pH units at pH 6.5 to less than 0.5 pH units at pH 7.6. The transmembrane electrical potential is approximately −0.17 and −0.14 V as calculated from the distributions of $TPMP^+$ and K^+ (valinomycin), respectively, at pH 6.5. The total electrochemical gradient for H^+ ions ($\Delta\mu_{H^+}$) is estimated to be −0.21 V ($TPMP^+$) and −0.18 V (K^+) while at pH 7.6 the values are −0.19 and −0.14 V, respectively (*see* Table IV). Calculations based on $TPMP^+$ distributions overestimate the transmembrane electrical potential. Mitochondria suspended in media of various [K^+] in the presence of valinomycin and the absence of an energy supply should exhibit a K^+ diffusion potential. When such a system is used to calibrate the transmitochondrial membrane potential obtained by $TPMP^+$ distribution, the latter agree with the K^+ (+ valinomycin) distribution measurements given in Table III (*22*). Thus the measurements using K^+ (+ valinomycin) should give the most accurate value for the transmembrane electrical potential.

The mitochondria are capable of synthesizing ATP up to [ATP]/[ADP][P_i] values of $9 \times 10^4 M^{-1}$ to $2 \times 10^5 M^{-1}$. Several laboratories (*22,*

Table III. The Transmembrane Electrical Potential and pH Gradient of Isolated Rat Liver Mitochondria[a]

External pH	*Transmembrane Electrical Potential (V)* *TPMP⁺*	*Valinomycin + K⁺*	*pH Gradient*	*[ATP]/[ADP][Pᵢ]*
6.5	−0.166	−0.136	0.81	$9.4 \times 10^5 M^{-1}$
7.0	−0.177	−0.140	0.68	$2.5 \times 10^5 M^{-1}$
7.6	−0.164	−0.112	0.47	$2.8 \times 10^5 M^{-1}$

[a] Rat liver mitochondria were suspended at the indicated pH, glutamate, malate, ADP, and P_i were added, and respiration was allowed to continue until the respiratory rate attained a minimum value (maximum ATP synthesis). The membrane potential, pH gradient, and [ATP]/[ADP][P_i] were measured (*22*).

30, 31, 32 33) have reported values for the chemical potential gradient. These values range from a high of −0.23 V to a low of −0.14 V depending on the experimental conditions and the measuring methods (*see* Table IV). The variation in measured $[ATP]/[ADP][P_i]$ is apparently large ($2.2 \times 10^3 M^{-1}$ to $2.7 \times 10^5 M^{-1}$) but in some of the reported experiments (*30, 31*) low total adenine nucleotide concentrations were used and ATP synthesis was limited by the [ADP] rather than the free energy of ATP synthesis. This results in a low measured $[ATP]/[ADP][_i]$ and a corresponding low ΔG_{ATP}.

Thermodynamic Relationships of Proton Translocation Across the Mitochondrial Membrane and ATP Synthesis. The translocation of 1 mol of protons across the mitochondrial membrane occurs with a change in free energy (ΔG_{H^+}) expressed in Equation 13 or

$$\Delta G_{H^+} = nF\,\Delta\bar{\mu}_{H^+} \tag{17}$$

where

$$\Delta\bar{\mu}_{H^+} = \frac{RT}{nF}(pH_e - pH_i) + E \tag{18}$$

Table IV gives a summary of the calculated ΔG_{H^+} values and the free energy of synthesis of ATP (ΔG_{ATP}). The ΔG_{H^+} values range from 13.4 kJ/mol (3.2 kcal/mol) to 22 kJ/mol (5.3 kcal/mol) while the ΔG_{ATP} values range from 54.4 kJ/mol (13 kcal/mol) to 69 kJ/mol (16.5 kcal/mol). As noted previously the lower values (*30, 31*) probably arise from the limiting adenine nucleotide concentrations used in those experiments. The ratio $\Delta G_{ATP}/\Delta G_{H^+}$ has been calculated and ranges from a low of 2.6 to high of 5.2. The free-energy ratio ($\Delta G_{ATP}/\Delta G_{H^+}$) is equal to the number of moles of protons which would have to be transported from the extramitochondrial medium to the intramitochondrial space in order to provide the negative free change required for ATP synthesis. Eliminating the erroneously low values for $[ATP]/[ADP][P_i]$ and using the corrected $TPMP^+$ distributions narrows this range from 3.1 to 5.2. Assuming a stoichiometric mechanism, the minimum value of H^+/ATP in mitochondrial oxidative phosphorylation would be 4 or 5.

The Free-Energy Relationships Between ATP Synthesis and Proton Translocation in *P. denitrificans*. It is important to examine any proposed intermediate of oxidative phosphorylation for its general applicability and to make the measurements under conditions that are as physiological as possible. *P. denitrificans* can be grown aerobically so as to have a respiratory chain very similar to that of mitochondria (*32*). It is prokaryotic, thus having a single cellular compartment surrounded by a plasma membrane. The respiratory chain is present in the plasma membrane, analogous to the mitochondrial inner membrane. As shown

Table IV. The Measured Proton Electrochemical Relationship to

Ref.	*Method*[a]	$\Delta\bar{\mu}_{H^+}$ *(mV)*
Padam & Rottenberg (*34*)	K^+	−0.15
Nicholls (*31*)	Rb^+	−0.23
Azzone et al. (*30*)	$TPMP^+$ and K^+	−0.18
Mitchell & Moyle (*33*)	K^+	−0.23
Holian & Wilson (*22*)		
pH 6.5	$TPMP^+$	−0.21
	K^+	−0.18
pH 7.0	$TPMP^+$	−0.22
	K^+	−0.18
pH 7.6	$TPMP^+$	−0.19
	K^+	−0.14

[a] The methods used were measurements of the distribution of (Rb^+) or (K^+) ions in the presence of valinomycin or the distribution of the lipid-soluble organic ion ($TPMP^+$).

[b] The ratio $\Delta G_{ATP}/\Delta G_{H^+}$ is the number of moles of H^+ which have to be transported in order to provide the negative free-energy change required for the synthesis of 1 mol of ATP.

in Table II under normal aerobic conditions the respiratory chain from the intracellular NAD couple to Cytochrome *c* is near equilibrium with ATP synthesis. These cells can be grown readily at an external pH of 7.6 and measurement of the transmembrane pH and electrical gradients for cells suspended at this pH are presented in Table V. The pH gradient indicated by the distribution of a weak acid (DMO) and a weak base (methylamine) are 0.00 and −0.31 pH units, respectively, the difference indicating that there is some binding of either one or both. Therefore, the ΔpH will be somewhere between these two extremes. The transmembrane electrical potential can be estimated from the distributions of both positively charged $TPMP^+$ and negatively charged SCN^-. These two ions give values of −0.040 and −0.023 V, respectively (*18*), establishing a maximum value for $\Delta\bar{\mu}_{H^+}$ of approximately −0.04 V. When the ΔG_{H^+} is calculated and compared with ΔG_{ATP} (*see* Table VI) it is apparent that more than 10 mol of H^+ would have to be transported from the extracellular to the intracellular space in order to provide the energy required for ATP synthesis.

It is reasonable to question whether H^+/ATP stoichiometries of four to greater than ten are consistent with chemiosmotic coupling in oxidative phosphorylation. Any stoichiometry is theoretically possible but practical considerations restrict the possibilities. The original fuel cell model for chemiosmotic coupling was based on the well-known stoichiometric coupling of one H^+ per reducing equivalent in the oxidation and reduction of compounds such as quinones. This mechanism requires a stoichiometry

Gradient of Suspensions of Mitochondria and Its ATP Synthesis

ΔG_{H^+}		$[ATP]/[ADP][P_i]$	ΔG_{ATP}		ΔG_{ATP}[b] / ΔG_{H^+}
kJ/mol	*(kcal/mol)*	*(M^{-1})*	*kJ/mol*	*(kcal/mol)*	
13.8	(3.3)	2.3×10^5	65.7	(15.7)	4.7
22.2	(5.3)	1.3×10^5[c]	58.6	(14.0)	2.6
18	(4.3)	2.2×10^3[c]	54.4	(13.0)	3.0
21.8	(5.2)	1×10^4[d]	57.7	(13.8)	2.6
20.5	(4.9)	9×10^5	67.4	(16.1)	3.3
17.6	(4.2)	9×10^5	67.4	(16.1)	3.8
21.3	(5.1)	2×10^5	65.7	(15.7)	3.1
17.1	(4.1)	2×10^5	65.7	(15.7)	3.8
18.4	(4.4)	2.7×10^5	69	(16.5)	3.7
13.4	(3.2)	2.7×10^5	69	(16.5)	5.2

[c] These values were measured at ADP concentrations less than $5 \mu M$ and are certainly low for this reason.
[d] Not a measured value.

of 1 H^+/equivalent/site in the respiratory chain and $2H^+$/equivalent/site in the respiratory chain and $2H^+$/ATP in ATP synthesis. Thus it is not consistent with the thermodynamic data. The modified chemiosmotic coupling mechanisms required to give stoichiometries greater than 1 H^+/ equivalent/site are based on H^+ transport systems analogous to chemical ion transport mechanisms involving chemical high-energy intermediates. This gives rise to an unwieldy situation in which energy transduction is proposed to generate a chemical high-energy intermediate which is used to pump protons. ATP synthesis occurs when the proton electrochemical gradient is used to form another chemical high-energy compound for ATP synthesis.

Table V. The Transmembrane Electrical and pH Gradients of *P. denitrificans*[a]

	Membrane Potential (V)		*pH Gradient*	
Condition	*$TPMP^+$*	*SCN^-*	*DMO*	*Methylamine*
Aerobic	$-0.041 \pm .006$	$-0.023 \pm .007$	$0.00 \pm .17$	$-0.31 \pm .11$
Anaerobic	$-0.057 \pm .007$	$-0.029 \pm .008$	$0.19 \pm .13$	$-0.53 \pm .16$
Aerobic	$-0.040 \pm .003$	$-0.023 \pm .002$	$0.11 \pm .19$	$-0.49 \pm .11$

[a] The cells were suspended in an aerobic medium at pH 7.6 with 6mM glucose as substrate. The indicated probes ($TPMP^+$, SCN^-, DMO, and methylamine) were added and the distributions were measured while the cells were aerobic, after 2 min anaerobiosis, and reaeration for 2 min. The apparent failure of ΔpH to return to the initial value arises from a 0.1–0.2 pH unit metabolism-dependent acidification of the external medium which occurred during the experiment. Data was taken from Ref. *20*.)

Table VI. Thermodynamic Relationship of Transmembrane Proton Translocation and Oxidative Phosphorylation in *P. denitrificans*[a]

$\Delta\mu_{H^+}$	ΔG_{H^+}		ΔG_{ATP}		$\frac{\Delta G_{ATP}}{\Delta G_{H^+}}$
(V)	*kJ/mol*	*(kcal/mol)*	*kJ/mol*	*(kcal/mol)*	
−0.04	4.4	(1.05)	50.2	(12.0)	> 10

[a] The data were taken from Table V and the $\Delta\bar{\mu}_{H^+}$ was calculated as given in Equation 17.

The most conceptually difficult aspect of the multiproton mechanisms derives from the requirement for an oxidation–reduction-driven proton pump with $1n$ H^+ translocated per reducing equivalent transferred across each energy transduction site and an ATP-driven proton pump with exactly $2n$ H^+ translocated per ATP where n may be 2 or any other number up to and greater than 5. There is little precedent for the existence of such an intricate system and even less for its functioning at high rates and with full reversibility. Chemical coupling, for all of its uncertainties and as yet uncharacterized intermediates, is at least as plausible an hypothesis for energy transduction in oxidative phosphorylation as is chemiosmotic coupling.

Summary

Mitochondrial oxidative phosphorylation is a very efficient process in which the reactions of the respiratory chain from the NAD couple to Cytochrome *c* are near equilibrium with ATP synthesis ($\Delta G \rightarrow 0$). Under phosphorylating conditions a transmembrane electrical potential of −0.11–−0.17 V (negative inside) and a pH gradient of 0.4–0.8 pH units (alkaline inside) is observed. Thus 3–5 mol of H^+ ions would need to move across the membrane in order to provide enough energy to synthesize 1 mol of ATP if these reactions are coupled. The prokaryotic bacterium, *P. denitrificans,* has a respiratory chain and coupling efficiency similar to that of mitochondria but has a much lower transmembrane electrical potential (−0.04 V) and a pH gradient of approximately 0 (external pH of 7.6). These data suggest that proton transport is not a primary intermediate in the mechanism of oxidative phosphorylation in either mitochondria or *P. denitrificans.*

Glossary of Symbols

ATP = adenosine triphosphate
ADP = adenosine diphosphate
P_i = inorganic phosphate

E = transmembrane electrical potential difference
E_h = oxidation–reduction potential for a given couple
E_m = half-reduction potential
ΔE = difference in oxidation–reduction potential
$-\Delta G$ = net negative free energy change
$\Delta G_{o\text{-}r}$ = free energy change for redox reaction
$\Delta G^{\circ\prime}$ = standard free energy change at pH 7.0; −7.6 kcal/mol at 1mM [Mg^{++}] for the hydrolysis of ATP
pH_i = $-\log\,[H^+]_i$; the subscript i designating the intracellular compartment
pK_a = $-\log K_a$
R = gas constant
T = absolute temperature
Z = valence of a given ion
F = faraday constant
DMO = 5,5-dimethyloxazolidine-2,4-dione
$TPMP^+$ = triphenylmethyl phosphonium ion
$\Delta\bar{\mu}_{H^+}$ = proton electrochemical gradient

Acknowledgment

This work was supported by U.S. Public Health Service Grant GM 12202.

Literature Cited

1. Mitchell, P. "Chemiosmotic Coupling in Oxidative and Photosynthetic Phosphorylation"; Glynn, Res. Ltd: Bodmin, Cornwall, England, 1966.
2. Mitchell, P. *J. Bioenergetics* **1972,** *3,* 5–24.
3. Wilson, D. F.; Erecinska, M.; Dutton, P. L. *Ann. Rev. Biophys. Bioeng.* **1974,** *3,* 203–230.
4. Erecinska, M.; Wilson, D. F. *Arch. Biochem. Biophys.* **1978,** *188,* 1–14.
5. Ohnishi, T. *Eur. J. Biochem.* **1976,** *64,* 91–103.
6. Ohnishi, T.; Lim, J.; Winter, D. F.; King, T. E. *J. Biol. Chem.* **1976,** *251,* 2105–2109.
7. Lindsay, J. G.; Owen, C. S.; Wilson, D. F. *Arch. Biochem. Biophys.* **1975,** *169,* 492–505.
8. Wilson, D. F.; Erecinska, M.; Owen, C. S.; Mela, M. "Dynamics of Energy Transducing Membranes"; Ernster, L., Estabrook, R. W., Slater, E. C., Eds.; Elsevier: New York, 1974; p. 221–231.
9. Erecinska, M.; Veech, R. L.; Wilson, D. F. *Arch. Biochem. Biophys.* **1974,** *160,* 412–421.
10. Benzinger, T.; Kitzinger, C.; Hems, R., Burton, K. *Biochem. J.* **1959,** *71,* 409–414.
11. Guynn, R.; Veech, R. L. *J. Biol. Chem.* **1973,** *248,* 6966–6969.
12. Slater, E. C. *Q. Rev. Biophys.* **1971,** *4,* 35–71.
13. Holian, A.; Owens, C. S.; Wilson, D. F. *Arch. Biochem. Biophys.* **1977,** *181,* 164–171.
14. Wilson, D. F.; Owen, C. S.; Holian, A. *Arch. Biochem. Biophys.* **1977,** *181,* 749–762.

15. Erecinska, M.; Wilson, D. F., Nishiki, K. *Am. J. Physiol.: Cell Physiol.* **1978,** *3*(2), C82–C89.
16. Wilson, D. F.; Stubbs, M.; Oshino, N.; Erecinska, M. *Biochemistry* **1974,** *13,* 5305–5311.
17. Wilson, D. F.; Erecinska, M.; Drown, C.; Silver, I. A. *Amer. J. Physiol.: Cell Physiol.* **1977,** *2*(3), C135–C140.
18. Erecinska, M.; Kula, T.; Wilson, D. F. *FEBS Lett.* **1978,** *87,* 139–144.
19. Waddell, W. J.; Butler, T. C. *J. Clin. Invest.* **1959,** *38,* 720–729.
20. Deutsch, C.; Kula, T. *FEBS Lett.* **1978,** *87,* 145–151.
21. Deutsch, C. J.; Holian, A.; Holian, S. K.; Daniele, R. P.; Wilson, D. F. *J. Cell Physiol.* **1979,** *99,* 79–94.
22. Holian, A.; Wilson,t D. F., submitted for publication in *Biochemistry.*
23. Bakeeva, L. E.; Grinius, L. L.; Jasaitis, A. A.; Kulicne, V. V.; Levitsky, D. O.; Lieberman, E. A.; Severina, I. I.; Skulachev, V. P. *Biochim. Biophys. Acta* **1970,** *216,* 13–21.
24. Skulachev, V. P. *Curr. Top. Bioenerg.* **1971,** *4,* 127–190.
25. Schuldiner, L.; Kaback, H. R. *Biochemistry* **1975,** *14,* 5451–5461.
26. Lombardi, F. J.; Reeves, J. P.; Short, S. A.; Kaback, H. R. *Ann. N.Y. Acad. Sci.* **1974,** *227,* 312–327.
27. Miller, A. G.; Budd, K. *J. Bacteriol.* **1976,** *132,* 741–748.
28. Altendorf, K.; Hirata, H.; Harold, F. M. *J. Biol. Chem.* **1975,** *250,* 1405–1412.
29. Hirata, H.; Altendorf, K.; Harold, F. M. *Proc. Nat. Acad. Sci. U.S.A.* **1973,** *70,* 1804–1808.
30. Azzone, G. F.; Pozzan, T.; Massari, S. *Biochim. Biophys. Acta* **1978,** *501,* 307–316.
31. Nicholls, D. G. *Eur. J. Biochem.* **1974,** *50,* 305–315.
32. Scholes, P.; Smith, L. *Biochim. Biophys. Acta* **1968,** *153,* 363–375.
33. Mitchell, P.; Moyle, J. *Eur. J. Biochem.* **1969,** *7,* 471–481.
34. Padam, E.; Rottenberg, H. *Eur. J. Biochem.* **1973,** *40,* 431–437.

RECEIVED October 17, 1978.

14

Displacement Photocurrents in Pigment-Containing Biomembranes: Artificial and Natural Systems

FELIX T. HONG

Department of Physiology, Wayne State University, School of Medicine, Detroit, MI 48201

A review of selected literature concerned with displacement photoelectric currents in pigment-containing biomembranes is presented. From model system studies, it appears that the displacement photocurrent is a manifestation of a chemical capacitance rather than the ordinary membrane capacitance. Analysis indicates that observed relaxation of the displacement photoelectric responses is influenced critically by the particular method of measurement. While the tunable voltage clamp (quasipotentiostatic) method provides relevant kinetic data, the open-circuit (galvanostatic) measurement preserves little kinetic information. Two different molecular models (oriented dipole vs. interfacial charge transfer) are discussed in light of current insight and with special reference to visual membranes and purple membranes of Halobacterium halobium.

Displacement (capacitative) photocurrents have been found in a wide variety of pigment-containing biomembranes. The best known example is the early receptor potential (ERP), which was discovered in 1964 by Brown and Murakami in the frog retina (*1, 2, 3*). The most salient feature of the ERP that distinguishes it from other bioelectric phenomena is the apparent lack of a latency (*4*). It is also remarkable for its resistance to anoxia and to drastically altered ionic environments (*1, 5, 6, 7, 8*). The ERP has two components (*see* Figure 1): a cornea-positive and temperature-insensitive *R*1 component and a cornea-negative *R*2 component, which is inhibited by low temperature (*9*).

0-8412-0473-X/80/33-188-211$06.75/1

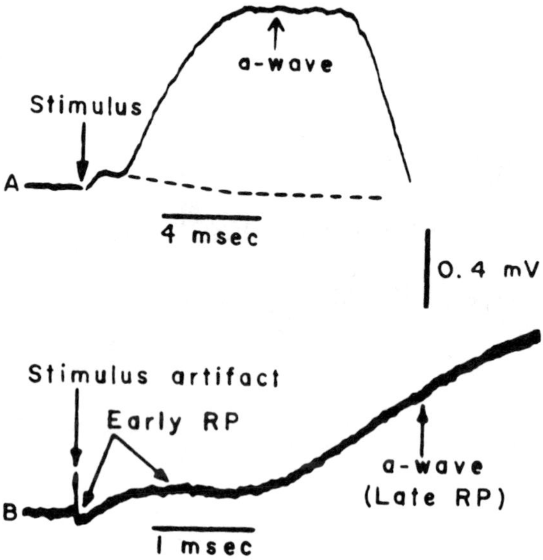

Vision Research

Figure 1. The ERP of a cynomolgus monkey. Responses were recorded by a tungsten microelectrode at about the level of the inner segments in the retina, with the reference electrode in the vitreous humor of the eye. The stimulus was an intense 20-μsec flash. In Record A*, the ERP intervenes between the stimulus flash and the rising phase of the a wave, which is part of the LRP. The time course of decay of the ERP, when isolated by cessation of artificial respiration, is shown by the dashed line. In Record* B*, the stimulus artifact consists only of the positive tip which is labeled. Thus the ERP is biphasic, and is followed by the LRP* (32).

The advent of techniques of forming model membrane systems makes it possible to study membrane phenomena under well-defined experimental conditions (*10, 11, 12, 13, 14*). Since Tien (*15, 16*) discovered a photoelectric effect in a chlorophyll-containing (artificial) lipid bilayer (synonymous with bilayer or bimolecular lipid membrane (BLM)), similar effects have been found and studied in lipid bilayers which contain other membrane-bound pigments. A fast photocurrent that possesses all the major characteristics of the ERP has been recorded in a lipid bilayer which contains a magnesium porphyrin and separates a redox gradient (*17, 18*). More recently, Montal and his co-workers succeeded in demonstrating an ERP-like signal in a rhodopsin-containing model membrane (*19, 20*), as well as in a model membrane that contains bacteriorhodopsin from *Halobacterium halobium* (*21*).

In this survey, we shall review selected data concerned with displacement photoelectric currents. Reviews on the photoelectric effects in biomembranes (*18, 22–29*) and on the ERP (*5, 30–37*) are available. The data will be analyzed in terms of an equivalent circuit scheme, which is generalized from model-system studies. We shall distinguish primary photoelectric events from the secondary ionic events. We further divide the primary effect into an ac (capacitative) photoelectric effect and a dc (resistive) photoelectric effect (*28*). Thus, the various descriptive terms such as fast photoelectric effect (fast photovoltage; FPV) (*38*), photoelectric potentials (PEP) (*39*), and displacement photocurrents (*40*) will be regarded as synonymous with the ac photoelectric effect. We shall describe two possible molecular models for the ac photoelectric effect (*18, 28, 37*), and discuss their validity in visual and in purple membranes in terms of available experimental data. Finally, we shall speculate on possible physiological significance of the ac photoelectric effect.

Classification of Photoelectric Effects

Operationally, a photoemf (photoelectromotive force) can be defined if stimulation by light generates an electric current across a pigment-containing membrane with two sides of the membrane kept equipotential. The term primary photoelectric effect refers to this current and the electrochemical gradient so generated. This definition thus excludes the secondary electrical events owing to dissipation of this gradient or a pre-existing gradient through light-controlled modulation of membrane ionic permeabilities. The latter are of considerable biological importance, but are beyond the scope of the present review (for a brief discussion and further references *see* Ref. *28*). The primary photoelectric effect can be divided further into a dc and an ac effect. A dc photoresponse leads to a net transfer of charges across the membrane, while an ac photoresponse results in no net transfer of charges. This classification is visualized best in terms of a generalized equivalent circuit, which was proposed initially to account for quantitative data in a model system of pigment-containing lipid bilayer (*see* Figure 2). The experimental basis of this scheme has been published elsewhere (*17, 18, 28*). In this generalized scheme, a parallel combination of ionic resistance, R_m, and membrane capacitance, C_m, represents a plain unmodified lipid bilayer. In addition, pigment photoreactions create a parallel electric pathway that includes a photoemf, E_p (with an internal resistance, R_p), which drives photoelectric current through another *RC* network: a transmembrane resistance, R_s, and a capacitance, C_p. The mechanism of charge transport across the membrane (through R_s) can be quite different from the ionic mechanisms encountered in nerve impulse generation. This dc current generally is not carried by the inorganic ions (Na^+, K^+, Cl^+, etc.) that are present in the aqueous phases in relative abundance (*18, 41*). Instead, the current may be carried either by mobile

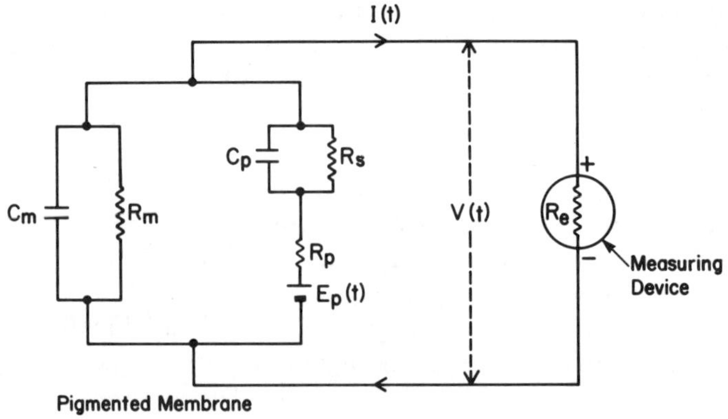

Figure 2. A generalized equivalent circuit for a pigmented membrane and a measuring device. The photoreactions are represented by a photoemf, $E_p(t)$*, with an internal resistance* R_p*. The photoemf drives a dc photoelectric current through a transmembrane pigment resistance,* R_s*, and an ac photoelectric current through a chemical capacitance,* C_p*. This photoelectric network is connected in parallel with the membrane ionic resistance,* R_m*, and the membrane capacitance,* C_m*. The measuring device has an access impedance, which is usually frequency dependent, but can be replaced by an effective access resistance,* R_e*, which is constant within the bandwidth of interest. The access impedance is the sum of the input impedance of the measuring device, the electrode impedance, and the resistance owing to intervening electrolyte solutions. Thus the relationship holds:* $V(t) = I(t)\ R_e$*. In an open-circuit measurement,* $I(t)$ *is so small that* $V(t)$ *is being read instead. The time course of* $I(t)$ *is dependent on the value of* R_e *(17).*

charged pigment molecules (*41*) or by a solid-state mechanism involving an electron-hole production (*22*) or a proton hopping in the pigment molecule (*42, 43*). At the two membrane–water interfaces, the continuity of a dc electric current requires that this specific charge-carrying function be relayed to the major inorganic ions in the aqueous phases through an interfacial electron (or proton) transfer reaction. This interfacial charge transfer is usually not strictly one way and a reverse interfacial transfer can occur as the system relaxes. Hong and Mauzerall (*17, 18*) have shown that such an interfacial charge transfer and the subsequent reverse transfer are tantamount to charging and discharging a capacitance that is connected in series to the photoemf. This capacitance is named chemical capacitance to distinguish it from the membrane capacitance that is connected in parallel. Historically, this interfacial charge transfer process has never been given serious consideration as a possible generating mechanism of the ac photoelectric effect. Instead, the ERP, for example, is regarded generally as a manifestation of a light-induced internal charge separation (displacement) of the visual pigment rhodopsin. These two molecular mechanisms will be described briefly later. For a detailed evaluation of these two mechanisms, interested readers are referred to Ref. *37*.

In a simple system such as the magnesium porphyrin–aqueous redox system studied by Hong and Mauzerall (*17, 18*), the resistive elements in the equivalent circuit can be treated as idealized elements with constant values, and all the *RC* parameters can be determined experimentally. The measured photocurrent can be fit to the computed time course with the amplitude of the photoemf as the only freely adjustable parameter (Figure 3A). Generally, the resistive elements need not be idealized elements. They can be voltage and/or time dependent. The secondary ionic events are incorporated into a single resistance, R_m. With this generalized model, we now can attempt to identify, at least tentatively, various signals in representative photobiological systems.

Photosynthetic membranes in plants contain chlorophyll–protein complexes as reaction centers, and electron transport chain components (*44, 45*). In higher plants, there are two reaction centers designated as photosystems I and II. Evidence indicates that these components are oriented and asymmetrically distributed across the photosynthetic membranes. Photons absorbed by the reaction centers initiate electron transfer reactions that are relayed through the electron transport chain. Eventually a proton gradient is generated across the photosynthetic membranes

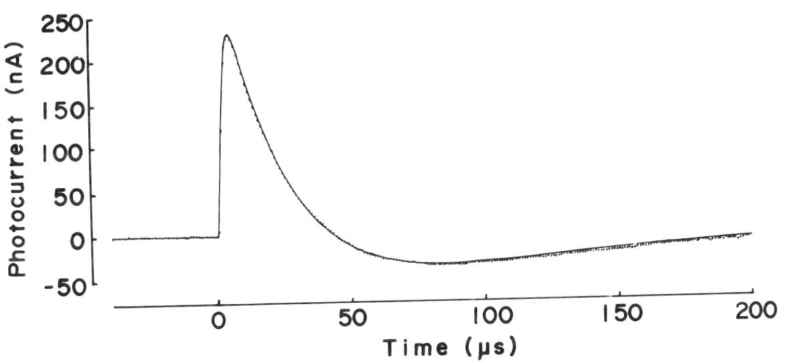

Proceedings of the National Academy of Sciences of the USA

Figure 3. Record **A.** *Photocurrent responses from a lipid bilayer that contains lipid-soluble magnesium* meso-*porphyrin di-*n-*amyl ester and separates a redox gradient with 10m*M *potassium ferricyanide and 0.5m*M *potassium ferrocyanide in one side (reference side) and 10m*M *potassium ferrocyanide in the other side. The aqueous phase also contains 1*M *NaCl buffered at pH 7.2. Temperature is 26°C. The light source is a dye laser pulse of 300-nsec duration and 590-nm output. The current was measured at an access impedance of 5.1 kΩ and an instrumental time constant of 1.5 μsec: (——), computed response based on the equivalent circuit shown in Figure 2, with input parameters determined experimentally; (· · ·), the measured response which was an average of 16 measurements. Notice that the area bound by the positive peak equals that bound by the negative peak (zero time integral). The dc photocurrent is estimated to be less than 1 nA and was not detected in this measurement (17).*

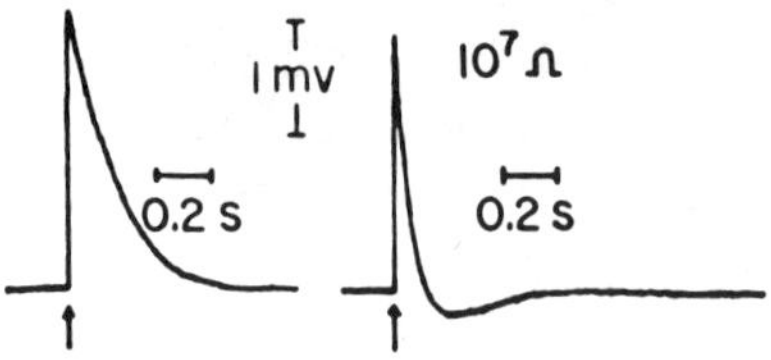

Biochimica et Biophysica Acta

*Figure 3. Record **B**. Open-circuit photovoltages across a lipid bilayer with a cyanine dye adsorbed to one interface. The light source is a xenon flash (8-μsec duration) delivered at time indicated by the arrows. The left record is the control with an access impedance of $10^9\Omega$. The right record shows the effect of adding a shunt resistor of $10^7\Omega$ (58).*

with the inside being more acidic than the outside of the photosynthetic membrane. Thus, photoreactions in the two photosystems can be represented by two photoemfs connected in series that drive a dc photocurrent through the membrane. The dissipation of this gradient is coupled to the synthesis of ATP (*46*) in accordance with Mitchell's chemiosmotic hypothesis (*47*) (secondary ionic event). As we shall see later in this chapter, an ac photoelectric effect also might be present. This ac photoelectric effect can be identified with the fast light-induced voltages in chloroplasts (*48, 49, 50*).

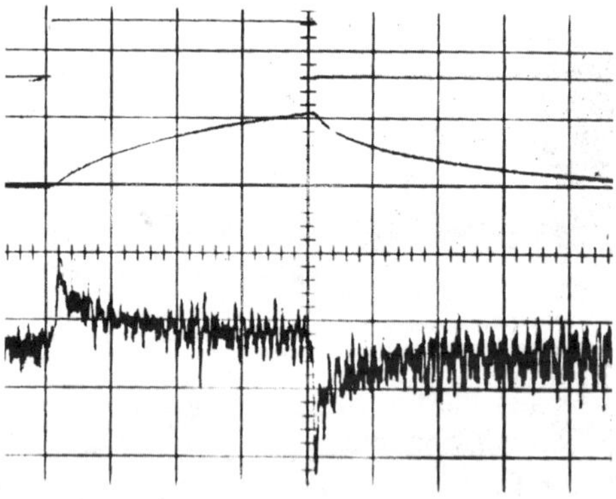

Biochimica et Biophysica Acta

*Figure 3. Record **C**. Photoelectric responses from a lipid bilayer containing retinal with 2mM potassium ferricyanide added to one aqueous phase. Scale: 1 mV/division; 5 pA/division; 2 sec/division. The light pulse* (top trace) *has a duration of 8 sec and a rise time of 5 msec. The photovoltage* (middle trace) *was measured by an electrometer (input impedance, $10^{16}\Omega$; rise time, 200 msec). The photocurrent* (bottom trace) *was measured by a picoammeter (input impedance, $10^7\Omega$; rise time, 20 msec)* (57).

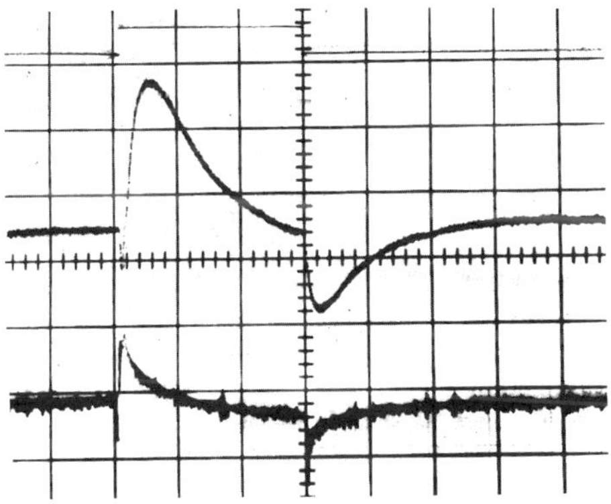

Biochimica et Biophysica Acta

*Figure 3. Record **D**. Photoelectric responses from a lipid bilayer containing Vitamin A acid with 2mM potassium ferricyanide added to one aqueous phase. Scale: 0.2 mV/division; 10 pA/division; 2 sec/division. The measurements of the light pulse* (top trace)*, the photovoltage* (middle trace)*, and the photocurrent* (bottom trace) *are similar to those in Record* C (57).

In the visual photoreceptor membranes, no dc photoelectric signal has ever been demonstrated convincingly. The two components *R*1 and *R*2 of the ERP are ac photoelectric signals (*4, 51*) and the photoemfs for these two components may be regarded as two consecutive photoemfs of opposite polarities (*37*). These two photoemfs drive electric current through a chemical capacitance, C_p, but not through a resistance, R_s (effectively, R_s = infinity). Montal et al. (*52*) have presented evidence of light-induced ionic channel formation in reconstituted rhodopsin lipid bilayer membranes. The electric current in this light-modulated channel must be regarded then as a secondary ionic event, according to the present classification, i.e. the current flow through R_m.

A photobiological system which has received much attention in the past few years is the purple membrane of *Holobacterium halobium* (*53*). The purple membrane contains only one single species of protein, bacteriorhodopsin, which contains retinal as the chromophore, similar to the visual pigment rhodopsin. However, it functions as a light-driven proton pump (*54*). The resulting proton gradient is utilized in the synthesis of ATP (*55*) similar to a photosynthetic membrane. The light-induced proton current across the purple membrane is a dc photoelectric current, while the passive dissipation of the proton gradient must be regarded as a secondary ionic event. The presence of an ac photoelectric effect was suggested first by Hong (*28*) and demonstrated experimentally by Trissl and Montal (*21*).

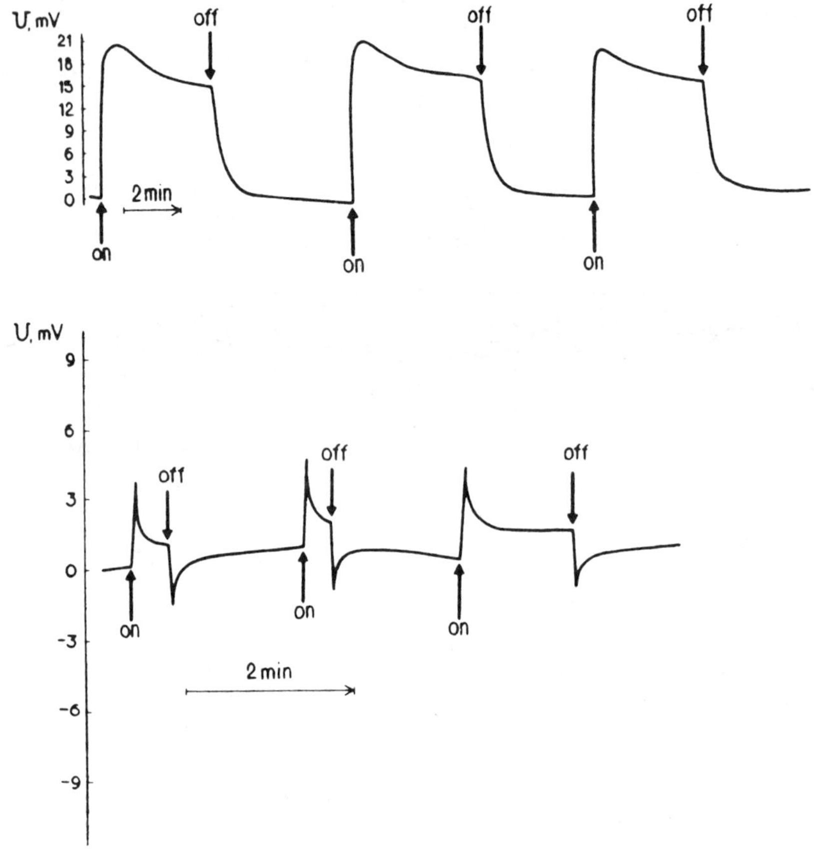

FEBS Letters

Figure 3. Record **E**. *Open-circuit photovoltages across a lipid membrane which contains oriented bacteriorhodopsin. The two aqueous phases have identical compositions: 0.15M KCl and 10mM Tris–HCl at pH 6.3:* (top trace), *the control photovoltage;* (bottom trace) *the photovoltage after adding* 2×10^{-7}M *of CCCP. Shunting with an external resistor has a similar effect* (59).

Methods of Measuring ac Photoelectric Responses

A systematic discussion of the methods of measuring the photoresponses from a typical photosystem as represented by the equivalent circuit in Figure 2 will facilitate our interpretation and classification of various photoelectric data. Let us consider the two common methods of electric measurements: galvanostatic (also known to electrophysiologists as the current clamp or open-circuit method) and potentiostatic (also known as the voltage clamp or short-circuit method). In an idealized galvanostatic measurement, the measuring device should have an infinite input impedance (relative to the source impedance of the system being measured) so that no appreciable current is drawn by the measuring device. In an idealized potentiostatic measurement, the two aqueous phases near each membrane–water interface should be maintained at a fixed potential at all time, and thus the access impedance, R_e, (sum of the input impedance of the amplifier, the electrode impedance, and the impedance caused by intervening electrolyte solutions) must be effectively zero relative to the membrane source impedance. It is instructive to consider the effect of varying the access impedance and the effect of a shunting path such as produced by adding ionophores to the membrane or produced by connecting a parallel resistor across the input of the amplifier.

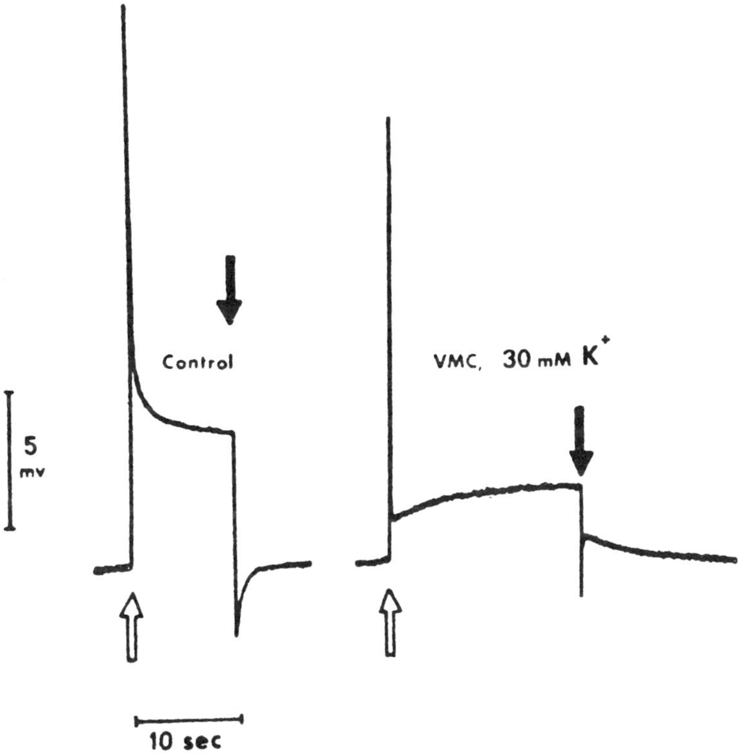

Plant Science Letters

Figure 3. Record F. Open-circuit photovoltages recorded intracellularly from an intact chloroplast of Peperomia metallica*. Open and closed arrows mark the "on" and the "off," respectively, of the rectangular pulses of white light. Left record is the control in the absence of valinomycin. Right record shows the effect of 1μM valinomycin and 30mM external potassium ions. The shunting effect of valinomycin shortens the relaxation time constants and reduces the amplitude of the steady-state photovoltage in accordance with the equivalent circuit shown in Figure 2 (67).*

In a steady-state galvanostatic measurement, the effect of shunting is a reduction in the amplitude of the photoelectric signal. Thus, a very "leaky" membrane presents a difficult condition for measuring a photovoltage. In contrast, the steady-state dc photoelectric current, as measured by a (quasi-) potentiostatic method, remains essentially constant when the ionic resistance (R_m) is decreased by several orders of magnitude (*18, 41*). In fact, this method can be used to differentiate between the primary photoelectric and the secondary ionic events. The effect of shunting on transient photoelectric signals is more complex, since both the amplitude and the relaxation time course of the photosignals may be affected. We shall consider a type of measurement named the tunable voltage clamp method, first introduced by Hong and Mauzerall (*17, 56*), in which a finite and nonzero access impedance is included in the data analysis. The galvanostatic and potentiostatic measurements can be regarded then as two extreme special cases with the access impedance approaching infinity and zero, respectively. In a photosystem in which all of the *RC* elements are constant (voltage independent and time invariant), it is possible to obtain analytical results explicitly. Here we shall summarize the results with respect to two types of light stimulation, namely, brief pulse stimulation, (in which the duration of the light pulse is much shorter than the intrinsic relaxation time of the membrane photosystem), and the long rectangular pulse stimulation (in which the duration of the light pulse is much longer than the intrinsic relaxation time of the photosystem) (*see* Figure 4). The analytical details can be found in Ref. *18*. Two important parameters deserve attention: a membrane discharging time constant (τ_m) and an intrinsic photoelectric relaxation time constant (τ_p);

$$\frac{1}{\tau_m} = \frac{1}{R_e C_m} + \frac{1}{R_m C_m} \tag{1}$$

and

$$\frac{1}{\tau_p} = \frac{1}{R_p C_p} + \frac{1}{R_s C_p} \tag{2}$$

Except for the case of an idealized potentiostatic measurement, the measured responses are always the results of interactions between τ_p and τ_m and other *RC* time constants, resulting in two apparent relaxation time constants, τ_s (the short apparent relaxation time constant) and τ_l (the long apparent relaxation time constant), which are dependent on the value of the access impedance in a way that is predictable by the equivalent circuit. Such a relationship has been confirmed quantitatively in magnesium porphyrin-containing lipid bilayers (*see* Figure 3A). In the limit of infinite access impedance (galvanostatic), τ_l approaches the

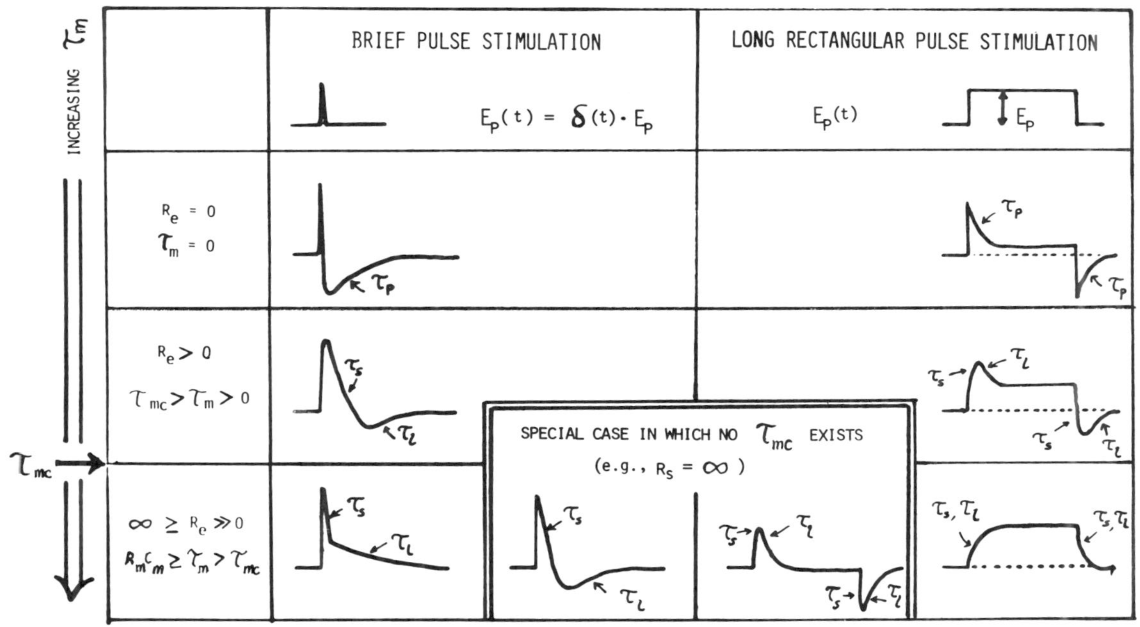

Figure 4. The dependence of time course of photoelectric responses on the membrane discharging time constant. Two types of stimulations are considered: brief pulse stimulation and long rectangular pulse stimulation: τ_p is the intrinsic relaxation time constant: τ_m is the membrane discharging time constant; τ_s and τ_l are the short and long apparent relaxation time constants, respectively; and τ_{mc} is a critical value of τ_m. See text for further explanation.

membrane time constant, R_mC_m, while τ_s is aproximately twice the value of τ_p (*18*). (Of course, in a galvanostatic measurement, the current through the measuring device is effectively zero and the voltage, which is proportional to the small residual current, is measured instead.) Frequently τ_s is masked completely because of inadequate time resolution (τ_p is usually in the microsecond range). In this case, the galvanostatic measurement could be devoid of relevant kinetic information, since it would measure nothing but the membrane *RC* relaxation. Failure to recognize this fact can lead to an erroneous interpretation of experimental data. In view of this distorting effect of a nonzero access impedance, one might prefer to consider an ideal potentiostatic measurement instead. However, a zero access impedance is very difficult, if not impossible, to achieve. Hong and Mauzerall (*18, 56*) have shown that an access impedance of 380 Ω is still far from effectively zero at 1 MHz, mainly because of the presence of C_p, which bestows the membrane system with a low source impedance (of the order of 10 kΩ at 1 MHz). Furthermore, a nonzero but small access impedance can be a blessing rather than a nuisance, since additional information that is inaccessible in an ideal potentiostatic measurement can be obtained. It has been suggested that the optimal measurement is that with the access impedance tuned to match the source impedance (hence the name tunable voltage clamp method).

Caution must be exercised in using an instrument to measure a (quasi-) short-circuit current. A crucial parameter is the input impedance of the instrument in addition to the electrode impedance. Since the source impedance of the membrane can be as low as 10 kΩ at 1 MHz, it is not good enough to have an input impedance of 10^7 Ω (that of a commercial picoammeter). It has been reported that the photocurrent response is the first time derivative of the photovoltage response (*19, 20*). However, this type of relationship possibly indicates that the measurement was not made under short-circuit conditions. It is true that current flow through a discharging capacitor is proportional to the first time derivative of the voltage across it. It is equally true that the discharging of a capacitor is accelerated when it is connected to a low-input impedance device (in effect, reducing the discharging time constant). Therefore, the first time derivative relationship holds only when both the current and the voltage are measured under conditions with the same access impedance (relative to the source impedance) and should break down otherwise. Without this precaution in mind, a current measurement could be misleading.

The diagram in Figure 4 indicates that as τ_m is changing from an idealized zero to its maximum value R_mC_m, the photoelectric relaxation time course also changes continuously, with increasing distortion owing to interaction with the other *RC* parameters, which are unrelated to the photoelectric event. In general, a critical value (τ_{mc}) of the discharging time constant exists. If $\tau_m < \tau_{mc}$, the brief pulse response is characterized by a biphasic waveform (a positive spike followed by an undershoot below the dark baseline), while the long rectangular pulse response is characterized by an "on-overshoot" and an "off-undershoot." In addition, a dc component also may be present. The net amount of charges transferred across the membrane is represented by the time integral of the current response. In a measurement with a short τ_m, the ac component is much enhanced and the time integral of the current response becomes almost zero (the amplitude of the dc component is negligible) (*see* Figure 3A). As τ_m becomes longer, the ac component decreases in amplitude progressively, owing to increasing distortion through interaction of various *RC* constants. Eventually, when τ_m exceeds the critical value τ_{mc}, the brief pulse response relaxes monotonically to the baseline without an undershoot while the long pulse response rises and decays monotonically. We shall refer to the "undershoot" in brief pulse responses and the "on-overshoot" and "off-undershoot" in long pulse responses as the characteristic high-pass-filter response, since this waveform is characteristic of a high-pass *RC* filter. Thus, when τ_m is reduced by the shunting effect of ionophores or a parallel shunt resistor, or by a decrease in access impedance, a characteristic high-pass-filter response might appear. In Figure 3C, using a picoammeter (input impedance $10^7\ \Omega$) in place of an electrometer (input impedance $10^{16}\ \Omega$) causes a sufficient reduction of the access impedance (and hence the discharging time constant) to reveal the characteristic high-pass-filter response in a lipid bilayer which contains retinal (*57*). Other examples are given by a cyanine-dye-containing lipid bilayer (*58*), which is shunted by a resistor of $10^7\ \Omega$ (*see* Figure 3B), and by a bacteriorhodopsin-containing model membrane (*59, 60, 61*), which is shunted by an ionophore CCCP (2,4,6-trichlorocarbonylcyanidephenylhydrazone) (*see* Figure 3E). Results similar to Figure 3E are found also in model membranes, which contain subchromatophore pigment–protein complexes from *Rhodospirillum rubrum* (*62, 63, 64*). Usually the appearance of a characteristic high-pass-filter response can be construed as the presence of a series capacitance connected to the photoemf, namely, a chemical capacitance. However the data presented in Refs. *62, 63,* and *64* require a special consideration. The membranes were made by fusions of pigment-containing lipid vesicles to a plain planar membrane. The effect of ionophores and a shunt resistor has been interpreted as a consequence of a peculiar structure in which bacteriorhodopsin–lipid vesicles remain intact but attached to a planar membrane (*see* Figure 12 in Ref. *63*). Thus there are three membrane capacitances: one of the planar membrane, one of the vesicle, and one of the fused region. The interaction among these three capacitances qualitatively explains the observed effect. In other words, the capacitance at the fused region fulfills the requirement of a series capacitance. However, this interpretation has an inherent difficulty. Here, it must be pointed out that the characteristic high-pass-filter response appears also in membranes made by a conventional method in which no fusion with vesicles took place (*59, 65*). The simplest interpretation is that the characteristic high-pass-filter response is a manifestation of an ac photoelectric effect (*28*) rather than an artifact specific to the particular method of making pigment-containing membranes.

There are, however, special cases in which a critical discharging time constant (τ_{mc}) does not exist (e.g., when R_s = infinity). In these cases, the characteristic high-pass-filter response persists even under open-circuit conditions (*see* e.g. Figures 3D and 3F cited here and Figure 4 in Ref. *66*). Here, the data obtained from the chloroplast of *Peperomia metallica* (*67*) shown in Figure 3F suggest the existence of an ac photoelectric signal.

It is apparent that the equivalent circuit shown in Figure 2 offers a simple and unified interpretation of ac photoresponses from different systems. For example, the difference of photoresponses shown in Figures 3C and 3D merely reflects the difference in relative magnitude of various *RC* parameters in the equivalent circuit (*18*); there is no need to postulate different models to explain the results (*57*).

Molecular Basis of a Chemical Capacitance

We have emphasized that the ac photoelectric response is a manifestation of a chemical capacitance, which is physically distinct from the membrane capacitance; the former is connected in series to the photoemf while the latter is connected in parallel. We shall briefly present molecular pictures to demonstrate that this is indeed the case. We shall summarize only the basic feature of two molecular models which are results of a calculation based on the Gouy–Chapman double-layer theory. The interested readers are referred to Refs. *18* and *37* for the details.

Interfacial Charge Transfer (ICT) Model. The basic feature of this model is that the membrane-bound pigment can reversibly exchange a charge (electron or proton) with the aqueous charge donors or acceptors upon light illumination. Since there are two interfaces and electric signals generated there tend to cancel each other, no external detection can be made unless an asymmetry exists. This asymmetry may be in the two aqueous phases (e.g., a redox or pH gradient) or in the membrane (e.g., an asymmetric and oriented distribution of photopigment in the membrane). We shall consider a simple case in which interfacial charge-transfer reactions occur at a single membrane–water interface (*see* the left interface in Figure 5). The interfacial charge-transfer reactions are usually faster than the transmembrane charge transport (dc effect) and thus the latter can be ignored during the brief transient of an ac photoresponse. The light-induced interfacial charge transfer and its subsequent relaxation thus causes formation and decay of a sheet of surface charge at the interface. A simple calculation (*18*) shows that the ionic atmosphere relaxation time is less than 1 nsec in an isotonic electrolyte solution, while it can be as long as 1 sec in the membrane phase. We assume that the interfacial charge recombination occurs on a time scale of 1 μsec to 1 msec. Thus, the ionic distribution in the aqueous phase is always in quasiequilibrium during the transient, while there is practically no charge screening in the membrane phase. This feature makes it possible to carry out the kinetic and the electrostatic calculations independently and separately. Also a constant field condition can be imposed in the membrane phase because of insignificant space charge build-up in the membrane phase. As a result of such a calculation, the potential profile and the charge-density profile across the membrane and the aqueous phases under short-circuit conditions are shown in Figure 5 (*see* the left half of the figure), which must be regarded as a snapshot during the brief transient of the ac photoresponse. As expected the charged pigment molecules are concentrated near the reacting interface. Because of the extremely small thickness (< 70 Å) of a membrane, the polarizing effect of the surface charge extends beyond the membrane phase and reaches the opposite aqueous phase. Thus, the positive surface charge is balanced by the excess negative charges at both diffuse double layers under short-circuit conditions. Linearization of the results permits a re-interpretation of the profiles in terms of three lumped capacitances: a geometric capacitance, C_g, owing to the membrane dielectric and two double-layer capacitances, C_d. Notice that the double-layer capacitance at the reacting interface has an opposite polarity to the remaining two capacitances. The only sensible way to connect a photoemf in order to generate such a charge distribution is as shown in Figure 5 (*see* the left half of the figure). Macroscopically, this ac photoresponse is characterized by a high-pass-filter circuit with a series capacitance, C_p. Here C_p is the result of a combination of three more fundamental capacitances: the geometric capacitance and the opposite double-layer capacitance combine in series; the result of such a combination then combines in parallel with the adjacent double layer capacitance. C_p is thus physically distinct from the membrane capacitance, which is a series combination of the same three fundamental capacitances.

Oriented Dipole Model. The basic feature of this model is that the membrane-bound pigment is oriented and distributed asymmetrically across the membrane, and that a light-induced conformational change

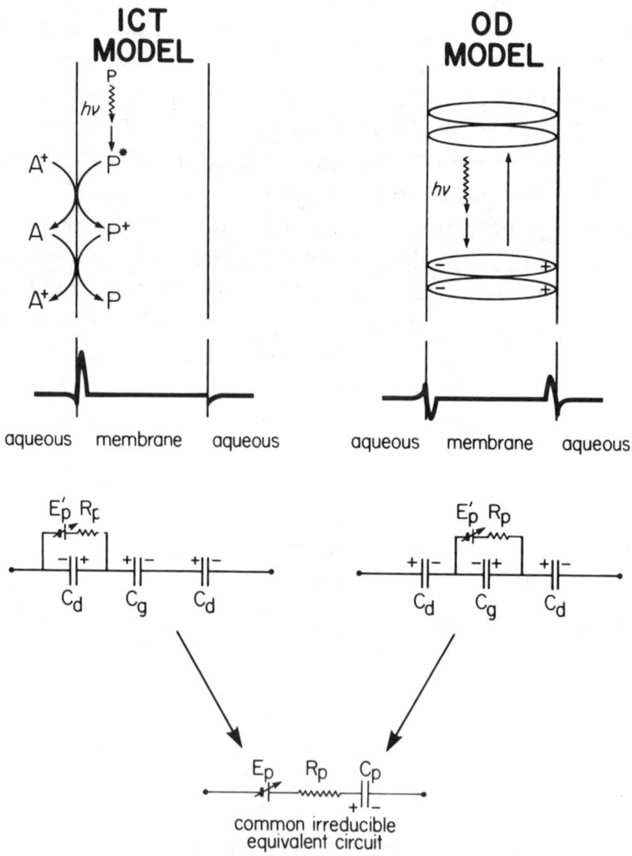

Bioelectrochemistry and Bioenergetics

Figure 5. Diagrams explaining the difference of an ICT model and an OD model. The upper half of the diagram shows the two models and their charge distribution profiles during an ac photoelectric transient. In the ICT model, P *is the ground-state photopigment, and* P* *is one of the intermediates.* P^+ *is a positively charged intermediate, a product of an interfacial charge-transfer reaction:* $P^* + A^+ \rightleftarrows P^+ + A$ *where* A^+ *and* A *are positive charge donor and acceptor, respectively. The OD model is self-explanatory. However, the diagram does not imply the actual shape of the pigment molecule and the location of the pigment in the membrane. Also the extent of charge separation is tentative and arbitrary. In both diagrams, photoreactions are indicated by wavy arrows, while dark thermal reactions are indicated by smooth arrows. The lower half of the diagram shows the respective equivalent circuit of the two models as a result of interpreting the Gouy–Chapman calculation. Notice that the equivalent circuits of the two models can be reduced to the same irreducible equivalent circuit, similar to that shown in Figure 2. The* C_p *is a combination of three fundamental capacitances: a geometric capacitance,* C_g*, and two double-layer capacitances,* C_d*.* See *text for further explanation (37).*

and the accompanying redistribution of internal charges cause formation of an array of oriented electric dipoles across the membrane, which subsequently decay according to the kinetics of the pigment photoreactions. In this model, the pair of separated charges remains correlated with each other as parts of the same molecule. In contrast, the separated charges in the ICT model become decorrelated rapidly with each other as the ionic cloud in the aqueous phase relaxes. We need only to consider a prototype of this model in which we assume that the pigment molecules span the membrane and that the charge movement extends across the entire thickness of the membrane. The possibility that the pigment might not span the membrane or the charge movement might cover only a fraction of the membrane thickness does not affect the general conclusion. Besides, the analysis could be modified easily to take this possibility into account. Using a similar technique as in the ICT model, such a calculation leads to a charge distribution and an equivalent circuit as shown in Figure 5 (*see* the right half of the figure). In the oriented dipole (OD) model, the photoemf is in series with the two double-layer capacitances, but is in parallel with the geometric capacitance, as far as the external measurement is concerned. This model also leads to the same irreducible equivalent circuit as the ICT model does; that is, both models contain a series capacitance C_p with different physical constructions. In any event, neither series capacitance is identical with the ordinary membrane capacitance (C_m). The displacement photoelectric response known as the ERP is not a manifestation of the membrane capacitance, as is commonly believed.

Mechanism for the Photoelectric Effect in Visual Membranes and in Purple Membranes: OD vs. ICT Model

Correlation of the ERP components with the rhodopsin photochemistry has been attempted since the discovery of the ERP. Most data pertain to the *R*2 component of the ERP, which may be correlated with the Metarhodopsin I-to-Metarhodopsin II reaction (*4, 68*). The major conformational change of rhodopsin occurs at the stage of Metarhodopsin I-to-Metarhodopsin II transition (*34, 36, 69*). Such a polypeptide macromolecule as rhodopsin might carry multiple charges (with one net negative charge (*70*)). Since rhodopsin molecules are oriented in the visual membranes (e.g., *see* Ref. *71*), it is reasonable to expect a light-induced transient change of electric dipole moments during this major conformational change from Metarhodopsin I to Metarhodopsin II. However, it has been established equally well that a net (presumably interfacial) uptake of one proton takes place in the same reaction (*72, 73*). It is therefore equally plausible for an interfacial proton transfer process to generate an ac photoelectric response. As these two mechanisms are not mutually exclusive, it will not be surprising if both mechanisms eventually turn out to be valid. It is also possible that one mechanism generates the *R*1 component, while the other mechanism generates the *R*2 component. Hong (*37*) has critically reviewed the literature of the ERP and found no solid experimental evidence that could rule out either mechanism. The major evidence that apparently contradicts an ICT mechanism is the lack of a dramatic pT effect on the *R*2 component (*34, 35*). The following consideration might shed some light on this problem (*37*). It is the rise time and the amplitude of the *R*2 component that were being correlated with the Metarhodopsin I-to-Metarhodopsin II transition (*see* Figure 11, Ref. *35*). The rate of formation of Metarhodopsin II depends on the availability of Metarhodopsin I and/or the availability of the aqueous protons. Therefore, it is possible that, over a considerable pH range of the medium, the rate of formation of Metarhodopsin II depends solely or mainly on the availability of Metarhodopsin I, which, in turn, depends on the availability of photons and factors other than pH of the medium. If one is willing to consider this possibility, one would expect the decay of the *R*2 component to be sensitive to the pH of the medium. The fact that this has not been observed might simply be a trivial consequence of the method used in the measurement since open-circuit measurements often leave out the intrinsic decay kinetics of an ac photosignal. The questions as to which mechanism is operative in the generation of the *R*2 as well as *R*1 component remains open.

In bacteriorhodopsin membranes, the direct demonstration of the presence of an ac photoelectric effect was made by Trissl and Montal (*21*). These authors deposited a layer of phospholipid and oriented bacteriorhodopsin onto a thin (6 μm) Teflon membrane. This method ensures that only capacitative photoresponses will be detected since the dc resistance of the membrane (that of the Teflon film) is as high as $10^{14}\ \Omega$. Using an open-circuit method and a millisecond flash of light, Trissl and Montal (*21*) found a photoelectric signal that rises in about 15 msec and decays in about 1 sec. If a 300-nsec dye laser pulse was used in an open-circuit measurement, the same time course was reproduced except that a small positive signal with unresolved latency was barely discernible (*40*). Based on additional data obtained by the tunable voltage clamp method, Hong and Montal (*40*) identified this latter one as a new (*B*1) component of the displacement photocurrent, while they named the former as a *B*2 component (*see* Figure 6).

In the open-circuit measurements of Trissl and Montal (*21*), the photosignal (*B*2) was not affected significantly by changing the pH or the ionic strength of the medium. Were it not for the arguments presented above, an interfacial charge-transfer mechanism could be elimi-

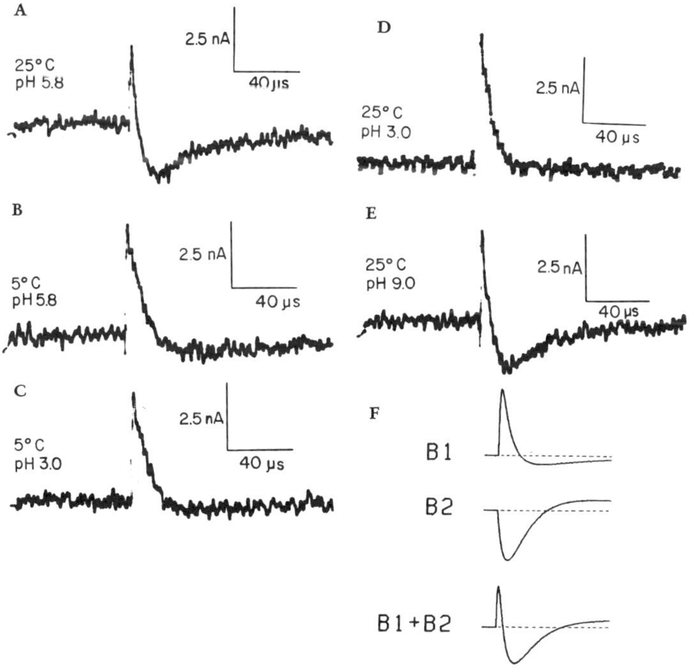

Biophysical Journal

Figure 6. Photoelectric responses from a bacteriorhodopsin model membrane as measured by the tunable voltage clamp method. The membrane *was made according to the method of Trissl and Montal* (21)*: an interfacial layer of purple membrane fragments was formed in one aqueous plhase* (2M *NaCl with imidazole buffer), which was apposed to a thin Teflon film* (6 *μm thick, an area of 0.16 cm²) by slowly raising the water level. The other aqueous phase contains only electrolyte solution. Records* A–E *show the temperature and pH effects on the photocurrent responses, which were taken from the same membrane in the sequence of presentation. The access impedance was 1 kΩ. The instrumental time constant was 15 μsec. The light source was a dye laser pulse with a duration of 300 nsec and an output at 590 nm. Both the temperature and the pH effects are reversible. Record* F *shows a schematic decomposition of a typical photocurrent response at room temperature and neutral pH as the sum of a* B1 *and* B2 *component. Each component has a time course similar to that shown in Figure 3A and was computed according to the equivalent circuit shown in Figure 2, with parameters chosen to illustrate the possibility of an unresolved second positive phase which may or may not be present in the sum, depending on the relationship of the time courses of the* B1 *and* B2 *components.* See *text for further explanation* (40).

nated (*21*). In fact, a remarkable pH effect on the B2 component is evident in a quasi-short-circuit (tunable voltage clamp) measurement with pulsed-laser stimulation (*see* Figures 6A–6E): the *B*2 peak decreases at low pH (pH 3) or low temperature (5°C), while the *B*1 peak actually increases at low pH or low temperature. However, the enhancement of the positive peak is an apparent effect resulting from the overlapping of *B*1 decay and *B*2 rise, as we can see in the decomposition of a typical signal (*see* Figure 6F).

Notice that the photosignals are not decomposed into two components with a single relaxation time constant, but rather into two components, each of which has a biphasic time course. This feature is required by the condition that the time–integral of an ac photocurrent eventually must go to zero. The diagram actually was computed in terms of the model presented in Figure 2 with parameters chosen to indicate the possible presence of a second slow positive phase of the photocurrent (curve fitting has not been attempted because of insufficient signal-to-noise ratio in the data). Thus, low temperature and/or low pH inhibit the *B*2 component but do not affect significantly the *B*1 component. The close resemblance of the *B*1 and *B*2 components to the *R*1 and *R*2 components of the ERP, respectively, in temperature effect is apparent. It is suggested that a tunable voltage clamp measurement of the ERP might reveal a pH dependence of the *R*2 component, which is inconspicuous under open-circuit conditions.

Using a 15-nsec laser pulse, Drachev et al. (*74*) recently have reported observation of several fast components of photoelectric signals in a bacteriorhodopsin model membrane under open-circuit conditions. Although these authors did not identify them as capacitative signals, we shall make a tentative identification with the ac photosignals reported by Hong and Montal (*40*). A first phase with a submicrosecond rise time and with a polarity opposite to the dc component can be identified with the *B*1 component. A second phase which rises in 20–50 μsec with a polarity identical to the dc component can be identified with the *B*2 component. A third phase of the same polarity as the second phase with a rise time of 6–12 msec has no counterpart in the data of Hong and Montal. As suggested by Drachev et al. (*74*), the second and third phases are generated by interfacial proton transfers at different sides of the membrane. The absence of a third component in the data of Hong and Montal (*40*) might be caused by the effect of the Teflon film which prevents the bacteriorhodopsin molecules from making contact with another aqueous phase. Drachev et al. (*74*) also observed a fourth phase with a relaxation time of 1 sec, which they correctly identified as the discharge of the photovoltage (cf: *see* Figure 4). These authors found that the first phase increases in amplitude as the pH decreases below 2, while the second and third phases are abolished. They also found that subsitution of D_2O for H_2O causes a marked increase of the amplitude of the first phase and a deceleration of the rise of the second and third phases. These results are consistent with the above identification. However, the increase in the amplitude of the first phase might be an apparent effect secondary to the suppression of the second and third phases. It is suggested that a decrease of temperature also might cause an apparent increase of the amplitude of the first phase. A tentative correlation of the photoelectric signals with the photochemistry of bacteriorhodopsin may be considered. The *B*1 component and the first phase of Drachev et al. may be correlated with the formation and decay of K_{590} (*40, 74*); the *B*2 component and the second phase may be correlated with the formation and decay of M_{412} (*40, 74*); and the third phase may be correlated with the transition $M_{412} - BR_{570}$ (*74*). Applebury et al. (*75*) have found that the formation of the *K* component has a rise time of 11 psec and a striking temperature independence that is consistent with a quantum mechanical tunneling mechanism. This explains the lack of a latency and the temperature independence of the *B*1 component, if the above correlation is correct.

The striking pH effect presented in Figure 6 strongly suggests that the *B*2 component might be generated by an interfacial proton transfer process. This view is in agreement with that of Drachev et al. (*74*) if the aforementioned identification of components is correct. However, further detailed kinetic analysis is required to establish this mechanism. On the other hand, the *B*1 component and the first phase of Drachev et al. (*74*) probably are generated by an OD mechanism. This could, of course, be accomplished by an internal charge separation accompaning a conformational change of bacteriorhodopsin upon illumination. However, an OD mechanism could be accomplished possibly without the involvement of a conformational change of the apoprotein moiety of bacteriorhodopsin. For example, an internal charge separation resulting from electron movement in the conjugate double-bond system of the chromophore, or intramolecular proton transfers without a major conformational change (the slight isotope effect of the $BR_{570} - K_{590}$ transition (*74*) supports this latter mechanism), could generate the *B*1 component. This latter view is shared by V. P. Skulachev (*76*), and is reinforced further by an observation of Kobamoto and Tien (*77, 78*). These authors have studied flash-light-induced photoelectric responses from a lipid bilayer which contains all-*trans*-retinal and separates a pH gradient under open-circuit conditions. Although the system differs somewhat from a visual or purple membrane (e.g., the asymmetry is in the two aqueous phases instead of the membrane itself), photoelectric signals resembling the *R*1 and *R*2 components of the ERP in temperature and pH effects were observed. The above discussion also is applicable to the generating mechanism of the ERP.

Conclusions and Perspectives

In this survey, primary photoelectric events in pigment-containing biomembranes are distinguished from secondary ionic events. The photoelectric events are divided further into an ac and a dc event based on a generalized equivalent circuit scheme. The ac photoelectric response is a manifestation of a chemical capacitance, which is physically distinct from the membrane capacitance. As predicted by the equivalent circuit, the time course of an ac photoelectric response is influenced critically by the discharging time constant of the membrane, which is related to the access impedance, and the membrane resistance and capacitance. The variation of time courses of a photoresponse under various experimental conditions could be accounted for in a unified manner, at least qualitatively, by changes of parameters in the equivalent circuit and the measuring device. Since a galvanostatic (open-circuit) measurement often preserves little kinetic information; all previous conclusions based on kinetic analyses of open-circuit data could be premature and must be re-evaluated. It is suggested that the tunable voltage clamp method might be used for this purpose.

Two major physiological functions of natural pigment-containing membranes are photosynthesis and vision. The photosynthetic membrane converts light energy into chemical energy for use in metabolism. In order for a light-induced charge separation to be useful as a photon-energy-conversion process, the separated charges must lead to the formation of an ionic gradient across the membrane. It is not good enough for charges to be separated across a membrane–water interface, and only allowed to recombine subsequently. In this regard, only the dc photoelectric effect contributes to light energy conversion, while the ac photoelectric effect could be regarded as a waste of energy.

In a visual membrane, photons serve to trigger the release of energy stored in advance across the membrane through active transport (*79*). No photon energy conversion is required. It is, therefore, a legitimate question to ask whether the ERP might serve any physiological function in visual transduction processes. However, the general conclusion is that the ERP is an epiphenomenon, a mere byproduct of visual transduction processes. This conclusion was reached mainly because of the small size of the ERP (*33*) and because of the observation that an externally applied voltage does not generate the same exciting effect as photons do (*80*), Hong (*18, 28, 37*) has argued that these reasons are not sufficient to rule out the possibility of an ERP-dependent gating mechanism; namely, the local electric field owing to the ERP might be intense enough to control and modulate permeability to specific ions such as Ca^{2+}. On the other hand, the discovery of a light-induced ionic channel formation in reconstituted rhodopsin membranes (*52*) tends to favor the possibility that the ERP is an epiphenomenon, a gating current that is the electrical manifestation of the conformational change of the visual pigment accompanying the gating action. The physiological role of the ERP in the visual transduction processes remains to be elucidated.

The purple membrane system is unique in the sense that the photopigment resembles the visual pigment rhodopsin, while the membrane performs a function similar to photosynthetic membranes. Because of the close resemblance of the ac photoelectric responses from a bacteriorhodopsin model membrane to the ERP, and because of the relative ease in the handling of bacteriorhodopsin molecules, the ac photoelectric responses from a bacteriorhodopsin membrane can be studied as a model of the ERP. Furthermore, the study of the photoelectric relaxation in a bacteriorhodopsin model membrane can provide direct electrokinetic data of charge movements. These types of information, when correlated with spectroscopic data, could be useful in elucidating the underlying molecular mechanism of light-induced proton translocation across the membrane. This latter knowledge could contribute toward the design of an efficient and practical solar cell in meeting the ever-increasing demand of energy consumption in the face of dwindling resources of fossil fuel.

Acknowledgments

The author wishes to thank B. Mobley, J. A. Sedensky, and W. Wojtkowski for critical reading of the manuscript. This research has been supported by Grant GM-25144 from the National Institutes of Health.

Glossary of Symbols

R_m = membrane (ionic) resistance
C_g = geometric capacitance
C_d = double-layer capacitance
C_m = membrane capacitance
R_s = transmembrane (photoelectric) resistance
E_p = photoemf
C_p = chemical capacitance
R_p = internal (photoelectric) resistance
R_e = access resistance
τ_p = photoelectric relaxation time constant
τ_m = discharging time constant
τ_s = short apparent relaxation time constant
τ_l = long apparent relaxation time constant
τ_{mc} = critical discharging time constant
ERP = early receptor potential
OD = oriented dipole
ICT = interfacial charge transfer

ATP = adenosine triphosphate
BLM = bimolecular lipid membrane, lipid bilayer
FVP = fast photovoltage
PEP = photoelectric potential

Literature Cited

1. Brown, K. T.; Murakami, M. "A New Receptor Potential of the Monkey Retina with no Detectable Latency," *Nature (London)* **1964,** *201,* 626–628.
2. Cone, R. A. "Early Receptor Potential of the Vertebrate Retina," *Nature (London)* **1964,** *204,* 736–739.
3. Brown, K. T.; Murakami, M. "Biphasic Form of the Early Receptor Potential of the Monkey Retina," *Nature (London)* **1964,** *204,* 739–740.
4. Cone, R. A. "Early Receptor Potential: Photoreversible Charge Displacement in Rhodopsin," *Science* **1967,** *155,* 1128–1131.
5. Pak, W. L. "Some Properties of the Early Electrical Response in the Vertebrate Retina," *Cold Spring Harbor Symp. Quant. Biol.* **1965,** *30,* 493–499.
6. Brindley, G. S.; Gardner-Medwin, A. R. "The Origin of the Early Receptor Potential of the Retina," *J. Physiol. (London)* **1966,** *182,* 185–194.
7. Pak, W. L.; Rozzi, V. P.; Ebrey, T. G. "Effect of Changes in the Chemical Environment of the Retina on the Two Components of the Early Receptor Potential," *Nature (London)* **1967,** *214,* 109–110.
8. Arden, G. B.; Bridges, C. D. B.; Ikeda, H.; Siegel, I. M. "Mode of Generation of the Early Receptor Potential," *Vision Res.* **1968,** *8,* 3–24.
9. Pak, W. L.; Cone, R. A. "Isolation and Identification of the Initial Peak of the Early Receptor Potential," *Nature (London)* **1964,** *204,* 836–838.
10. Mueller, P.; Rudin, D. O.; Tien, H. T.; Wescott, W. C. "Reconstitution of Cell Membrane Structure *in vitro* and Its Transformation into an Excitable System," *Nature (London)* **1962,** *194,* 979–980.
11. Mueller, P.; Rudin, D. O.; Tien, H. T.; Wescott, W. C. "Methods for the Formation of Single Bimolecular Lipid Membrane in Aqueous Solution," *J. Phys. Chem.* **1963,** *67,* 534–535.
12. Bangham, A. D. "Membrane Models with Phospholipids." In "Progress in Biophysics and Molecular Biology"; Butler, J. A. V., Noble, D., Eds.; Pergamon: Oxford, New York, Frankfurt, 1968; Vol. 18, pp 29–95.
13. Takagi, M.; Azuma, K.; Kishimoto, U. "A New Method for the Formation of Bilayer Membranes in Aqueous Solution," *Annu. Rep. Biol. Works, Fac. Sci., Osaka Univ.* **1965,** *13,* 107–110.
14. Montal, M.; Mueller, P. "Formation of Bimolecular Membranes from Lipid Monolayers and a Study of Their Electrical Properties," *Proc. Natl. Acad. Sci. USA* **1972,** *69,* 3561–3566.
15. Tien, H. T. "Light-induced Phenomena in Black Lipid Membranes Constituted from Photosynthetic Pigments," *Nature (London)* **1968,** *219,* 272–274.
16. Tien, H. T. "Photoelectric Effects in Thin and Bilayer Lipid Membranes in Aqueous Media," *J. Phys. Chem.* **1968,** *72,* 4512–4519.
17. Hong, F. T.; Mauzerall, D. "Interfacial Photoreactions and Chemical Capacitance in Lipid Bilayers," *Proc. Natl. Acad. Sci. USA* **1974,** *71,* 1564–1568.
18. Hong, F. T. "Charge Transfer Across Pigmented Bilayer Lipid Membrane and Its Interfaces," *Photochem. Photobiol.* **1976,** *24,* 255–189.
19. Trissl, H.-W.; Darszon, A.; Montal, M. "Rhodopsin in Model Membranes: Charge Displacements in Interfacial Layers," *Proc. Natl. Acad. Sci. USA* **1977,** *74,* 207–210.

20. Trissl, H.-W. "Light-Induced Conformational Changes in Cattle Rhodopsin as Probed by Measurements of the Interfacial Potential," *Photochem. Photobiol.* **1979**, *29*, 579–588.
21. Trissl, H.-W.; Montal, M. "Electrical Demonstration of Rapid Light-Induced Conformational Changes in Bacteriorhodopsin," *Nature (London)* **1977**, *266*, 655–657.
22. Tien, H. T. "Electronic Processes and Photosensitization in Bilayer Lipid Membranes," *Photochem. Photobiol.* **1972**, *16*, 271–290.
23. Tien, H. T. "Bilayer Lipid Membranes (BLM): Theory and Practice"; Marcel Dekker: New York, 1974.
24. Tien, H. T.; Chen, V. K.-H. "Interfacial Photochemistry of Lipid Bilayer Membranes," In "Progress in Surface and Membrane Science"; Cadenhead, D. A., Danielli, J. F., Rosenberg, M. D., Eds.; Academic: New York, 1974; Vol. 8, pp 119–160.
25. Tien, H. T. "Electronic Processes and Photoelectric Aspects of Bilayer Lipid Membranes," *Photochem. Photobiol.* **1976**, *24*, 97–116.
26. Berns, D. S. "Photosensitive Bilayer Membranes as Model Systems for Photobiological Processes," *Photochem. Photobiol.* **1976**, *24*, 117–139.
27. Tien, H. T. "Photoelectric Bilayer Lipid Membrane: A Model for the Thylakoid Membrane," *Brookhaven Symp. Biol.* **1976**, *28*, 105–131.
28. Hong, F. T. "Photoelectric and Magneto-Orientation Effects in Pigmented Biological Membranes," *J. Colloid Interface Sci.* **1977**, *58*, 471–497.
29. Tien, H. T. "Light Transduction by Pigmented Bilayer Lipid Membranes," *Bioelectrochem. Bioenerg.* **1978**, *5*, 318–334.
30. Brown, K. T.; Watanabe, K.; Murakami, M. "The Early and Late Receptor Potentials of Monkey Cones and Rods," *Cold Spring Harbor Symp. Quant. Biol.* **1965**, *30*, 457–482.
31. Cone, R. A. "The Early Receptor Potential of the Vertebrate Eye," *Cold Spring Harbor Symp. Quant. Biol.* **1965**, *30*, 483–491.
32. Brown, K. T. "The Electroretinogram: Its Components and Their Origins," *Vision Res.* **1968**, *8*, 633–677.
33. Pak, W. L. "Rapid Photoresponses in the Retina and Their Relevance to Vision Research," *Photochem. Photobiol.* **1968**, *8*, 495–503.
34. Arden, G. B. "The Excitation of Photoreceptors," In "Progress in Biophysics and Molecular Biology"; Butler, J. A. V., Noble, D., Eds.; Pergamon: Oxford, New York, Frankfurt, 1969; Vol. 19, pp 373–421.
35. Cone, R. A; Pak, W. L. "The Early Receptor Potential," In "Handbook of Sensory Physiology, Vol. 1: Principles of Receptor Physiology"; Loewenstein, W. R., Ed.; Springer-Verlag: Berlin, Heidelberg, New York, 1971; pp 345–365.
36. Ostroy, S. E. "Rhodopsin and the Visual Process," *Biochim. Biophys. Acta* **1977**, *463*, 91–125.
37. Hong, F. T. "Mechanisms of Generation of the Early Receptor Potential Revisited," *Bioelectrochem. Bioenerg.* **1978**, *5*, 435–455.
38. Hagins, W. A.; McGaughy, R. E. "Membrane Origin of the Fast Photovoltage of Squid Retina," *Science* **1968**, *159*, 213–215.
39. Smith, T. G.; Brown, J. E. "A Photoelectric Potential in Invertebrate Cells," *Nature (London)* **1966**, *212*, 1217–1219.
40. Hong, F. T.; Montal, M. "Bacteriorhodopsin in Model Membranes: A New Component of Displacement Photocurrent in the Microsecond Time Scale," *Biophys. J.* **1979**, *25*, 465–472.
41. Hong, F. T.; Mauzerall, D. "The Separation of Voltage-Dependent Photoemfs and Conductants in Rudin–Mueller Membranes Containing Magnesium Porphyrins," *Biochim. Biophys. Acta* **1972**, *275*, 479–484.
42. Nagle, J. F.; Morowitz, H. J. "Molecular Mechanisms for Proton Transport in Membranes," *Proc. Natl. Acad. Sci. USA* **1978**, *75*, 298–302.

43. Kozlov, I. A.; Skulachev, V. P. "H^+-Adenosine Triphosphatase and Membrane Energy Coupling," *Biochim. Biophys. Acta* **1977**, *463*, 29–89.
44. Trebst, A. "Energy Conservation in Photosynthetic Electron Transport of Chloroplasts," *Annu. Rev. Plant Physiol.* **1974**, *25*, 423–458.
45. Anderson, J. M. "The Molecular Organization of Chloroplast Thylakoids," *Biochim. Biophys. Acta* **1975**, *416*, 191–235.
46. Jagendorf, A. T.; Uribe, E. "ATP Formation Caused by Acid–Base Transition of Spinach Chloroplasts," *Proc. Natl. Acad. Sci. USA* **1966**, *55*, 170–177.
47. Mitchell, P. "Chemiosmotic Coupling and Energy Transduction"; Glynn Research: Bodmin, U.K., 1968.
48. Ebrey, T. G. "Fast Light-Evoked Potential from Leaves," *Science* **1967**, *155*, 1556–1557.
49. Witt, H. T. "On the Analysis of Photosynthesis by Pulse Techniques in the 10^{-1} to 10^{-8} Second Range," In "Nobel Symp. V: Fast Reactions and Primary Processes in Chemical Kinetics"; Claesson, S., Ed.; (Almqvist and Wiksell: Stockholm) Interscience: New York, London, Sydney, 1967; pp 261–316.
50. Witt, H. T. "Coupling of Quanta, Electrons, Fields, Ions and Phosphorylation in the Functional Membrane of Photosynthesis, Results by Pulse Spectroscopic Methods," *Q. Rev. Biophys.* **1971**, *4*, 365–477.
51. Hagins, W. A.; McGaughy, R. E. "Molecular and Thermal Origins of Fast Photoelectric Effects in the Squid Retina," *Science* **1967**, *157*, 813–816.
52. Montal, M.; Darszon, A.; Trissl, H.-W. "Transmembrane Channel Formation in Rhodopsin-Containing Bilayer Membranes," *Nature (London)* **1977**, *267*, 221–225.
53. Stoeckenius, W. "The Purple Membrane of Salt-Loving Bacteria," *Sci. Amer.* **1976**, *234*(6), 38–46.
54. Oesterhelt, D.; Stoeckenius, W. "Functions of a New Photoreceptor Membrane," *Proc. Natl. Acad. Sci. USA* **1973**, *70*, 2853–2857.
55. Danon, A.; Stoeckenius, W. "Photophosphorylation in *Halobacterium halobium*," *Proc. Natl. Acad. Sci. USA* **1974**, *71*, 1234–1238.
56. Hong, F. T.; Mauzerall, D. "Tunable Voltage Clamp Method: Application to Photoelectric Effects in Pigmented Bilayer Lipid Membranes," *J. Electrochem. Soc.* **1976**, *123*, 1317–1324.
57. Schadt, M. "Photoresponse of Bimolecular Lipid Membranes Pigmented with Retinal and Vitamin A Acid," *Biochim. Biophys. Acta* **1973**, *323*, 351–366.
58. Huebner, J. S. "Photovoltages of Bilayer Lipid Membranes in the Presence of Cyanine Dyes," *Biochim. Biophys. Acta* **1975**, *406*, 178–186.
59. Drachev, L. A.; Kaulen, A. D.; Ostroumov, S. A.; Skulachev, V. P. "Electrogenesis by Bacteriorhodopsin Incorporated in a Planar Phospholipid Membrane, *FEBS Lett.* **1974**, *39*, 43–45.
60. Drachev, L. A.; Frolov, V. N.; Kaulen, A. D.; Liberman, E. A.; Ostroumov, S. A.; Plakunova, V. G.; Semenov, A. Yu.; Skulachev, V. P. "Reconstitution of Biological Molecular Generators of Electric Current: Bacteriorhodopsin," *J. Biol. Chem.* **1976**, *251*, 7059–7065.
61. Herrmann, T. R.; Rayfield, G. W. "The Electrical Response to Light of Bacteriorhodopsin in Planar Membranes," *Biophys. J.* **1978**, *21*, 111–125.
62. Drachev, L. A.; Kondrashin, A. A.; Samuilov, V. D.; Skulachev, V. P. "Generation of Electric Potential by Reaction Center Complexes from *Rhodospirillum rubrum*," *FEBS Lett.* **1975**, *50*, 219–222.
63. Drachev, L. A.; Frolov, V. N.; Kaulen, A. D.; Kondrashin, A. A.; Samuilov, V. D.; Semenov, A. Yu.; Skulachev, V. P. "Generation of Electric Current by Chromatophores of *Rhodospirillum rubrum* and Reconstitu-

tion of Electrogenic Function in Subchromatophore Pigment–Protein Complexes," *Biochim. Biophys. Acta* **1976**, *440*, 637–660.
64. Barsky, E. L.; Dancshazy, Z.; Drachev, L. A.; Il'ina, M. D.; Jasaitis, A. A.; Kondrashin, A. A.; Samuilov, V. D.; Skulachev, V. P. "Reconstitution of Biological Molecular Generators of Electric Current: Bacteriochlorophyll and Plant Chlorophyl Complexes, *J. Biol. Chem.* **1976**, *251*, 7066–7071.
65. Dancshazy, Z.; Karvaly, B. "Incorporation of Bacteriorhodopsin into a Bilayer Lipid Membrane; A Photoelectric–Spectroscopic Study," *FEBS Lett.* **1976**, *72*, 136–138.
66. Ullrich, H.-M.; Kuhn, H. "Photoelectric Effects in Bimolecular Lipid-Dye Membranes," *Biochim. Biophys. Acta* **1972**, *266*, 584–596.
67. Vredenberg, W. J.; Buychev, A. A. "Changes in the Electrical Potential Across the Thylakoid Membranes of Illuminated Intact Chloroplasts in the Presence of Membrane-Modifying Agents," *Plant Sci. Lett* **1976**, *7*, 101–107.
68. Gedney, C.; Ward, J.; Ostroy, S. E. "Isolation and Study of Rhodopsin and Cone Responses in the Frog Retina," *Am. J. Physiol.* **1971**, *221*, 1754–1759.
69. Montal, M.; Korenbrot, J. I. "Rhodopsin in Cell Membranes and the Process of Phototransduction," In "The Enzymes of Biological Membranes"; Martonosi, A., Ed.; Plenum: New York, 1976; Vol. 4; pp 365–405.
70. Blasie, J. K. "Net Electric Charge on Photopigment Molecules and Frog Retinal Receptor Disk Membrane Structure," *Biophys. J.* **1972**, *12*, 205–213.
71. Liebman, P. A. "*In situ* Microspectrophotometric Studies on the Pigments of Single Retinal Rods," *Biophys. J.* **1962**, *2*, 161–178.
72. Matthews, R. G.; Hubbard, R.; Brown, P. K.; Wald, G. "Tautomeric Forms of Metarhodopsin," *J. Gen. Physiol.* **1963**, *47*, 215–240.
73. Falk, G.; Fatt, P. "Rapid Hydrogen Ion Uptake of Rod Outer Segments and Rhodopsin Solutions on Illumination," *J. Physiol. (London)* **1966**, *183*, 211–224.
74. Drachev, L. A.; Kaulen, A. D.; Skulachev, V. P. "Time Resolution of the Intermediate Steps in the Bacteriorhodopsin-Linked Electrogenesis," *FEBS Lett.* **1978**, *87*, 161–167.
75. Applebury, M. L.; Peters, K. S.; Rentzepis, P. M. "Primary Intermediates in the Photochemical Cycle of Bacteriorhodopsin," *Biophys. J.* **1978**, *23*, 375–382.
76. Skulachev, V. P., personal communication, 1978.
77. Kobamoto, N.; Tien, H. T. "Light-Induced Electrical Effects in a Retinal Bilayer Lipid Membrane," *Biochim. Biophys. Acta* **1971**, *241*, 129–146.
78. Kobamoto, N.; Tien, H. T. "The Effect of Temperature on the Biphasic Photoresponses of an All-*trans*-Retinal Bimolecular Lipid Membrane," *Biochim. Biophys. Acta* **1972**, *266*, 56–66.
79. Hagins, W. A. "The Visual Process: Excitatory Mechanisms in the Primary Photoreceptor Cells, *Annu. Rev. Biophys. Bioeng.* **1972**, *1*, 131–158.
80. Toyoda, J.; Nosaki, H.; Tomita, T. "Light-Induced Resistance Changes in Single Photoreceptors of *Necturus* and *Gekko*," *Vision Res.* **1967**, *9*, 453–463.

RECEIVED October 17, 1978.

15

A Surface Chemical Model of Salt, Acid, and "Water" Taste

J. A. DESIMONE, G. L. HECK, and S. K. DESIMONE

Department of Physiology, Medical College of Virginia,
Virginia Commonwealth University, Richmond, VA 23298

We show that certain phenomena demonstrated by electrophysiological and psychophysical studies of taste reception in mammals can be given a unified surface chemical interpretation. Many investigators postulate that taste reception commences with a taste cell membrane conformational change which subsequently results in taste nerve stimulation. A role for charged phospholipids in salt and acid taste also has been suggested. We present evidence based on studies of charged phospholipid monolayers that surface pressure changes in taste cell membranes may serve as the conformational changes leading to salt, acid, and the so-called "water response." The latter is the sour–bitter taste perceived when a sodium chloride-adapted tongue is placed in contact with dilute salt solutions or water. We show that the variation of the surface pressure of phospholipid monolayers provides a basis for the water response as well as the acid and salt responses.

Unlike the auditory and visual systems, the precise nature of the gustatory stimulus–receptor relation is unknown. Broadly speaking, certain compounds, when placed in contact with the tongue, initiate a chain of events leading to stimulation of the taste nerves. The perceived response is usually complex and many investigators believe it to be a superposition of certain primary tastes. In man these usually are held to be salty, sour, sweet, and bitter although this is still a matter of considerable controversy (*1, 2, 3*). The notion of four primary tastes often is considered to imply four separate and independent receptor types sensitive to one

0-8412-0473-X/80/33-188-239$05.50/1

quality. There is evidence for and against this view of receptor segregation. Evidence for separate receptors on the basis of quality comes from inhibition and cross-adaptation studies. Evidence against comes from electrophysiological measurements on taste nerves and taste receptor cells.

A surface-active plant extract, gymnemic acid, is capable of blocking sweet taste responses without substantially affecting the other taste qualities (*4, 5*). This suggests that sweetness is somehow distinct and may be an independent taste modality. In cross-adaptation studies the taste response to a given substance such as NaCl can be attenuated or adapted in time following prolonged exposure. The result is that responsiveness to all salty-tasting substances is attenuated, but the other taste qualities are unaffected (*6*). This also suggests independent mechanisms based on quality. Electrophysiological studies on the taste nerves of mammals such as the cat, rat, and hamster, however, indicate that few taste fibers are specific for a given quality (*7, 8*). Indeed, multiple sensitivities are the rule rather than the exception. Thus specificity of the taste response in mammals is not associated apparently with quality-specific peripheral anatomical structures such as the individual taste cells comprising the taste bud or the nerves innervating these cells.

To complicate matters further various species including man are capable of responding to water as an apparent taste stimulant (*9, 10, 11, 12, 13*). Responses are detected both electrophysiologically in mammals (*12*) and by behavioral studies in many (*13*). The response to water depends to some degree on the nature of the substance in contact with the taste cells prior to water. It is therefore apparent that it is not water per se to which the cells respond but to dilution from the previous adapting state. Bartoshuk (*13*) has shown that human subjects adapted to $0.01M$ NaCl recognize the more concentrated solutions as salty and the more dilute solutions as sour–bitter. Figure 1 shows that in man water after NaCl tastes as if an acidic substance were present on the tongue. In the cat a similar U-shaped response curve is obtained when neural response is plotted as a function of the salt concentration present. The position of the minimum depends on the adapting NaCl concentrations (*see* Figure 2).

It is our purpose to show that surface chemical principles can provide a mechanism for a salt dilution response mediated by a distinctly acidic stimulus. We also show that the same principles provide mechanisms for salt and acid responses. There is reason to believe that charged phospholipids may be important surface-active groups on the apical border of taste cells (*14, 15*). We propose that taste cells can respond to changes in surface pressure occurring in the membrane in contact with the oral cavity. If this is the case, the nonmonotonic character of the surface pressure of charged phospholipid monolayers as a function of the salt con-

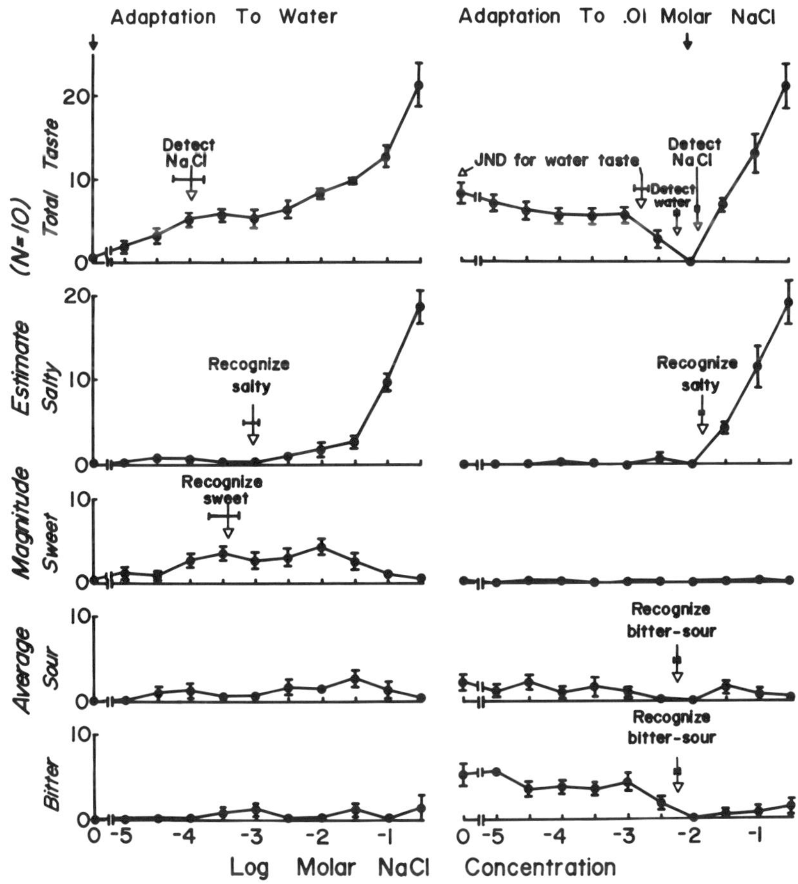

American Psychological Association

Figure 1. The taste of water and NaCl under conditions of (left) *adaptation to water and* (right) *adaptation to 0.01M NaCl. Water following NaCl is decidedly bitter–sour* (13).

centration in the subphase suggests a simple mechanism for the water response. Our experimental work utilizes two approaches: (1) equilibrium oil–water interface studies and (2) quasiequilbrium studies at the air–solution interface. In the latter studies equilibrium is achieved by forming the film on one subphase and then transferring it to another. This permits us to study the rate of change of the surface tension as the surface salt concentration changes continuously. This technique simulates the approach of a salt solution to the receptor surface as it occurs in vivo. The theoretical analysis assumes that the development of the diffuse double layer outside the phospholipid monolayer is diffusion controlled and that the theory of the surface pressure of weakly acidic surfactants developed by Payens (*16*) applies here.

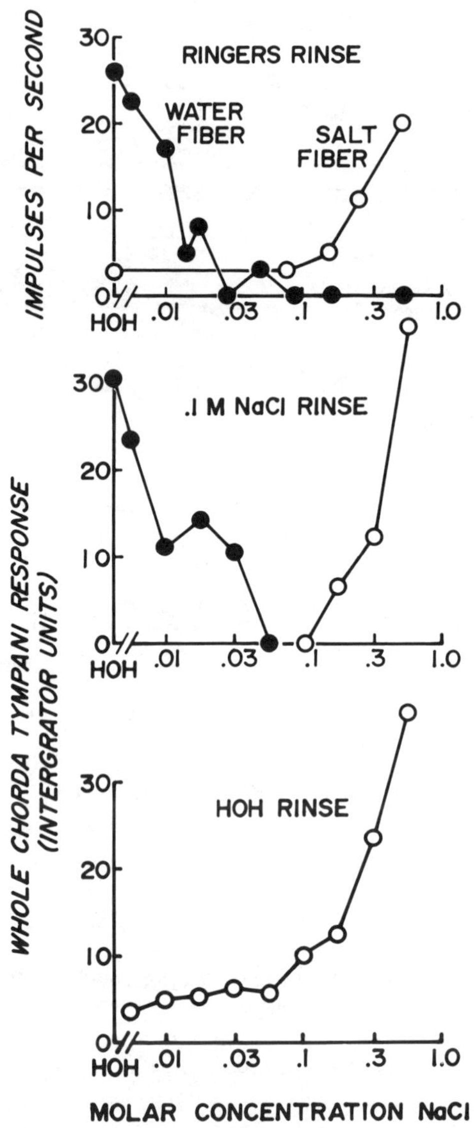

Plenum Press

Figure 2. The response of the afferent taste nerves in the cat to increased and decreased NaCl concentration relative to adapting conditions consisting of: (top) *Ringers solution;* (middle) *0.1*M *NaCl; and* (lower) *water.* (33)

The Model

We assume that the interface between the outer apical taste cell surface and the aqueous content of the oral cavity can be modeled as a charged phospholipid monolayer. While admittedly crude, there is justification for this. Kurihara et al. (*17*) have found that taste papillae are richer in phospholipid than the surrounding lingual epithelium. Beidler (*15*) has shown that the response of rat tongues to acid resembles a titration curve with a pK of about 1.8 indicating a strong acidic group, possibly a phosphate. Finally Kamo et al. (*14*) have been able to model many of the characteristics of the taste cell intracellular potential using millipore filters impregnated with phospholipid extracted from taste papillae. If we assume that, like a monolayer of phosphatidic acid, taste receptors are characterized by a fixed negative surface charge density, there will be a tendency for cations to be attracted and anions to be repelled from the surface. Thus cations should be the principal determinants of taste responses and this has been proven experimentally (*18*). Accordingly the surface charge, σ, is given by

$$\sigma = \mathrm{e} \sum_i z_i \Gamma_i \qquad (1)$$

where e is the protonic charge and z_i is the valence of the ith surface-bound ion whose surface concentration is Γ_i. Since the surface-bound ions are in equilibrium with the ions in solution, each Γ_i depends generally on the concentration of mobile ions at the surface. The affinity of surface phosphate groups is, of course, highest for hydrogen ions, but other ions also can be bound (*19*). Indeed the difference in response among the alkali cations implies some direct ion–site interaction. Since we wish to focus attention on the dilution or water response of taste receptors, we shall assume that only hydrogen ions are capable of direct surface binding. Sodium ions therefore affect the surface only through electrostatic screening.

In the special case of phosphatidic acid, Equation 1 is

$$\sigma = -\mathrm{e}\Gamma_1 - 2\mathrm{e}\Gamma_2 \qquad (2)$$

where Γ_1 and Γ_2 are the surface concentrations of the monovalent and the divalent forms, respectively, of the surface-active phosphatidic acid anions. Using the law of mass action we obtain

$$\sigma = -\mathrm{e}\Gamma \frac{K_1/\mathrm{He}^{-\phi} + 2K_1K_2/\mathrm{H}^2\mathrm{e}^{-2\phi}}{1 + K_1/\mathrm{He}^{-\phi} + K_1K_2/\mathrm{H}^2\mathrm{e}^{-2\phi}} \qquad (3)$$

where Γ is the surface concentration of phosphatidic acid in all its forms, K_1 and K_2 are the dissociation constants of the two acidic protons, H is the hydrogen ion concentration in the bulk phase, and ϕ is the normalized surface potential, $e\psi/kT$. Here ψ is the potential, k is Boltzmann's constant, and T is the temperature. As is now well established, the surface hydrogen ion concentration will differ generally from the hydrogen ion concentration in the bulk (*16, 20*). Because the ion concentration at the surface depends on the surface potential, the surface pH may vary at constant-bulk pH. This can occur with variations in the ionic strength of the medium at constant-bulk pH. Thus we envisage a class of taste receptors whose primary affinity is for hydrogen ions. It is obvious that such a surface may act as an acid receptor, but we believe it also plays a key role in water and salt responses as well.

The surface potential can be found self-consistently by obtaining an additional expression for the charge density by solving the appropriate Poisson–Boltzmann problem. For semi-infinite planar geometry and assuming only uni–univalent salts, we obtain

$$\sigma = \sqrt{\frac{N\epsilon kTc}{500\pi}} \sinh \frac{\phi}{2} \tag{4}$$

where ϵ is the dielectric constant, N is Avogadro's number, and c the molar salt concentration. Eliminating σ between Equations 3 and 4 permits us to obtain ϕ for any specific ionic strength.

Changes in the ionic strength also can be expected to change the surface pressure at constant area. Payens (*16*) has shown that the increase in surface pressure owing to the dissociation of hydrogen ions from a monolayer of weakly acidic molecules is given by

$$\Delta\Pi_e = \int_0^{\psi} \sigma d\psi \tag{5}$$

where $\Delta\Pi_e$ is the electrostatic contribution to the surface pressure. It also happens that the expression coincides with Davies' result (*21*) for strong acids and bases. We envisage, therefore, that like a charged monolayer, the apical membrane of a taste receptor suffers changes in surface potential and pressure with changes in ionic strength. Furthermore, we postulate that changes in pressure may be propagated in the plane of the membrane in accordance with Singer's proposal for a fluid mosaic structure (*22*) and may serve as the direct stimulus to the receptor.

The use of Equation 5 implies an equilbrium charging process whereby the monolayer is reversibly allowed to dissociate to its final state when placed in contact with a reservoir of fixed composition. With Equation 4, the result is

$$\Delta\Pi_e = \sqrt{\frac{N\epsilon k^3T^3c}{125\pi e^2}}\left(\cosh\frac{\phi}{2} - 1\right) \quad (6)$$

If the surface anions are weakly dissociable, $\Pi\Delta_e$ is generally a nonmonotonic function of c. The reason for this was given by Payens (*16*). At low ionic strength the high surface potential causes the surface hydrogen ion concentration to rise. This results in neutralization of the fixed anions and hence a low surface pressure. As the ionic strength is raised, the surface hydrogen ion concentration falls; the film therefore charges and the surface pressure rises. The process continues until the film approaches complete dissociation. Further increases in ionic strength now serve only to screen the surface charges and the pressure now falls again. We see here the essence of the surface chemical basis of the water response. If we postulate that the receptor responds only to decreases in surface pressure, then the nonmonotonic character of $\Delta\Pi_e$ shows it can respond to very dilute as well as very concentrated salt solutions in accord with the electrophysiological and psychophysical results presented earlier. Furthermore, it gives an explanation for the perceived sour component to the water response, i.e. dilution corresponds to a decrease in local pH.

Taste reception in vivo begins with diffusion of the tastant through the unstirred boundary layer projecting from the taste pore into the homogeneous fluid mixture in the oral cavity. We model this process experimentally by studying the diffusion of sodium chloride solutions through an unstirred region toward a spread film of phosphatidic acid while monitoring the surface tension. The analysis assumes that sodium chloride diffuses electroneutrally throughout the zone except for a region of a few Debye lengths out from the surface. Within this zone we assume that the surface equilibrates rapidly with the solution of varying composition just outside the zone. We accordingly can use the same analysis that holds for an equilibrium double layer but with the following difference: the surface concentration of NaCl, $c(t)$, varies owing to diffusion through the unstirred zone

$$c(t) = c_0\left(1 - \frac{4}{\pi}\sum_{n=0}^{\infty}\frac{(-1)^n}{2n+1}\, e^{\frac{-(2n+1)^2\pi^2\xi}{4}}\right) \quad (7)$$

where $\xi = Dt/l^2$. Here D is the diffusion coefficient of NaCl, t is the time, l equals the diffusion path length, and c_0 is the NaCl concentration in the stirred region. Using this method we can demonstrate the nonmonotonic character of the surface pressure from a single experiment. This is because the film can be brought from a subphase of water through a series of salt concentrations to a final concentration. If the final concentration is sufficiently high, the time course will pass through an extremum as we will show.

Experimental

Chemicals. Deionized, distilled water was used throughout. All chemicals were reagent grade and sodium chloride was recrystallized using the method of Betts and Pethica (*23*). L-α-Phosphatidic acid, dipalmitoyl (PA), was obtained from Sigma Chemicals. The PA was checked for purity using two-dimensional, thin-layer chromatography. Ninety-eight percent of the PA was found as a single spot. In a few cases atomic absorption spectroscopy revealed the presence of calcium in the PA. It was removed by successive washing with HCl solutions (*24*). For purposes of spreading at the oil–water interface, acid-washed PA was dissolved in 60% 2-propanol to a concentration of approximately $10^{-4}M$. An accurate determination of the concentration was obtained by phosphate analysis (*25*).

Equilibrium Studies at the Oil–Solution Interface. Surface tension measurements were made using the Wilhelmy plate method with a Cahn Electrobalance Model RG. The plate was a glass microscope cover slip, acid washed and made hydrophobic with a coating of carbon black. Experiments were done in a 10-cm diameter glass crystallizing dish. Approximately 25 mL of heptane was layered over 100 mL of aqueous solution. The interfacial tension was taken. If the interfacial tension did not agree satisfactorily with the value obtained using the drop-volume method, a fresh set-up was prepared. Phosphatidic acid was introduced into the oil–solution interface using a micrometer syringe. The drop in surface tension was recorded after allowing a few minutes for spreading and equilibration. The experiments were done at a room temperature of 22°–23°C.

Film Transfer Studies at the Air–Solution Interface. For these studies a Teflon trough with two adjacent compartments was constructed. Each compartment is 1-cm deep and has a surface area of 97.1 cm^2. Initially a Teflon brace rests on the left-hand compartment. The brace has the same dimensions as the compartment and serves to make it deeper. Solutions are placed in each of the adjacent compartments. A film of phosphatidic acid is spread from 9:1 chloroform–methanol on the surface of the left-hand compartment. With the brace held fixed, the trough is moved beneath it to the left. This transfers the film plus about 1 mm of the original subphase onto the adjacent subphase. Changes in surface tension are recorded as the new subphase equilibrates by diffusion toward the film. Studies without a film were done to test the ability of the apparatus to follow the diffusion of NaCl through a thin layer of transferred water. Between 0 and 2*M* NaCl the surface tension increases linearly with concentration, i.e. changes in surface tension are proportional to changes in concentration. A measure of the thickness of the solution tranferred was obtained by determining the extent to which the salt solution is diluted by the transfer of water. After transfer the final solution was on the average 91.5% as concentrated. NaCl concentrations were determined by flame photometry.

Results

Figure 3 shows the results of the oil–solution interface studies. Increasing the NaCl concentration in the subphase up to 0.1*M* causes film expansion. At 1*M* NaCl the film is again more condensed but the precision of the experiments is less than that at lower concentration. The film expansion is in accord with the results of Payens (*16*). The condensation at 1*M* is in qualitative agreement with theory, but quantitative agreement is poor. In Figure 4 we display the response of the surface pressure to variations in salt concentration at fixed area. Each point on a given curve represents a different experiment. (Film transfer studies are designed to overcome this problem.) In both Figures 3 and 4, the solid lines are drawn according to the relation

$$\Pi = \frac{kT}{A - A_0} + \Delta\Pi_e \tag{8}$$

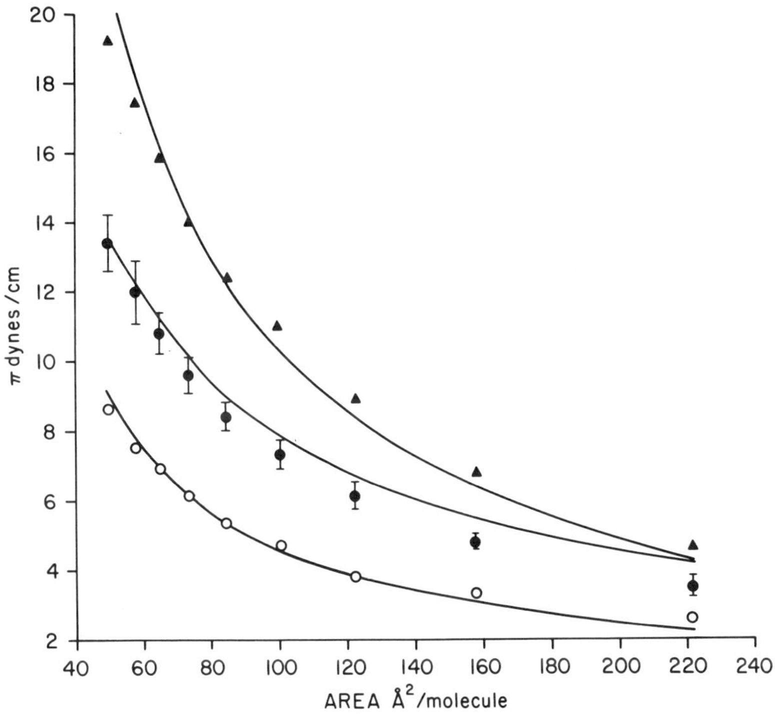

*Figure 3. The surface pressure of phosphatidic acid at the heptane–solution interface as a function of the area per molecule. The lower curve (○) is on a water subphase; the center curve (●) is on 0.001*M *NaCl and bars indicate the standard error in three experiments. The upper curve (▲) is on 0.1*M *NaCl. The solid lines are drawn according to Equation 8.*

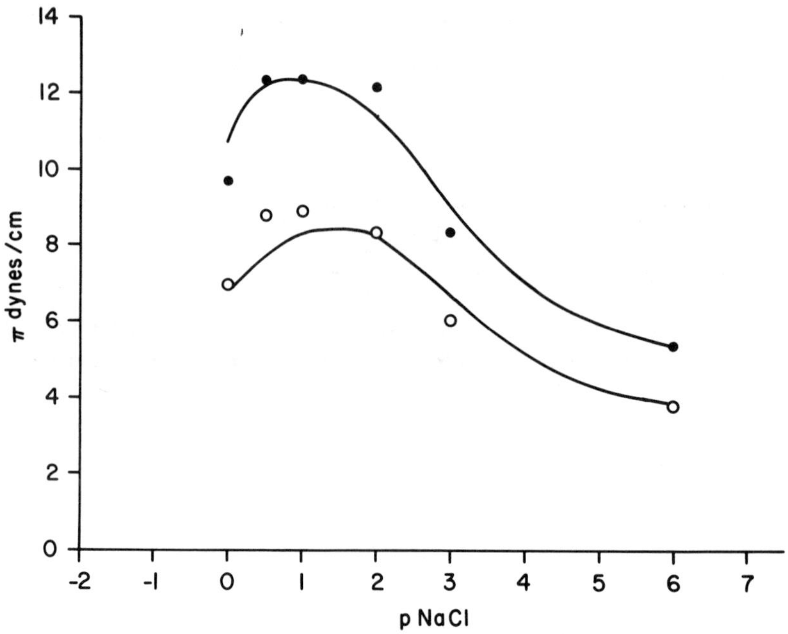

Figure 4. The variation of the surface pressure of phosphatidic acid as a function of NaCl concentration for two areas. Each point falls on a line of constant area in a series of experiments displayed as in Figure 3. The nonmonotonic character of Π is seen. The solid lines are drawn from Equation 8: (●), A = 85 Å²; (○), A = 123 Å².

where Π is the surface pressure, $(\gamma_0 - \gamma)$, γ_0 is the surface tension of the clean interface, and γ is the surface tension of the surface with film. A is the molecular area; A_0 is a constant found to be 2Å^2 and $\Delta\Pi_e$ is the electrostatic part of Π given by Equation 6. The values of the intrinsic pK's were pK_1 = 2.8 and pK_2 = 6, and the pH of the solutions, though unbuffered, never differed appreciably from 6. The data are in overall agreement with the effects predicted by Payens (*16*). The rather low value of A_0, however, probably results from neglecting the contribution of van der Waals (VDW) cohesive effects and perhaps other important local effects such as specific sodium binding and discreteness of charge.

Figure 5 shows the effect of pH on the surface pressure at 0.1*M* NaCl. The curves resemble titration curves with inflection points near pH = 7.5. Abramson et al. (*26*) report an apparent pK_2 value of 7.93 for PA micelles in 0.1*M* NaCl. Träuble and Eibl (*27*) indicate values between 7.5 and 8.0 for 0.5*M* NaCl. It is clear that the film responds to decreases in pH with a lowered surface pressure. The same response is observed upon decreasing the NaCl concentration at a constant pH. Both effects result in decreases in the local pH. Therefore one might expect "water-sensitive" taste nerve fibers to be acid responsive as well, if the basis of the water response is a local decrease in pH following dilution of the adapting salt solution. Cohen et al. (*12*) report that the water fibers of the cat respond to acid as well as dilution but they do not respond to increases in NaCl concentration. If the taste cell responds to decreases in surface pressure only, these facts can be given a consistent interpretation.

Figure 6 shows a typical time course for the diffusion of 2*M* NaCl through a water layer transferred from the adjoining trough compartment. The response is concentration dependent. With the aid of Equation 7 these data can be used to estimate the diffusion coefficient *D* of NaCl. We therefore verify that the concentration-dependent change in surface tension is caused by diffusion. The method is limited by the accuracy with which the diffusion path length can be measured. We used the change in concentration of the final solution as a measure of the volume of water transferred from the water compartment to the salt compart-

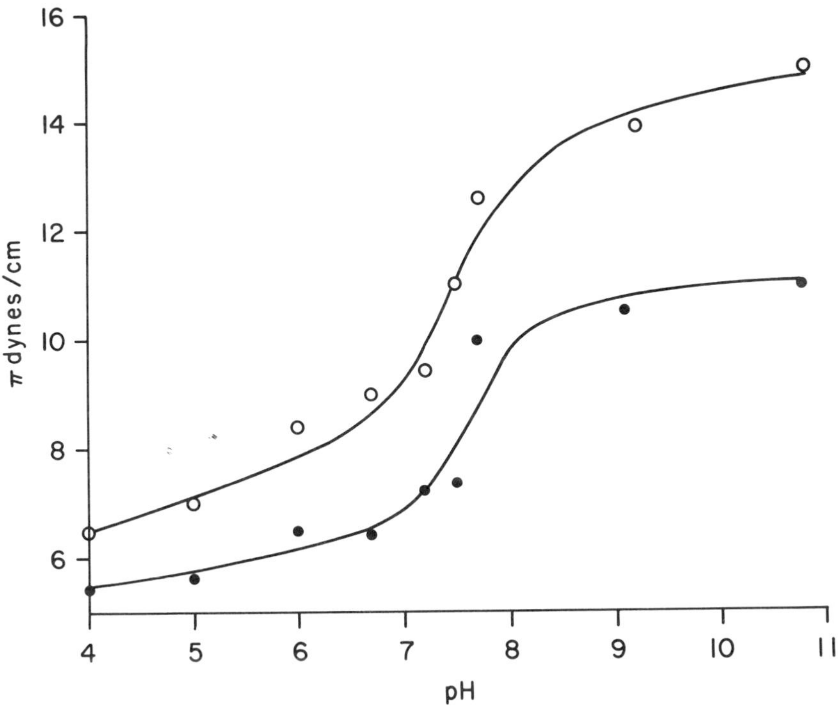

*Figure 5. The surface pressure of phosphatidic acid as a function of the pH of the subphase at two areas. All solutions contained 0.1*M *NaCl and were adjusted in pH using 1m*M *sodium phosphate buffers. The curves show an inflection point between 7.5 and 8.0.* (○), A = *85Å²*; (●), A = *123 Å²*.

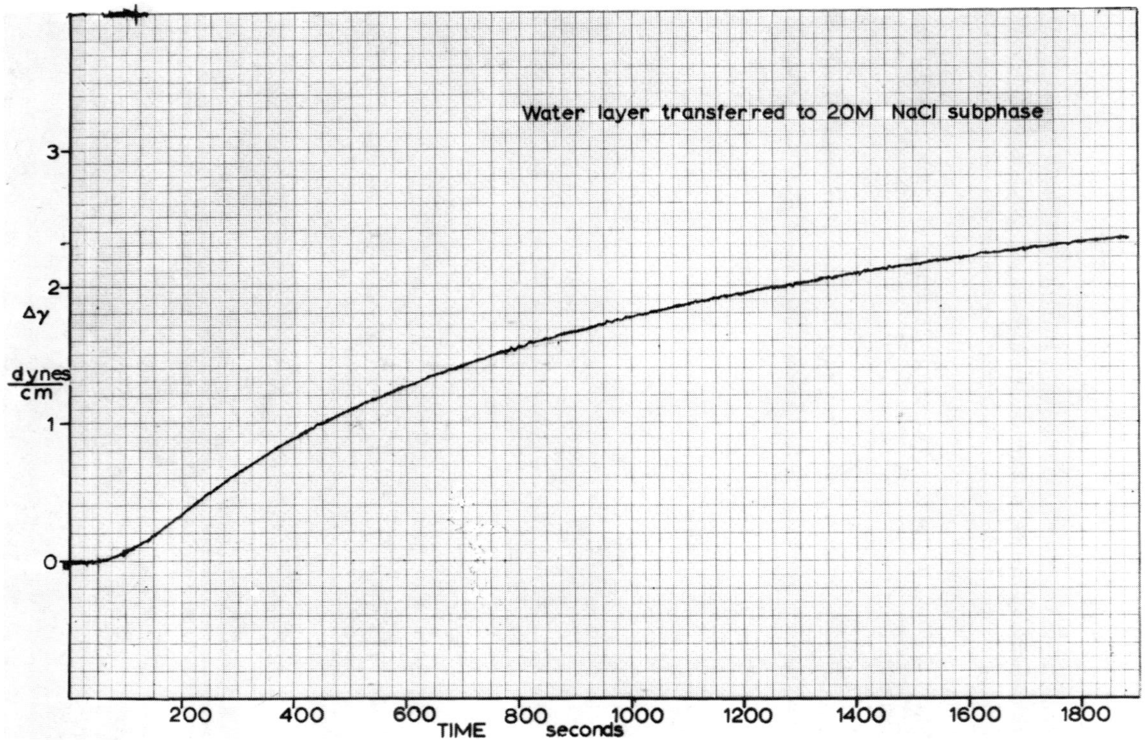

Figure 6. The surface tension increase following the transfer of a layer of water about 1 mm thick to a new subphase of 2M NaCl. The time course, the estimated diffusion coefficient of NaCl, and the linear relation between surface tension and NaCl concentration indicate that the change is a measure of the salt's diffusional relaxation.

ment. We estimate on this basis a diffusion thickness of 1.1–1.2 mm. At 2*M* we obtain the range of *D* values, 0.97×10^{-5}–1.2×10^{-5} cm^2/sec. At 0.5*M* we obtain a range of 1.2×10^{-5}–1.4×10^{-5} cm^2/sec. In making the computation we have ignored activity corrections which are necessary to correct for the concentration dependence of *D*. These estimates of *D* are in accord with the more accurately determined values reported in the literature which cluster about 1.2×10^{-5} cm^2/sec (*28*). We conclude that the relaxation process observed is in fact caused by diffusion.

In film transfer studies PA was spread onto the air–water interface in the left-hand compartment until a surface pressure change of 1.5–2.0 dynes/cm was observed. This corresponds to an area of about 55 Å^2/molecule. The film plus a layer of water about 1 mm thick then was transferred onto a salt subphase as described above. Figures 7, 8, and 9 show the change in surface tension recorded for 0.1*M*, 0.5*M*, and 2*M* NaCl, respectively. Each curve shows a rapid decline in surface tension. For the 0.1*M* curve the decline is monotonic. However for the 0.5*M* and 2*M* curves a minimum is observed. The position of the minimum can be predicted by combining the thermodynamic relations for the change in surface tension owing to electrostatic charging and the theory of diffusion. Recall that from the equilibrium studies the maximum surface pressure occurs between 0.1*M* and 1*M* NaCl for the film areas examined. We therefore might expect to find a minimum surface tension for the cases involving salt solutions with concentrations above 0.1*M*.

The predicted change in surface tension is shown in Figure 10. We construct the curves by finding the local concentration of salt for any normalized time ξ from the solution to the diffusion equation. We then use this value as the equilibrium concentration to calculate the surface tension difference, $\Delta\gamma$, according to Equation 6. The theoretical curves are for an area of 50 Å^2/molecule. From the theoretical curve for 2*M* NaCl we estimate a value of $\xi = 0.125$ at the minimum. This corresponds to 126 sec using $D = 1.2 \times 10^{-5}$ cm^{2}sec/ and $l = 0.11$ cm. For the 0.5*M* curve the value of ξ at the minimum is between 0.250 and 0.275. This translates into 252 to 277 sec. These estimates are in excellent agreement with observation. From the position of the minimum it is possible to estimate the salt concentration at the surface tension minimum from the time course of diffusion in the absence of the PA film. For the 2*M* curve the salt concentration at $t = 125$ sec is between 0.2 and 0.25*M*. An estimate of 0.2*M* can be obtained from the diffusion course of a 0.5*M* solution without PA at the interface. The theoretical value is about 0.2*M*. Thus the position and concentration of the minimum agree well with the theory; however the value of the surface tension drops do not. The theory overestimates the drop in surface tension. The reason is probably the

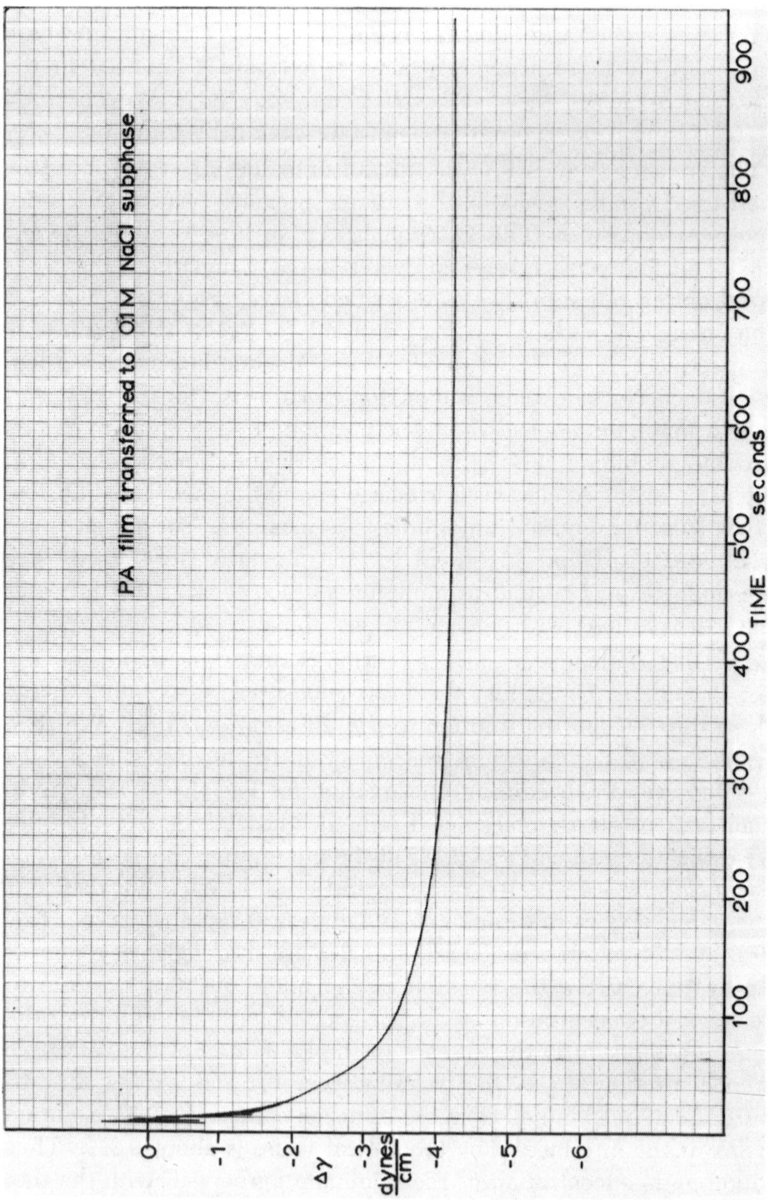

Figure 7. The change in surface tension following the transfer of a phosphatidic acid monolayer at an area of 55 Å² per molecule to a subphase of 0.1M NaCl

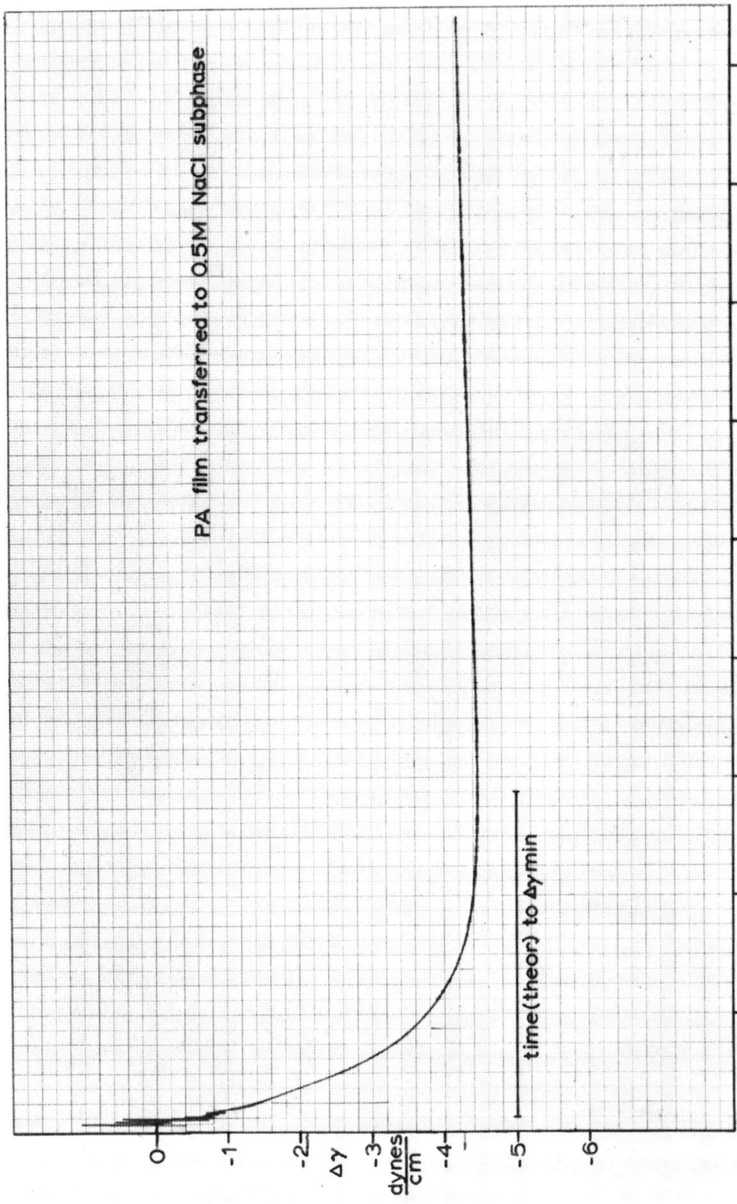

Figure 8. The same as Figure 7 except that the subphase concentration is 0.5M NaCl

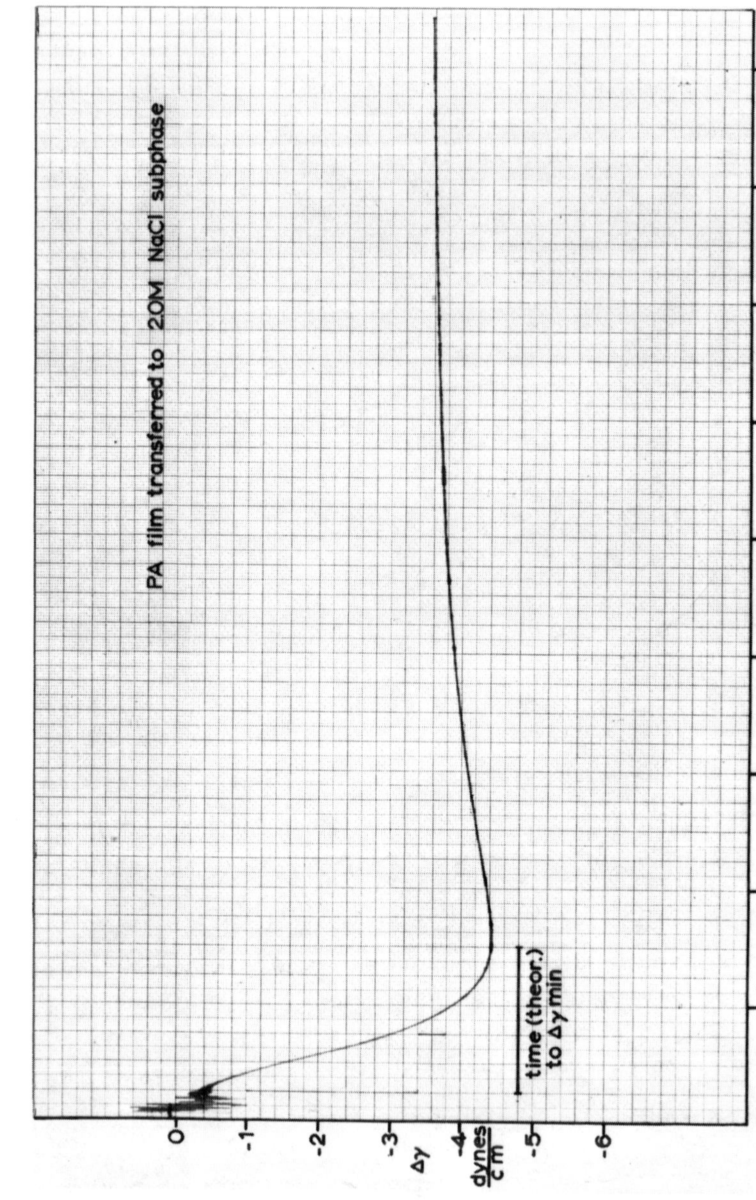

Figure 9. The same as Figure 7 except that the subphase concentration is 2M NaCl

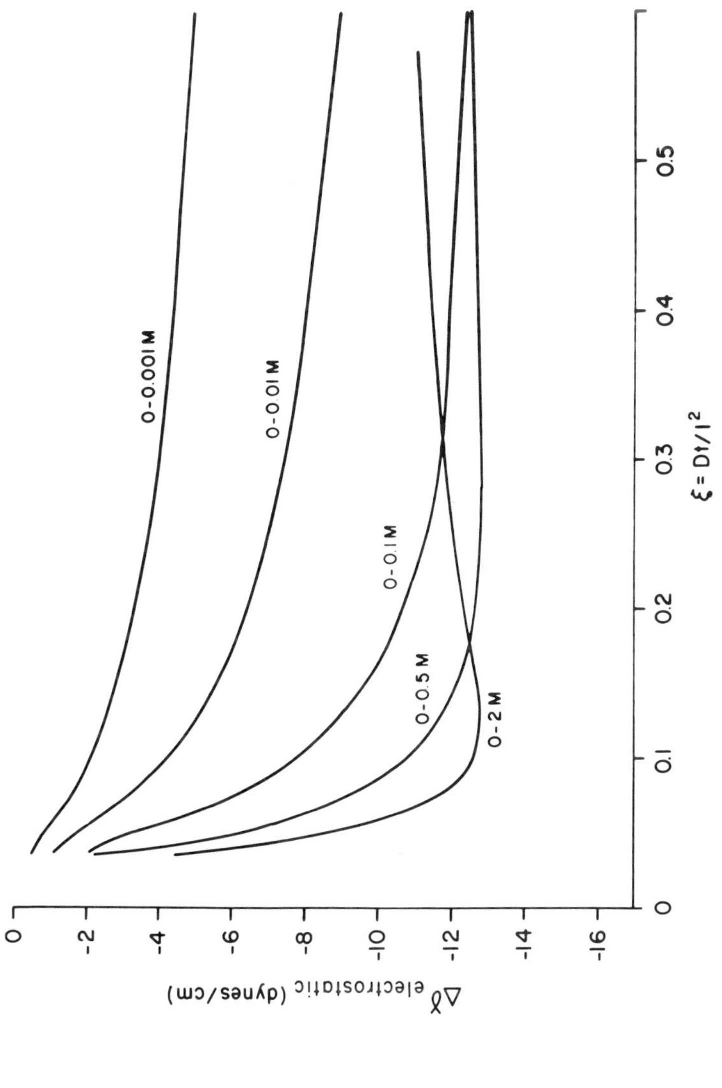

Figure 10. The theoretical change in surface tension owing to diffusion of NaCl (at the indicated concentrations) toward a monolayer of phosphatidic acid at 50Å²/molecule

neglect of cohesive pressure owing to VDW interactions among the hydrocarbon chains. Such effects are usually more important at the air–solution interface than at the oil–solution interface.

Discussion

The film transfer studies show that surface pressure changes in charged phospholipid monolayers can occur as rapidly as salt reaches the film by diffusion. Assuming quasiequilibrium behavior for events occurring within the electrical double layer gives reasonable agreement between experiment and theory in the case of phosphatidic acid. It readily is shown that the initial drop in surface tension that occurs when PA is transferred from water to salt solutions depends on the salt concentration. This drop also depends on the ionic character of the film because dipalmitoyl lecithin, a zwitterionic species, does not show it. This is expected because the basis of the drop is the ionization of the film as the salt diffuses toward the surface. The minimum in the time course for the cases involving $0.5M$ and $2M$ NaCl can be explained also on the basis of electrostatic screening and we accordingly observe agreement between theory and experiment. However it is not possible to rule out sodium chloride penetration into the monolayer as a possible contribution to the rise in surface tension when using the more concentrated salt solutions.

From the standpoint of taste reception we have shown that the dilution or water response may be a local acid response. This would account for the sour–bitter taste experienced when progressively more dilute sodium chloride solutions are placed on the tongue (*13*). The fact that water-sensitive nerve fibers are usually acid responsive as well has been shown in the cat (*12*). Nerve fibers in the cat which respond to dilution often respond to very hypertonic salt solutions also. This too is in keeping with the behavior of model phospholipid monolayers if we assume that decreased surface pressure initiates a neural response. Thus for sufficiently high salt solutions the dilution-sensitive receptors may augment the response of salt receptors. The latter do not give a dilution response but respond only to increasing salt concentrations. This behavior is expected from very acidic, pH-insensitive surface anions (*16*) which seem to characterize the salt receptor (*29*).

The particular mechanism detailed here to link a local acid response to that water response seen in some cases in chemoreception, is but one possible instance of a coupling among proton concentration at the receptor membrane, surface charge density, and surface pressure. This mechanism involves only the particular association of the proton to an anionic site in a negatively charged phospholipid. Other monovalent anions may associate with membrane components; some divalent cations such as calcium probably form strong associations. Specific adsorption of anions is highly implicated in taste. Those surface pressures and potentials resulting from simultaneous adsorptions of a number of ions are complex and can be evaluated for only specific aggregates of appropriate association parameters. A few generalizations are, however, possible.

Water responses after a particular tastant will depend both upon the pH attained at the receptor membrane during stimulation with that tastant, and the magnitude and sign of the surface charge density. In terms of a surface model, a water response after acid stimulation signifies that the receptor membrane has attained a surface charge density with a positive sign. This occurs since the decrease in surface proton concentration in response to water must result in decreased proton association. Reduced proton association will lower surface charge density and elicit a response owing to a drop in the surface pressure only if the net surface charge is positive. This is in contrast with the mechanism presented earlier where a decrease in proton association increases surface charge density in a membrane with net negative charge.

Specific associations by any species of ion will modulate surface charge density directly, and the water response will consequently vary. In particular, adsorption of anions such as benzoate should decrease the water response to acid. Similarly, a chemoreceptor which shows no water response to NaCl might be induced to produce water responses when anion association is high, e.g. following stimulation with sodium benzoate.

The effects of both surface charge and pH on conductance changes in nerve excitation have been investigated by Bass and Moore (*30, 31*). They demonstrate that changes in conductance with voltage changes can be correlated with those local pH changes which can be expected to be induced by Wien dissociation in the presence of large electric fields. These fields induce increased proton association with anionic basic sites, and pH transiently increases in the membrane. These changes are associated with conductance changes in the membrane. This view is entirely compatible with the hypothesis that a change in the surface pressure is the link between local ionic changes and conductance changes in neural membranes.

The mechanisms discussed here emphasize the role of surface charge density in modulating the composition of the ionic double layer and consequent reflexive modulation of surface charge density. This density is modulated through neutralization of surface charge in response to changes in specific ion composition of the double layer. Consequent to these changes in surface charge density is a change in surface pressure. Divalent ions will be affected much more strongly than monovalent ions by the electrostatic effects of changes in surface charge density. Those ions, such as calcium, which can be expected to show specific adsorption at anionic sites should dramatically affect both surface pressure and surface potential. Katchalsky (32), in studies on polyelectrolytes, notes the ability of divalent ions to act as bridges between anionic sites on different host molecules so as to cause precipiation. Such bridging would change the surface pressure of membranes; the electrostatic effects caused by adsorption of the divalent cation would be augmented by nonelectrostatic effects.

The specification of surface pressure as a link between membrane events and neural response predicts a strong but not exclusive dependence on electrical events at the membrane. This dependence, however, is predicted to reverse as the surface charge density goes through zero; the surface pressure depends on the absolute magnitude and not the sign of the surface charge density. This stands in contrast to the surface potential which, in general, is predicted to be a monotonic function of surface charge density. In particular, reference to water response mechanisms (both the increases and decreases in proton association) are predicted to produce water responses, but only if they occur at opposite signs of surface charge density.

Acknowledgments

The authors wish to thank K. Owens and C. Richard for assistance with the TLC determinations. This work was supported by NIH Grant NINCDS, 1R01-N-13767 to J. A. D. G. L. H. acknowledges support from NIH Training Grant T32-HL-07110.

Glossary of Symbols

A = molecular area
A_0 = constant representing "characteristic" molecular area
c = molar salt concentration
D = diffusion coefficient
e = protonic charge
K_i = intrinsic dissociation constant
k = Boltzmann's constant
l = diffusion thickness
N = Avogadro's number
T = temperature
t = time
z = valence

Greek Letters

Γ = surface concentration

γ = surface tension
γ_0 = surface tension of clean interface
ϵ = dielectric constant
$\xi = Dt/l^2$ = normalized time
Π = surface pressure
Π_e = electrostatic contribution to the surface pressure
σ = surface charge density
$\phi = e\psi/kT$ = normalized surface potential
ψ = surface potential

Literature Cited

1. Pfaffmann, C.; Bartoshuk, L. M.; McBurney, D. H. In "Handbook of Sensory Physiology"; Beidler, L. M., Ed.; Springer-Verlag: New York, 1971; Vol. IV, Part 2, p 76.
2. Price, S.; DeSimone, J. A. *Chem. Senses Flavor* **1977**, *2*, 427.
3. Dethier, V. G. *Science* **1978**, *201*, 224.
4. Warren, R. M.; Pfaffmann, C. *J. Appl. Physiol.* **1959**, *14*, 40.
5. Kurihara, Y. *Life Sci.* **1969**, *8*, 537.
6. Smith, D. V.; McBurney, D. H. *J. Exp. Psychol.* **1969**, *80*, 101.
7. Pfaffmann, C. *J. Cell. Comp. Physiol.* **1941**, *17*, 243.
8. Ogawa, H.; Sato, M.; Yamashita, S. *J. Physiol. (London)* **1968**, *199*, 223.
9. Dethier, V. G. "The Hungry Fly," Harvard University: Cambridge, MA, 1976.
10. Rees, C. J. C. *Proc. R. Soc., Lond. Ser. B.* **1970**, *174*, 469.
11. Zotterman, Y. *Acta Physiol. Scand.* **1949**, *18*, 181.
12. Cohen, M. H.; Hagiwara, S.; Zotterman, Y. *Acta Physiol. Scand.* **1955**, *33*, 316.
13. Bartoshuk, L. M. *J. Comp. Physiol. Psychol.* **1974**, *87*, 310.
14. Kamo, N.; Miyake, M.; Kurihara, K.; Kobatake, Y. *Biochim. Biophys. Acta* **1974**, *367*, 1.
15. Beidler, L. M. "Olfaction and Taste II," Hayashi, T., Ed.; Pergamon: New York, 1967; p 506.
16. Payens, Th. A. J. *Philips Res. Rep.* **1955**, *10*, 425.
17. Kurihara, K.; Koyama, N.; Kurihara, Y. "Olfaction and Taste IV"; Schneider, D., Ed.; Wissenschaftliche Verlagsgesellschaft MBH: Stuttgart, 1972; p 234.
18. Beidler, L. M. *J. Gen. Physiol.* **1954**, *38*, 133.
19. Nir, S.; Newton, C.; Papahadjopoulos, D. *Bioelectrochem. Bioenerg.* **1978**, *5*, 116.
20. Ninham, B. W.; Parsegian, V. A. *J. Theor. Biol.* **1971**, *31*, 405.
21. Davies, J. T. *Proc. R. Soc. London, Ser. A* **1951**, *208*, 224.
22. Singer, S. J. "The Molecular Organization of Biological Membranes"; Rothfield, L. I. Ed.; Academic: New York, 1971; p 145.
23. Betts, J. J.; Pethica, B. A. *Trans. Faraday Soc.* **1956**, *52*, 1581.
24. Rathbone, L. *Biochem. J.* **1902**, *85*, 461.
25. Turner, J. D.; Rouser, G. *Anal. Biochem.* **1970**, *38*, 423.
26. Abramson, M. B.; Katzman, R.; Wilson, C. E.; Gregor, H. P. *J. Biol. Chem.* **1964**, *239*, 4066.
27. Träuble, H.; Eibl, H. *Proc. Nat. Acad. Sci. USA* **1974**, *71*, 214.
28. Jost, W. "Diffusion," Revised ed., Academic: New York, 1960; p 477.
29. DeSimone, J. A.; Heck, G. L.; DeSimone, S. K., unpublished data.

30. Bass, L. J.; Moore, W. J. "Structural Chemistry and Molecular Biology"; Rich, A., Davidson, N., Eds.; W. H. Freeman and Co.: San Francisco, 1968; p 356.
31. Bass, L.; Moore, W. J. *Prog. Biophys. Mol. Biol.* **1973**, *27*, 143.
32. Katchalsky, A. *Biophys. J.* **1964**, *4*, 9.
33. Bartoshuk, L. M. "Drinking Behavior: Oral Stimulation, Reinforcement, and Preference"; Weijnen, A. W. M., Mendelson, J., Eds.; Plenum: New York, 1977; p 329.

RECEIVED October 17, 1978.

16

The Modulation of Synaptic Membrane Excitability by Cholesterol

WILLIAM D. NIEMI

Departments of Physiology and Neurology, College of Physicians & Surgeons, Columbia University, New York, NY 10032

PHILIP C. SU

Neurology Service, Veterans Administration Hospital, Northport, NY 11768 and the Division of Neurology, State University of New York at Stony Brook, NY 11794

The introduction of cholesterol into the innervated membrane of the eel electroplaque by using lecithin cholesterol liposomes greatly alters the chemical and electrical excitability of the membrane and eventually causes a block in synaptic transmission. The amplitude and maximum rate of rise of the action potential decrease until only local responses are elicited. At this point endplate potentials still can be elicited via nerve stimulation, but they are subthreshold in amplitude and have prolonged time courses. With further incubation these also disappear. Using bath-applied carbamylcholine, a decrease is seen in the chemosensitivity of the postsynaptic membrane and this decrease can be reversed partially after a 1-hr incubation with lecithin liposomes.

The living cell membrane is viewed no longer as a static, lipid bilayer coated with protein but now is considered to be a dynamic, three-dimensional liquid crystalline matrix containing heterogeneous islands of protein that can undergo both micro- and macroscopic movement. Man's view of the membrane has progressed from the Gorter–Grendel bilayer of phospholipid to the Danielli–Davson protein–phospholipid sandwich and finally to the present "fluid-mosaic" model (*1*). The mem-

0-8412-0473-X/80/33-188-261$05.00/1

brane's composition and state of fluidity are recognized now as important factors governing membrane function and influencing cellular behavior. Rheological and physiological properties of the cell membrane are dependent on the microviscosity of the lipids in the membrane (*2, 3, 4*). The apparent microviscosity of the membrane is governed by a number of factors. These include temperature (*5*), the cholesterol-to-phospholipid ratio (*6*), fatty-acid chain length and degree of saturation (*7*), and calcium-ion concentration (*8*). This chapter deals mainly with the effects of cholesterol on membrane excitability. In artificial phospholipid-membrane systems, cholesterol increases the apparent microviscosity of the membrane (*9*) and decreases the membrane's permeability to cations (*10*), glucose (*11*), anions (*12*), protons (*13*), and azobenzene dyes (*14*). The model membrane's ability to transport water goes down with the incorporation of cholesterol (*15*). In addition to its effects on passive transport, cholesterol can inhibit the ability of certain phospholipids to activate ATPases which are associated with the active transport of cations across cell membranes (*16*).

The fact that cholesterol can alter the electrical potential and capacitance of phospholipid bilayers (*17*) encouraged us to investigate what effects it might have on an excitable cell such as the densely innervated electric eel electroplaque. In this study we used the now-well-established technique of altering the cholesterol:phospholipid ratio of membranes by using cholesterol–lecithin liposomes (*2, 18, 19, 20, 21*). Cells which are incubated in a suspension of such liposomes will exchange cholesterol with the liposomes until the cell membrane's cholesterol:phospholipid ratio approaches and equilibrates with that in the liposome (*19*). This process is thought to involve a translocation of the cholesterol from the liposome to the cell membrane during physical contact (*21*), although an alternative hypothesis has been proposed which envisions the fusion of the liposome with the cell membrane (*22*). The end result in a cell membrane with a low cholesterol content is to increase its cholesterol:phospholipid ratio and this usually is accompanied by an increase in the apparent microviscosity of the membrane (*23*). It was interesting to us to see if this could influence synaptic transmission at cholinergic synapses.

The electroplaque from the Sachs' organ of the electric eel was chosen as the model for the cholinergic cell used in this study because it contains a dense innervation of more than 50,000 synapses on its caudal surface and is pharmacologically similar to the mammalian neuromuscular junction. It is a large, rectangular cell (those used in this study were approximately 3 $\times$ 10 mm) that possesses both chemically and electrically excitable membranes and is very amenable to electrophysiological studies because of its high density of ionic channels (*24*) and lack of contractility. The extrasynaptic membrane of the electroplaque, like that of the vertebrate muscle, responds to electrical stimuli in an all-or-none fashion after a certain threshold is reached (usually between −50 and −60 mV). The resulting potential change, the action potential (AP), is caused by a simultaneous increase in sodium permeability and decrease in potassium permeability (K^+ inactivation) during the rising phase of the action potential (*24*). The chemosensitive membrane of the electroplaque undergoes desensitization in the continued presence of agonists such as carbamylcholine (*25*) in a manner similar to that of a frog muscle endplate (*26*).

Experimental

The eel electroplaque cells were isolated as previously described (*27*). The cells were divided into three groups: a control (untreated) group, a group that was treated with lecithin liposomes, and a group that was treated with either liposomes containing cholesterol (Δ^5-cholesten-3-ol, Sigma grade 99%+, standard for chromatography) or with liposomes containing chloesterol hydrogen succinate (Sigma). The liposomes were prepared by mixing 40 mg of lecithin (Sigma Type III-E from egg yolk, prepared chromatographically per a modification of the procedure of Singleton et al. (*28*)) and 20 mg of cholesterol (or of cholesterol hydrogen succinate) and evaporating the hexane off at room temperature under nitrogen. Ten mL of eel Ringer's solution (*29*) then was added to the mixture which was subjected to intermittent sonication under nitrogen for 60 min at 4°C in a Braunsonic 1510 with a 4-mm probe. The peak power of the sonicator was not allowed to exceed 50 W. The liposome suspension was centrifuged for 60 min at 35,000 g. The middle translucent layer was removed carefully with a long needle and a Buchler pump and liposomes from this layer were incubated with isolated electroplaque cells. Lecithin liposomes were made in a similar fashion except that the cholesterol was left out. Single cells were placed in 10-mL beakers containing 3 mL of the liposome suspension and were shaken gently at room temperature for periods of up to 60 min. A treated electroplaque was mounted in an Ussing-type lucite chamber so that the cell separated two pools of eel Ringer's solution. The posterior (caudal) face of the electroplaque was perfused with carbamylcholine-containing solutions while the anterior, noninnervated face was bathed in oxygenated eel Ringer's solution. The membrane potential across the innervated membrane was monitored continuously using a glass microelectrode which penetrated the noninnervated face just opposite of the "window" of the innervated side. Thus the glass microelectrode faced the inside surface of that portion of the innervated membrane that was being perfused with the carbamylcholine solution (*see* Figure 1). The membrane conductance was measured with intermittent hyperpolarizing pulses of current from a Grass photoelectric constant current unit. Slow membrane potential changes were recorded on a Varian chart recorder and the rapid potential changes were photographed from a 564 Tektronix storage oscilloscope. Direct stimulation of the innervated membrane was achieved

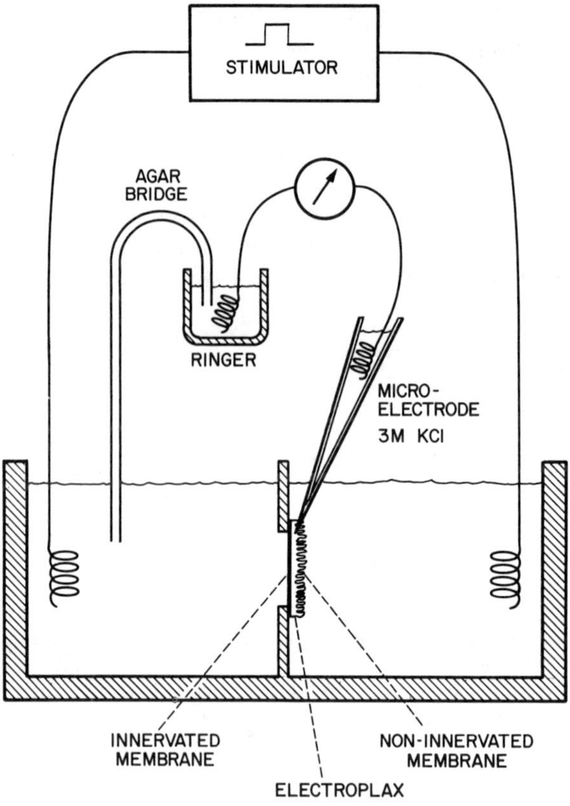

Figure 1. Diagram of the lucite chamber used to hold the eel electroplax. The chamber consisted of 2 halves between which a 1 × 3-mm window and spacer were placed. The cell was held against the window by a nylon mesh and it separated the left and right pools. The microelectrode (held in a micromanipulator) was used to penetrate the noninnervated face just opposite the area of the innervated membrane that was being bathed in the other pool. The potential was monitored with a high-impedance, negative-capacitance preamplifier and displayed on an oscilloscope.

Table I. The Effects of Cholesterol Enrichment on the Properties of the Electrically Excitable Membrane of the Eel Electroplaque

	AP (mV)	*R(V/sec)*	*F(V/sec)*	*OS (mV)*	*RP (mV)*	*n*
Control[a]	131 ± 11	401 ± 55	122 ± 20	46 ± 5	86 ± 9	3
Lec	127 ± 10	397 ± 32	115 ± 18	44 ± 4	83 ± 8	3
Lec–chol	90 ± 20	202 ± 31	71 ± 22	10 ± 6	81 ± 6	3

[a] Control cells were incubated in normal eel Ringer's solution for the same amount of time that the treated cells were incubated in the liposomes. The cells were incubated for 1 hr in either lecithin liposomes (lec), 4 mg/mL, or lecithin–cholesterol liposomes (lec–chol), 4 mg/mL:2 mg/mL. AP = action potential; R = maximum rate of rise in volts per second; F = maximum rate of fall in volts per second; OS = overshoot of the AP; RP = resting potential; and n = number of cells. The values are expressed as the mean ± SD.

with depolarizing current pulses which passed through the cell from one pool to the next while indirect (neural) stimulation was achieved by passing hyperpolarizing current pulses from one pool to the next. This resulted in the stimulation of the nerve twigs attached to the posterior face. The perfusion rate of the drug solution across the innervated face of the electroplaque was 5 mL/min and the size of the Mylar window covering the innervated side of the cell was 1 × 3 mm. The amount of cholesterol that was taken up by the cell was determined by using ^{3}H-cholesterol (Amersham) in some of the experiments. After thoroughly rinsing the cell, followed by dissolution in Scintisol, the samples were placed in a liquid scintillation counter.

Results

The Effect of Cholesterol on Synaptic Transmission. After the cells were incubated for 1 hr in cholesterol–lecithin liposomes containing radio-labelled cholesterol, each electroplaque had taken up, on the average, approximately 150 μg of cholesterol. This uptake represents a 50% increase in the total cholesterol content of a normal electroplaque (*30*).

Electroplaques which had undergone the above treatment showed a significant decrease in the amplitude and the maximum rate of rise of action potentials (AP) of the postsynaptic membrane (*see* Table I). The amplitude of the AP prior to treatment was 131 ± 11 (mean ± SD) and after 1 hr of incubation in the cholesterol–liposome suspension it was reduced to 90 ± 20 mV. The maximum rate of rise of the AP was reduced by about 50% while the maximum rate of fall of the AP was reduced to about 60% of the control value. The overshoot of the AP after treatment was only 10 mV as compared with 46 mV for the control. In some experiments the complete abolition of the AP occurred during the incubation period (*see* Figure 2) and these were not included in Table I. As seen in Figure 2, in some cells significant effects were seen after only 10 min of incubation in the cholesterol–lecithin liposomes. After 35 min of incubation only local responses were elicited from the innervated membrane with direct stimulation, and only subthreshold potentials were elicited with indirect (neural) stimulation (*see* Figure 2d). After 1 hr of incubation no endplate potentials were elicited with neural stimlulation. Thus a block of synaptic transmission had occurred.

The Effect of Cholesterol on the Chemosensitivity of the Postsynaptic Membrane. To test whether the block in synaptic transmission was caused by the loss of either the neuronal AP or the chemosensitivity of the postsynaptic membrane, we bath-applied carbamylcholine (a cholinergic agonist which is not hydrolyzed to any significant extent by acetylcholinesterase) to the innervated membrane and measured the maximum amount of depolarization which occurred in 2 min (*see* Table II). Figure 3 shows that a significant decrease in the cell's response to 40μM carb-

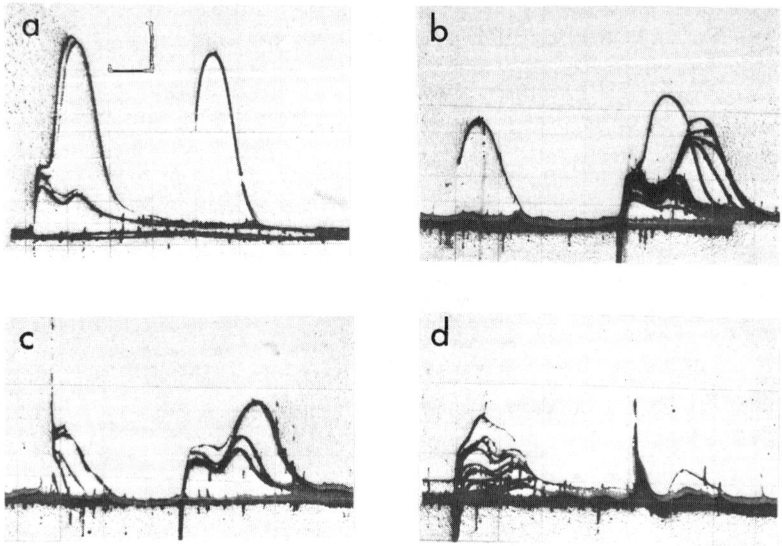

Figure 2. The effect of cholesterol liposomes on directly and indirectly elicited action potentials of the innervated membrane of the eel electroplaque. (a) Control cell in normal eel Ringer's solution. The potentials at the left are the result of indirect (neural) stimulation. The smaller ones are endplate potentials while the larger ones are action potentials. To the right is a single directly elicited action potential. (b) Same cell after a 10-min incubation in a cholesterol–lecithin liposome suspension; a single directly elicited action potential is seen at the left and indirectly elicited ones are at the right. (c) Same cell 15 min later. Directly elicited action potentials are at the left and indirectly elicited ones are at the right. (d) Same cell 35 min later. The indirectly elicited potentials are to the left; only subthreshold potentials are seen. To the right only endplate potentials can be elicited even though a depolarizing pulse is sent across the cell. The recordings are dc and the calibration is 25 mV and 2 msec per division.

Table II. The Chemosensitivity of the Postjunctional Membrane of the Innervated Face of the Eel Electroplaque after Exposure to Lecithin Liposomes Containing Either Cholesterol or Cholesterol Hydrogen Succinate

Carb Conc	*Untreated*[a]	*Lec*	*Chol–Lec*	*CHS–Lec*
20 μM	15.3 ± 2.4 (3)	10.3 ± 1.5 (3)	7.0 ± 2.5 (3)	12.4 ± 1.9 (3)
40 μM	32.9 ± 3.5 (3)	30.2 ± 2.5 (3)	11.8 ± 2.2 (3)	16.3 ± 6.7 (3)
60 μM	41.5 ± 9.6 (3)	39.0 ± 2.6 (3)	20.8 ± 8.5 (3)	—
100 μM	50.6 ± 2.6 (3)	46.3 ± 4.5 (3)	30.6 ± 6.1 (3)	31.0 ± 7.7 (3)

[a] Isolated eel electroplaques were incubated for 1 hr in one of the above designated solutions: normal eel Ringer's (untreated) or eel Ringer's containing either 4 mg/mL lecithin liposomes (lec), liposome containing 4 mg/mL lecithin plus 2 mg/mL cholesterol (chol–lec), or 4 mg/mL lecithin plus 2 mg/mL cholesterol hydrogen succinate. The values tabulated are the maximum amounts of depolarization produced by the indicated carbamylcholine (carb) solution in 2 min. The number of cells treated is indicated in the parentheses. The values are expressed as the mean ± SD.

amylcholine has occurred after treatment with cholesterol–lecithin liposomes. The control cell (Cell a) shows a 26-mV depolarization while the treated cell (Cell b) shows only a 7-mV depolarization. The treated cell's response to 1m*M* carbamylcholine is also quite different from the control. The control cell undergoes a rapid 50-mV depolarization which then reverses itself in a few minutes by 25 mV while the treated cell undergoes a slower 30-mV depolarization which does not reverse in the continued presence of 1m*M* carbamylcholine. Figure 4 shows a group of concentration–response curves for treated and untreated cells. The maximum depolarization produced in 2 min by various concentrations of carbamylcholine is plotted for four different cells. Cells that were treated with either eel Ringer's solution or with the lecithin liposomes showed a normal sensitivity to carbamylcholine, while cells treated with either cholesterol–lecithin liposomes or with cholesterol hydrogen succinate liposomes showed a reduced response to carbamylcholine as well as a reduced maximum depolarization at high agonist concentrations. The dose–response curve does not shift simply to the right but undergoes a decrease in maximum response indicating that some unsurmountable inhibition has occurred.

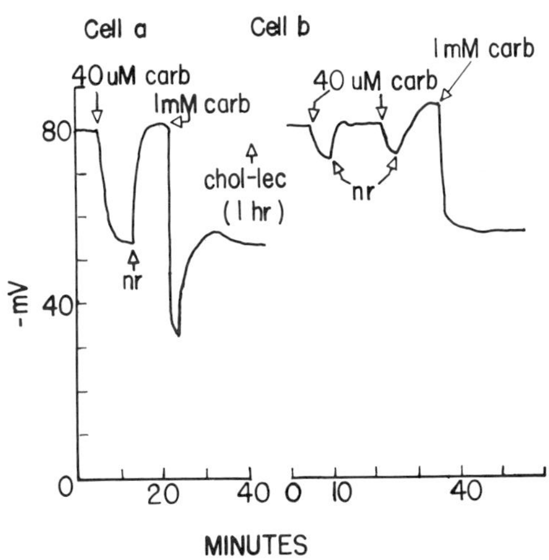

Figure 3. Two electroplaque cells from the same electric organ were treated in parallel, one with eel Ringer's and the other with a suspension of cholesterol-lecithin liposomes containing 2 mg/mL cholesterol and 4 mg/mL lecithin; they then were mounted into lucite chambers for electrophysiological recording. Cell a had an initial resting potential of −80 mV and it depolarized by 26 mV upon exposure to 40μM carbamylcholine and by 50 mV with 1mM carbamylcholine whereupon it began to show desensitization. Cell b, which had been exposed to the cholesterol–lecithin liposomes for 1 hr, was mounted and it gave only a 7-mV depolarization with 40μM carbamylcholine while 1mM carbamylcholine gave a slow 35 mV of depolarization. NRT = tris-buffered eel Ringer's.

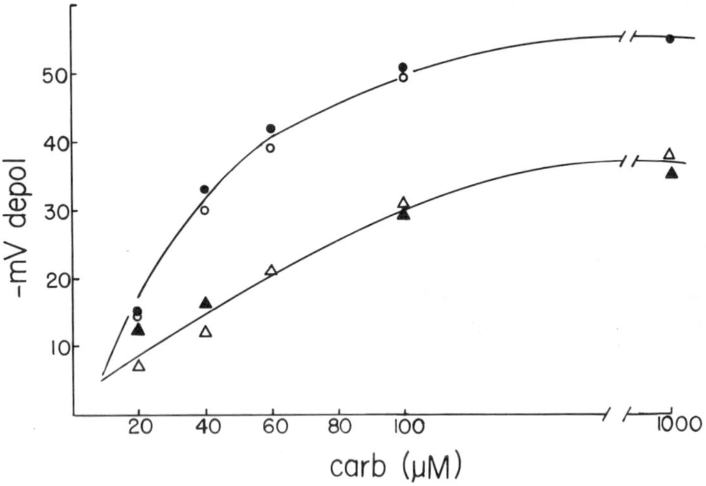

Figure 4. The concentration–response curves of four different cells using carbamylcholine as an agonist perfused onto the innervated face. The values depicted represent the maximum depolarization attained in 2 min: (●) eel Ringer's; (○), incubated 1 hr in lecithin loposomes; (▲), 1 hr in cholesterol–lecithin liposomes; (△), 1 hr in liposomes containing cholesterol hydrogen succinate and lecithin.

Results similar to the above are seen when other cholinergic agonists are used in place of carbamylcholine. In Figure 5 the response of a single electroplaque after treatment with cholesterol–lecithin liposomes to an agonist which is more potent than carbamylcholine, i.e. 3,3′-bis(α-(trimethylammonium)-methylazobenzene) dibromide (Bis-Q) (*31*) is shown using 1μM Bis-Q. The initial response to this concentration of Bis-Q consists of a 60-mV depolarization which is accompanied by a significant increase in conductance (as noted by the decrease in the amplitude of the hyperpolarizing pulses) and this is reversed by flushing with normal eel Ringer's solution. If cholesterol–lecithin liposomes then are introduced into the pool bathing the innervated face one sees a markedly reduced response to 1μM Bis-Q 30 min later. The cell experiences a 15-mV depolarization which reverses itself by 10 mV in the continued presence of the Bis-Q. The membrane potential then reaches a steady-state level some 12 mV depolarized from the initial resting level. If one now challenges the cell with 1mM carbamylcholine, the response is a 3-mV depolarization and no noticeable change in conductance.

Attempts to reverse the block caused by cholesterol using lecithin liposomes in an attempt to remove the cholesterol from the membrane were only partially successful. The best that could be achived in some cells after a 1-hr incubation in lecithin–liposomes was a 50% recovery in the postsynaptic membrane sensitivity to carbamylcholine. This resistance to reversal of inhibition may be the result of the limited access that the liposomes have to the electroplaque membrane (*see* the Discussion and Conclusions section).

Direct effects such as the actual physical screening of acetylcholine receptor sites by the liposomes seem to be ruled out by the experiment depicted in Figure 6. When either lecithin liposomes or cholesterol–lecithin liposomes are added simultaneously with the agonist solution, no inhibition of depolarization is observed (*see* Figure 6). In all cases the rates of depolarization and repolarization also are not affected. The liposomes have to be in contact with the cell some time before any significant inhibition occurs.

Discussion and Conclusions

The reduction in chemosensitivity and the block of synaptic transmission which occurs in the eel electroplaque after incubation with cholesterol–lecithin liposomes is most likely the result of an alteration of the cell membrane's fluidity owing to a change in its normal cholesterol:phospholipid ratio. This membrane fluidity is dependent upon the state of the phospholipids in that membrane. The fatty-acid chains of

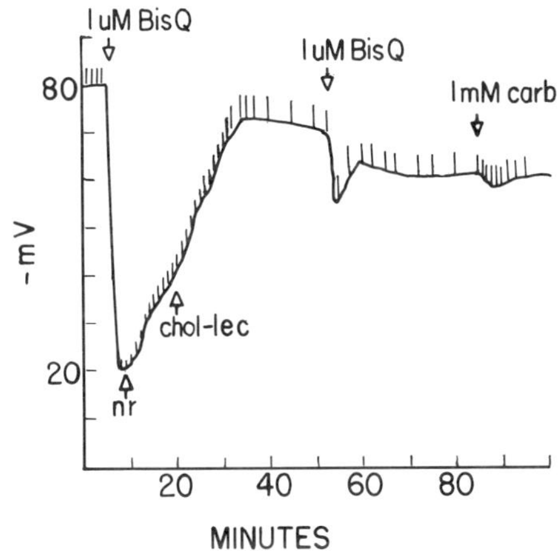

Figure 5. A normal eel electroplaque was mounted into the chamber and its innervated face perfused with 1μM Bis-Q resulting in a depolarization of 59mV. During its recovery it was perfused with a suspension of cholesterol–lecithin liposomes and challenged again 30 min later with 1μM Bis-Q and this time the cell only depolarized 18 mV. After being exposed to 1μM Bis-Q for 30 min it was challenged with 1mM carbamylcholine and gave only a 3-mV depolarization. The hyperpolarizing blips are pulses of constant current.

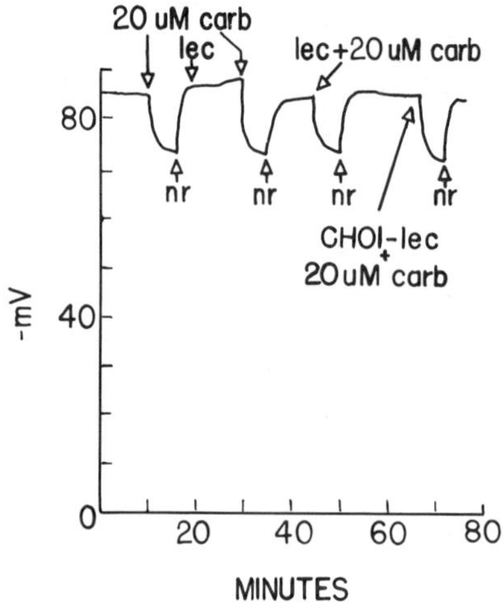

Figure 6. A single electroplaque whose innervated face was perfused sequentially with the following solutions revealed no immediate effect of the liposomes on chemosensitivity: 20μM carbamylcholine; lecithin liposomes (4 mg/mL); and 20μM carbamylcholine lecithin liposomes plus 20μM carbamylcholine cholesterol–lecithin liposomes (2 mg:4 mg/mL). The abbreviation nr = normal eel Ringer's.

the phospholipids can be flexible and interact depending upon their chain length and degree of saturation. This flexibility and "cooperativity" could be disrupted by the insertion of cholesterol into the membrane. In artificial phospholipid membranes an increase in cholesterol content causes the phospholipids to lose their ability to interact and prevents them from undergoing a transition to the gel state at low temperatures (*32*). In synaptic transmission rapid protein conformational changes as well as cooperativity between the membrane's constituents are both vital in the gating of ionic currents. An increase in microviscosity could reduce the rate and efficiency with which the acetylcholine receptor could open membrane channels while a decrease in microviscosity (increase in fluidity) could reduce cooperativity between the receptor and its ion channels. If one lowers the temperature in frog muscle from 22° to 3°C the average "shot amplitude" from acetylcholine noise measurements increases to 0.51 μV from 0.22 μV (*33*). If this lowered temperature is accompanied by an increase in microviscosity, then it is apparent that slight increases in microviscosity do not compromise receptor function. Also, it has been our experience that lowered temperatures do not result in a reduction of the AP amplitude in the electroplaque. Thus it is possible that the results we see are the result of an increase in membrane fluidity. Gage et al. (*34*) have shown that long-chain alcohols such as octanol, which increase membrane fluidity also decrease open-channel lifetime resulting in a reduction of the amplitude of depolarization and a decrease in the single-channel conductance. Thus noise measurements and single-ion-channel recordings on cholesterol-enriched membranes need to be done.

An effect of cholesterol on the amplitude and spontaneous firing rate of the AP recorded from Aplysia neurons has been observed (*35*). A decrease in the AP amplitude and spontaneous rate of firing was seen in the Aplysia after incubation with cholesterol–lecithin liposomes. This effect could be reversed with lecithin liposomes. Complete reversal was not seen in our preparation possibly because of the complex surface geometry of the electroplaque. The electroplaque is highly invaginated and has a dense nerve net over its posterior surface. Getting cholesterol into the innervated membrane may be easier than getting it out owing to the fact that the liposomes can make contact with only a fraction of the electroplaque's surface.

Most of the discussion has so far dealt with depolarization. We did not attempt to quantitate cholesterol's effects on repolarization because in the electroplaque repolarization is complex, involving both ion-channel closure and the re-establishment of its ionic gradients through active transport (*36*). In our experiments, cholesterol did not seem to have any significant effect on the electroplaque's ability to repolarize, indicating to us that it did not inhibit active sodium transport. However, this needs to be pursued further with more definitive experiments.

Glossary of Symbols

AP = action potential
R = maximum rate of rise of the AP
F = maximum rate of fall of the AP
OS = overshoot of the AP
RP = resting potential
mV = millivolts
V = volts
sec = second
lec = lecithin liposomes
chol–lec = cholesterol–lecithin liposomes
Bis-Q = 3,3′-bis(α-(trimethylammonium)-methylazobenzene) dibromide
CHS = cholesterol hydrogen succinate
carb = carbamylcholine

Literature Cited

1. Singer, S. J.; Nicolson, G. L. *Science* **1972**, *175*, 720.
2. Shinitzky, M.; Inbar, M. *J. Mol. Biol.* **1974**, *85*, 603.

3. Rottem, S.; Hubbell, W. L.; Hayflick, L.; McConnell, H. M. *Biochim. Biophys. Acta* **1970**, *219*, 104.
4. Sha'afi, R. I.; Rodan, S. B.; Hintz, R. L.; Fernandez, S. M.; Rodan, G. A. *Nature (London)* **1975**, *254*, 525.
5. Szabo, G. *Nature (London)* **1974**, *252*, 47.
6. Papahadjopoulos, D.; Cowden, M.; Kimelberg, H. *Biochim. Biophys. Acta* **1973**, *330*, 8.
7. Huang, L.; Jaquet, D. D.; Haug, A. *Can. J. Biochem.* **1974**, *52*, 483.
8. Torch, W. C.; Abood, L. G. *Int. J. Neurosci.* **1973**, *5*, 143.
9. Cogan, U.; Shinitzky, M.; Weber, G.; Mishida, T. *Biochemistry* **1973**, *12*, 521.
10. Papahadjopoulos, D.; Nir, S.; Ohki, S. *Biochim. Biophys. Acta* **1972**, *266*, 561.
11. Demel, R. A.; Bruckdorfer, K. R.; Van Deenen, L. L. M. *Biochim. Biophys. Acta* **1972**, *255*, 321.
12. Papahadjopoulos, D.; Watkins, J. C. *Biochim. Biophys. Acta* **1967**, *135*, 639.
13. Scarpa, D.; DeGier, J. *Biochim. Biophys. Acta* **1971**, *241*, 789.
14. Horton, W.; Schuff, A. R.; McClure, D. W. *Biochim. Biophys. Acta* **1973**, *318*, 225.
15. Finkelstein, A.; Cass, A. *Nature (London)* **1967**, *216*, 717.
16. Kroes, J.; Ostwald, R. *Biochim. Biophys. Acta* **1971**, *249*, 647.
17. Ohki, S. *Biophys. J.* **1969**, *9*, 1195.
18. Bruckdorfer, K. R.; Edwards, P. A.; Green, C. *Eur. J. Biochem.* **1968**, *4*, 506.
19. Bruckdorfer, K. R.; Graham, J. M.; Green, C. *Eur. J. Biochem.* **1968**, *4*, 512.
20. Cooper, R. A.; Handl, J. H. *J. Clin. Invest.* **1968**, *47*, 809.
21. Murphy, J. R. *J. Lab. Clin. Med.* **1962**, *60*, 86.
22. Papahadjopoulos, D.; Mayhew, E.; Poste, G.; Smith, S. *Nature* **1974**, *252*, 163.
23. Borochov, H.; Shinitzky, M. *Proc. Nat. Acad. Sci. U.S.A.* **1976**, *73*, 4526.
24. Nakamura, Y.; Nakajima, S.; Grundfest, H. *J. Gen. Physiol.* **1965**, *49*, 321.
25. Larmie, E. T.; Webb, G. D. *J. Gen. Physiol.* **1973**, *61*, 263.
26. Katz, B.; Thesleff, S. *J. Physiol.* **1957**, *138*, 63.
27. Schoffeniels, E.; Nachmansohn, D. *Biochim. Biophys. Acta* **1957**, *26*, 1.
28. Singleton, W. S.; Gray, M. S.; Brown, M. L.; White, J. L. *J. of Amer. Oil Chem. Soc.* **1965**, *42*, 53.
29. Webb, G. D.; Hamrell, B.; Farquharson, D.; Niemi, W. D. *Biochim. Biophys. Acta* **1973**, *297*, 313.
30. Bartels, E.; Rosenberg, P. *J. Neurochem.* **1972**, *19*, 1251.
31. Bartels, E.; Wasserman, N. H.; Erlanger, B. F. *Proc. Nat. Acad. Sci. U.S.A.* **1971**, *68*, 1820.
32. Chapman, D. "Lipid Dynamics in Cell Membranes," in "Cell Membranes Biochemistry, Cell Biology, & Pathology"; Weissmann, G., and Claiborne, R., Eds.; HP Publishing Co., Inc.: New York, 1975; Chapter 2, p. 21.
33. Katz, B.; Miledi, R. *Nature (London), New Biology* **1971**, *232*, 124.
34. Gage, P. W.; McBurney, R. N.; Van Helden, D. *J. Physiol.* **1978**, *274*, 279.
35. Stephens, C. L.; Shinitzky, M. *Nature* **1977**, *270*, 267.
36. Karlin, A. *Proc. Nat. Acad. Sci., U.S.A.* **1967**, *58*, 1162.

RECEIVED November 20, 1978.

17

Increased Ionization of Catecholamines in the Presence of Imidazole and Related Compounds

M. B. ABRAMSON and M. CHOI

The Saul R. Korey Department of Neurology, Albert Einstein College of Medicine, Bronx, NY 10461

Spectrophotometric measurements of dopamine and norepinephrine show that at physiological pH the presence of imidazole or related proton acceptors causes an increase in the ionization of the phenolic OH. The absorbance at 295 nm permits an estimation of the concentration of the ionized species. While the pK_a' *of norepinephrine in 0.1*M *sodium phosphate is 8.9, in 0.1*M *imidazole, histidine, or tris buffer at pH 7.4, it is 8.6, 8.7, or 8.4, respectively. Increasing the concentration of imidazole or tris causes greater lowering of the* pK_a'*. The monophenolic compounds, normetanephrine, or tyramine did not show this increased ionization except at pH levels* > 9*. Using codispersions of imidazole in egg lecithin in phosphate buffer with a concentration of 3m*M *imidazole, the* pK_a' *of norepinephrine was 8.6. This increased ionization may be important in some biological systems.*

It has been assumed that the active molecular form of the catecholamines which is involved in various processes in the nervous system is the cation. This view is based upon the fact that at physiological pH about 97% is present as the cation, since the p*K* of the first phenolic hydroxyl is roughly 8.8 and remains essentially unionized. In a recent study in this laboratory (*1*), the uptake of norepinephrine by synaptosomes from rat brain increased with pH up to approximately pH 8.2, and presumably the active species was the zwitterion form of the molecule with negative phenolate and positive amine. This chapter describes experiments which show that at a constant pH, the ionization of the phenolic group of the

0-8412-0473-X/80/33-188-273$05.00/1

catecholamine can be increased as a result of the interaction with certain *N* bases in the medium. This presents the view that in some actions of the catecholamines in the nervous system, a reaction may occur with a specific agent such as histidine in the medium or in the membrane surface at the uptake or receptor sites which could then alter the charged state of the catecholamine. The different behavior observed for the catecholamines and such related compounds as tyramine, octopamine, or normetanephrine could result from the inability of these latter compounds to form a zwitterion by ionization of a hydroxyl group at neutral pH. In another area of interest, the active site of some hydrolytic enzymes is observed to include a histidine residue (2). It is possible that the imidazole functions by catalyzing a proton transfer in the enzyme–substrate complex.

In order to relate the reactions we describe to biological phenomena, we include some experiments in which imidazole or some related compound is incorporated with egg lecithin and then dispersed in water forming multilamellar structures. This provides a system possessing some of the characteristics of the biological membrane.

Experimental

Materials. Dopamine, L-norepinephrine, and normetanephrine were from Regis Co.; for spectroscopic measurements solutions of these compounds were prepared daily at concentrations of either $1 \times 10^{-3}M$ or $2 \times 10^{-3}M$ in $0.01M$ HCl. Tris(hydroxymethyl)aminomethane base was from Fisher Scientific; imidazole was from Eastman Chemical; histamine was from Sigma, and L-histidine was from Calbiochem. Egg lecithin from Sylvana Co. was used for lipid preparations. In preparing the buffer media, solutions of tris base or Na_2HPO_4 were brought to the desired pH with HCl. Solutions of $0.10M$ imidazole, histamine, or histidine were prepared in a similar manner. All pH measurements were made with either a Corning Model 12 or Orion 601A pH meter using a Thomas or Fisher combined glass and reference electrode. The water used was deionized, distilled, and redistilled from Pyrex.

Lipsomes. Lipid codispersions were made by dissolving a weighed amount of imidazole or related compound in absolute ethanol, adding lecithin in alcohol and mixing. The solvent was removed by a stream of nitrogen followed by several hours in vacuum. The required volume of water was added and the tube was shaken until the lipid was freed from the glass surface. A clear dispersion then was formed by exposure to ultrasonics using either an MSE or Branson sonifier. In most instances, a 2–10 min exposure at 1-min intervals sufficed to reduce the turbidity to low levels. Aliquots of this system were added to $0.10M$ sodium phosphate to give phosphate concentrations from $0.025–0.05M$ and the desired pH was established by the addition of HCl or NaOH.

Spectroscopic Measurements. Most of the spectra reported here were obtained using a Perkin–Elmer No. 576 spectrophotometer equipped with a memory unit for the storing and correcting of a nonhorizontal baseline. The cell compartment was temperature controlled and all spectra were obtained at 25°C. Additional spectra were obtained using a Cary Model 308 spectrophotometer. In obtaining the spectra, 3 mL of the buffer or test solution was added to each of two matched 10-mm quartz cuvets and the baseline was scanned from 320–260 nm and stored in the memory unit. Whenever the baseline departed significantly from the horizontal, the samples were discarded. 0.01*M* HCl (30 μL) was added to the solution in the reference cell and 30 μL of either 1–2m*M* dopamine or norepinephrine in 0.01*M* HCl was added to the sample cuvet and rapidly stirred. The spectra were obtained within 2 min after mixing, and repeated to detect whether there were any changes resulting from oxidation.

Spectroscopic studies of the dispersed lipids were run with cuvets in the compartment for turbid samples. Equal 3-mL portions of the codispersion in phosphate buffer were added to two matched cuvets and the baseline spectrum was obtained and stored in the memory unit of the instrument. Whenever the baseline showed differences between the contents of the two cuvets, small changes in the concentrations were made until a satisfactory baseline was obtained. When this could not be done, new samples were taken. The spectra then were obtained as before.

Results

Spectral Changes. The absorption spectra of aqueous solutions of dopamine and norepinephrine show prominent bands in the regions of 280 and 240 nm. In this study, our attention was directed only to hte changes in the longer wavelength band. At pH levels above 7, the maximum of this band shifts to the red and a shoulder increases in intensity in the 295-nm region (*see* Figure 1). At pH 10, the maximum is at 295 nm (3). These changes are associated with the deprotonation of one phenolic OH with increasing pH. Above pH 10, the second phenolic OH is deprotonated and additional changes are observed in the spectra. When the pH is above 8, the spectra are complicated further by the formation of oxidation products with absorbances in the 300-nm region.

Using 0.10*M* sodium phosphate buffer systems with $2 \times 10^{-5}M$ norepinephrine or dopamine at pH 7.4, the maximum is at 279.6. However, this differs in 0.10*M* tris buffer at pH 7.4, as these compounds then give spectra with the maximum at 281 nm and increased absorbance at approximately 295 nm (*see* Figure 2, Curves 1 and 2). These spectral shifts to the red are similar to those observed with phosphate buffers at a somewhat higher pH although both systems are at the same pH. With the increasing pH of the tris buffer, there is a further increase in this red shift. This change in spectra was dependent upon the concentration of tris as shown by the following experiment.

Several solutions of tris were prepared with concentrations ranging from 0.02*M*–0.10*M*, all at pH 7.4. The ionic strength was maintained at 0.10 by the addition of NaCl. The shift in the wavelength of the maxi-

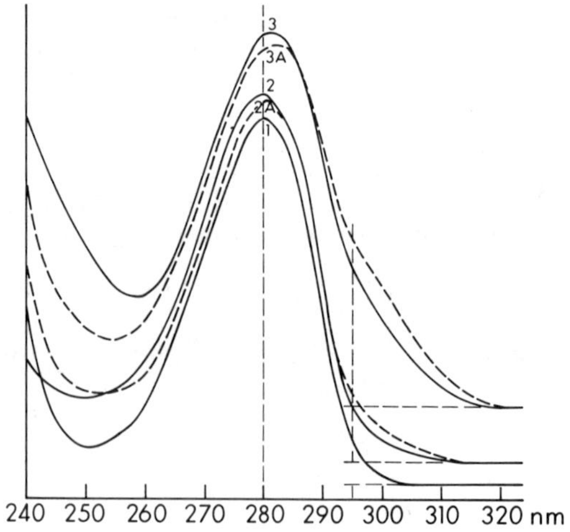

Figure 1. Effect of pH on spectra of 2 × 10⁻⁵M norepinephrine. Spectra 1, 2, and 3 are in 0.10M sodium phosphate at pH: 1, 6.60; 2, 7.45; 3, 8.27. Spectra 2A and 3A are in 0.10M imidazole at pH: 2A, 7.41; 3A, 8.23. Note the shift in maximum to longer wavelength and the increased absorbance in the region of 295 nm with increased pH in the imidazole solutions. Spectra are displaced vertically for clarity.

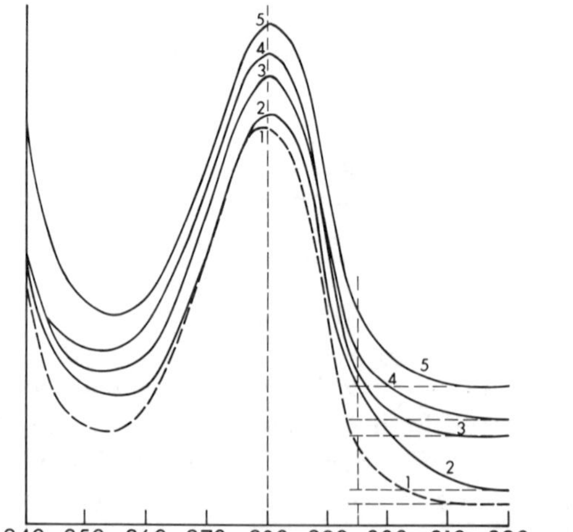

Figure 2. Spectra of 2 × 10⁻⁵M norepinephrine at pH 7.4 in 0.10M solutions of 1, sodium phosphate; 2, tris; 3, imidazole; 4, histamine; and 5, histidine. The change in wavelength of maximum and the increased absorbance at 295 nm in Solutions 2–5 compared with 1 can be seen.

mum and the absorbance near 295 nm increased with the tris concentrations. Similar changes were observed with mixed buffers prepared with tris and sodium phosphate at pH 7.4 with total buffer concentration equal to 0.10*M*. To determine whether these spectral changes resulted from oxidation, a number of spectra were obtained in the absence of oxygen. Using a Thurnberg cell with the buffer medium in the main portion of the cell and the acidified solution of catecholamine in the side arm, the solutions were deaerated by a stream of argon and the catecholamine solution was added and mixed under the argon atmosphere. Spectra obtained in this way showed the same characteristics as those obtained immediately after mixing in the normal atmosphere. Other experiments were performed to determine whether oxidation was involved in the observed changes. In numerous instances, the buffer solutions with either tris or phosphate were prepared with added 0.1m*M* Na_2EDTA and 1m*M* sodium metabisulfite to minimize the oxidation of the catecholamine. No difference was observed in the spectra taken with or without these added antioxidants. To further monitor the importance of oxidation during the course of these studies, the spectra were scanned immediately after adding the acidified catecholamine to the buffer and were completed within 2 min after mixing. The spectrum then was repeated and the absence of a change in the 280–295-nm region indicated negligible oxidation.

These studies were extended to other biologically relevant bases by using solutions of imidazole and some of its derivatives as the medium in place of the tris buffer. Spectra in imidazole resembled those obtained in tris at the same concentration and pH. Here, however, the displacement of the maximum towards the red and the broadening of the band on the long wavelength side were somewhat less than was seen with tris. Histamine and histidine also showed similar but decreased effects (*see* Figure 2). Although the results reported here deal chiefly with norepinephrine, numerous experiments with dopamine showed similar effects.

To obtain a better understanding of the interaction of tris or imidazole with the catecholamines, we studied the spectra of tyramine or normetanephrine in these systems. Tyramine and normetanephrine possess a single phenolic hydroxyl with higher pK_a values than that of the first phenolic OH in the catecholamines. At pH 7.4, the spectra of normetanephrine in 0.1*M* solutions of sodium phosphate, tris, imidazole, or histamine are alike with maxima at 278 nm and a weak shoulder in the 285-nm region (*see* Figure 3, Curves 1 and 1A). However, in 0.1*M* tris at pH 8.4 and higher, normetanephrine shows increased intensity of the shoulder at 284 and 295 nm (*see* Curves 2 and 2A). Larger differences are evident at pH 9.4 (*see* Curves 3 and 3A).

To study the effect of *N* bases when present on a membrane surface, we incorporated some of these compounds with egg lecithin into a multilamellar dispersed system. Figure 4 shows spectra of 2×10^{-5} nore-

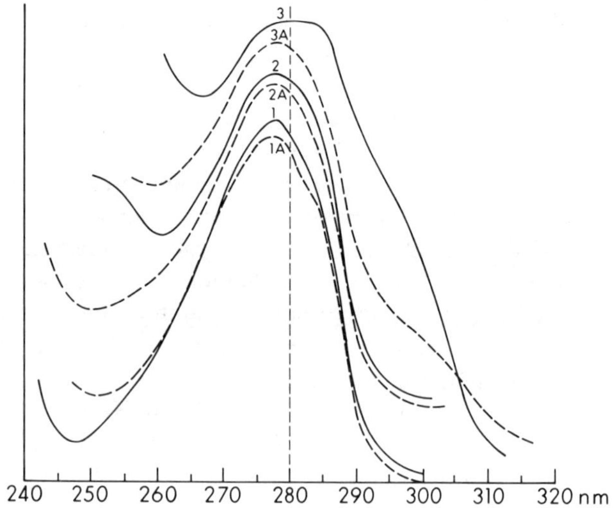

Figure 3. Spectra of 1 × 10^{-5}M normetanephrine in either: 1, 2, and 3; 0.10M tris; or 1A, 2A, and 3A; 0.1M sodium phosphate, at 1, 1A, pH 7.40; 2, 2A, pH 8.4; 3, 3A, pH 9.4. Significant changes between the spectra in the two buffer systems are evident only at high pH.

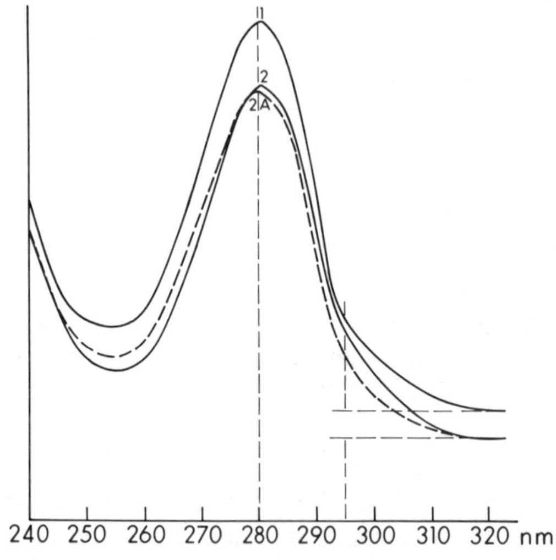

Figure 4. Effect of either imidazole or histamine in codispersions with egg lecithin on the spectra of 2 × 10^{-5}M norepinephrine: (1) dispersion of 1.2 mg/mL lecithin and imidazole to give a concentration of 3.4mM imidazole in 0.025M sodium phosphate at pH 7.74; (2) dispersion of 1.4 mg/mL lecithin with histamine to give a concentration of 1.9mM histamine in 0.025M sodium phosphate at pH 7.80; and (2A) 0.025M sodium phosphate alone at pH 7.80.

pinephrine in 0.025*M* sodium phosphate at pH 7.75. In Curve 1, a codispersion of egg lecithin and imidazole is added. The solution then contained 1.2 mg/mL lecithin and 3.4m*M* imidazole. Curve 2 depicts the spectrum in a codispersion of 1.4 mg/mL lecithin and 1.9m*M* histamine. Curve 2A is for 0.025*M* sodium phosphate without lipid dispersion. The shift in the wavelength of the maximum and the increased absorbance at 295 nm in Curves 1 and 2 compared with 2A can be noted. In another experiment in which a lecithin dispersion not containing any of the *N* bases was used, there was no change in the spectrum of norepinephrine. In still another experiment, an attempt was made to incorporate tris into lecithin for codispersion, without success. Nevertheless, a dispersion of lecithin with 3m*M* tris was prepared. This did not alter the spectrum of norepinephrine in 0.025*M* sodium phosphate at pH 7.4.

Changes in pK_a. We interpret these changes in the spectra of the catecholamines in the presence of tris and other *N* bases studied as the result of an effective decrease in their pK_a values in these media. These apparent pK_a' values were estimated by the following procedure. At pH 6.6, the catecholamines can be considered to be fully protonated, and the absorbance at 280 nm gives a measure of the concentration of the cationic species present. We calculated the absorbance at 295 nm of the same concentration of catecholamine solution in the zwitterion form by using the observed absorbance at 280 nm at pH 6.6 multiplied by the ratio of the extinction coefficients for the zwitterion at 295 nm to the cation at 280 nm as given by Kappe and Armstrong (3). Using the same stock solution of acidified catecholamine and the same final concentration of catecholamine in all experiments in a series, we measured the absorbance at 295 nm under different conditions. This absorbance includes contributions from both forms of the catecholamines and therefore needed correction for the small absorbance of the unionized species at 295 nm. The absorbance at 295 nm (A_{295}) of a sample gave the fraction α present in the zwitterion form according to the following

$$A_{295} = (1 - \alpha)\,\mathrm{B} + \alpha \mathrm{C}$$

where *B* is the absorbance at 295 nm of the fully protonated form and *C* is for the zwitterion alone. The pK' was obtained in the customary manner

$$\mathrm{p}K' = \mathrm{pH} + \log \frac{1 - \alpha}{\alpha}$$

The pK_a' values obtained in this way in solutions of tris, imidazole, histamine, and histidine are shown at several pH levels in Table I. Tris produces the greatest decrease of the pK_a' of norepinephrine. Also, the lowering of the pK_a' is less at higher pH levels.

Table I. Apparent pK_a for 2 $\times 10^{-5}M$

*0.10*M *Tris*		*0.10*M *Imidazole*		*0.10*M *Histamine*	
pH	pK_a'	*pH*	pK_a'	*pH*	pK_a'
6.60	7.91	6.61	8.19 ± 0.10		
7.02	8.08 ± .02	7.01	8.45 ± .05	7.02	8.30
7.42	8.36 ± .03	7.41	8.58 ± 0.17	7.33	8.54 ± .04
7.82	8.49 ± .02	7.75	8.74 ± 0.05	7.80	8.73 ± .02
8.19	8.57 ± .01	8.14	8.81		

To determine whether such interactions of catecholamines with *N* bases would take place on membrane surfaces, imidazole, histamine, or histidine was mixed with egg lecithin to form codispersions in phosphate buffer. Spectra then were obtained with a concentration of catecholamine in the final system the same as that used in the previous series of experiments, and the pK_a' values were calculated in the same manner. Table II gives these values. As seen from the table, systems containing imidazole in concentrations as low as 2.45m*M* gave pK_a' values for norepinephrine lower than those obtained in phosphate buffer systems alone. The addition of dispersed lecithin without imidazole did not produce this decrease in the pK_a' values. Histamine and histidine, which were more difficult to incorporate into membranes, showed smaller changes in the pK_a values. In these membrane systems, the effect of pH on the observed pK_a' values

Table II. Apparent pK_a for 2 $\times$ $10^{-5}M$ Norepinephrine in the Presence of Various Lecithin Codispersions

Dispersion	*Medium*	*pH*	pK_a'
Lecithin–Imidazole			
*2.45m*M *imidazole*		7.00	8.50 ± 0.15
1.45 mg/mL lecithin	0.025*M* PO_4	7.40	8.51 ± 0.15
*3.4m*M *imidazole*		7.00	8.50 ± 0.06
1.21 mg/mL lecithin	0.025*M* PO_4	7.39	8.61 ± 0.05
		7.78	8.72 ± 0.02
		8.17	8.84 ± 0.03
*15m*M *imidazole*		7.07	8.25 ± 0.16
2.2 mg/mL lecithin	0.033*M* PO_4	7.40	8.41 ± 0.15
		7.79	8.53
Lecithin–Histamine			
*1.0m*M *histamine*	0.025*M* PO_4	7.40	8.80 ± 0.03
1.4 mg/mL lecithin		7.77	8.60 ± 0.01
		8.20	8.92 ± 0.01
Lecithin–Histidine			
*1.3m*M *histidine*	0.025*M* PO_4	7.00	8.78 ± 0.10
1.33 mg/mL lecithin		7.38	8.85 ± 0.03
		7.82	8.81 ± 0.02
		8.15	8.68 ± 0.16

Norepinephrine in Different Media

0.10M Histidine		0.10M Sodium Phosphate	
pH	pK_a'	*pH*	pK_a'
		7.04	8.93
7.35	8.63 ± .02	7.46	8.87 ± 0.01
7.41	8.65	7.87	8.85 ± 0.02
7.47	8.70 ± .03	8.27	8.91 ± 0.01

is not as definite as in the systems in solution. Changes in the structures of the dispersed lipids associated with pH changes could make the effects more complex.

Discussion

The results of this study show that such *N* bases as tris, imidazole, or compounds containing the imidazole ring could accept protons from a weak acid group and increase the ionization of the latter. In the instance of tris with $pK_a = 8.3$, in the pH range studied, a relatively high concentration exists in the deprotonated or base form which is able to accept a proton. Although this effect may occur also in interactions of similar *N* bases with other molecules which possess a weak oxygen acid, in most instances it is difficult to directly detect any increased ionization. The fact that the catecholamines show distinctly different spectra for the protonated and deprotonated phenolic hydroxyl makes it possible to study this effect.

Several studies have been reported dealing with the spectra of catecholamines at different pH levels. Spectroscopic methods possess the advantage in that they are not influenced by changes in ionogenic groups present on the alkyl amine chain, but do indicate the ionization of the phenolic groups. Lewis (*4*) used both potentiometric and spectroscopic procedures to determine the ionization constants of a number of amines. He showed that reasonably good agreement was obtained for the pK_a values by the two methods, provided that there was a sufficient difference between the ionization of the phenolic and amine groups. Another study of the ionization of some phenolic amines by Kappe and Armstrong (*3*) also showed the applicability of spectroscopic methods for the determination of the ionization properties. In these studies the only buffer systems used were glycine or glycylglycine. Spectra taken in our study showed that these compounds had no effect on the ionization of the catecholamines. In neither of these studies was there any attempt made to determine whether there was an interaction of the catecholamines with any molecule present in the solution.

In a later study, Martin (*5*) calculated the microconstants for a number of equilibria involving the various forms of the phenolic and catecholic alkyl amines. His work also was based on spectroscopic and potentiometric measurements. Additionally, he determined a ratio for the concentrations of the zwitterion and the neutral molecular species. This showed that for the catecholamines, the zwitterion form was present in a higher concentration than the neutral molecule. Still, another potentiometric study by Antikainen and Witikainen (*6*) showed some interesting relations between the structures of the catecholamines and their acidic properties. They showed that the presence of an alkyl side chain increases the acid strength of the phenolic groups, and that changes in the side chain did not alter the ionic properties of the phenolic group but did alter the ionization of the protonated amine so that the ionization of this group in adrenaline is less than for norepinephrine. An exception to the aforementioned effect is that the presence of a methyl group in the α position of the side chain increases the acid strength of the ring hydroxyl. These subtle changes in ionization can enter into the biological activity of the naturally occurring catecholamines as well as some pharmacological agents with related structures by altering the concentration of the zwitterionic form present at physiological pH.

There are some studies of these ionic characteristics of the amines and their physiological activities in various systems. Lewis (*4*) examined the effect of the amines on smooth muscle from rabbit uterus or intestines. His findings indicated that there was no direct relationship between the pharmacological activity and the pK_a of the amines. However, the amines which did stimulate muscle had closely similar ionization and at a fixed pH no conclusion can be drawn relating ionization and activity. Another study of the effect of pH on the physiological activity of epinephrine (*7*) was performed by measuring the depression of the spontaneous activity of an isolated segment of ileum from rabbit. The effect of the epinephrine decreased as the pH increased from 7.4 to 8.6. In this range, there is a significant decrease in the concentration of the cationic form which is suggested as the active form. They pointed out, however, that at an elevated pH and a prolonged incubation period, especially in an oxygenated medium, there could be loss of epinephrine as a result of oxidation. We have reported similar effects for dopamine oxidation (*8*). Our results of the pH dependence of the initial uptake of norepinephrine by rat brain synaptosomes (*1*) revealed that the pH-uptake profile has a maximum near pH 8.2 at 37°C. This suggests that the zwitterion form of the amines most likely is the active species for this process.

The possibility of the involvement of the phenolic OH group in biological systems is supported further by the present spectroscopic study of the chemical interaction of this group with a *N* base in aqueous media.

The pK_a for imidazole at 25°C is 7.14 (*9*). In the pH range studied in this work, more than 50% of the imidazole is present as the free base with an electron pair on the *N* which may accept a proton from a weak acid whose p*K* is not too much greater than that of imidazole, leading to a decrease in the apparent p*K* of the weak acid. In this respect, with increasing pH, the concentration of the base form of imidazole increases and hence, we would expect to find a further decrease in the pK_a' of the catecholamine. Our data, however, shows a small increase (Table I). This effect may be explained in part by the fact that at higher pH, the contribution of the imidazole reaction becomes a smaller fraction of the observed change in spectra because of the change resulting from the pH increase alone. Oxidation effects also add uncertainty to the measurements at higher pH. The effect of imidazole as a proton acceptor in the ionization of the catecholamines is not a unique example of this reaction. There are numerous instances of hydrogen bonding and proton-transfer reactions in aqueous or aprotic systems involving imidazole (*10*). Furthermore, for enzymatic systems, it is well established that in a serine protease such as α-chymotrypsin, the active site imidazole (Hist 57) serves uniquely as a proton donor and a proton acceptor simultaneously (*7*).

The experiments performed with lecithin dispersions containing imidazole or related compounds present some interesting results. The total concentration of *N* base used in these systems ranged from 1.2m*M*–15m*M*. Simple solutions of imidazole at these concentrations in the absence of lecithin showed no effect on the pK_a' of norepinephrine. It is possible that, if some imidazole is released from the lecithin codispersions to the surrounding solution, its concentration would be too low to have an effect. The fact that we observed an effect when dispersions containing the *N* base were used indicates that the reaction probably takes place at the lamellar surface of the codispersion. In an experiment in which lecithin was added to phosphate buffer and then imidazole was added to give a concentration of 2m*M*, no change was observed in the spectrum of norepinephrine. This indicates that because the increased deprotonation was observed in the experiments with codispersions, the reaction probably occurred at the surface of the lipid at the sites where the *N* base was exposed to aqueous medium.

Measurements of the ionization characteristics of catecholamines in earlier studies (*3, 4, 5, 6*) led to the widely held view that at physiological pH the predominant form of these neurotransmitters was the cation. This implies that in the nervous system the active form is probably cationic. However our earlier work (*1*) which showed that the zwitterion form was taken up preferentially by synaptosomes presented questions concerning the relatively low concentration of this species at pH 7.4. The significance of our present work resides in the fact that the findings described above show that the possibility exists for the interaction of the phenolic hydroxyl group of the catecholamines with various general bases present in biological systems. The influence of the ionization of the phenolic group of norepinephrine on its uptake by rat brain synaptosomes may be attributed to a specific interaction of the hydroxyl group with a base present at the uptake site to form the zwitterion. It may be interesting to speculate that the phenolic OH group is involved with a base present at the receptor site of the post-synaptic membrane leading to the formation of the zwitterion species.

Acknowledgments

The authors wish to thank F. Gaskin and D. O. Shah for making available their spectroscopic equipment and for their interest in this work.

Literature Cited

1. Choi, M.; Abramson, M. B. *Biochim. Biophys. Acta* **1978**, *540*, 337–345.
2. Tanizawa, K.; Bender, M. L. *J. Biol. Chem.* **1974**, *249*, 2130–2134.
3. Kappe, T.; Armstrong, M. D. *J. Med. Chem.* **1965**, *8*, 368–374.
4. Lewis, G. P. *Br. J. Pharmacol. Chemother.* **1954**, *9*, 488–493.
5. Martin, R. B. *J. Phys. Chem.* **1971**, *75*, 2657–2661.
6. Antikainen, P. J.; Witikainen, U. *Acta Chem. Scand.* **1973**, **27**, 2075–2082.
7. Reynolds, R. C.; Hardman, H. F. *Eur. J. Pharmacol.* **1972**, *20*, 249–255.
8. Abramson, M. B.; Choi, M.; Katzman, R. *Fed. Proc., Fed. Am. Soc. Exp. Biol.* **1976**, *35*, 1647.
9. Paiva, A. C. M.; Juliano, L.; Boschow, P. *J. Am. Chem. Soc.* **1976**, *98*, 7645–7648.
10. Robinson, B. H. In "Proton-Transfer Reactions," Calder, E. F., Gold, V., Eds.; Chapman and Hall: London, 1975; p. 121–152.

RECEIVED October 17, 1978.

18

Protein Phosphorylation and Bioelectrogenesis

E. SCHOFFENIELS

Department of General and Comparative Biochemistry, University of Liège, 17 place Delcour, B-4020 Liège, Belgium

High-molecular-weight proteins extracted from the walking nerves of Crustacea with SDS undergo a cycle of phosphorylation–dephosphorylation that is influenced by electrical stimulation, the ionic composition of the medium, and compounds such as local anaesthetics, diphenylhydantoin, veratridine, and TTX. It is suggested that the ion-gating mechanism in axonal membranes is controlled by the net state of specific protein phosphorylation.

For the past few years I have entertained the idea that phosphorylation of specific proteins in nerve membranes could be associated directly with the production of an action potential. The rationale behind this idea and the first experimental evidence were presented at the Second International Meeting on Torpedo in 1976 (*1*). Briefly stated, it is obvious that electrical activity in conducting membranes is a dissipative process, therefore irreversible: the analysis of the ionic currents and of the thermal events contemporary to the conductance changes unmistakably reveal that a chemical has to be used up as in any irreversible process occurring in a biological system (*2, 3*). For more than 20 years Nachmansohn has proposed repeatedly that acetylcholine in conjunction with adequate enzymes could be the specific operating substance controlling the configuration of a receptor protein (*4*). More recently, calcium ions have been included in the process (*5*). However there are still many uncertainties as to the exact nature of the role of acetylcholine and calcium ions. Moreover, since the enthalpy of hydrolysis in acetylcholine in the presence of a complex buffering medium is $+7$–$+12$ Kcal mol^{-1} (*6*), it is incompatible with the heat absorption (i.e. $\Delta H < 0$ according to the sign convention of the biochemists) observed by Hill

0-8412-0473-X/80/33-188-285$05.00/1

and his associates (*7*) during the descending phase of the action potential, an event that according to Nachmansohn should be controlled by the acetylcholine hydrolysis.

These facts lead to the idea that another energy-producing process must be evoked to explain the dissipative nature of the impedance variation cycle (IVC) described for the first time by Cole and Curtis in 1939 (*8*). I already have proposed that IVC should be controlled by enzymes (*9*). I would like to report on a phosphorylating system located in nerves and the properties that are affected by the electrical stimulation, the ionic composition of the medium, local anaesthetics, TTX, veratridine, and more generally by those compounds known to affect the electrical activity of nerves in vivo.

Two types of methodology have been applied. Membranes are prepared from the nerves of the walking legs of various species of crabs (*Carcinus moenas, Eriocheir sinensis,* and *Maja squinado*) and the phosphorylation of this membrane fraction is studied in various incubating medium using $\{\gamma - {}^{32}P\}$ ATP as the substrate in concentrations as low as $10^{-8}M$. (This work is done in collaboration with G. Dandrifosse, is reported elsewhere (*11*).) Since one assumes that an enzymatically controlled dephosphorylation process is taking place, membrane fragments also are used to study the properties of the phosphoprotein phosphatase(s) that exist in our preparation. This work, carried out in collaboration with P. Wins and G. Dandrifosse, also will be published elsewhere.

In a second type of experiment intact nerves isolated from the above species of crustacea are submitted to electrical stimulation or conditions known to affect electrical activity, and phosphorylation of the proteins separated by gel electrophoresis is examined.

Experimental Methods

Animals. The experiments were done on nerves isolated from the walking legs of three species of Crustacea: *Eriocheir sinensis, Carcinus moenas,* and *Maja squinado.* However most of the experiments were done with *E. sinensis* that had adapted to fresh water. The saline used as incubation medium was either sea water for the sea-water-adapted animals or 50% sea water in the case of *E. sinensis.* Both media were supplemented with 1m*M* phosphate.

All of the experiments were done using nerves isolated from the same side of the animal and were compared with control nerves isolated from the opposite symmetrical legs. The temperature of the medium was 18° ± 1°C.

Electrical Activity Measurement. A single nerve was stimulated through external silver–silver chloride electrodes using a Grass–Stimulator or a Tectronix pulse generator complemented with a stimulus isolation unit.

The electrical activity was recorded with external silver–silver chloride electrodes on a Tectronix oscilloscope. The maximum amplitude of the biphasic potential was generally close to 6 mV. The stimulating and recording electrodes were placed at both ends of the nerve thus leaving a large portion of nerve in between (up to 6 cm) dipping into a beaker containing the saline. In order to keep that portion of the nerve in the liquid, a small hook made of glass was hooked into the middle of the nerve. Electrical stimulation was performed at 10 sec^{-1} during 20-min periods.

After stimulation the nerve is placed in 250 μL of a stop solution brought to pH 7.2 and containing 8% sodium dodecyl sulfate (SDS), 25m*M* mercaptoethanol, 10m*M* NaH_2PO_4, and 6mM EDTA. Up to 12 nerves are pooled in both experiment and unstimulated control. The nerves then are ground with a small pestle in the presence of sea sand which has been washed carefully with hydrochloric acid and distilled water. The suspension is transferred into a small dialyzing bag and dialyzed overnight at room temperature against 40 mL of the above stop solution that is changed four times during the process.

SDS–Polyacrylamide Gel Electrophoresis. A known volume of solution is collected from the dialyzing bag. One aliquot is used for protein determination according to Ref. *10*. To the known volume 10 μL of tracking dye (bromophenol) and 50 μL of a saturated solution of sucrose is added.

Two concentrations of polyacrylamide have been used. In the first set of experiments the stacking gel had a concentration of 5% while that of the resolving gel was 10%. In other experiments reported elsewhere the concentrations were 3.2% and 5%, respectively.

The electrophoresis was run for about 4 hr at a constant current of 4 mA per cylindrical gel with a running buffer of 50m*M* tris HCl, 2m*M* EDTA, 375m*M* glycine, and 0.1% SDS adjusted to pH 8.3. The gel was stained for protein with 0.06% Coomassie blue (G 250) in 45% methanol–7% acetic acid for at least 4 hr. It was destained overnight in 7% acetic acid with several changes.

The gels were sliced according to the positions of some bands (*see* Figure 1) and the radioactivity was measured by liquid scintillation counting (Lumagel) after dissolving the gel in 0.5 mL H_2O_2 overnight in an oven at 50°C. The results are expressed in counts per 10 min after correcting for the protein concentration.

In some experiments the sub-bands 1, 2, or 3 were not separated, therefore, when the results are given for a letter without a numeral, they refer to the total activity of the slice identified by the letter.

Chemicals. SDS, sea sand, glycine, sucrose, bromophenol blue, EDTA, methanol, NaH_2PO_4, and ammonium persulfate were bought from Merck. Mercaptoethanol, tris, tetrodotoxin (TTX), tetracaine, and ouabain were supplied by Sigma Chemical Co. Bis-acrylamide, acrylamide, Temed, and Coomassie blue (G 250) were bought from Serva.

^{32}P as orthophosphate (sodium salt) was prepared as neutral sterile and injectable solution by I.R.E., Fleurus, Belgium, with a specific activity of 100 mCi/mg phosphorus.

Experimental Procedure. The labeling of endogenous ATP was obtained according to three different experimental procedures.

1. Radioactive phosphate is added to the saline in which the nerves are submitted either to electrical stimulation or to the action of a compound. The maximum activity used is 10 mCi per 80 mL of saline. The experiments performed in this condition are referred to as incubation in ^{32}P.
2. Nerves are incubated for 30 min in a radioactive saline. Afterwards these are submitted to electrical stimulation or to the action of a compound in a cold saline. This type of experiment is referred to as preincubation in ^{32}P.
3. About 2.5 mCi of radioactive phosphate are injected into the intact animals. The nerves subsequently are isolated at different time intervals.

Results

Before investigating in detail the various means of controlling the protein phosphorylation in nerves it was necessary to perform some elementary control experiments. Accordingly after obtaining a constant reproducible pattern of electrophoresis (*see* Figure 1) we would demonstrate that after incubating nerves in a saline solution containing ^{32}P as orthophosphate, the radioactivity was distributed among all of the bands. Since we suspected that part of this radioactivity could be the result of contamination by low-molecular-weight components (including orthophosphate) and occurring not only during the electrophoresis but also during the destaining process, we have decided to submit our material to a thorough dialysis prior to electrophoresis. This has improved considerably the results, and after dialysis the contamination by the destaining solution is abolished completely. However all of the bands still remained labeled but the highest activity is located from Band A to D (*see* Figure 1).

Also, if it may be assumed reasonably that the radioactivity detected along the electrophoretic pattern is the result of some sort of phosphorylation of proteins, it was imperious to demonstrate that we were dealing with proteins. Using pronase completely destroys the electrophoretic pattern, while ribonuclease has no effect on it. From these results we conclude that we are dealing with proteins. However at this stage we have no idea as to the type of bonding phosphate involved, but we may be sure that any low-molecular-weight phosphorylated compound that is adsorbed loosely on the protein has been removed during dialysis and electrophoresis.

Another question is whether or not inorganic phosphate, as we used it, is the substrate of reaction. Though no inhibitor of oxidative phosphorylation has been used so far in our experiments, we have, together

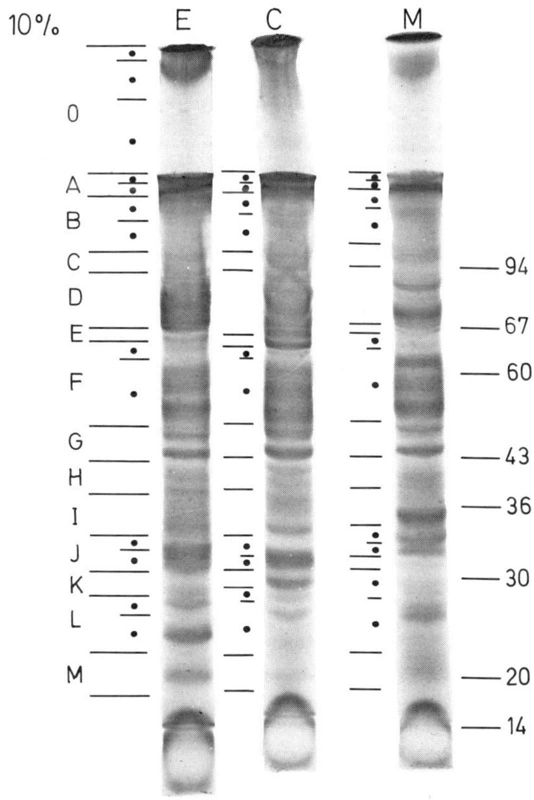

Figure 1. SDS gel electrophoresis pattern obtained with nerves of E. sinensis *(E),* C. moenas *(C), and* M. squinado *(M) extracted in SDS solution as indicated in the text. The lines and letters on the LHS indicate the position and numbers of cutting. The dots mean 1, 2, or 3, e.g.* O_1, O_2, O_3; A_1, A_2. *The apparent molecular weight* ($\times 10^{-3}$) *indicated on the RHS has been obtained with a Pharmacia calibration kit. Bands* O_1, O_2, *and* O_3 *correspond to the stacking gel (concentration of 5%). Bands A to M: resolution gel at a concentration of 10%.*

with some results reported below, indirect evidence that ATP is indeed the substrate. A membrane fraction of nerves never is phosphorylated when inorganic phosphate is used as substrate, while with the concentration of $\{\gamma - {}^{32}P\}$ ATP as low as $10^{-8}M$ a significant phosphorylation of the membrane proteins is observed. Moreover this process is affected by agents known to interfere with bioelectrogenesis such as local anaesthetics, veratridine, TTX, etc. (*11*).

As indicated by the results of Table I electrical stimulation affects the labeling of the proteins. Since we observe a decrease in the radioactivity this could well mean that (a) the fluxes of orthophosphate are

Table I. Effect of Electrical Stimulation on Protein Phosphorylation in Nerves of *E. sinensis* and *M. squinado* (Counts per 10 min): Incubation Was in ^{32}P; Gel Concentration Was 10%.

	E. sinensis		M. squinado	
	Control	*Stimulated (10 sec⁻¹; 20 min)*	*Control*	*Stimulated (10 sec⁻¹; 20 min)*
A_1	432	279	218	70
A_2			183	80
B_1	504	176	251	18
B_2			749	231
C	74	59	558	352
D	546	299	1262	492

Table II. Protein Phosphorylation in Nerves of *E. sinensis* (Counts per 10 min): Effect of 5×10^{-5} Ouabain; Gel Concentration Was 10%.

	Incubation in ^{32}P [a]		*Preincubation in ^{32}P (30 min)* [b]	
	Control	*Ouabain*	*ONS*	*OS*
A_1	855	290	66	58
A_2	229	88	65	75
B_1	333	93	121	9
B_2	492	348	1390	445
C	461	180	—	—
D	2074	1281	910	804

[a] In this experiment the nerves are incubated for 30 min in a radioactive saline solution with or without ouabain. No stimulation.
[b] ONS = incubation in ouabain without stimulation; OS = incubation in ouabain with stimulation. In this experiment the nerves were preincubated in radioactive saline solution for 30 min and then transferred in a cold saline solution containing ouabain. Frequency of stimulation 10 sec^{-1}, duration 20 min.

Table III. Protein Phosphorylation in Nerves of *E. sinensis* (Counts per 10 min). Effect of Electrical Stimulation (10 sec^{-1} During 20 min) in a Cold Saline after Preincubation for 30 min in Radioactive Saline Solution. Gel Concentration Was 10%.

	Control	*Stimulated*
A	1135	760
B	996	1138
C	414	59
D	1916	2080

affected by the electrical activity of the nerves; (b) the conductance of the membrane is controlled by the net phosphorylation state of specific proteins; and (c) the metabolic changes associated with repetitive nerve firing induces changes in ATP concentration or specific radioactivity. However, note that (a) and (c) do not necessarily exclude (b).

That a decrease in phosphate influx could explain part of the data is illustrated by the results of Table II obtained in the presence of ouabain (c (control) vs. o (ouabain); unstimulated). Ouabain is an agent known to affect the phosphate influx in squid giant axons (*11*) although it has been demonstrated with the same biological material that electrical stimulation has no effect on the phosphate influx (*12*).

However, if the nerves are preincubated in radioactive phosphate for 30 min prior to the application of ouabain thus permitting the biosynthesis of radioactive ATP in the nerve, the situation is rather different. If one applies ouabain to both control nonstimulated (ONS) and stimulated nerves (OS), only Bands B_1 and B_2 are affected. Therefore, independently of a possible effect on the phosphate influx, electrical stimulation modifies the phosphorylation pattern.

The electrical stimulation performed on nerves that have been preincubated in a radioactive saline solution also shows a modification of the phosphorylation pattern (*see* Table III). These results hardly could be explained solely through an increase in phosphate efflux, assuming that electrical activity affects the movements of labeled phosphate in Crustacean nerves in the same way that it affects garfish olfactory and rabbit vagus nerves (*13*).

Electrical activity certainly alters the energy metabolism of the nerves and one therefore should expect a change in ATP concentrations and in ATP specific radioactivity. In order to further check out this possibility, crabs have been injected with radioactive phosphate and the nerves have been isolated after 2- and 4-hr periods. Table IV shows that electrical stimulation increases the labeling of Bands A, B, and D, which apparently is in contradiction with the results of Table I. Though other interpretations are possible, it is reasonable to assume that the phos-

Table IV. Protein Phosphorylation in Nerves of *E. sinensis* (Counts per 10 min). Crabs Injected with ^{32}P as Indicated. Frequency of Stimulation 10 sec^{-1} for 20 min. Gel Concentration Was 10%.

	120 min		*240 min*	
	Control	*Stimulated*	*Control*	*Stimulated*
A	413	925	3918	6703
B	832	1079	2807	5803
C	303	249	1107	577
D	1082	1647	3686	8098

phorylation of proteins proceeds via at least two pools of ATP having different size and/or turnover numbers, one of which is being preferentially used during the electrical stimulation. When stimulation occurs in the radioactive saline solution, the specific activity of the ATP pool being used is lower than that of the pool that is used in resting conditions. Thus this explains the results of Table I. In this interpretation one assumes that electrical activity increases the rate of the phosphorylation–dephosphorylation cycle. On the other hand, a preincubation of 30 min in a radioactive saline solution (*see* Table III) or the injection of radioactive phosphate into the animals 2 or 4 hr prior to the experiment (*see* Table IV) allows the specific activity of the pool used during electrical stimulation to be higher, thus leading to an increase in the labeling of some proteins after electrical stimulation. Therefore, depending on the specific activity of the pool supplying ATP, the radioactivity of the protein should be lower or higher than that of the unstimulated control. Results obtained with crabs injected with radioactive phosphate 24 or 48 hr prior to the experiments show that the patterns of labeling of stimulated and control nerves are identical; thus they are in agreement with the idea that a steady state has been reached with respect to the specific activity of the different phosphorylated compounds.

The results presented so far do not bring any evidence that (a) the proteins—the phosphorylation of which is affected by electrical stimulation—are located in the axonal membrane and that (b) they are specifically involved in the ion-gating mechanism.

Until we have isolated and characterized some of the proteins making up the electrophoretic pattern (a problem we are now working on), we have to use indirect evidence to support further our hypothesis.

Since electrical activity is produced by an extra influx of sodium and an added loss of potassium ions, one may expect a temporary change in intramembraneous ion concentration and therefore a differential effect of sodium and potassium ions on the phosphorylation process. Table V

Table V. Protein Phosphorylation in Nerves of *E. sinensis* as Percent of Control in SDS. Nerves Preincubated 30 min in ^{32}P. Gel Concentration Was 10%. Explanations in Text.

	KCl + ATP *(10^{-3}M)*	*KCl*	*NaCl*
A_1	101	444	4
A_2	181	163	12
B_1	[a]	[a]	[a]
B_2	[a]	[a]	[a]
C	[a]	[a]	[a]
D	48	43	43

[a] Not measurable.

Table VI. Effect of TTX ($8 \times 10^{-8}M$) on Protein Phosphorylation in Nerves of *C. moenas* (Counts per min). Preincubation in TTX = 20 min followed by Incubation in TTX + ^{32}P. Gel Concentration Was 10%.

	Control	*TTX*
A_1	1014	2176
A_2	1822	1857
B_1	1439	1092
B_2	1625	1242
C	1123	860
D	1215	1261

shows that this is indeed the case with the proteins located in Band A_1, i.e. the high-molecular-weight components in our electrophoretic pattern (*see* Figure 1).

The experiment is done as follows. Nerves are incubated 30 min in a saline solution containing radioactive inorganic phosphate. They are divided into six groups of 12 symmetrical nerves. After the incubation period one group is placed in 250 μL of stop solution (control group) while the symmetrical group of nerves is incubated in a medium containing 120mM KCl + ATP $10^{-3}M$. After 10 min of exposure to this condition, the nerves are place in 250 μL of stop solution. The same type of procedure is applied to the other groups of symmetrical nerves except that the experimental group is incubated 10 min in KCl or in NaCl. It is obvious that K ions favor the phosphorylation process of the proteins located in Band A_1, while Na ions enhance the dephosphorylation process. When the specific radioactivity of endogenous ATP is lowered by adding a large amount of exogenous cold ATP to the K medium, the radioactivity located on the protein Band A_1 is lower than with K ions alone.

There is now direct electrophysiological evidence that compounds such as TTX, veratridine, and local anaesthetics, etc. modify in a very specific manner the gating processes responsible for the generation of an action potential.

In order to investigate further the possibility that protein phosphorylation could be involved directly in bioelectrogenesis, these compounds have been tried. With nerves isolated from *E. sinensis,* diphenylhydantoine at saturation decreases markedly the labeling of proteins in Bands A_1 and B_1 and has a less pronounced effect on the labeling of the other bands. Veratridine and tetracaine also alter markedly the phosphorylation pattern.

As shown by the results of Table VI obtained on nerves isolated from *C. moenas,* TTX increases the phosphorylation in Band A_1 and decreases it in Bands B_1, B_2, and C.

The results presented so far show that electrical stimulation and the compounds known to affect the process of bioelectrogenesis interfere with the phosphorylation of proteins mainly located in a region of the SDS gel electrophoresis where the molecular weights are higher than 60.000 (*see* Figure 1). Proteins in the A_1 band seem to be the most affected.

A decrease in the reticulation of the gel in order to improve the resolution in that region was indicated. After a few trials, it appeared that a 5% concentration was adequate: there was indeed a much better separation of the high-molecular-weight components. Results obtained as a result of this improved resolution are discussed elsewhere (*15*).

Discussion

Since the phosphorylation of endogenous proteins was reported first by Johnson et al. (*16*) there has been a large accumulation of data showing that it is a rather general process controlling most significantly biological events (*16, 17, 18*). Among the processing reactions that modify the structure of a protein and therefore its biological activity, covalent modifications such as phosphorylation or glycosylation are certainly of the utmost importance. Phosphorylation of serine or threonine residues are transient processes fluctuating in response to various stimuli. Hence the net state of protein phosphorylation fluctuates accordingly together with their biological activity. Examples may be found in two reviews recently published (*17, 18*). Surprisingly enough besides our suggestion made some years ago (*1, 19, 20*), protein phosphorylation never has been proposed as a means of controlling the IVC responsible for the action potential in axonal membranes.

However it generally is accepted that proteins are the molecular structures that control the impedance of the conducting membranes. One of the most pertinent arguments to support this is the high specificity exhibited by the conducting membrane towards the cations carrying the action current. So far there is little direct evidence as to the molecular structure controlling the ion movement. Are we dealing with an oligomeric structure or with a single large protein? At any rate it may be useful to coin a name for the protein(s) controlling the cation movements during nerve activity and I suggest calling that class of proteins "conductin," leaving the possibility of specifying sodium conductin, potassium conductin, or calcium conductin if it turns out that different molecular species control the sodium, potassium, and calcium currents.

The results presented in this chapter, if they do not demonstrate that the net state of phosphorylation of some specific proteins determines the conductance of the axonal membrane, point nevertheless very strongly to the possibility that such a mechanism indeed exists. They are significant enough when taken together to encourage us to explore further this approach in trying to solve the molecular basis of bioelectrogenesis in nerves.

When considering the results of the electrophoresis it is the highest-molecular-weight components that are affected consistently either by the electrical stimulation or by tetracaine, TTX, etc.

If one assumes that the major proteins making up the IVC are high-molecular-weight components, the results obtained are in reasonable agreement with the idea that in resting conditions the sodium conductin is phosphorylated while at the peak of the action potential it is dephosphorylated thus corresponding to the high conductance state of the membrane to sodium ions. Therefore the electrical stimulation should increase the turnover of phosphate from phosphoproteins. Depending on the specific activity of the pool supplying the ATP the radioactivity of the protein should be lower (*see* Table I) or higher (*see* Tables III and IV) and that of the unstimulated control. Results obtained with crabs injected with radioactive phosphate 24 or 48 hr prior to the experiments show that the patterns of labeling of stimulated and control nerves are identical; thus they are in agreement with the idea that a steady state has been reached with respect to the specific activity of the different phosphorylated compounds.

On the other hand, since a local anesthetic blocks the action potential without depolarization, its action on the protein could be explained by the inhibition of the dephosphorylation process leading to a slowing down of the phosphate turnover. As for the effect of TTX, an agent known to block the sodium current by preventing the increase in conductance, its effect could be explained in terms of an increase in the net state of sodium conductin phosphorylation, a situation typical of rest.

This is obviously mere speculation. Nothing is known about the proteins that form the pathways for the ion movements. Until we know if we are dealing with oligomeric structures or a single, large protein, it will be difficult to exclude the possibility that some of the lower-molecular-weight components might be protomeric units dissociated by our experiment conditions, such as SDS extraction.

Moreover until we have isolated and characterized thoroughly the conductins it is obviously impossible to ascertain that the conducting state of the membrane is indeed controlled by the net state of phosphorylation of these intrinsic proteins. The only strong argument so far in favor of our hypothesis is that compounds that are supposed to act in a specific manner on the generation of an action potential also act on the phosphorylation process not only on the intact nerve but also on membranes prepared from nerves (*11*).

The alternative to our interpretation is that repetitive nerve firing profoundly affects the energy metabolism and therefore the phosphorylation state of proteins that are not necessarily involved directly in bioelectrogenesis. The sole merit of our results then would be to demonstrate that most of the compounds used to study the ionic currents such as TTX, veratridine, and the like do not have the specificity one likes to believe they have.

Finally it may be argued that some of the results observed may be an expression of a change in phosphate influx. We do not have information regarding the control of phosphate fluxes in Crustacea nerve. However in squid giant axon the influx is independent of electrical stimulation but is inhibited by ouabain (*12, 13*). This is obviously the case with *E. sinensis* as shown by the results of Table II; the incubation of nerves in a radioactive saline solution shows a decrease in the labeling of all the bands with respect to the control in the presence of ouabain. Now if the nerves are incubated first in a radioactive saline solution in order to label at least part of the endogenous ATP and then stimulated in a cold saline solution in the presence of ouabain, only Bands B_1 and B_2 show a decrease in radioactivity with respect to the unstimulated control also in contact with ouabain.

If one compares these results with those obtained under the same conditions but without ouabain (*see* Table III), a reasonable conclusion is that the phosphorylation of the various bands must proceed from different pools of ATP that have not yet obtained the same level of specific radioactivity.

Other interpretations are certainly as valid as previous proposals but it is too early to comment further on this situation before knowing more about the number of ATP pools as well as their size and turnover rate.

Acknowledgments

Many thanks to my colleague G. Dandrifosse for many helpful discussions and comments.

Thanks also are extended to G. Pirard for her able technical assistance. This work was aided by Grant No. 2.4544.76 from the Fonds de la Recherche Fondamentale Collective.

Literature Cited

1. Schoffeniels, E. Phosphorylation and Bioelectrogenesis, 2nd International Meeting on Torpedo, Arcachon, France, Sept. 1976.
2. Margineanu, D. G.; Schoffeniels, E. *Proc. Natl. Acad. Sci. USA* **1977**, *74*, 3810–3813.

3. Neumann, E. *Adv. Biol. Med. Phys.* **1977,** *16,* 235–239.
4. Nachmansohn, D. *Harvey Lect.* **1955,** *49,* 57–99.
5. Neumann, E. In "Biochemistry of Sensory Function"; Jaenicke, L., Ed.; Springer–Verlag: Berlin and New York, 1974; pp. 456–514.
6. Sturtevant, J. M. *J. Biol. Chem.* **1972,** *247,* 968–969.
7. Abbott, B. C.; Hill, A. V.; Howarth, J. V. *Proc. R. Soc., Ser. B* **1958,** *148,* 149–187.
8. Cole, K. S.; Curtis, O. *J. Gen. Physiol.* **1939,** *22,* 649.
9. Schoffeniels, E. *Arch. Int. Physiol. Biochim.* **1970,** *78,* 205–223.
10. Schacterle, G. R.; Pollack, R. L. *Anal. Biochem.* **1973,** *51,* 654–655.
11. Schoffeniels, E.; Dandifosse, G. *Proc. Natl. Acad. Sci. USA* **1980,** *77,* 812.
12. Caldwell, P. C.; Lowe, A. G. *J. Physiol.* **1970,** *206,* 271–280.
13. Ibid., **1966,** *186,* 24P.
14. Ritchie, J. M.; Straub, R. W. *J. Physiol.* **1978,** *274,* 539–548.
15. Schoffeniels, E. In "Epithelial Transport in the Lower Vertebrates"; Lahlou, B., Ed.; Cambridge Univ. Press: 1979.
16. Johnson, E. M.; Ueda, T.; Maeno, H.; Greengard, P. *J. Biol. Chem.* **1972,** *247,* 5650–5652.
17. Greengard, P. *Science* **1978,** *199,* 146–152.
18. Williams, M.; Rodnight, R. *Progr. Neurobiol.* **1977,** *8,* 183–250.
19. Schoffeniels, E.; Margineanu, D. G. *Bull. Acad. R. Med. Belg.* **1977,** *132,* 127–135.
20. Schoffeniels, E.; Margineanu, D. G. "Power Dissipation in Bioelectrogenesis: A Molecular Approach," *Biosc. Comm.* **1978,** *4,* 205–216.

RECEIVED October 17, 1978.

19

The Permeability of Adsorbed- and Spread-Membrane Protein (Spectrin–Actin) Films to Ions

MARTIN BLANK, L. SOO, and R. E. ABBOTT

Department of Physiology, Columbia University, 630 W. 168 St., New York, NY 10032

The surface properties of spectrin and actin films suggest that these erythrocyte-membrane proteins are present in vivo as an interacted network that also may constitute a significant ion-permeability barrier. Using polarography, we have shown that the ionic permeability of adsorbed and spread spectrin and actin monolayers at the mercury/water interface at different pH's is consistent with an isoelectric point of 5.5. At pH 7.4, the spectrin and actin layers are relatively impermeable to anions (with an estimated permeability of about 10^{-4} cm/sec for an adsorbed film) in the range where the erythrocyte membrane is very permeable. Our observations suggest that the specialized anion carrier molecules in the membrane penetrate an anion-depleted spectrin and actin layer.

In the last few years there have been many advances in analyzing the components and the structure of the red cell membrane (*1, 2*). Many lipid and protein components have been localized in the structure (*3*) and several individual proteins have been isolated, e.g. spectrin (*4*) and actin (*5*). It is now well established that the spectrin component of the erythrocyte membrane is derived from the cytoplasmic side, contains actin, and accounts for about a third of the membrane protein. It is relatively easy to extract and appears ". . . to form an anastomosing network beneath the erythrocyte membrane" (*5*).

The recently obtained surface-packing information and the surface rheological properties of spectrin–actin (S + A) monolayers support the

0-8412-0473-X/80/33-188-299$05.00/1

view that it is an important functional component of the membrane that is present as an interacted cross-linked network on the inner face (*6*). Such a layer in the cell membrane may constitute a significant ion-permeability barrier. To examine this possibility we have obtained information about ion transport rates through this layer.

We have been studying the transport of molecules and ions across monomolecular films of lipids and proteins for some time, and we now have a reasonable understanding of the physical factors that influence the rate (*7*). (It should be mentioned that our early studies of monolayer permeability provided us with sufficient insight into monolayers to predict the correct magnitudes of other important properties, such as the self-diffusion coefficient of lipids (*8*) in membranes and the flipping frequency (*9*) of lipid molecules between layers.) More recent studies of monolayer permeability to ions have provided information about effects of surface charge density, electrolyte concentration, layer thickness, etc., the fundamental factors governing electrical resistance in very thin membranes and in interfacial regions (*10, 11*). These results can be summarized and explained in terms of electrical double-layer theory (*12*).

Experimental

Preparation of S + A. The large number of proteins that are present in the erythrocyte membrane can be separated into a number of well characterized components. The components associated with SDS polyacrylamide gel Bands 1 and 2 and called spectrin can be obtained without the aid of surface-active agents (*see* Figure 1). It is possible to extract this material from erythrocyte ghosts by using low salt and also by incubating with EDTA. The extraction treatment generally produces an additional low-molecular-weight component (Band 5 on SDS polyacrylamide gels) that has been identified as actin. This preparation constitutes about one third of the membrane protein and the major components of the endofacial layer of the erythrocyte membrane.

In our experiments, we used standard techniques in preparing spectrin and actin. Human erythrocytes were obtained from freshly outdated blood bank blood. Following centrifugation and removal of the plasma and buffy coat, the erythrocytes were washed with cold isotonic buffer (30m*M* sodium phosphate (pH 7.4), 117m*M* NaCl, and 2.8m*M* KCl). The washed, packed cells were hemolyzed in 12 vol cold 8m*M* sodium phosphate, pH 7.4, for 30 min at 4°C, and ghost membranes isolated by the method of Dodge et al. (*13*). The washed ghosts were stored frozen at − 20°C overnight and then extracted with 10 vol 0.1m*M* sodium EDTA, pH 8.0, for 20 min at 37°C. Centrifugation at 78×10^3 g (30 min, 4°C) yielded a clear supernatant (typical protein concentration about 0.2 mg/mL) which was S + A (Bands 1, 2, and 5) as judged by the Coomassie Brilliant Blue bands seen after SDS–polyacrylamide gel electrophoresis (PAGE) (*see* Figure 1). This solution was used to form spread films at an air/water interface or was diluted to nanomolar concentrations for experiments on adsorbed films.

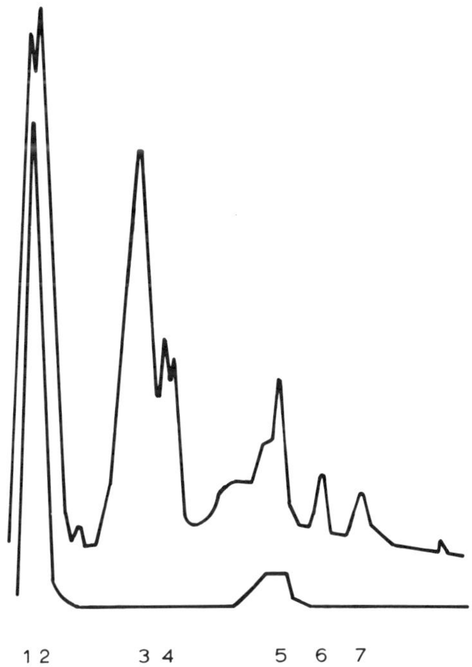

Figure 1. Two tracings of the densities of the bands seen with proteins from the erythrocyte membrane after SDS treatment and PAGE. The numbers correspond to the major peaks on the upper tracing—a sample of proteins from the whole membrane. The lower tracing is of an S + A preparation used in these experiments.

Some experiments were done with bovine serum albumin (BSA) in order to compare our measurements on S + A films with those of a known substance. The BSA was obtained from Sigma and used without further treatment.

Trough Measurements. The experiments were performed with a Langmuir-type trough and barriers made of Teflon. A Teflon barrier was attached to an apparatus which varied the area at a rate of 10 cm^2/sec. The original area for spreading the film was 300 cm^2 and the experiments were performed at a temperature of 25°C and in the pH range of 3–8. The surface tension was measured with a Sanborn (Model 311A) transducer and an attached sand-blasted platinum plate.

The Teflon apparatus and a glass rod were cleaned with detergent, rinsed several minutes in running cold water and finally in doubly distilled water. The surface tension was recorded after filling the trough with 0.1*N* NaCl solution, sweeping the surface, and setting up the dipping plate. Precise amounts of S + A spreading solution were delivered from Lang–Levy pipets and spread down the glass rod to the surface. The glass rod then was rinsed several times with 0.1*N* NaCl solution. After waiting 5 min, the movable Teflon barrier was started with a speed of 1.58 mm/sec or 1.58 cm^2/sec and the surface-tension readings were recorded automatically.

When S + A films were spread in the polarographic cell, film thicknesses were estimated assuming that the partial molar volume of a protein on a surface is equal to that in solution, 1.3 cm^3/g. Combining this figure with our value for the surface concentration of a close-packed S + A film, 1.4×10^{-7} g/cm^2, we obtain a thickness of 18 Å per monolayer (*6*).

Ion Transport Through Monolayers. The movements of ions through oriented monolayers was studied by determining the effect of an adsorbed or spread monolayer on the polarographic reduction of an ion at a mercury/water interface (*10*). The permeating ions are shown in Table 1 along with the pH ranges in which they were used. (The complex ions of Cu^{++} assume different charges depending on the anions present.)

To overcome the experimental limitations in connection with the need for soluble substances as films, we extended the technique devised by Pagano and Miller (*14*), which utilized the polarographic technique in conjunction with the conventional surface film balance. This makes it possible to form films at an air/water interface under conditions where the surface concentration and charge are well characterized, and then to study the ionic permeability of these films at a mercury/water interface. We therefore have studied the permeability of both adsorbed protein films and spread films formed directly in the polarography cell.

In both types of experiments the results were interpreted in terms of a transport rate constant, k_c, using a method described earlier (*10*). At polarizations where the rate of electroreduction is high, the diffusion current across an electrode covered by a monolayer is given by the diffusion of the depolarizer to the surface and by the rate of crossing the monolayer. This problem has been treated formally like the kinetic current, where the electrode processes are replaced by the crossing of the monolayer, and the solution of the differential equation for the dropping mercury electrode is the same as given by Koutecky (*15*), who tabulated $X = (k_c t^{1/2}/D^{1/2})$ as a function of the reduced current I/I_0, where I and I_0 are the diffusion-controlled instantaneous currents in the presence and absence of a film, respectively. Knowing the time, t, and D, the diffusion coefficient, we used his tabulated data for calculating k_c from the reduced current.

Table I. The pH Range of Ion Probes

pH Range	*Cation Used*	*Anion Used*
3–6.5	Cu^{++} complex in NO_3^- solution	Cu^{++} complex in Cl^- solution
6	Cu^{++} complex in NO_3^- solution	Cu^{++} complex in EDTA solution
5.5–9	Tl^+	Iodoacetate

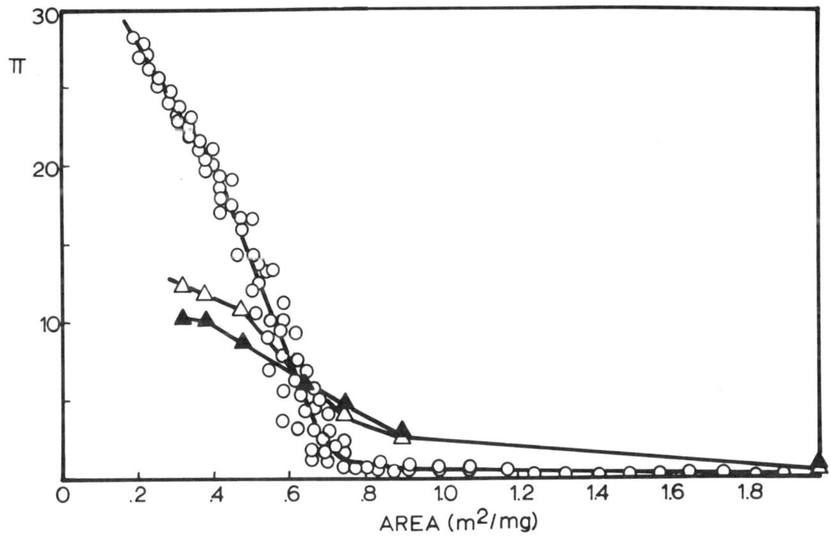

*Figure 2. The surface pressure, Π (in dynes/centimeter), vs. the surface area of spread S + A monolayers (meters²/milligram) at 25°C. The curves refer to: (○), compressed monolayers on 0.1*M *NaCl; (△), monolayers formed by additions of S + A at constant area on 0.1*M *NaCl; (▲), monolayers formed by additions of S + A at constant area on 0.1*M *phosphate buffers.*

Results

When we deposit the S + A from a 0.1m*M* EDTA solution on to a 0.1m*M* saline surface at a pH of 5.8 (approximately the isoelectric point of S + A) and compress the film on a standard surface balance, we obtain the surface pressure data shown in Figure 2. If we extrapolate to zero surface pressure, the intercept is about .7 m^2/mg. On the same abscissa, we also have plotted the surface pressure–surface concentration curves obtained by incremental spreading of S + A at a constant area. The curves are really quite different from the compressed films and the extrapolated area at zero surface pressure is considerably larger. These results indicate that the structure of an S + A film is very dependent upon the conditions under which it is formed, an observation that we shall underscore in the ion transport measurements.

When the S + A films are formed directly in the polarography cell at the air/water interface and their permeabilities to ions are determined at different pH's, the values of k_c are shown in Figure 3 as functions of the surface concentration. As observed previously in the case of adsorbed films (*16*), at pH 3 the S + A film is relatively impermeable to cations and at pH 6.4, to anions. The values of k_c are also about the same order

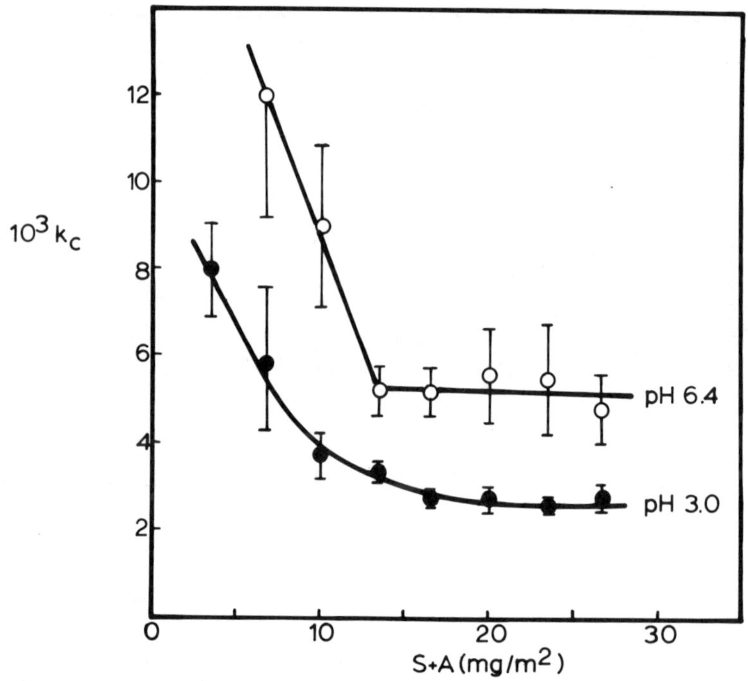

Figure 3. The permeability, k_c (in centimeters/second), of spread S + A films to Cu^{++} ions as a function of the surface concentration of S + A (in milligrams/meter²). At pH 3.0 the data apply to the permeation of cations while at pH 6.4 they apply to anions.

of magnitude for both types of films (*see* Figure 4). However, there is an enormous difference between the thickness of the impeding layers. In the spread films (*see* Figure 3), there are over ten monolayer thicknesses on the k_c plateau regions, while for the adsorbed films (*see* Figure 4) there is not even one complete monolayer present. One can estimate (*16*) that it would take about 17 sec at the highest S + A concentration to achieve a complete monolayer, a drop time that we cannot achieve.

Although the spread S + A monolayer is apparently much looser than the adsorbed layer, it is still much more compact than a spread serum albumin film. Ion transport measurements at pH 3 equivalent to those in Figure 3 are shown in Figure 5 for BSA films. Apparently, one must go to about 100 times the surface concentration (or thickness) of S + A layers in order to observe a comparable value of k_c for BSA.

The pH dependence of adsorbed S + A films (when compared at the same S + A solution concentration and mercury drop times) show the expected selectivity around pH 5.5. While the earlier results (*16*) up to a pH of about 6 suggested a much steeper rise in selectivity on the alkaline side, the more expanded set of observations show greater symmetry. The decrease in selectivity at pH 9 may be associated with the swelling of the monolayer or the slow degradation of the protein (*see* Figure 6).

Discussion

The Structure of S + A Interfacial Films. In this chapter we have presented data that relate to four different types of film. In Figure 2, the standard Π-A curve with the intercept at .7 m^2/mg is obtained by spreading a known amount of S + A at almost zero surface pressure, and slowly compressing. The other curve, usually called a Π-C curve, is done at constant area with increments of S + A deposited into the already existing film. The Π-A curve reflects a film where the molecules are spread out more effectively, whereas the Π-C film contains many molecules that have been incorporated under a significant surface pressure

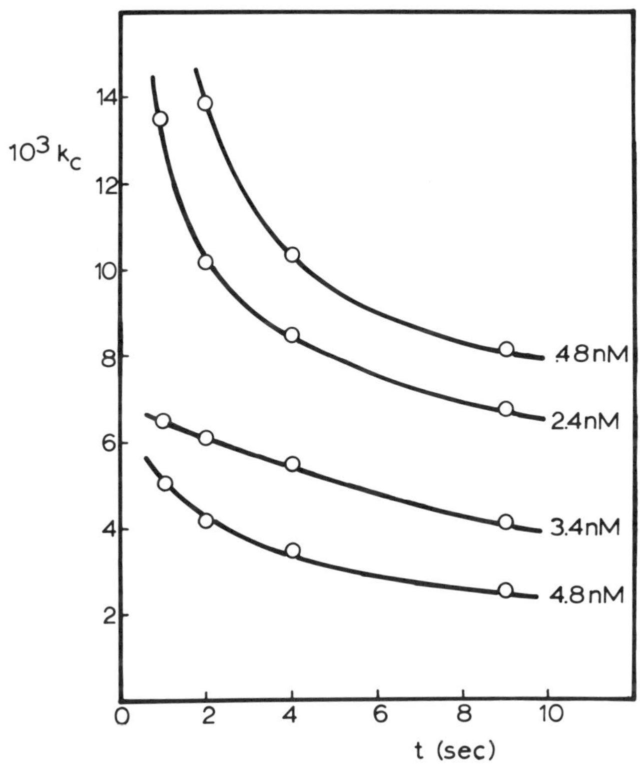

Figure 4. The permeability, k_c (in centimeters/second), of adsorbing S + A monolayers at pH 3 to Cu^{++} ions as a function of time (in seconds). The concentrations listed refer to the S + A in solution (asuming a molecular weight of 200,000).

and so have not been able to spread out to the same extent. These two curves are similar to published data for BSA films (*17*), and lead us to conclude that S + A films are subject to the same packing restrictions on a surface.

The third type of film is the one studied in Figure 3. It is spread like in the Π-A experiment, but at such high concentration that the effect is like many layers of a Π-C film. The S + A molecules do not have the opportunity to spread out at the interface, and because of the many layers in these films, the S + A molecules do not experience the same asymmetry between hydrophilic and hydrophobic regions as is present in a single layer. These spread films undoubtedly contain very poorly packed S + A molecules, and the very high permeabilities bear this out.

Still another type of film is used in Figure 4. This fourth type is adsorbed at the mercury/water interface and probably is closest to the Π-A film at higher areas. The adsorbing S + A molecules always arrive at an expanding interface and the film never reaches the full monolayer concentration. Therefore, there is ample room for the molecules to spread and readjust their orientations at the surface. This film is therefore very different in structure from the one used in Figure 3, and the permeabilities indicate this as well.

The S + A film that is of greatest interest to us, of course, is the complex structure on the inner face of the red cell membrane. Using our

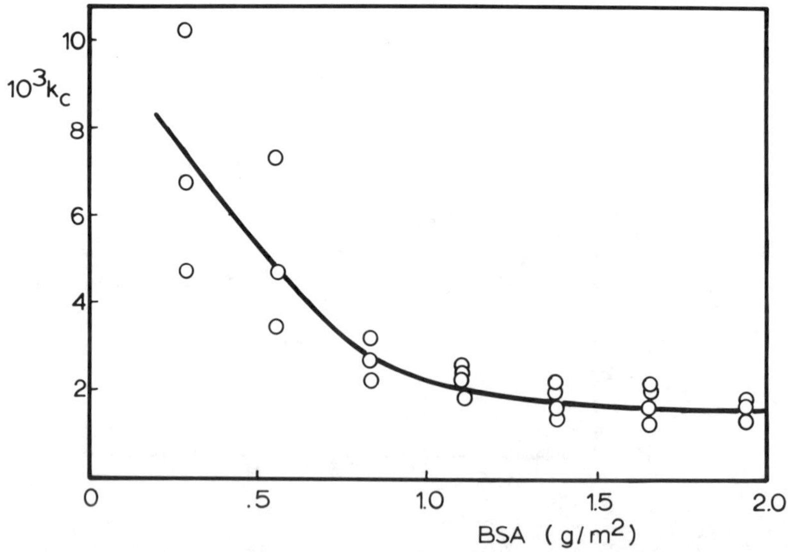

Figure 5. The permeability, k_c (in centimeters/second), of spread BSA films to Cu^{++} ions at pH 3 as a function of the surface concentration of S + A (in grams/meter²)

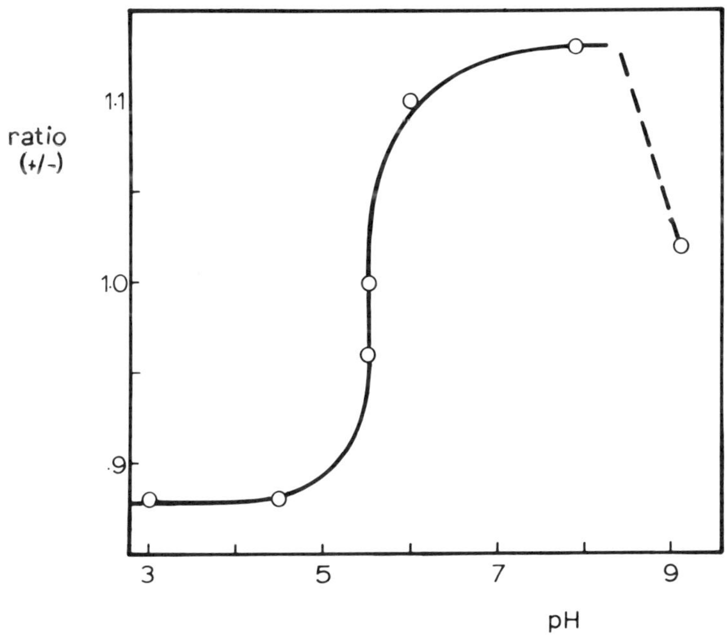

Figure 6. The ratio of the reduced current for the permeation of adsorbed S + A monolayers by cations to the reduced current for anions as a function of the pH. The ratio (+/−) is a measure of the selectivity of the S + A monolayer for cations relative to anions.

surface measurements, we calculate that this layer is probably about two monolayers thick (*6*). In addition to the thickness, the natural S + A film is very different from the four films we have discussed already. It is in contact with the membrane lipids on one side and a concentrated hemoglobin solution on the other. Furthermore, it appears likely that some of the other membrane proteins, in particular the Band 3 protein, may extend through this layer.

It is difficult to decide which type of surface film the natural S + A film resembles most closely. The number of S + A molecules present in the membrane is closer to the adsorbed film and the packing is believed to be more orderly (*2*). Also, the ease with which the S + A molecules can be removed from the membrane by low ionic strength media, suggests a loose and more accessible structure. For these reasons we feel that the Π-A type of film below 10 dyn/cm pressure resembles most closely the natural S + A film, and that the data of Figure 4 are more relevant to the membrane state than those of Figure 3.

It appears somewhat unreasonable that this less concentrated (in two dimensions) but more ordered surface structure is less permeable than

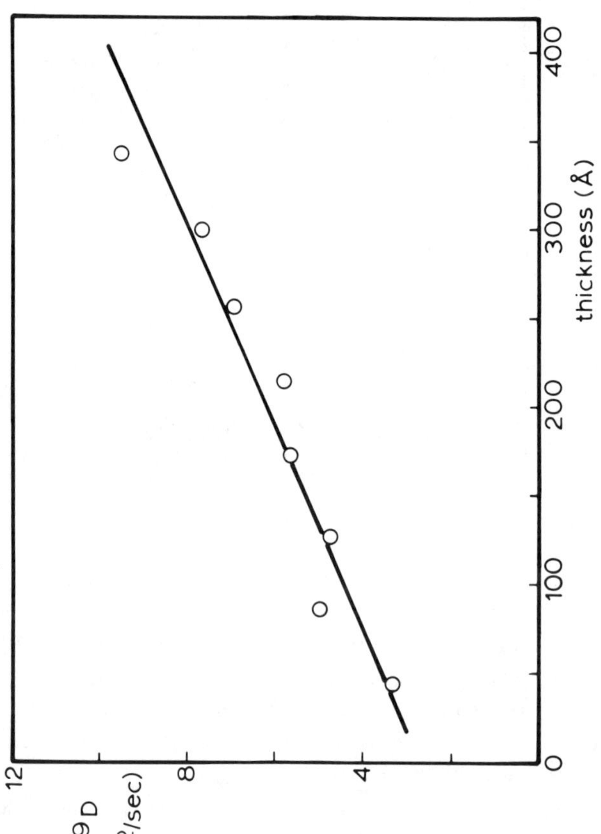

Figure 7. The diffusion coefficient, D (in centimeters²/second), of Cu^{++} ions through an S + A film at pH 3 as a function of the estimated thickness of the film (in angstroms)

the many-layered one formed by the rapid spreading method at constant area. Apparently the major factor controlling the flow of ions through the layer is the surface charge (*12*), and this is probably optimized during a more orderly adsorption process, as opposed to rapid spreading. Protein dissolution in the spread films at higher pressure probably contributed to the greater permeability, but recent experiments indicate that only about 30% of the S + A layer dissolves under the conditions of the experiments.

The Permeability of S + A Films. As already mentioned, the k_c values in Figures 3 and 4 refer to different types of films of S + A and also to different thicknesses. The magnitude of k_c for spread S + A films (*see* Figure 3) reaches a plateau value of about 10^{-3} cm/sec at relatively high surface coverages. For the adsorbed films (*see* Figure 4), the k_c decreases with time and S + A concentration, but shows no plateau. This is to be expected, since the form of the current–time curve indicates that the S + A film is still below the saturated monolayer stage (*10*). Higher concentrations of S + A and longer drop times cannot be reached experimentally, but we can estimate the time at which the S + A monolayer would be complete and also the approximate magnitude of k_c at that point. Extrapolating from the behavior of BSA under the same conditions, we have determined the time of formation of a complete monolayer as a function of the BSA concentration, and have extrapolated those results to the time required for the formation of a S + A monolayer at the highest S + A concentration studied. Using this information, the shortest time (as seen in Figure 3) for a full monolayer of S + A is 17 sec, and k_c would be about 10^{-4} cm/sec on extrapolation (*16*).

The permeabilities of the two types of S + A films differ by an order of magnitude, so extrapolating to the same layer in the red cell membrane, involves a choice between a resistance of the membrane layer to ion flow that is either negligible or significant. For reasons given in the previous section, the adsorbed type of film is probably more closely related, in which case $k_c \sim 10^{-4}$ cm/sec and significant. This, of course, would mean that the special anion carrier molecules (Band 3 in the SDS gel) in the membrane would have to extend through the S + A layer in order to be effective. There is some evidence that the S + A layer in vivo does constitute a significant ion-permeability barrier. Lepke and Passow (*18*) have shown that internal trypsin can cause a slight increase in anion equilibrium exchange in erythrocytes. In any case, the S + A layer in vivo represents a significant anion-depleted (cation-rich) zone in the membrane, which could cause unusual effects in ion transport kinetics (*19*).

The Variation of Permeability with Film Thickness. In early studies of monolayer permeability to water and gases a very strong dependence of the transport rate on the thickness of the film was found

(7). These observations suggested that as one increases the film thickness there is a transition of monolayer properties to bulk properties. In a more recent study of the viscoelasticity of spread S + A films of different thicknesses (6), we found that when the surface properties were converted to equivalent bulk values (by introducing the film thickness) there was also a variation with thickness. The viscosity coefficient and elastic modulus became independent of thickness at about 80 Å, or four monolayers, while the residual yield decreased with thickness over the entire range of the measurements. These results were interpreted to mean that the molecular interactions that account for the properties must become less effective within a layer as the number of layers increases and there are more interactions between layers.

We can examine our permeability data in the same way and calculate a diffusion coefficient upon multiplying the transport rate coefficient (in centimeters per second) by the estimated film thickness (in centimeters) to yield a diffusion coefficient (in cm^2/sec). The results of this calculation, shown in Figure 7, indicate that the diffusion coefficient increases with the thickness of the S + A film over the entire range of measurement. This result is directly in line with the variation of the residual yield in our rheological studies (6), where the original layer had the strongest orientation and interplanar interactions, and all subsequent layers showed less orientation and weaker interactions. The subsequent layers apparently become more permeable as a result of this change in structure.

Acknowledgment

This work was supported by Research Grant PCM 76–11676 from the National Science Foundation.

Literature Cited

1. Steck, T. L. *J. Cell Biol.* **1974,** *62,* 1.
2. Kirkpatrick, F. H. *Life Sci.* **1976,** *19,* 1.
3. Zwaal, R. F. A.; Roelofsen, B.; Colley, C. M. *Biochim. Biophys. Acta* **1973,** *300,* 159.
4. Marchesi, V. T.; Steers, E. *Science* **1968,** *159,* 203.
5. Tilney, L. G.; Detmers, P. *J. Cell Biol.* **1975,** *66,* 508.
6. Blank, M.; King, R. G.; Soo, L.; Abbott, R. E.; Chien, S. *J. Colloid Interface Sci.* **1978,** *69,* 67.
7. Blank, M. *Prog. Surf. Membr. Sci.* **1979,** *20,* 789.
8. Blank, M.; Britten, J. S. *J. Colloid Sci.* **1965,** *20,* 789.
9. Blank, M.; Britten, J. S. *Chem. Phys. Lipids* **1973,** *10,* 286.
10. Miller, I. R.; Blank, M. *J. Colloid Interface Sci.* **1968,** *26,* 34.
11. Sweeney, G. D.; Blank, M. *J. Colloid Interface Sci.* **1973,** *42,* 410.
12. Britten, J. S.; Blank, M. *Bioelectrochem. Bioenerg.* **1977,** *4,* 209.
13. Dodge, J. T.; Mitchell, C.; Hanahan, D. J. *Arch. Biochem.* **1963,** *110,* 119.
14. Pagano, R.; Miller, I. R. *J. Colloid Interface Sci.* **1973,** *45,* 126.

15. Koutecky, J. *Collect. Czech. Chem. Commun.* **1953,** *18,* 597.
16. Blank, M.; Soo, L.; Abbott, R. E. *J. Electrochem. Soc.* **1979,** *126,* 1487.
17. Mussellwhite, P. R.; Palmer, J. J. *J. Colloid Interface Sci.* **1968,** *28,* 168.
18. Lepke, S.; Passow, H. *Biochim. Biophys. Acta* **1976,** *455,* 353.
19. Blank, M.; Britten, J. S. *Bioelectrochem. Bioenerg.* **1978,** *5,* 528.

Explanatory Note

The authors note that their article in the *Journal of Membrane Biology,* **1979,** *47,* 185–193, provides more detailed information about the packing of S + A at an interface.

RECEIVED October 17, 1978.

20

Influence of Fluxes on Stationary-State Electric Potential and Concentration Profiles

V. S. VAIDHYANATHAN

Department of Biophysical Sciences, 114 B, Cary Hall, School of Medicine, State University of New York at Buffalo, Main Campus, Buffalo, NY 14214

The influence of fluxes on concentration and electrical potential profile, as expressed by a set of coupled nonlinear differential equations, is analyzed. These equations are derived using a correction to the limiting expressions for chemical potentials and the molecular expressions for frictional force. From general theory and numerical considerations a simplified set of expressions are obtained enabling one to compute stationary-state membrane potential from fluxes and properties of surrounding electrolyte solutions. Application of the theory to various giant axon systems yield reasonable values of resting potentials.

Isothermal transport of solutions containing three or more permeant ions across a diffusion barrier, such as a biological membrane, is of significant biophysical interest (*1*). The central problem which must be solved is to compute the value of the electric potential difference between the two solutions adjacent to the membrane phase in terms of known physical parameters of the system on an acceptable physical basis. A large number of papers on this subject deal with the problem of finding solutions of familiar Nernst–Planck Equations (*2*). These equations are nonlinear and solutions of these equations in closed analytic form have not been obtained yet, in spite of attempts by many over a period of seven decades. In certain specialized circumstances, solutions associated with the names of Planck, Henderson, and Schlogl are available (*3*). In the opinion of the author, the assumptions involved in obtaining these solutions are not applicable for real membrane systems of interest in biology. (One of the referees has expressed the opinion that transport

0-8412-0473-X/80/33-188-313$06.00/1

in biological membranes is accomplished by discrete macromolecular complexes of one kind or another, and therefore any continuous treatment is unrelated to biology and should not be done. In spite of this remark, the continuous nature of present treatment has some merits in the opinion of the author.) A widely accepted equation in membrane physiology is the Goldman equation (*4*), which involves the assumption that electric potential profile is linear. This assumption enables one to evaluate the formal integral of Equation 7 of this chapter. However, this assumption requires validity of microscopic electroneutrality and is not aesthetically attractive when one must reconcile with the Poisson Equation. In spite of these, the discussion of observed phenomena in biological literature usually is limited to analysis based on equilibrium Nernst potential and sometimes involving activity coefficient correction. (The literature in this field is vast (*see* e.g. Ref. *6*), and specific references are avoided consciously in this chapter.) In spite of a considerable amount of available experimental information on electrical potential difference, boundary concentrations of ions, and the influence of divalent ions on fluxes of monovalent (*5*) ions and permeabilities, our understanding of the theoretical basis for experimentally observed variation of fluxes and electrical potential as a function of the properties of a membrane barrier and surrounding solutions is far from satisfactory (*6, 7*). Attempts to explain the uneven distribution of monovalent ions in solutions of either side of a biological membrane on the basis of the Donnan Effect also is present in the literature.

The well-studied excitation phenomena of nerve axon systems (*8*) clearly indicate that the state of affairs present in biological systems are definitely not in an equilibrium state. Under nonequilibrium conditions, fluxes that are present alter the equilibrium concentration profiles and the electric potential profile through the Poisson Equation. Thus, any theory that attempts to explain the influence of fluxes must explicitly exhibit the flux dependence of electrical and concentration profile gradients (*9*). In this chapter, the influence of fluxes on concentration profiles of both ionic and nonionic species are assumed to be described by the basic set of coupled nonlinear differential equations (*see* Equations 6a–6e.) These equations are sufficiently general enough to include the influence of any chemical reactions occurring in the diffusion barrier, though analysis of this chapter is restricted to stationary states in the absence of chemical reactions. The justification for validity of assumed basic differential equations is presented later in this chapter.

This chapter focuses on the inclusion of ion–neutral molecule interaction contribution to chemical potentials of permeant ionic species, thus reducing the problem to be solved by the Taylor Series. Certain exact results such as Equations 8, 9, and 10 can be obtained. Equation 10 for example is the generalization of Maxwell's well-known osmotic balance equation.

Certain results of mathematical interest based on these equations have been published elsewhere (*10*). These technical details are not very interesting to the experimenter. With this in view, a simplified theory is presented in the section dealing with theory which enables one to compute membrane potential from Debye–Hückel parameters of the surrounding solution and the observed stationary-state fluxes of permeant ions and neutral molecules. As shown in the section on theory, Equation 17, which relates membrane potential difference to gradients of electric potential profile at specified locations, is derived. Methods of evaluating the various quantities are presented.

Application of the theory to various axon systems leading to the calculation of resting potentials is presented in the "Sample Calculation" section. Though admittedly they are approximate and somewhat ambitious, calculated values are in reasonable agreement with observed values. Provided that one has values of stationary-state fluxes of individual permeant ionic and neutral species—independently measured—properties of solutions and the value of the electric potential which also is measured independently, it is possible to apply the theory to a clean physical system and to verify the extent to which theory agrees with experimentation. Such flux data are scant, and the data available for axon systems do not satisfy the zero net electric current condition. In spite of these deficiencies and the approximate validity of simplified equations, the reasonable agreement between theory and experiments for various axon systems is evidently not the result of coincidence. Certain symmetry arguments and roots of polynomial considerations enable one to compute charge-density profiles. One computes membrane thickness to be $\sim$ 57 Å and the thickness of diffuse double-layer regions to be $\sim$ 21 Å, if the region lacking microscopic electroneutrality is of the order of 100 Å.

A measure of the activity coefficient correction factor, denoted by a constant parameter, λ^2, (assumed to be constant in this chapter), plays an important role in determining the stationary-state membrane potential, as determined by the ionic flux term, ZR. When λ^2 equals zero, the activity coefficient of all solute ions equals unity. The ion–neutral molecule interaction energy contribution to the chemical potential of permeant ions is denoted by H*. An approximate expression for λ^2 as equalling $-[\eta_1^2 + \eta_2^2]/(\eta_1^2\,\eta_2^2)$, where η_1^2 and η_2^2 are the Debye–Hückel parameters of Solutions 1 and 2 surrounding the membrane phase, is derived.

This chapter shows that the stationary-state membrane potential evolves in a natural self-consistent manner, determined by bulk solution properties of electrolytes and by the fluxes as determined by membrane properties. These also determine the extent of the region lacking microscopic electroneutrality. It is worthwhile recalling that the classic Debye–Hückel approach assumes the existance of regions lacking electroneutrality of a spherical shape having a radius of $1/\eta$. This chapter is for physical chemists who are interested in various aspects of membrane-transport problems. The results presented in this chapter indicate that nature is simple because of its complexity. Experimentally it is known that the permeability coefficient of a specified ion and molecule across a specific membrane is affected by electrical potential difference and chemical reactions. Experimental observations that the flux of one kind of species is influenced by forces on another kind of molecules is understood by the formal expression (*see* Equation 8) of this chapter. In Debye–Hückel theory, the limiting expression for the chemical potential of ionic species α is given by the relation (*11*)

$$\mu_\alpha = \mu_\alpha{}^\circ(T, P) + kT \ln C_\alpha + Z_\alpha e\phi - (e^2\eta/\epsilon) \quad (1)$$

where C_α is the concentration of ion α in solution, ϵ is the dielectric coefficient of solution, Z_α is the signed valence charge number of ion α, e is the electronic charge, k is the Boltzmann constant, and T is absolute temperature. The expression $\mu_\alpha{}^\circ$ is the composition-independent part of the chemical potential and η is the Debye–Hückel ionic atmosphere parameter. The approximations involved regarding the potential of mean force in the derivation of Equation 1 are well known.

Derivation of Basic Equations

The experimental verification of Equation 1 for extremely dilute electrolyte solutions clearly indicates the need for the last term in Equation 1, while the usual derivation of Nernst–Planck Equations contain only the first three terms. (However, the last term is incorporated when one expresses the Nernst–Planck Equation with activity). Assume that the expressions for chemical potentials of ionic species σ and nonionic species j in an inhomogeneous diffusion medium may be written as (*12*)

$$\sigma(x) = \mu_\sigma{}^*(T, P) + Z_\sigma e\,\phi(x) + kT \ln C_\sigma(x) + \sum_\eta C_\eta(x)\, H_{\sigma\eta} + \sum_j C_j(x)\, H_{\sigma j} \quad (2a)$$

$$\mu_j(x) = \mu_j{}^*(T, P) + kT \ln C_j(x) + \sum_K C_k(x)\, H_{jk} + \sum_\sigma H_{j\sigma}\, C_\sigma(x) \quad (2b)$$

In Equations 2a and 2b $\mu_\sigma{}^*(T, P)$ and $\mu_j{}^*(T, P)$ are the composition- and position-independent part of chemical potentials of ionic species σ and uncharged species j, in standard states. The term x is a position variable, defined along the axis of transport, normal to the yz plane of the diffusion barrier. The expression $\phi(x)$ is the electric potential felt by a unit charge placed at x and $C_\sigma(x)$ and $Cj(x)$ are the concentrations of species σ and j, respectively, at x, expressed in moles/cubic centimeter. $H_{\sigma\eta}$ are the molecular integrals over the potential energy of these interactions, weighted by probability function of the σ and η-th kind of species. $H_{\sigma}j$ and Hkj are similar molecular integrals of ion–nonelectrolyte interactions and dispersion force interactions. The leading terms of Equation 2 are familiar limiting expressions for chemical potentials in dilute solutions that express entropic contributions. The additional terms express approximately the potential energy contributions to chemical potentials of the species in question. The assumptions that $H_{\sigma\eta}$ may be regarded as position independent and may be expressed as $Z_\sigma Z_\eta e^2 H$, and that Hj_σ can be expressed as $Z_\sigma eH^*$, enables one to write the gradients of chemical potentials by using the Poisson Equation as

$$\mu_\sigma'(x) = kT\,[d \ln C_\sigma/dx] + Z_\sigma e\,\{F'(x) + H^* \sum_j C_j'\}$$

$$\mu_j'(x) = kT\,[d \ln C_j/dx] + \sum_K H_{jk}C_k' + H^* e \sum_\eta Z_\eta C_\eta'$$

$$F'(x) = \phi'(x) - \lambda^2\phi'''(x)$$

$$\lambda^2 = (H\epsilon/4\pi)$$

$$\phi''(x) = -\,(4\pi e/\epsilon) \sum_\sigma Z_\sigma C_\sigma(x) \tag{3}$$

In Equation 3, as well as in the following equations, the first derivative with respect to position variable x is denoted by a single prime. Higher-order derivatives are denoted by the appropriate number of primes. The approximation that the positional dependence of the dielectric coefficient, ϵ, may be neglected has been incorporated. This is a serious assumption whose validity to biological membrane systems is questionable. An empirical expression stating that $\epsilon(x) = \epsilon(o) - \epsilon_2 dx + \epsilon_2 x^2$, where $\epsilon(o)$ is the dielectric constant of the bulk solution, d is the distance between the two solutions where inhomogeneity exists, and ϵ_2 is the second Taylor coefficient of $\epsilon(x)$, can be utilized. In this chapter, for simplicity, ϵ is regarded as x-independent.

The frictional coefficient of ionic species σ and ζ_σ and of the nonionic species j and ζj in the diffusion barrier can be written as

$$\zeta_\sigma(x) = [kT/D_\sigma(x)] = r_\sigma + C_\sigma\,\zeta_{\sigma\sigma} + \sum_{\eta\neq\sigma} C_\eta(x)\zeta_{\sigma\eta} + \sum_j C_j(x)\zeta_{\sigma j} \tag{4a}$$

$$\zeta_j(x) = [kT/D_j(x)] = r_j + C_j\zeta_{jj} + \sum_{k\neq j} C_k\zeta_{kj} + \sum_{\sigma} C_\sigma\zeta_{\sigma j} \tag{4b}$$

where D_σ and Dj are the diffusion coefficients of species σ and j, respectively, in the diffusion barrier. The terms r_σ and rj are the contributions to ζ_σ and ζj, respectively, of the immobile molecules of the membrane phase that arise from intermolecular interactions. The term $\zeta_{\sigma\eta}$ is the partial frictional coefficient and represents the contribution per mole of species η to ζ_σ.

Under stationary-state conditions, with no chemical reactions, the expressions for gradients of chemical potentials satisfying the Gibbs–Duhem Equation (*13, 14*) can be written for isothermal conditions also

$$\mu_\sigma'(x) = -\ [J_\sigma\zeta_\sigma(x)/C_\sigma(x)] + \sum_{\eta} J_\eta\zeta_{\sigma\eta} + \sum_{j} J_j\zeta_{\sigma j}$$

$$\mu_j'(x) = -\ [J_j\zeta_j(x)/C_j(x)] + \sum_{\sigma} J_\sigma\zeta_{\sigma j} + \sum_{\kappa} J_k\zeta_{jk} \tag{5}$$

where J_σ and Jj are the fluxes of molecules σ and j, expressed in moles per centimeters squared per second in the stationary state. In the absence of chemical reactions, the fluxes are independent of the position variable. In the presence of chemical reactions, the positional dependence of flux of species participating in chemical reactions is given by an equation of continuity. Substitution of equations 4 into Equation 5 and equating this with Equation 2 yields the basic set of coupled differential Equations (*see* Equations 6a–6e) upon rearrangement. This determines the concentration profiles of all species and the electrical potential profile in the diffusion barrier.

Specifically, for a system containing n kinds of ionic species and one kind of permeant uncharged species (solvent), one has

$$C_\sigma' + [Z_\sigma e/kT]C_\sigma(x)\ [F'(x)\ + H^*C_j'] + R_\sigma - S_{\sigma\sigma}C_\sigma + \sum_{\eta\neq\sigma} S_{\sigma\eta}C_\eta + S_{\sigma j}C_j = 0 \tag{6a}$$

$$C_j' + (H_{jj}/kT)C_jC_j' + R_j - S_{jj}C_j + \sum_{\sigma} S_{j\sigma}C_\sigma = [H^*\epsilon/4\pi kT]\ C_j(x)\phi'''(x) \tag{6b}$$

$$R_\sigma = (J_\sigma r_\sigma/kT)\,; \qquad R_j = (J_j r_j/kT) \tag{6c}$$

$$S_{\sigma\eta} = (J_\sigma\zeta_{\sigma\eta}/kT)\,; \qquad S_{j\sigma} = (J_j\zeta_{\sigma j}/kT) \tag{6d}$$

$$S_{\sigma\sigma} = \{\sum_{\eta\neq\sigma} S_{\eta\sigma} + S_{j\sigma}\}\,; \qquad S_{jj} = \sum_{\sigma} S_{\sigma j} \tag{6e}$$

In the absence of chemical reactions in the diffusion barrier, $S_{\sigma\eta}$ and $S_{\sigma j}$ are constants independent of x. When one neglects the composition-dependent (hence position-dependent) part of diffusion coefficients, and ignores H^*, (i.e., $H^* = 0$), one obtains from Equation 6a

$$C_\sigma'(x) + (Z_\sigma e/kT)C_\sigma(x)F'(x) + (J_\sigma/D_\sigma) = 0 \tag{7}$$

When λ^2 is set equal to zero, $F'(x) = \phi(x)$ and one has the familiar Nernst–Planck Equation. Thus, Nernst–Planck equations do not contain the effect of solvent molecules and its flow on ion fluxes since ion–neutral molecule interactions are absent. Coupling between fluxes of ionic and nonionic species, as expressed by phenomenological relations and experimental observations of anomalous osmosis, streaming potentials, etc., are ignored in the familiar Nernst–Planck formulations. Together with the Poisson Equation, the set of $(n + 1)$ equations (*see* Equations 6a and 6b) form the required $(n + 2)$ equations to solve for the $(n + 1)$ unknown concentration profiles and the electric potential profile $\phi(x)$. In the absence of chemical reactions, the matrix of constant coefficients $S_{\sigma\eta}$ and $S_{\sigma j}$ is singular. The influence of fluxes on the electric potential profile and concentration profiles are expressed implicitly by differential Equations 6a–6e and explicitly by the formal solutions of Equation Set 8 that are

$$C_\sigma(x) = C_\sigma(\mathrm{o}) \exp\left[-U_\sigma(x)\right] - I_\sigma(x)$$

$$C_j(x) = C_j(\mathrm{o}) \exp\left[U_j(x)\right] - I_j(x)$$

$$U_\sigma(x) = (Ze/kT)\left[\Delta F(x) + H^*\Delta C_j\right] - S_{\sigma\sigma}x$$

$$U_j(x) = (H^*/4\pi kT)\,\phi''(x) + S_{jj}x - (H_{jj}/kT)\Delta C_j$$

$$I_\sigma(x) = \exp\left[-U_\sigma\right]\int_{\mathrm{o}}^{x}\{R_\sigma + \sum_{\eta\neq\sigma} S_{\sigma\eta}C_\sigma + S_{\sigma j}C_j\}\exp(U_\sigma)\,dx$$

$$I_j(x) = exp\,(U_j)\int_{\mathrm{o}}^{x}\{[R_j + \sum_{\sigma} S_{j\sigma}C_\sigma]\exp(-U_j)\}dx$$

$$\Delta F(x) = F(x) - F(\mathrm{o})$$

$$\Delta C_j(x) = C_j(x) - C_j(\mathrm{o}) \tag{8}$$

Equation Set 8 expresses the influence of chemical reactions occurring in the diffusion barrier on the flux and concentration profile of Species α, in which Species α does not participate. The dependence of permeability of a specified species, say β, on the electrical potential difference, fluxes of other species, chemical reactions occurring in which species β does not participate, and on concentrations of other species also are expressed formally by Equation Set 8.

Thus, though Equation Set 6 are more complicated than the familiar Nernst–Planck Equations, they are expected to be more realistic in their applicability in describing the situations of biological membrane systems. The concentrations of ions in solutions of biological systems, for example in axoplasm and typical bathing solutions, are of the order of 0.5 mol/L. This value is too high for the activity coefficients of ions to be regarded as unity.

Two other exact relations (*see* Equations 9 and 11) can be obtained from Equation Set 6. Summing up the $(n + 1)$ equations of Set 6, the collective motion of all permeant species is describable by the following equation

$$\begin{aligned} A'(x) + C_j'(x) + [H_{jj}C_jC_j'/kT] + (R + R_j) \\ = [\epsilon/4\pi kT]\phi''(x)F'(x) \\ + (H^*\epsilon/4\pi kT)\ (d/dx)\ C_j(x)\phi''(x) \\ A(x) = \sum_\sigma C_\sigma(x); \qquad R = \sum_\sigma R_\sigma \end{aligned} \tag{9}$$

Equation 9 may be integrated to yield

$$\begin{aligned} \Delta A(x) + \Delta C_j(x) + (R + R_j)x + (H_{jj}/2kT)[C_j(x)^2 - C_j(\text{o})^2] \\ = (\epsilon/8\pi kT)\{\phi'(x)^2 - \phi'(\text{o})^2 - \lambda^2[\phi''(x)^2 - \phi''(\text{o})^2]\} \\ + (H^*\epsilon/4\pi kT)\{C_j(x)\phi''(x) - C_j(\text{o})\phi''(\text{o})\} \end{aligned} \tag{10}$$

Equation 10 is the generalization of Maxwell's osmotic balance equation (*15*).

Multiplication by Z_σe of Equation 6a for the ionic species σ, and summing up all ionic species, one obtains the third exact equation, by using the Poisson Equation

$$\begin{aligned} \phi'''(x)[1 + \eta^2(x)\lambda^2] = \eta^2(x)[\phi'(x) + H^*C_j'] + ZR \\ + (4\pi\text{e}/\epsilon)\sum_\sigma C_jS_{\sigma j} + \sum_\sigma\sum_\eta C_\sigma S_{\eta\sigma}\ (Z_\eta - Z_\sigma) \\ \eta^2(x) = (4\pi\text{e}^2/\epsilon kT)\sum_\sigma C_\sigma(x)Z_\sigma^2 \\ ZR = (4\pi\text{e}/\epsilon)\sum_\sigma Z_\sigma R_\sigma \end{aligned} \tag{11}$$

ZR has the same dimenison as ϕ''' viz, electrostatic units per centimeter to the fourth. R_σ has the dimension of moles per centimeters to the fourth. Fluxes observed in squid axon systems (*16*) are of the order of 10^{-12} mol/cm^2/sec. The membrane thickness is of the order of 50–100 $\times$ 10^{-8} cm. The observed permeability coefficients for univalent ions are of the order of 10^{-8} cm/sec. If one assumes a linear concentration profile and Fick's Law, and identifies the permeability coefficient of ionic species to be approximately $P_\sigma = (D_\sigma/h)$, where D_σ is the effective value of the diffusion coefficient of in the membrane phase and h is the thickness of the diffusion barrier, and using the above-mentioned values of fluxes, it follows that the diffusion coefficients for univalent ions in most membranes should be of the order of 10^{-14} cm^2/sec. This value is much smaller in comparison with the values of the order of 10^{-5} cm^2/sec obtained for univalent ions in aqueous solutions (*17*). The ion–water partial frictional coefficients computed from experimental data of diffusion coefficients are of the order of 10^{-8} erg–cm–sec/mol, while ion–ion partial frictional coefficients (*10*) again computed from experimental diffusion data in aqueous solutions are of the order of 10^{-7} erg–cm/mol. From the ion concentration values in axoplasm and bathing solutions and fluxes, one computes that the terms R_σ of Equation Set 6 is of the order of 10–200 mol–cm^{-4}, while the terms $C_\eta S_{\sigma\eta}$ are computed to be of the order of 10^{-5} mol–cm^{-4}. Thus, the third and fourth terms of right-hand side (RHS) of Equation 11 can be neglected with negligible error. The neglect of the third and fourth terms in comparison with the magnitude of the R_σ term is not generally valid. Such neglect is permissable only for the giant squid axon system, for which computations can be made with the available data. This justifies applying the theory to test experimental data in such dirty systems as squid axon membranes.

When one cannot neglect ion–solvent interactions, H^* is nonzero. When H^* is nonzero, and when $C_j' = 0$ and $J_j = 0$, it follows from Equation 6b that ϕ''' should be a constant independent of position variable x. In fact, it should equal $-(4\pi/H^*\epsilon)\sum_\sigma J_\sigma\zeta_{\sigma j}$. If $\sum_\sigma J_\sigma\zeta_{\sigma j}$ equals zero, then ϕ''' also equals zero and a linear profile is possible. Generally, in the case when flux and concentration gradient of a neutral permeant (solvent) molecule j vanishes, $\phi(x)$ can at best be a cubic function of the position variable. This conclusion is subject to the validity of the assumption that the dielectric coefficient is independent of the position variable. When $J_j = 0$, $C_j' \neq 0$, it follows that

$$C_j(x) = C_j(\text{o}) \exp\{[H^*\epsilon/4\pi kT]\Delta\phi'' + S_{jj}x - (H_{jj}/kT)\Delta C_j\} \tag{12}$$

These considerations suggest that one could expand $\phi(x)$ in a Taylor series

$$\Delta\phi(x) = \sum_{i=1} \phi_i x^i;\ \phi_i = (1/i!)[d^i\phi/dx^i]_{x=0} \tag{13}$$

retaining as many terms as one desires.

Simplified Theory

Physical considerations require that in regions of axoplasm (Solution 1) and bathing (Solution 2) solutions, at distances far from membrane solution interfaces, electroneutrality should be preserved. If one denotes the extent of the region lacking electroneutrality as d, (inclusive of diffuse double-layer regions δ_1 and δ_2 and a membrane of thickness h, $d = h + \delta_1 + \delta_2$), then the above requirement suggests that

$$\phi''(\mathrm{o}) = \phi''(d) = 0$$

$$\phi''(x) = 0 \text{ for values of } x < 0 \text{ and } x > d$$

$$\phi''(x) \neq 0 \text{ for values of } 0 < x < d \tag{14}$$

The bulk average concentrations measured in Solutions 1 and 2 will correspond to concentrations of species at $x = 0$ and $x = d$, respectively. Another physical consideration that should be valid in the resting state is that the zero net electric current condition should be satisfied. To satisfy the conditions stipulated in this equation, the charge-density profile must have at least one extremum value, if it is different from zero. To satisfy the macroscopic electroneutrality, there must be a minimum of two extremum values in the charge-density profile. The charge-density profile ($\phi''(x)$ one expects when $Cj' = 0$ and $Jj = 0$, is presented schematically by Plot A of Figure 1. Equation 14 reduces Equation 10 to

$$\phi'(d)^2 - \phi'(\mathrm{o})^2 = [8\pi kT\xi/\epsilon]$$

$$\xi = \Delta A(d) + \Delta C_j(d) + (R + R_j)d + (H_{jj}/2kT)\ \delta C_j \Delta C_j$$

$$\delta C_j(d) = C_j(d) + C_j(\mathrm{o}) \tag{15}$$

For axon systems, considerations of magnitudes of R_σ and $S_{\sigma\eta}C_\eta$ terms suggest that Equation Set 11 can be approximated with negligible error as

$$\eta_{\mathrm{o}}^2\phi'(\mathrm{o}) + ZR = (1 + \eta_{\mathrm{o}}^2\lambda^2)\phi'''(\mathrm{o})$$

$$\eta_2^2\phi'(d) + ZR = (1 + \eta_2^2\lambda^2)\phi'''(d) \tag{16}$$

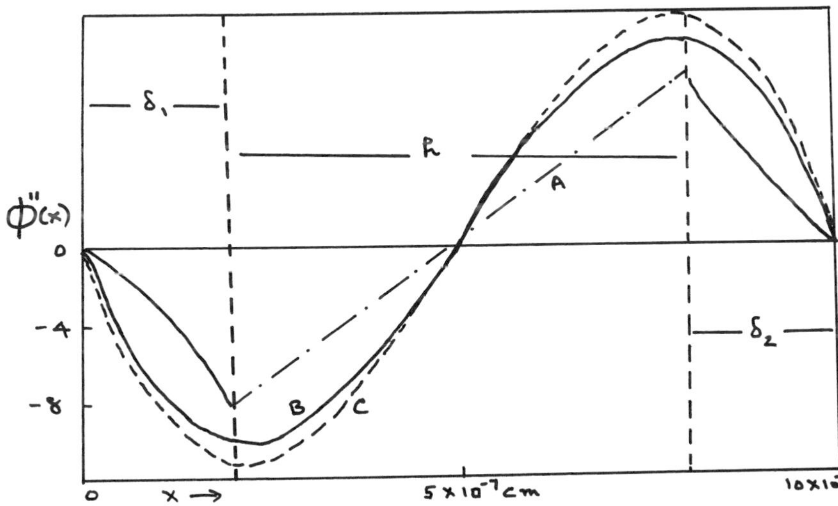

Figure 1. Charge-density profile computed from membrane potential (60 mV = 1.98824 × 10⁻⁴ esu/cm) and the value of d $= 10^{-6}$ *cm for a membrane system using Taylor expansion coefficients. The* y *axis is in* 10^{-8} *esu/cm³. Plot* A *is schematic; Plot* B *is obtained when* $\lambda^2 = 0$*; and Plot* C *is obtained using* $\lambda^2 = -0.193 \times 10^{-14}$ *cm²*.

The terms η_0^2 and η_2^2 are the values of Debye–Hückel parameters of the axoplasm and bathing solutions of axon systems, respectively. Computed values of these and the values of $A(o)$ and $A(d)$ from the concentrations of ions reported by Hurlbut (*16*) for various axon systems are presented in Table I.

The assumption of the validity of Equation 13 with the additional assumption that terms of the order of ϕ_7 and higher can be neglected, lead to an extremely simple expression for the stationary-state membrane potential difference $\phi(d)$

$$\phi(d) = (d/2)\delta\phi' + (d^3/120)\delta\phi''' \tag{17a}$$

$$\delta\phi' = \phi'(d) + \phi'(o) \tag{17b}$$

$$\delta\phi''' = \phi'''(d) + \phi'''(o) \tag{17c}$$

Equation 17 is subject only to the validity of the Taylor expansion approach and approximation involved in the neglect of ϕ_7 and higher-order terms. Equation 17 is a simple expression. Provided that one has knowledge of the five quantities, namely, d, $\phi'(o)$, $\phi'(d)$, $\phi'''(o)$, and $\phi'''(d)$, $\Delta\phi(d)$ can be computed. Also, one observes that the stationary-state membrane potential is determined by bulk solution ion concentra-

Table I. Values of $A(o)$, $A(d)$, and the Debye–Hückel Parameters for Various Giant Axon Systems Computed from Concentrations of Axoplasm and Bathing Solutions Reported in Ref. *16*

Axon	A*(o)*	A*(*d*)*[a]	η_o^2	η_2^2 [b]
Lologo Forbesii	778	1184	4.2918	6.5315
Loligo Pealii	858	992	4.7253	5.4633
Dosidicus Gigas	1026	1110	5.6509	6.1233
Sepioteuthis Sepioda	716	1160	3.9498	6.4543
Cuttlefish	784	1184	4.3249	6.5315
Lobster	626	1066	3.5087	5.9748
Crayfish	564	484	3.1113	2.6700

[a] The values of $A(o)$ and $A(d)$ are in 10^{-6} mol–cm^{-3}.
[b] The values of η_o^2 and η_2^2 are in 10^{-14} cm^{-2}.

tions through $A(d)$ (cf Equation 15), the Debye–Hückel parameters, and the membrane properties that determine the values of d, $(R + Rj)$, and ZR, as expressed in Equation Set 16. Only four of the five unknowns of Equation 17 are not determined yet, since one has the expression valid for any value of ZR

$$\phi'(d) = [\eta_o^2/\eta_2^2]\,\phi'(o) + (L/\eta_2^2)$$

$$L = (1 + \eta_2^2\lambda^2)\phi'''(d) - (1 + \eta_o^2\lambda^2)\phi''(o) \qquad (18)$$

When ξ equals zero, it follows from Equation Set 15 that $\phi'(d) = \pm \phi'(o)$. Equation Set 15 will be satisfied by the relations

$$\phi'(d) = (\eta/2) + [4\pi kT\xi/\epsilon\eta]$$

$$\phi'(o) = -(\eta/2) + [4\pi kT\xi/\epsilon\eta]$$

$$\eta = \Delta\phi'(d) = \phi'(d) - \phi'(o) \qquad (19)$$

From experiments one can measure stationary-state concentrations of ions and solutes in Solutions 1 and 2 and can measure the values of fluxes of permeant species. From additional information regarding diffusion coefficients of permeant species in the diffusion barrier, one could compute R_σ, Rj, and ZR. Thus, ξ is known and one knows the ratio of $[\phi'(d)/\phi'(o)]$. From Equation Set 18, $[L/\phi'(o)]$ is now known. Equation Sets 13 and 14 yield, when ϕ_7 and higher-order terms are negligible,

$$\phi'(d) = \phi'(o) - d^3\,[2\,\phi_4 + 5\,\phi_5 d + 9\,\phi_6 d^2]$$

$$\phi'''(o) = 6\,\phi_3 = -12\,\phi_4 d - 20\,\phi_5 d^2 - 30\,\phi_6 d^3$$

$$\phi'''(d) = 12\,\phi_4 d + 40\,\phi_5 d^2 + 90\,\phi_6 d^3$$

$$\Delta F'(d) = -\,d(12\lambda^2 + d^2)[2\,\phi_4 + 5\,\phi_5 d + 9\,\phi_6 d^2] - 12\,\lambda^2 d^3\,\phi_6 \quad (20)$$

If in a membrane system under the stationary state, the condition that $\Delta F'(d) = 0$ is satisfied and, in addition, ϕ_6 equals zero, one has

$$\lambda^2 = -(d^2/12) \quad (21a)$$

This relation suggests that since d is definitely positive, then λ^2, which is a measure of the activity coefficient correction factor, is negative. This is in agreement with the Debye–Hückel expression (*see* Equation 1). Equation 21a thus gives an approximate relation between d and λ^2. In general, from Equation Set 16 one has

$$\lambda^2 = -(1/\eta_o{}^2) + [\phi'(o)/\phi'''(o)] + ZR/[\eta_o{}^2\phi'''(o)] \quad (21b)$$

$$= -(1/\eta_2{}^2) + [\phi'(d)/\phi'''(d)] + ZR/[\eta_2{}^2\phi'''(d)] \quad (21c)$$

Sample Calculation

From analysis to be presented in the section dealing with a charge-density profile, one could assume that under the stationary state, $\Delta\phi''' = \phi'''(d) - \phi'''(o) = 0$. If in a membrane system the term ZR equals zero and $\Delta\phi'''$ equals zero, derives from Equation Set 16 that

$$\Delta\phi' = -\,[\Delta\eta^2/\eta_o{}^2\eta_2{}^2]\,\phi'''(o)$$
$$\delta\phi' = \{\lambda^2 + [\delta\eta^2/(2\,\eta_o{}^2\eta_2{}^2)]\}\,\phi'''(o)2 \quad (22)$$

Since, from the values listed in Table I, $\Delta\eta^2$ does not equal zero for most axon systems if $\Delta\phi' = 0$, it follows from Equation 22 that $\phi'''(o)$ should equal zero. From Equation Set 20 one concludes that all Taylor coefficients, ϕi, with $i = 1, 2, 3, 4, 5$, and 6, vanish. Thus, the difference in electric potential $\Delta\phi(d)$ should equal zero when the flux term $ZR = 0$ and $\Delta\phi''' = 0$. Under these conditions, ξ also should equal zero.

Consider now the case where ZR does not equal zero and $\Delta\phi''' = 0$. From Equation Set 16, one has the relation

$$\eta_o{}^2\eta_2{}^2\,\Delta\phi' + \Delta\eta^2\,\phi'''(o) = \Delta\eta^2 ZR \quad (23)$$

If in the membrane system, the conditions where $\xi = 0$ and $\Delta\phi' = 0$ are satisfied, one concludes that

$$\phi'''(o) = \phi'''(d) = ZR$$
$$\phi'(o) = \phi'(d) = \lambda^2 ZR$$
$$\phi_6 = 0, \quad \phi_5 = (ZR/10d^2), \qquad \phi_4 = -(ZR/4d)$$
$$\phi_3 = (ZR/6) \tag{24}$$

Since $\Delta F'(d)$ equals zero when $\Delta\phi' = 0 = \Delta\phi'''$, Equation 21a is valid. Hence $\phi_1 = \phi'(o) = -(ZR\ d^2/12)$. The stationary-state membrane potential is given now by the expression

$$\Delta\phi(d) = -ZR(d^3/15) \tag{25}$$

Thus, if one has knowledge of *ZR* from experiments, it follows from Equation Sets 24 and 25 that the Taylor expansion coefficients of the electric potential profile and the stationary-state electric potential difference between the two solutions can be computed from a knowledge of *d*. The sign of the electric current contribution term *ZR* is opposite to that of the membrane potential. If the potential difference is 60 mV, corresponding to 1.98824×10^{-4} esu/cm, and *d* is 1×10^{-6} cm, it follows that *ZR* should equal -2.98236×10^{15} esu/cm^4. Since ξ should equal zero when $\Delta\phi' = 0$, one can compute $(R + Rj)$ from Equation 15 from a knowledge of the boundary solution concentration and *d*. Conversely, if one has knowledge of permeant species, boundary solution properties, and an estimation of the diffusion coefficients of permeant species in the diffusion barrier, one can compute the stationary-state membrane potential $\Delta\phi(d)$. If one assumes that $\xi = 0$ in the resting state, and ignores the flux contribution of water, $Rj = 0$ assuming that $\Delta Cj = 0$, it follows from the values of concentrations listed in Ref. 16 for the Loligo Forbeseii axon system that $R = \sum_\sigma R_\sigma$ should equal -406 mol/cm^4 and that $\sum_\sigma Z_\sigma R_\sigma$ should correspond to about -66 mol/cm^4 in the resting state.

Calculation of Resting Potentials of Axon Systems

The parameter λ^2, which expresses deviation from the unity of activity coefficients of all solute ions, is a thermodynamic quantity, and is not expected to be strongly dependent on *ZR*. An expression for λ^2 in terms of bulk solution properties can be obtained realizing that under the stationary state $\phi'(o)$ is unique and assuming that λ^2 is independent of ZR. From Equation Sets 15 and 18, a quadratic equation for $\phi'(o)$ is derived easily.

$$\phi'(o)^2 - 2\,L\,\eta_o^2\,\phi'(o)/[\delta\eta^2\,\Delta\eta^2] - L^2/[\delta\eta^2\Delta\eta^2]$$
$$+ [\eta_2^4/\delta\eta^2\,\Delta\eta^2](8\pi kT\xi/\epsilon) = 0$$
$$\delta\eta^2 = \eta_2^2 + \eta_o^2; \qquad \Delta\eta^2 = \eta_2^2 - \eta_o^2 \tag{26}$$

Since $\phi'(o)$ is unique under the stationary-state condition, one derives from Equation Set 26 that

$$L^2 = (8\pi kT\xi/\epsilon)\,\delta\eta^2\Delta\eta^2$$

$$\phi'(o) = L\eta_o^2/(\delta\eta^2\Delta\eta^2)$$

$$\phi'(d) = L\eta_2^2/(\delta\eta^2\Delta\eta^2) \tag{27}$$

By approximating that $\phi'''(o) = \phi'''(d) \neq 0$, one obtains

$$L = \phi'''(o)\lambda^2\Delta\eta^2$$

$$\phi'(o) = \lambda^2\eta_o^2\phi'''(o)/(\delta\eta^2)$$

$$\phi'(d) = \lambda^2\eta_2^2\phi'''(o)/(\delta\eta^2) \tag{28}$$

From Equation Sets 16 and 28 one now can derive that

$$\lambda^2 = -[\delta\eta^2/(\eta_o^2\eta_2^2)] \tag{29}$$

Equation 29 is exact when $ZR = 0$ and $\phi'(o)$ is unique. When $\Delta\phi''' \neq 0$, one derives from Equation 16 and 27 that

$$ZR\,\delta\eta^2\Delta\eta^2 = (1 + \eta_o^2\lambda^2)\eta_2^4\phi'''(o) - (1 + \eta_2^2\lambda^2)\eta_o^4\phi'''(d) \tag{30}$$

In Table II, the values of λ^2 (computed by using Equation 29), the values of d, and the extent of the region lacking electroneutrality (computed by using Equation 21a) are presented for various axon systems. In the same table, the computed values of $\Delta\phi(d)$ and the observed resting potentials of different axon membrane systems are presented. One

Table II. Calculated and Observed Values of Resting Potentials for Various Axon Systems Using Equations 15, 17a, 21a, and 29

Axon	$[8\pi kT\Delta A/\epsilon]$[a]	$-\lambda^2$[b]	d[c]	$\Delta\phi_{calculated}$[d]	$\Delta\phi_{observed}$[e]
Loligo Forbesii	3.1584	0.3861	21.525	−77.5	−60
Loligo Pealeii	1.0054	0.3947	21.762	−73.4	−60
Dosidicus Gigas	1.0424	0.3400	20.199	−94.14	—
Sepioteuthis Sepiodea	3.4540	0.4081	22.130	−75.89	−56
Cuttlefish	3.1117	0.3843	21.475	−76.07	−62
Lobster	3.3012	0.4524	23.299	−75.15	−75
Crayfish	−0.6223	0.6959	28.898	−74.7	−87

[a] The values of $(8\pi kT\Delta A/\epsilon)$ are in 10^6 erg–cm^{-3}.
[b] The values of λ^2 are in 10^{-14} cm^2.
[c] The values of d are in 10^{-8} cm.
[d] The values of potentials are in millivolts.
[e] 60 mV = 1.98824×10^{-4} esu/cm.

may observe that the values computed from simplified theory are in reasonable agreement with experiments.

In calculating the Debye–Hückel parameters for these systems, the temperatures of 9°C for lobster and of 14°C for loligo pealii as well as a value of 80 for the dielectric coefficient of bulk aqueous solutions that enter the calculations were used. For all other systems, it was assumed that $\epsilon = 78.5$ and that the 19°C temperature were used. Negative root values of Equation Set 27 for $\phi'''(o)$ (assuming that $\Delta\phi''' = 0$), were used. In computing $A(o)$ and $A(d)$ of Table I, it was assumed that the equivalent concentration of monovalent unspecified impermeable ions are present in axoplasm and bathing solutions to preserve macroscopic electroneutrality. This needs to be done since the values of concentrations of sodium, chlorine, and potassium, reported in Ref. 16, do not satisfy electroneutrality conditions. Contributions from Cj and fluxes to ξ were ignored, since values of fluxes under resting-state conditions satisfying zero net electric conditions were not available.

Evidently the values presented in Table II are approximate. The values of flux and permeability of water across the membrane systems are not available. The neglect of positional dependence of the dielectric coefficient is a serious source of error. This is important when considering solutions of the differential equations presented in the first section of this chapter, but not so important in the simplified theory since only bulk solution properties enter in computation of $\Delta\phi(d)$. Using the bulk solution dielectric coefficient underestimates the value of d, which is about 20×10^{-8} cm, while in reality d should be much larger.

The values of permeant ion fluxes (influx and outflux) are reported for the loligo forbesii, loligo pealii, cuttlefish, and lobster systems. If the net flux of a specified ion required in the calculation is assumed to equal the difference between reported values of outflux and influx, the computed values (e.g., for loligo forbesii, $J_{Na} = -7$, $J_K = 20$, and $J_{Cl} = -1 (\text{x } 10^{-12}$ mol–cm^{-2} sec^{-1}) do not satisfy the zero net electric current condition. Using these and the estimated values of diffusion coefficients from permeabilities, one computes $R_{Na} = -200$, $R_K = 44.44$, and $R_{Cl} = 13.7$ mol/cm^4. Computed values of ZR using these values is -0.657×10^{16} esu/cm^4 while the value in the resting state should equal -0.298×10^{16}, corresponding to a potential of 60 mV and $h = 1 \times 10^{-8}$ cm. The situation for other axon systems are worse. Again the computed value of R should be about -66 mol/cm^4, while the values from fluxes yield -169.25 mol/cm^4 for loligo forbesii. A real test of the theory should come from actual measurements of fluxes of permeant ions and electric potential difference and permeability of a clean physical system in which neglecting the last term of Equation 11 is justified.

Analysis from the Charge-Density Profile

The charge-density profile $\phi''(x)$ satisfies Equation Set 14. If $\phi''(x) \neq 0$ for values of x, then $0 < x < d$ so that there is an extremum point, ϕ_3, and ϕ_4 cannot vanish. In order that the diffusion barrier and adjacent diffuse double-layer regions are macroscopically electrically neutral, there must be a minimum of two extremum points in $\phi''(x)$ such that the simplest case is when ϕ_5 does not vanish. From Equation Set 20, it follows that when there is one extremum point in $\phi''(x)$, occurring at perhaps $x = x_2$ (though not physically permitted), one obtains $\phi'''(o) = -\phi''(d) = -12\ \phi_4 d$ and $d = 2x_2$.

If ϕ_6 and higher-order terms of Equation 13 are negligible, the inflection point x_1 and the extremum points x_2 of the function $\phi''(x)$ are given by

$$x_1 = -\ (\phi_4/5\phi_5)$$

$$x_2{}^2 + (2\phi_4/5\phi_5)x_2 + (\phi_3/10\phi_5) = 0$$

$$x_2 = x_1\{1 \pm (3)^{-1/2}\}$$

since

$$5\phi_5 d = -\ 2\phi_4$$

$$\phi_3 = -\ 2\phi_4 d/3 \text{ when } x_1 = d/2 \qquad (31)$$

Thus, when the inflection point of $\phi''(x)$ occurs at midpoint, x_2 equals 0.2113 and 0.7887 d. If extremum in charge density occurs at membrane–solution interfaces, one has the membrane thickness h, as the distance between two extreme values of $\phi''(x) = 0.5774\ d$. When d equals 1×10^{-6} cm on this premise, the membrane should have a thickness of about 57×10^{-8} cm and the diffuse double layers should have a thickness of 21×10^{-8} cm on both sides.

When the ϕ_1, ϕ_3, ϕ_4, ϕ_5, and ϕ_6 terms of Equation 13 are not negligible, while ϕ_7 and higher-order terms are negligible, and $\phi''(x)$ has one unique inflection point at $x = x_1$, one has

$$x_1{}^2 + (\phi_5/3\phi_6)\ x_1 + (\phi_4/15\phi_6) = 0$$

$$x_1 = -\ (\phi_5/6\phi_6)\,; \qquad 5\phi_5{}^2 = 12\phi_4\phi_6 \qquad (32)$$

Generally, when there are two real inflection points of $\phi''(x)$ at $x = x_1$ the relation is

$$x_1 = -(\phi_5/6\phi_6) \pm (\phi_5/6\phi_6)\{1 - [12\phi_4\phi_6/5\phi_5^2]\}^{1/2}$$
$$5\phi_5^2 \geqslant 12\phi_4\phi_6 \qquad (33)$$

For the symmetrical case, when the two inflection points are spaced equally apart from the midpoint of the diffusion barrier, $d/2$, it follows that $\phi_5 = 3\phi_6 d$. In this situation, one derives from Equation Sets 16, 18, and 20 that

$$\phi'''(d) = -\phi'''(o) = 12\phi_4 d + 10\phi_5 d^2$$

$$\phi'(d) = \phi'(o) - 2d^3(\phi_4 + \phi_5 d)$$

$$\phi'(o)\Delta\eta^2/2d = \phi_4[\eta_2^2 d^2 + 12 + 6\lambda^2\delta\eta^2] + \phi_5 d[\eta_2^2 d^2 + 10 + 5\lambda^2\delta\eta^2]$$

$$\phi'(d)\Delta\eta^2/2d = \phi_4[\eta_o^2 d^2 + 12 + 6\lambda^2\delta\eta^2] + \phi_5 d[\eta_o^2 d^2 + 10 + 5\lambda^2\delta\eta^2]$$

$$-(\Delta\eta^2/2d)\, ZR = \phi_4\{\eta_o^2\eta_2^2(d^2 + 12)\lambda^2) + 6\,\delta\eta^2\} + \phi_5 d\{\eta_o^2\eta_2^2(d^2 + 10\,\lambda^2) + 5\,\delta\eta^2\}$$

$$\Delta\phi(d) = (d/2)\,\delta\phi' = (d^2/\Delta\eta^2)\,\{\phi_4[24 + \delta\eta^2(d^2 + 12\,\lambda^2)] + \phi_5 d\,[20 + \delta\eta^2(d^2 + 10\,\lambda^2)]\}$$

$$-(2\pi kT\xi/\epsilon)/[d^2(\phi_4 + \phi_5 d)] = \Delta\phi(d) \qquad \mathbf{(34)}$$

Equation Set 34 can be used to compute the Taylor expansion coefficients of $\phi(x)$. In the simpler case, when $\phi''(x)$ has only one unique inflection point, $\phi_3 = -(5/2)\phi_6 d^3$, $\phi_4 = (15/4)\phi_6 d^3$, $\phi_5 = -3\phi_6 d$, and $\phi' = -(3/2)\phi_6 d^5$. In addition

$$\Delta\phi = \phi'(o)d - (3/4)\phi_6 d^6 = -(8\pi kT\xi)/(3\epsilon\phi_6 d^4) \qquad (36)$$

If one has knowledge of $\Delta\phi$ and ζ from experiments, one can compute ϕ_6 and all other quantities. With the assumption that $d = 1 \times 10^{-6}$ cm, $\xi = \Delta A$, and $\Delta\phi = 1.98824 \times 10^{-4}$ esu/cm for the loligo forbesii axon system, one obtains $\phi_6 = 5.295 \times 10^{33}$, $\phi'(o) = -3.7725 \times 10^3$ esu/cm^2, $\phi'(d) = 4.1702 \times 10^3$, and $\phi_3 = 13.2378 \times 10^{15}$ esu/cm^4. Using Equation Sets 16 and 18, one computes that $\Delta F'(d) = 211$ esu/cm^2, $\lambda^2 = 4.867 \times 10^{-14}$ cm^2, and $ZR = 1.1942 \times 10^{17}$. Apart from demonstrating the feasibility of such calculations, these values justify the approximation of neglect of $\Delta F'(d)$ since the computed magnitude of this quantity is only about 3% of the value of $\Delta\phi'(d)$. In Figure 1, the charge-density profile computed from a membrane potential of 60 mV for various values of λ^2 (assuming that $\phi''(x)$ has one inflection point) is presented. The term d is assumed to equal 1×10^{-6} cm. When $\phi'''(d) = -\phi'''(o)$, then $\Delta\phi'$ — $(4\pi kT\ \Delta\ Ad/\epsilon\Delta\phi)$ and $(2 + \lambda^2\delta\eta^2)\phi'''(d) = \eta_2{}^2\phi'(d) - \eta_0{}^2\phi'(o)$.

The Significance of λ^2

The essential difference between the simplified theory of this chapter and the Nernst–Planck Equations is the presence of the λ^2 term. The condition of electroneutrality at $x = 0$ and at $x = d$, combined with neglecting H_{jj}, ΔC_j, and R_j, reduces Equation 10 to the same result that one would obtain from the Nernst–Planck Equation when $R_\sigma = (J_\sigma/D_\sigma)$. The long-range nature of coulombic interactions and the low dielectric coefficient of the biological membrane system forbid assuming that λ^2 equals zero. Equation Set 16 expresses the role of λ^2 in determining the relation existing between $\phi'(o)$, ZR, and $\phi'''(o)$. The role of λ^2 is stressed further by Equation 9. By approximating that $C_j' = 0 = R_j$, one has

$$A'(x) + R = (\epsilon/4\pi kT)\ \phi''(x)F'(x) \tag{37}$$

From Equations 14 and 37, it follows that $A'(o) = A'(d) = -R$. From Equation 37,

$$(\epsilon/4\pi kT)\{\phi'''(x)F'(x) + \phi''(x)F''(x)\} = A''(x)$$

$$\lambda^2\phi'''(o)^2 - \phi'(o)\phi'''(o) + (4\pi kT/\epsilon)A''(o) = 0$$

$$\lambda^2\phi'''(d)^2 - \phi'(d)\phi'''(d) + (4\pi kT/\epsilon)A''(d) = 0 \tag{38}$$

Under stationary-state conditions $\phi'(o)$, $\phi'''(o)$, $\phi'(d)$, and $\phi'''(d)$ have unique values. Hence one has

$$\phi'(o) = 2\ \lambda^2\ \phi'''(o)$$

$$\phi'(d) = 2\lambda^2\ \phi'''(d)$$

$$\phi'(d)^2 = [16\pi kT/\epsilon]\ \lambda^2 A''(d)$$

$$\phi'(o)^2 = [16\pi kT/\epsilon]\ \lambda^2 A''(o)$$

$$\Delta A = 2\lambda^2[A''(d) - A''(o)] \tag{39}$$

It is evident from Equation Sets 24 and 39 that λ^2 relates $\phi'''(x)$ with $\phi'(x)$. The term λ^2 also relates the derivatives of $A(x)$ with $\Delta A(x)$. From Equation Sets 16 and 39 one has

$$ZR = \phi'''(d)\ [1 - \eta_2{}^2\lambda^2] = \phi'''(\mathrm{o})\ [1 - \eta_o{}^2\lambda^2]$$

$$\Delta\phi'''(d) = \lambda^2\Delta\eta^2\phi'''(\mathrm{o})/[1 - \eta_2{}^2\lambda^2]$$

$$\phi'(d)[1 - \eta_2{}^2\lambda^2] = \phi'(\mathrm{o})\ [1 - \eta_o{}^2\lambda^2] \tag{40}$$

When λ^2 equals zero, it follows that $\Delta\phi' = 0$, a condition that cannot be met as long as $\Delta A + Rd$ does not equal zero, as demanded by Equation 10. In the approximation that terms of ϕ_7 and higher-order terms of Equation 13 can be neglected, use of Equation 20 enables one to evaluate the following

$$\phi_6 = (d^2 + 24\ \lambda^2)\ \Delta\phi'''/12d^5$$

$$\phi_5 = (\phi'''(\mathrm{o})/10d^2) - (\Delta\phi'''/10d^4)\,(d^2 + 20\,\lambda^2)$$

$$\phi_4 = -(\phi'''(\mathrm{o})/4d) + (\Delta\phi'''/8d^3)\,(d^2 + 40\ \lambda^2)$$

$$\Delta\phi(d) = \phi'''(\mathrm{o})\,(d/60)\,(120\ \lambda^2 + d^2) + \Delta\phi'''d\ \{5\ \lambda^2 + (23/120)\,d^2\}$$

$$[2\pi kT\ \Delta A/\epsilon]\ (1 - \eta_2{}^2\lambda^2)^2 = \phi'''(\mathrm{o})^2\Delta\eta^2\lambda^6(2 - \lambda^2\delta\eta^2) \tag{41}$$

If ξ, ZR, and $\Delta\phi$ for a diffusion barrier system are known and $\eta_2{}^2$, $\eta_o{}^2$, and d are also known, then $\phi'''(\mathrm{o})$ and λ^2 may be regarded as unknowns and these can be evaluated from the above equations. Also, the Taylor expansion coefficient, $A''(\mathrm{o})$, can be evaluated now by using Equation Set 39. $A'(\mathrm{o}) = -R$. Additional differentiation with respect to the x of Equation Set 38 yields

$$A'''(x)\,(4\pi kT/\epsilon) = \phi''''(x)F'(x) + 2\ \phi'''F'' + \phi''F'''$$

$$\{3\ \lambda^2\phi'''(\mathrm{o}) - \phi'(\mathrm{o})\}\phi''''(\mathrm{o}) + (4\pi kT/\epsilon)A'''(\mathrm{o}) = 0$$

$$\lambda^2\phi'''(\mathrm{o})\phi''''(\mathrm{o}) = -(4\pi kT/\epsilon)A'''(\mathrm{o}) \tag{42}$$

Such equations together with the constraint that $\sum_{1=2}^{\eta} i(i - 1)A_i d^{(i-2)}$ can be used to evaluate the Taylor coefficients of $A(x)$. For example, when $ZR = 0$ and $\lambda^2 = 0$, and if $\Delta A(x) = A_1x + A_2x^2 + A_3x^3$, then $A_1 = -R$; $A_2 = -[\Delta A\eta_o{}^2\eta_2{}^2]/\delta\eta^2$ and A_3d equals $(2/3)\Delta A\eta_o{}^2\eta_2{}^2/(\delta\eta^2)$.

Discussion

In biological literature, the resting potentials of various axon systems often are computed with the Nernst equilibrium expression, which invariably does not tally with the observed potentials (*16*). Abundant support can be given for the statement that the state of affairs of the axon systems at resting state is a stationary-state condition with non-vanishing individual fluxes and is not an equilibrium condition. Historically, Teorell (*1*) defined these dynamic effects as a diffusion effect on the ionic distributions. Our objective has been to crystalize the qualitative ideas expressed by Teorell. Many schools of thought have recognized these problems, and have attempted to solve the problem of the influence of fluxes on stationary-state membrane potentials by using the familiar Nernst–Planck Equations. The only satisfactory solution of the Nernst–Planck Equation is the result of the work by Planck and is restricted to a two-ion symmetrical electrolyte (*2*). In spite of many attempts, a solution of the Nernst–Planck Equations in closed analytic form valid for many ion systems is not available. In the opinion of the author, the Nernst–Planck Equations are too simple to be justified on the basis of physical arguments and too complicated to be solved mathematically.

The approach that this chapter took was initiated with the philosophy that a more complicated set of equations than the one presented in this chapter may be simpler to solve and may be applied realistically to biological membrane systems. As may be noted from Equation 17, which is a simple though evidently approximate result, one concludes that nature is simple because of its complexity.

The results of Equation 10 may be recognized as a generalization of Maxwell's osmotic balance relation with strain owing to the electric field (*15*). Subject only to the validity of assumed Equation Sets 2, 4, and 6, the results of Equation Sets 8, 9, and 11 are exact. The results of Table II indicate the applicability of the approach of this chapter to biological membrane transport problems. The conclusion that $A'(o) = A'(d) = -R$, obtained from Equation 37 (which is exact), can be reconciled with the basic Equation 6a, neglecting the H* term in a primitive manner by stating that

$$(Z_\sigma e/kT)\,C_\sigma(x)F'(x) = S_{\sigma\sigma}\,C_\sigma - \sum_{\eta\neq\sigma}^{n} S_{\sigma\eta}C_\eta \qquad (43)$$

A three-permeant-ion system, together with the zero net electric current condition, $\sum_\sigma Z_\sigma J_\sigma = 0$, yields two relations for the ratio of fluxes of two permeant species, say (J_β/J_α), in terms of partial frictional coefficients and the concentrations of the three species in one boundary solution. Assuming that ion–ion interactions, which depend on valence charge numbers, and contributions to partial frictional coefficients are invariant, these two relations can be justified by only specific values of boundary solution concentrations.

A touchy problem is the maintenance of constant composition of axoplasm and bathing solutions, with fluxes present in the membrane under the stationary state. This important problem, has been discussed by Lerch and Wolf (*18*) and has been avoided consciously by us. The situation is somewhat akin to the constant rate of flow of river water into a large lake. The flow of water at distances far from the mouth of the river is zero, and the flux of water in the river is finite and nonzero. Thus, in the interfacial-region fluxes must dissipate, and in this region the stationary state cannot be maintained. This problem and the variation of the dielectric coefficient as a function of position variable x are complicated mathematically, and thus are avoided in this chapter.

Although specific attention to experimental information of axon systems is described in this chapter in order to relate simplified theory to experiments, it is evident that the approach is applicable to an arbitrary membrane diffusion system, with or without impermeant molecules and ions in solution. For axon systems, it can be shown quite simply that the R_σ terms are considerably larger in magnitude than the neglected $S_{\sigma\eta}C_\eta$ terms. It would be ideal to compute the stationary-state electric potential from observed fluxes and bulk solution composition. However, the fluxes reported for axon systems do not satisfy the zero electric current condition and the flux of water across axon membrane systems is not available. In other systems and situations, neglecting the last term in the RHS of Equation 11 may not be reasonable. Finally, this chapter presents certain preliminary results based on a novel idea. The treatment of this chapter is approximate but hopefully further attempts along these lines will yield more fruitful results.

Glossary of Symbols[a]

k = Boltzmann Constant
e = proton charge
T = temperature in Kelvin scale
x = position variable defined normal to yz plane of diffusion barrier
d = dielectric coefficient of a solution
h = membrane thickness
$C_\sigma(x)$ = concentration of ionic Species σ at x
J_σ = net flux of ion σ in moles centimeters squared per second
Z_σ = signed valence charge number of Ion σ
D_σ = diffusion coefficient of Ion σ
$A(x)$ = sum of the concentrations of all ions at Location x
A_i = Taylor expansion coefficient of $A(x)$
$S_{\sigma\sigma}$, $S_{\sigma\eta}$ = constant terms defined in Equation Set 6

R_σ = constant parameter approximately equalling (J_σ/D_σ)

$R = \sum_\sigma R_\sigma$

$ZR = (4\pi e/\epsilon) \sum_\sigma Z_\sigma R_\sigma$

Greek Letters

ϵ = distance between locations where the difference in the electric potential is measured; extent of the region lacking electroneutrality.

ζ_σ = frictional coefficient of Ion σ

$\zeta_{\sigma\eta}$ = partial frictional coefficients

P_σ = permeability coefficient of an ion

$\Delta_{\phi(d)}$ = difference in electric potential, assumed equal to measured electric potential difference

ϕ_i = Taylor expansion coefficient of $\phi(x)$

η_0^2, η_2^2 = Debye–Hückel parameters of solutions

$\xi = \Delta A(d) + \Delta C_j(d) + d(R + R_j)$

[a] Greek subscripts are used to denote ionic species and roman subscripts are used to denote nonionic species. The symbol δ is used to indicate the sum of values of specified quantity at two locations. The symbol Δ is used to denote differences in values at two locations of a specified quantity. A single prime indicates a first derivative with respect to x of a quantity. Higher derivatives are denoted by the appropriate number of primes.

Literature Cited

1. Teorell, T. "Membrane Phenomena," *Discuss. Faraday Soc.* **1956,** *21,* 9.
2. MacInnes, D. A. "Principles of Electrochemistry"; Dover Publications: New York, 1961.
3. Lakshminarayaniah, N. "Transport Phenomena in Membrances"; Academic: New York, 1969.
4. Goldman, D. E. *J. Gen. Physiol.* **1943,** *27,* 37.
5. Blaustein, M. P.; Russell, J. P. *J. Membr. Biol.* **1975,** *22,* 285.
6. Vaidhyanathan, V. S. In "Topics in Bioelectrochemistry and Bioenergetics"; Milazzo, G., Ed.; Wiley: New York, 1976; Vol. I, 288–378.
7. Vaidhyanathan, V. S. In "Electrical Phenomena at the Biological Membrane Level"; Roux, R., Ed.; Elsevier: New York, 1977; p. 113.
8. Kobatake, Y.; Inoue, I.; Ueda, T. In "Advances in Biophysics"; Kotani, M., Ed.; Vol. 7, 43–89.
9. Shinagawa, Y. *J. Theor. Biol.* **1977,** *64,* 551.
10. Vaidhyanathan, V. S. *Bull. Math. Biol.* **1979,** *41,* 365.
11. Hill, T. L. *Discuss. Faraday Soc.* **1956,** *21,* 31.
12. Kirkwood, J. G. "Collected Papers of Kirkwood"; Gordon & Breach: New York, 1967; p. 26.

13. Vaidhyanathan, V. S. *J. Theor. Biol.* **1971**, *33*, 1.
14. Bearman, R. J. *J. Chem. Phys.* **1958**, *21*, 1278.
15. Agin, D. In "Foundations of Mathematical Biology"; Rosen, R., Ed.; Academic: New York, 1971; Vol. 1.
16. Hurlbut, W. P. In "Membranes and Ion Transport"; Bittar, E., Ed.; Academic: New York, 1971; Vol. 2.
17. Tyrell, H. J. V. "Diffusion and Heat Flow in Liquids"; Butterworths: London, 1961.
18. Lerche, D.; Wolf, H. *J. Bioelectrochem. Bioenerg.* **1975**, *4*, 293, 304.

RECEIVED October 17, 1978.

CELLS AND TISSUES

21

Membrane Impedance As a Probe for Interfacial Electrochemical Control of Living Cell Function

ARTHUR A. PILLA

Bioelectrochemistry Laboratory, Orthopedic Research Laboratories and Department of Applied Chemistry and Chemical Engineering, Columbia University, 630 W. 168th Street, New York, NY 10032

The concept of electrochemical information transfer at cell surfaces and junctions has been proposed as a means to modulate cell regulation. Potential-dependent interfacial phenomena involving specific adsorption and membrane transport and their possible interaction is described quantitatively by kinetic analysis. This leads to experimentally testable transient impedance expressions. Results show that the cell surface exhibits adaptive electrochemical kinetic changes as a function of its microenvironment and probably its cycle stage and functional state. This allows the range of relaxation times which may be involved in the kinetics of cell regulation to be estimated. Utilization of this approach to encode low-level pulsating current for the modulation of the kinetics of cellular response to an existing environment has been successful in vitro (cell division, protein synthesis, and Ca^{2+} kinetics) and in vivo (healing of recalcitrant bone fractures).

It now is established clearly that cellular function, both in vitro and in vivo, is influenced by weak (in the microamperes per centimeters squared range) electric currents (*1, 2*). The role of electromagnetism in living systems has received considerable attention for more than 100 years (*3*). However recent work in the skeletal system, wherein recalcitrant bone fractures are healed using both dc and pulsating currents,

0-8412-0473-X/80/33-188-339$05.25/1

has served to focus considerable attention on the role of electricity in cellular phenomena. Interestingly, this modern era began because of the discovery of piezoelectric-like phenomena in wood (*4*) and the desire of bone physiologists to understand how bone tissue adaptively responds to mechanical input (*5*). Piezoelectricity appeared to offer a rational approach to the problem since the suspected electrical link could thereby be provided. Thus bone, when stressed, exhibits an electrical relaxation which could be the basis of a signal triggering cell activity. However, many recent studies have shown that the piezoelectric or strain-generated voltages observed when bone tissue is stressed are properties of its nonliving organic matrix collagen (*6*). In fact, absolutely dead bone, i.e., no living cells, exhibits the same electric relaxation. At least one study has shown that a suspension of living cells (no collagen matrix) responds functionally to periodic stress input in an identical manner as the same cell type in a matrix (*7*). It therefore appears that direct mechanical perturbation of the cell (probably via specific membrane deformation) is the underlying mechanism. As will be shown in this study, the basic phenomenon could very well be a change in electrochemical properties at specific membrane sites.

However more physically meaningful present approaches may appear to be, the piezoelectric concept has led to considerable recent activity in which the electric signal was applied for bone repair and remodeling via nonpiezoelectric means. Thus bone repair has been enhanced using dc current applied via electrodes in direct contact with the cells and tissues involved (*8*). It is evident that under these conditions electrolysis or faradaic effects, which modify the chemical composition of extracellular medium, may play a central role (*9*). In addition, the redistribution of minority concentration ions (e.g., K^+, Ca^{2+}, and Mg^{2+}) owing to transference number effects can significantly change a cell's ionic boundary conditions with possible biochemical consequences (*10*). Direct effects of electric currents at the cellular level have been observed recently (*11, 12, 13*). In this case cell-function modification appears to depend upon current waveform parameters and it appears that a relatively high specificity is involved.

For all of the above cases it is clear that there must be a central linking mechanism by which certain specific cell regulatory processes are either triggered into action or experience some rate modification. Since a cell's environment is largely electrochemical consisting of ions and dipoles, it recently has been proposed that electrochemical surface interactions could play an important role in the modulation of cell function (*14, 15, 16, 17*). Thus the concept of electrochemical information transfer in vivo was created to provide a working model having unifying and predictive capabilities for using electrical currents to modify cell behavior.

Basically this approach specifically recognizes the potential, or electrical charge, and relaxation-time dependence of specific interactions involving membrane sites which can direct cell development and function. A charged species can enter into very specific interaction with a membrane surface site via molecular steps such as dehydration, displacement, and spatial modification (*18, 19*). These events are the essence of specific adsorption (binding) or phase penetration (partitioning) processes which are electrochemical in nature since they occur in the highly charged and structured environment at the membrane–fluid junction. Quantitative knowledge of these dynamic electrochemical surface interactions allows their chronic manipulation for possible cell function modification via a change in the rate of a regulatory process. The electrochemical information transfer concept has been quantitated in preliminary studies (*20*). This chapter extends the modeling approach and discusses relevant in vitro results which lend credence to the basic approach.

Membrane Surfaces, Cell Regulation, and Electrochemical Information Transfer

A cell's first contact with its environment, whether in isolation or in a tissue structure, is via its cytoplasmic membrane. One of the most important properties of this membrane's intra- or extracellular fluid interface is that it is electrified. This characteristic arises because of the charged functional groups which are present at the hydrophilic interfaces of the lipids, proteins, and carbohydrates making up the basic molecular structure of the cell membrane. These regions exhibit charge separation wherein the electroneutrality condition no longer holds. The structure of this electrified interface is potential dependent and is determined by the particular ion and/or dipole interactions which are present. These in turn depend upon the specific molecular structures, and all interactions thereof, present in the membrane phase at a given state in cellular activity. The most generally accepted working model of membrane structure is known as the fluid mosaic model (*21*). This approach depicts the membrane as a two-dimensional solution of globular integral proteins in a fluid solvent of lipids in bilayer configuration. Because of the molecular structure of lipids and proteins, it is evident that even basic thermodynamic considerations require both hydrophobic and hydrophilic interactions. It is the latter which are important for this study since highly polar groups are in contact with the intra- and extracellular ionic environment. The charged-species interactions which occur here appear to be fundamentally important for cell function. Biochemical reactions on the cell surface involve charged reactants (*22*) and the availability and specificity of these charges are known to influence, e.g., membrane-bound enzyme activity (*7, 23, 24*). The surface potential is seen by an ionic species involved in membrane transport (*25, 26, 27, 28*) and appears to be primarily important in partitioning kinetics (*20*). Underlying all of the above is the emerging evidence that fundamental changes in membrane structural parameters such as configuration, fluidity, and lipid–protein interaction, etc., occur if specific surface charges are modified (*38*). That these are related functionally to cell activity is without doubt, even if only the distribution of proteins in the membrane is altered.

It is because of the above-mentioned influences of surface-charge configuration upon membrane structure (and therefore function) that the possible application of electrochemical information transfer to cell regulation was proposed. In order to more clearly illustrate the manner whereby low-level dynamic current injection may result in a real time modification of the structure of the cell surface which may be important to cell regulation, it is convenient to consider the diagram shown in Figure 1. Here the charged interface on one side of a membrane is shown in highly schematic form. By analogy with the electrode–electrolyte interface (*14*), two distinct types of electrochemical interactions occur. The first involves all of the nonspecific electrostatic interactions of water dipoles and hydrated (or partially hydrated) ions. The structure of this portion of the interfacial regions is established merely via the availability of entities with appropriate charge to satisfy minimum free-energy requirements commensurate with charge separations of molecular dimensions. For this, the concentration of water dipoles in tissue electrolytes is overwhelmingly larger than the concentration of even the Na^+ or Cl^- ions. Therefore, at hydrophilic surfaces a layer of water dipoles, configured approximately as shown in Figure 1, exists provided that the dielectric structure mediates the interaction of hydrated ions. The interfacial electrochemical situation just described is expected to exist in the aqueous electrolyte media of most living systems and would probably not vary significantly as the ionic environment changes owing to, e.g., functional and/or developmental activities. This already has been observed for one living membrane system (*20, 23*), and is most probably caused by the nonspecific nature of these interactions.

The mechanism of charge acceptance by the above depicted portion of the membrane structure is relatively simple, if low-level currents are used. The necessary ionic and molecular rearrangements probably would be restricted to a first approximation, to slight dipole bending and/or a small change (according to the quantity of injected charge per pulse) in the population of (mostly) hydrated cations in the plane of closest approach (*see* Figure 1). Such processes are infinitely rapid over the frequency range of interest both for analysis and perturbation (10^{-1}–10^8

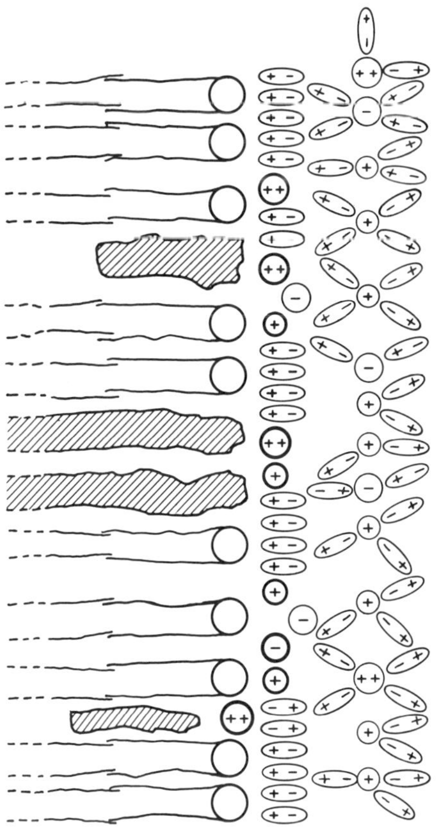

Figure 1. Schematic of the charged interfacial structure which may exist at one membrane–fluid interface. Purely electrostatic interactions involve water dipoles (elliptical structures containing + and −) in closest proximity to (mostly) lipid polar heads, followed by a plane of hydrated or partially hydrated ions (outer Helmholtz plane). This portion of the interfacial structure is responsible for electrostatic charging response to low-level current injection. Specific adsorption is depicted by dehydrated ions in the plane of closest approach to membrane structures (i.e. displacing water dipoles).

rad/sec). In addition, these reversible electrochemical responses appear to involve primarily aqueous-phase structural modifications not resulting (for low-level current only) in specific membrane structure changes. Therefore no functional consequence would be expected if only this process is perturbed by the chronic use of such-encoded pulsating current. Note that the net change in charge configuration at the level of energy input considered here is not sufficient for electrophoretic or dielectrophoretic responses to be elicited. Heating is also not a factor. In fact this study considers only electrochemical information in the context of the kinetic modulation of a cell's normal response to its given functional and microenvironmental situation. Responses to conditions introduced via electrolysis or pulsating current with high enough energy input for local heating or electrokinetic phenomena to result are considered to be abnormal modulation. While this may be of some ultimate therapeutic value, cell responses via these modalities are not considered in this study.

In approximately the same kinetic range as the interfacial phenomena depicted above, the lipid asymmetry exhibited by most living membranes (*38, 39, 40, 41, 42*) results in classical dielectric behavior (*44*) owing to charge separation across what is essentially a bilayer of insulating substances. As for nonspecific electrostatic interactions, the mechanism of low-level charge acceptance by this pathway is expected to be confined to the two lipoprotein interfaces with little or no bulk membrane structure modification. In fact it is most probable that the changes in the nonspecific electrical double-layer portions of the interfacial regions functionally occur in conjunction with the dielectric membrane response. This will be discussed in more detail below.

Based on all of the above information it is interesting to consider the manner by which the transient response of the relatively nonspecific dielectric and electrostatic interactions are coupled. In order to perform this in a quantitative manner it is most convenient to express this behavior (as well as that of all other membrane-charging pathways) in the frequency, as opposed to the time, domain. Under these conditions the experimentally testable system function of impedance can be obtained. Therefore all variables in this study will be given in terms of the complex frequency variable of the Laplace transformation (*46*). This complex variable, usually denoted by s, has a real, σ, and an imaginary, $j\omega$, part which define the axes of the Laplace plane. In the context of this study Laplace transformation allows the pulse response of a living membrane to be expressed in terms of its frequency content. In this manner it is the frequency functionality and not the waveshape which is used for comparative purposes. With this in mind it is useful to recall that all current which is seen by the membrane is defined by the sum of charging portion $i_c(s)$ and a phase-transfer portion $i_p(s)$. The interactions considered above contribute to $i_c(s)$ and are written as

$$\eta(s) = \frac{i_c(s)}{s}\left(\frac{1}{q_{E_1}} + \frac{1}{q_{E_2}} + \frac{1}{q_{E_3}}\right) \tag{1}$$

where $\eta(s)$ is the total voltage change experienced by both interfacial regions and the lipid bilayer structure; q_E is a coefficient representing the variation of charge with potential, wherein Subscript 1 refers to the extracellular interface, 2 refers to the lipid bilayer, and 3 is the intracellular interface, respectively. Equation 1 shows that for low-level perturbations (linear region), the membrane response is purely capacitive. The q_E coefficients are indeed capacitors whose physical significance are understood by examining Figure 1. By analogy with the electrode–electrolyte interface the value of both q_{E_1} and q_{E_3} is expected to be in the range of 10 $\mu F/cm^2$ since the two charged planes are only a few angstroms apart. In contrast, q_{E_2} is most often of the order of 0.5 $\mu F/cm^2$ since lipid bilayer thickness is 70–100 Å. If the majority of the surface area of both membrane interfaces exhibits electrostatic interactions only, then the transient membrane response for this case is dominantly that owing to the dielectric lipid bilayer structure since $q_{E_1} \simeq q_{E_3} >> q_{E_2}$.

The aforementioned discussion shows that low-level pulsating currents of duration less than approximately 50 μsec (for typical extracellular fluid resistivity) would primarily excite nonspecific dielectric membrane responses. If it is considered that the lipid bilayer portion of the membrane structure is bounded at both interfaces by the electrostatic environment shown in Figure 1, then the ratio of dielectric and electrostatic responses will be given by their relative associated capacitance values.

If dielectric relaxation occurs as described above, it is expected that chronic application of this type of electrochemical information to a given cell or tissue would not trigger functional responses. Using pulsating currents of less than 50–100 μsec duration has been ineffective in modifying cellular dedifferentiation (*16, 17*), DNA uncoiling (*11*), and Ca^{2+} utilization (*12*). Thus it appears that relatively high-frequency, low-level (no thermal effects) currents are not interpreted by the cell as useful electrochemical information. This, however, must take into account signal periodicity which, dependent upon the kinetic situation, may allow waveform configurations containing relatively narrow pulse widths.

Further examination of Figure 1 shows that there may be regions wherein, e.g., a divalent cation has competed effectively with electrostatically interacting water dipoles for specific membrane sites. This type of interaction involves such aqueous-phase steps as dehydration, displacement, and specific binding. In addition, molecular structure in the membrane phase, at least at the binding site, depends on the surface concentration, Γ, of specifically adsorbed species (*29*), and may exhibit longer-range structural modifications. The highly charged interfacial environment in which specific adsorption occurs allows its potential dependence to be considered. If the specifically adsorbed species does not undergo membrane transport itself then its kinetics contribute to $i_c(s)$. Thus (*16, 20*)

$$i_c(s) = q_{\Gamma_i}\, a\eta(s) \left(\frac{1}{A_i(s) + B_i(s)}\right) \tag{2}$$

where q_{Γ_i} and a_η are coefficients representing the dependence of surface charge on the surface concentration of adsorbed species i, Γ_i, and the potential dependence of adsorption, respectively; and

$$A_i(s) = \frac{1}{(D_i)^{1/2} C_i(s)^{1/2} \coth (\delta (D_i/s)^{1/2})} \tag{3}$$

in which C_i and D_i are the aqueous concentration and diffusion coefficient of the adsorbing species, respectively, and δ is the thickness of the diffusion layer at the convection limit. Also

$$B_i(s) = s\Gamma_i/(1 + s\Gamma_i/v_i) \tag{4}$$

where v_i is the exchange rate constant for adsorption.

Equation 2 was derived by considering that the adsorption of a single species exhibits finite kinetics (*20*). This is embodied in $B_i(s)$ which is written assuming that the adsorption isotherm (of any type) may be linearized. Therefore, if specific adsorption is linearly perturbed, its kinetics are represented by a single time constant wherein the equivalent resistance of adsorption R_A (*see* Figure 2) is given by

$$R_A = 1/a\ v_i\ q_{\Gamma_i} \tag{5}$$

indicating that, as expected, this quantity is inversely proportional to the exchange adsorption rate, i.e. it is a kinetic parameter. The equivalent capacitance of adsorption, C_A (*see* Figure 2), is given by

$$C_A = a\ \Gamma_i\ q_{\Gamma_i} \tag{6}$$

showing that the quantity of charge stored at those portions of the interface at which specific adsorption occurs is directly proportional to the surface concentration, Γ_i. In addition the binding of species of low aqueous concentration (e.g. Ca^{2+}, K^+, Mg^{2+}, etc.) may cause a transient aqueous concentration gradient to develop. This contributes to the overall relaxation process via $A_i(s)$ which describes in a general manner the impedance of aqueous diffusion, Z_D (*see* Figure 2). Here, if diffusion is semi-infinite, as most likely is the case for the typical cell–fluid interface situation, then Z_D behaves as a uniform R_DC_D transmission line (telegraphist's equation behavior) in which

$$R_D = 1/a\ C_iD_i\ q_{\Gamma_i} \tag{7}$$

showing that the kinetics of aqueous transport is inversely proportional to the diffusion coefficient, and

$$C_D = aC_iq_{\Gamma_i} \tag{8}$$

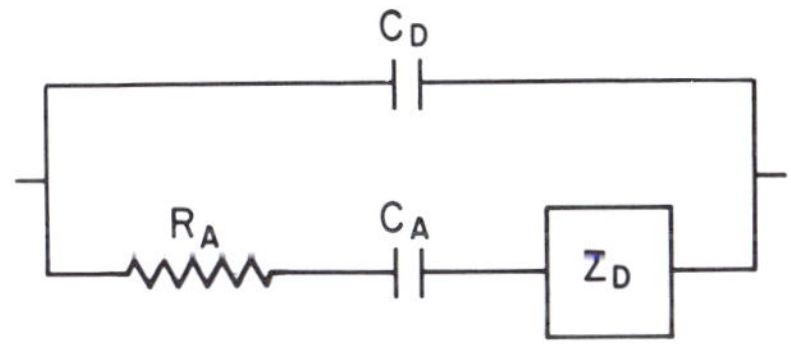

Figure 2. Electrical equivalent circuit illustrating one possible functional response of a living membrane to the charging portion of injected transient current. Here C_D is the observable dielectric and electrostatic response mostly caused by the lipid bilayer structure; R_A and C_A represent adsorption (binding) kinetics and surface concentration, respectively, of the adsorbing species; and Z_D is the aqueous mass transport impedance, which is present in case adsorption kinetics are fast enough to cause a transient concentration gradient.

which indicates that the charge involved in this process is a direct function of the initial bulk aqueous concentration of the transporting species. These parameters physically represent the time delay involved in the propagation of a concentration wave to (or from) the interfacial sites at which adsorption occurs. The mass transport and kinetic steps of specific adsorption are physically in series, since perceptible aqueous concentration gradients will exist only if the rate of adsorption is sufficiently rapid.

Equations 1 and 2 represent the total charge acceptance of a membrane independent of the existence of phase transfer ($i_p(s)$). In principle there can be more than one adsorbing species whose potential dependences are such that low-level current could excite several binding processes simultaneously. If the adsorptions are not interactive then Equation 2 can be repeated for the number of species involved. For each process neither the coefficients nor the surface concentration will necessarily be identical.

The derivation of the charging behavior of a living membrane given above is based on a kinetic model, the resting conditions of which do not have any special functionality a priori. The analytical expressions obtained in Equations 1 and 2 are represented, for convenience only, by the electrical equivalent circuit shown in Figure 2. The configuration (topology) of the circuit elements is that for the assumed model. In other words, the physical restraints dictated by the proposed kinetics can be represented only by the configuration shown. This is in contrast to the impedance response of an unknown "black box" for which an infinite number of equivalent electric circuits can be generated, as is common in classical circuit synthesis studies.

The process of specific adsorption is expected to exhibit a somewhat longer relaxation time than dielectric or electrostatic interactions. This is caused by the number of aqueous and membrane steps that may be involved, leading to a finite adsorption rate (τ_A) in the frequency range of interest. In comparison, electrostatic-charge acceptance appears infinitely rapid, being limited only by the resistance of the electrolyte pathway. A specific adsorption process has been detected (*20*) for the toad urinary bladder membrane. Interestingly the value of τ_A for this system is about 100 μsec. If this value is indicative of specific binding processes at living-cell membranes, then it can be anticipated that chronic application of pulsating currents having 100–200-μsec durations could have functional consequences. Thus the electrochemical information received by the cell under these conditions is a net change in the surface concentration of a specifically adsorbed species. If a divalent cation is involved, one of the results may be a modification in the activity of a membrane-bound enzyme (*24*). The transport of ions through a membrane, either to a site at the intracellular interface or to other cytoplasmic components, contribute to the phase-transfer portion, $i_p(s)$, of the total current. The kinetics of this process is a function of the potential change, $\eta(s)$, and the variation of concentration in both aqueous phases and within the membrane for each species involved. Aqueous concentration fluxes contribute to the phase-transfer impedance via functional relationships similar to that given in Equation 3. Of major importance, however, is the manner in which membrane fluxes must be taken into account in evaluating overall transient response (*29, 30, 31*). Within the membrane, mass transport is driven by concentration and electric field gradients. This can be taken into account by considering the Nernst–Planck equation for linear perturbation conditions. For this case a linearized potential profile applies in the perturbed equations. With these provisions it is possible to obtain analytical solutions of the transport equation for a variety of boundary conditions (*20, 32*). For example, the two limiting conditions for considering all membrane-phase concentration changes (ΔC_m) to occur on a single plane or uniformly throughout the membrane, result in substantially different transient responses. For the former, the phase-transfer impedance $Z_p(s)$ is given by

$$Z_p(s) = 1/I_E\ (1 + (I_e B(s) + I_m Z_M(s))) \tag{9}$$

in which $1/I_E$ is the reciprocal of the potential variation of $i_p(s)$ representing phase-transfer (partitioning) kinetics; I_e and I_m represent the variation of $i_p(s)$ with aqueous, C_e, and membrane concentration, C_m, respectively; $B_e(s)$ is the aqueous transport impedance given by an expression similar to Equation 3 for Species e developing a concentration gradient in the extracellular fluid only; and $Z_M(s)$ is the membrane mass transport impedance, given by

$$Z_M(s) = \frac{1}{nFC_m D_M(b\overline{V} + 1/\delta_M + (b^2\overline{V}^2 + s/D_M)\delta_M/3)} \tag{10}$$

where D_M is the diffusion coefficient within a membrane of thickness δ_M, $\overline{V}$ is the voltage field, and $b = nF/RT$.

The transport behavior described by Equation 9 is written for each species considered to contribute to phase-transfer current. It clearly shows that this is a three-step process in which both aqueous and membrane mass transport are functionally coupled to phase-transfer kinetics via I_E. Obviously any one of these can be rate limiting, but for one living membrane (*20, 33*) partitioning appears to be the slow step. In fact there is increasing evidence for many cell systems that membrane transport may be penetration controlled, leading to the supposition that a potential-dependent surface process may be regulatory for this relaxation step. A convenient representation for Equation 9 is given by the aperiodic electrical equivalent circuit shown in Figure 3a where R_p represents phase-transfer kinetics in series with membrane transport whose time constant is $\tau_M = R_M C_M$. Examination of Equation 10 shows that R_M is a function of V and D_M, i.e. transport kinetics, while C_M is related to the resting-membrane concentration. If R_p is limiting, it is highly unlikely that $B(s)$ will contribute significantly to phase-transfer relaxation. Therefore electrochemical information for this step will be related most likely to membrane surface and bulk phenomena. Impedance measurements for the toad urinary bladder show that this is true (*20, 33*).

If the boundary conditions for membrane transport are such that concentration variations are uniform throughout, $Z_p(s)$ is still given by Equation 9. However $Z_M(s)$ now has a different frequency response. Thus

$$Z_M(s) = \frac{1}{nFC_m D_M\ (\overline{V}b + 2/\delta_M)} \tag{11}$$

which generates purely resistive behavior over the frequency range of interest. Membrane transport of this type can be represented by the equivalent circuit of Figure 3b. This behavior is identical in function to that which exists for a diffusion layer of membrane thickness in the presence of a voltage field. Since the added step involving membrane-phase concentration, C_m, is not present, transport is easier and therefore may be more rapid. From a mechanistic point of view, this is an important distinction since it provides a means to assess the possible number of coupled (or regulatory) steps in membrane transport that are potential dependent. Studies of the toad bladder membrane in which membrane structure was altered by chemical cross-linking provide evidence that these transport pathways can be present in a living system (*33*).

The above discussion provides one possible approach for an evaluation of the kinetics of interfacial and/or transport processes at living membranes and how they may relate to cell regulation. In principle it is

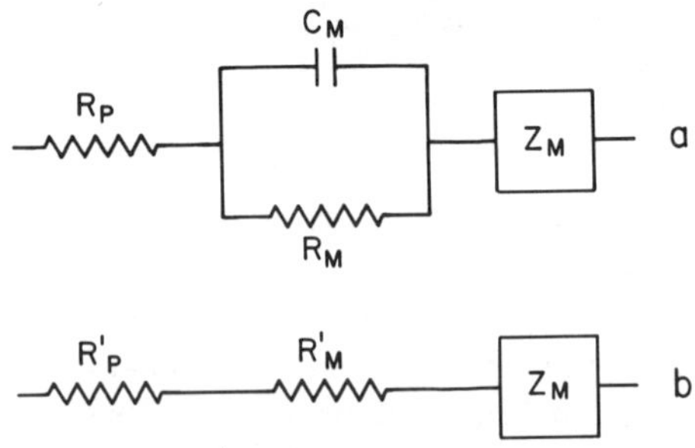

Figure 3. Electrical equivalent circuits representing membrane structural effects on the phase-transfer portion of injected current: (a) phase transfer (partitioning) kinetics (R_p) *coupled to a membrane transport impedance* (R_M, C_M) *allowing a residence membrane concentration of the transporting species; (b) the effect of cross-linking on the membrane transport step.* Z_M *for both cases represents aqueous mass transport impedance.*

possible to provide a large variety of dynamic models corresponding to suspected membrane-related mechanisms. The models are multiparameter, but experimental verification is often possible if an impedance technique is used in which a very wide frequency range is accessible. The Laplace plane real-axis impedance approach developed for electrode kinetic studies (*34, 35, 36*) has been used already with success for one living membrane study (*20, 33*). The experimental approach has been described previously, however for the sake of completeness a brief description will be given here. The isolated toad urinary bladder is subjected to a current perturbation and its voltage response is recorded. Both input and response waveforms are observed in the time domain using transient recording techniques with a 100-msec time resolution. These data then are transformed digitally into the frequency domain using real-axis ($s = \sigma$) Laplace transformation and the impedance evaluated over the range of 10^{-2}–10^{7} rad/sec. The number of relaxation processes present and the value of the associated parameters are obtained by functional analyses over relevant frequency ranges. The first step is to isolate the predominant relaxation observable at the highest frequencies studied. This is the dielectric membrane capacitance (*see* Equation 1). Isolation of this parameter from the total impedance allows for the detection of remaining processes. In this manner the functional response shown in Figure 4 has been obtained for the toad urinary bladder system. The kinetics associated with dielectric membrane charging and specific adsorption contribute to the charging current, $i_c(s)$. The transport pathway for this system is related, as expected, to passive Na^+ transport and shows partitioning (phase transport) to be overwhelmingly rate determining.

Since the transient response of a membrane is evaluated, an impedance picture is obtained over a time infinitesimally small compared with the cell's normal function cycle. In this manner the measured impedance can be used as a probe for the presence and variation of electrochemical membrane phenomena as a function of a cell's stage in cycle, functional activity, and microenvironment. In this manner, as will be shown below, the adaptive impedance responses can be used to configure a family of current waveforms for both the study and manipulation of cellular mechanisms which take into account the kinetics and irreversibility of nonfaradaic electrochemical reactions at cell surfaces.

Electrochemical Information in Practice

The analysis given above and its relation to membrane structure (function) can be used to formulate a predictive approach for using pulsating currents for cell function modification. Thus, knowledge of surface adsorption and membrane transport time constants and the manner in which they are coupled provides a rationale for the initial choice of waveform parameters. The first step is to decide, from physiological reasoning if possible, whether or not only surface or surface-coupled transport is important. As the analysis has shown, this determines the required frequency content (duration) of the applied current waveform. It is of course difficult to make this choice in an a priori manner since details concerning surface regulation in cell function are sparse. One possible reason for this is that the majority of studies

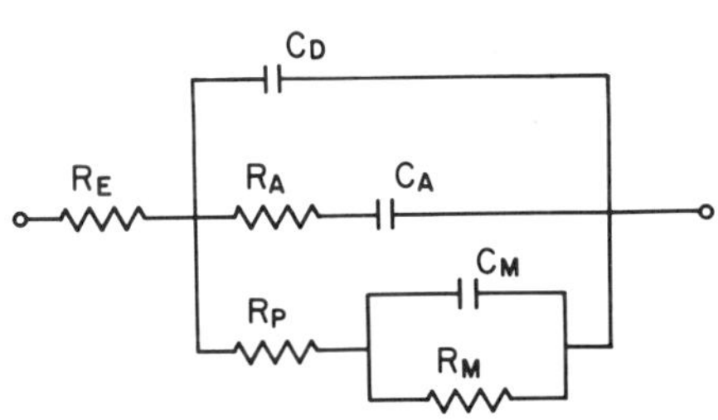

$$Z_1 = R_E C_D = 24 \pm 7 \mu sec$$

$$Z_2 = R_A C_A = 69 \pm 18 \mu sec$$

$$Z_3 = \frac{R_P R_M C_M}{R_P + R_M} = 3.4 \pm 2.9 msec$$

Figure 4. Functional impedance behavior obtained for the toad bladder epithelial cell. Three relaxation pathways corresponding to nonspecific dielectric (C_D), specific adsorption (R_A and C_A), and membrane mass transport (R_p, R_M, and C_M) processes. The relative time constants shown provide an indication of values for duration encoding of electrochemical information signals.

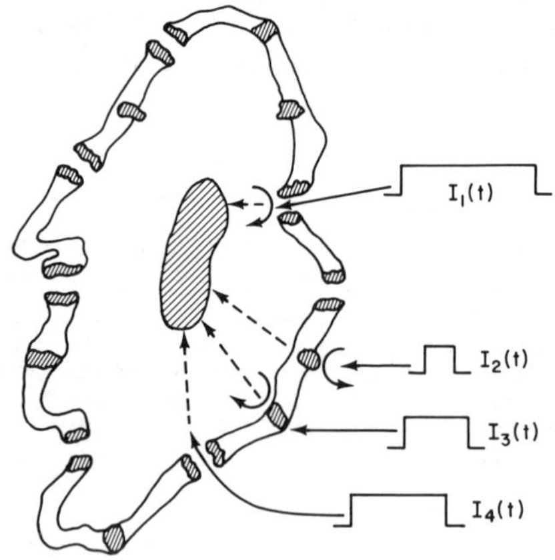

Figure 5. Schematic of possible membrane-related response to currents of varying waveform duration. Surface processes are represented by curved arrows and membrane transport by straight arrows traversing the membrane structure. Note that the number of coupled processes increases as pulse width increases ($I_1(t)$) to($I_4(t)$).

concerning environmental effects in cell behavior do not include an attempt to electrochemically modify the rate of the cell process (e.g. protein synthesis) concerned. In fact to the author's knowledge only three studies are available in which the rate effect of current injection on cell function has been examined (*11, 12, 13*). Therefore it is proposed that the use of pulsating current with designed modification of parameters in basic cell studies may provide mechanistic details not otherwise available.

The manner in which waveform duration could be selective in modifying cell function is illustrated schematically in Figure 5. Here, current pulses of progressively longer duration are shown impinging on the cytoplasmic membrane. As shown, the narrower the pulse width, the more selective it is for extracellular surface phenomena, i.e. membrane transport may not be excited. In order of the increasing pulse duration, the number of coupled processes increases. Thus the progression can be from extra- to intracellular surface via membrane transport. As expected from the impedance analysis, the transport process has the longest relaxation time if a residence membrane phase concentration exists (*see* Equations 10 and 11). Obviously part of the selectivity of action of interface specific current depends upon whether there is one dominant surface regulatory process and whether it is potential dependent. For a cell in a given phase of activity, e.g. collagen production for an osteoblast, it is reasonable to expect regulation of a very specific type and that the microenvironment plays a dominant role in the number of cells involved, as well as in their rate of collagen production. This aspect is very important because electrochemical information probably acts only as a trigger for the required biochemical sequence. Thus, as seen in Figure 5, the pulse having the longest duration will be seen by the whole cell surface and can excite all time constants which fall within its effective frequency content. Therefore a functional response can be expected only if the cell is programmed initially (e.g. by trauma) and if the cell surface is involved directly (via specific adsorption) or indirectly (via transport) in regulation.

It is important also to consider that as a cell's stage in its cycle changes, or as a repair process evolves, either the relaxation time of the relevant surface process may be modified or the regulatory mechanism itself may be different. Under these conditions, membrane impedance would be expected to undergo some degree of modification. For example, membrane structure modification by protein cross-linking significantly changes transport kinetics (*33*). The functional change is in fact rather dramatic resulting in a purely resistive response for this pathway as opposed to that shown for either case in Figure 3. To illustrate this adaptive change in membrane impedance, model calculations were performed using the experimentally determined impedance function of the toad urinary bladder membrane (shown in Figure 4 along with the time constants of the three isolated relaxations). Phase transfer (entry) is the rate-determining step for passive Na^+ transport for this system. Membrane transport is a two-step coupled process involving partitioning (R_p, *see* Figure 4) and transfer kinetics (R_M and C_M, *see* Figure 4) for this system. Entry becomes more difficult as the mucosal (bladder lumen) Na^+ concentration decreases, illustrating that perhaps Na^+ regulates its own partitioning rate (*20*). The experimental observation in this case is an increase in R_p. The effect of Na^+-concentration provoked variation in R_p on the overall impedance of this membrane is shown in Figure 6. The effect of environmental Na^+ concentration on the number of relaxation processes and their value is substantial. As R_p progressively increases (lower to upper curve) the specific adsorption process (R_A and C_A, *see* Figure 4) becomes uncoupled from the transport process (i.e. a distinct break appears in the slope of the impedance curve at intermediate frequencies). In addition it is evident that for a waveform of fixed duration and amplitude, the quantity of charge injected per pulse has less and less influence on the specific adsorption process as R_p decreases. Therefore it is evident that to obtain the proper net change in cellular boundary conditions, adaptive impedance changes must be taken into account.

The possible kinetic separation of the real time responses of a cell's surfaces and junctions means that rate modulation selectivity can be

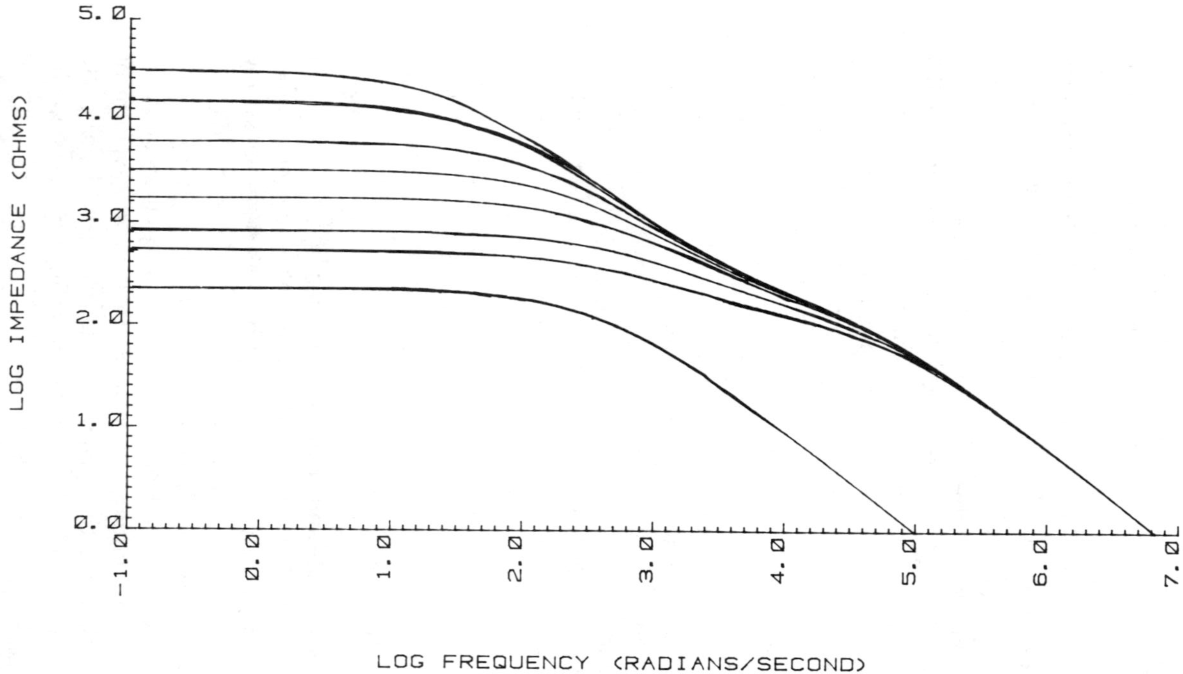

Figure 6. Model calculation showing the variation of impedance with mucosal Na^+ concentration for the toad bladder dynamic behavior illustrated in Figure 4. Entry kinetics (R_p) *is the variable for this set of curves which show that, as phase transfer becomes more difficult* (lower to upper plot), *the number and value of time constants perturbable by injected current increases.*

considered. For example if specific adsorption is excited (by a current waveform of appropriate frequency content) and this involves a regulatory enzyme, then its average activity could be increased by affecting a chronic change in the surface concentration of the bound entity at the relevant sites. Specific adsorption selectivity depends primarily upon the duration of each waveform, however selectivity for a particular binding process necessitates both its potential dependence and the net change required in real time (and over a sufficient time) to modulate the kinetics of the desired cell mechanism to be taken into account. This means that the amplitude of the current waveform probably will be restricted to a relatively small range for optimal selectivity. To accomplish this the relation between amplitude and repetition rate, which essentially determines the net charge injection per unit time, must be considered (*47*). Generally the larger the repetition rate (duty cycle) the more selective the waveform is in terms of amplitude. However this depends upon whether the perturbing waveform is kinetically unipolar or kinetically bipolar, as will be seen below.

While current can be applied in a variety of ways, the most unambiguous approach is to induce current electromagnetically. In this manner spatial distribution can be controlled and the effects of electrolysis which change the cell's chemical environment can be avoided (*9, 16*). This type of current induction is always bipolar and care must be taken that (especially for narrow pulses) there is no cancellation of effect if the cell membrane responds in a linear or near-linear manner. This problem can be avoided, and even used to advantage, if the frequency content of the two polarities are sufficiently different. For most in vitro and clinical studies using pulsating electromagnetic fields the induced current waveforms are of two distinct types—RSP and RPT as shown in Figure 7. The primary pulse duration in each case (τ) is chosen on the basis of impedance analysis to excite only extracellular surface phenomena or membrane transport coupled with intra- and extracellular interfacial processes. Thus τ for the RSP signal is between 100 and 350 μsec, while that for the RPT signal is usually between 2 and 20 msec. In this manner interface selectivity on the basis of relaxation times is possible. The opposite polarity portion of each pulse type (corresponding to the collapse of the magnetic field) is significantly different and allows for the possibility of a sequential introduction of electrochemical information. The opposite polarity portion of the RSP signal is of substantially lower amplitude (at least 5X) and wider width (usually 10X) than τ. Therefore this portion of the RSP waveform is of sufficient width to excite both surface and transport processes. The fact that they are juxtaposed in time allows for the possibility of a substantially different net change in cell boundary conditions than if the main polarity waveform is used alone.

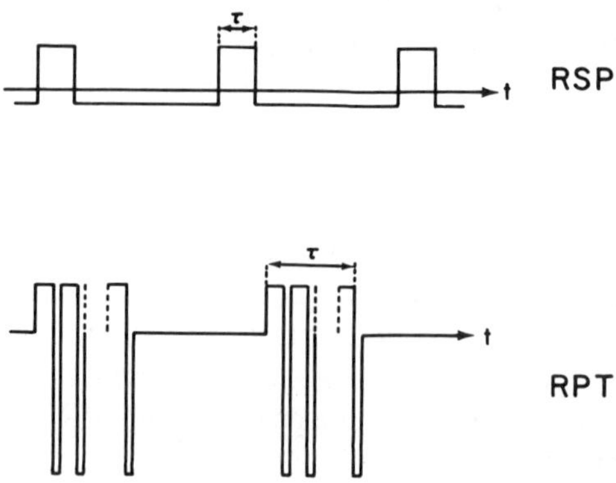

Figure 7. Schematic of the electromagnetically induced current waveform used for in vitro cellular dynamic studies and in vivo clinical studies. The RSP pulse has juxtaposed opposite polarity portions of significantly different but interface-selective widths. The RPT pulse is interface selective in one polarity only, the opposite polarity being of short enough duration to excite only nonspecific dielectric and electrostatic membrane interface relaxations provided kinetic irreversibility exists.

In contrast, the opposite polarity portions of the RPT waveforms are narrow enough (< 30 μsec.) to excite only the nonspecific dielectric and electrostatic relaxations. In this manner only a single polarity (of width defined by the envelope of the pulse train) may be functionally significant for the cell if sufficient nonlinearity is present.

The current waveforms described above are by no means the only choices, but they are designed to be interface and transport selective for a given set of regulatory steps. The extension of this approach is to provide a more sophisticated coding in terms of pulse sequences as determined by adaptive impedance changes (*see* Figure 6). Each time one of the aforementioned waveforms is seen by the cell, it receives electrochemical information. However, since the responses elicited are not of the all-or-none type (as in excitable membrane activity) it is apparent that repetition rate or signal frequency will ultimately determine if, and the rate at which, the cell will respond. An important part of proper encoding of electrochemical information is therefore repetition rate. Generally the more complex the code (e.g. the RSP waveform) the greater the role that signal frequency plays. This has been observed in studies of the dedifferentiation of the amphibian erythrocyte, where the rate of appearance of cells exhibiting DNA uncoiling is strongly dependent on the repetition rate of the RSP signal and relatively independent of that of the RPT waveform (*11*) . The reason for these differences has been attributed to the possibility that excitation of membrane transport relaxation is required before functional response occurs. If this is so then the RSP waveform contains what appears to be antagonistic portions and the net required change in cellular boundary conditions is achieved only within a narrow range of repetition rates (for the same level of induced current).

The requirement that the membrane transport time constant be satisfied for effective electrochemical information transfer has been observed in other cell systems as well. For example the rate of Ca^{2+} uptake by isolated chondrocytes or embryonic chick bone explants aappears to be modulated by current injection (*12*) only if the primary waveform width is sufficiently long and if it is not juxtaposed with an opposite polarity waveform of sufficiently low frequency content (e.g. RSP, *see* Figure 7). Also, clinical studies involving recalcitrant bone fractures show that optimal healing is dependent on pulse width and repetition rate (*37*).

Conclusions

This study was undertaken in order to provide further quantization for the concept of electrochemical information transfer and for its use for the kinetic modulation of regulatory processes in tissue growth and repair. Thus, if potential-dependent membrane processes are involved in cell regulation, it was first necessary to establish experimentally verifiable kinetic models and to ascertain whether or not these could be detected for a living cell. Experiments involving impedance determination of the toad bladder epithelial cell system have shown that potential dependent membrane transport regulatory processes are present (*20*). Also changes in this impedance function as a result of normal microenvironment changes mimicking bladder Na^{+} content indicates that an adaptive electrochemical response may be operative. These results show that the information obtained by using transient impedance approaches is significantly greater than that obtained when using steady-state (dc) approaches. However the application of transient electrochemical information for cell function manipulation (i.e. the reverse of the above) may require adaptive impedance changes to be taken into account. The degree to which this must be performed depends significantly on the particular cell process and/or tissue involved. For example, it appears that the cell population at the site of a nonunion (pseudarthrosis) of bone is relatively uniform and that the cells are probably relatively synchronous with respect to cycle stage (*45*). Under these conditions, the application of relatively constant parameter pulsating current may be sufficient for the moludation of the kinetics of this (essentially dormant) repair process. This appears to be the case for the clinical studies involving otherwise incurable bone fractures performed up to the present time (*37*).

Obviously a more complete embodiment of electrochemical information would require detection of function-related impedance changes for cells involved in developmental or repair processes. Studies are now in progress wherein the adaptive impedance variations involved in protein synthesis of isolated skeletal cells are being evaluated. An important result of this study is that the encoding of pulsating current for optimal electrochemical response could possibly be made to follow the cellular changes involved in complex repair processes. It is indeed intriguing to consider that, as mechanisms are uncovered with the aid of membrane-specific kinetic modification, not only fracture healing, but limb and organ regeneration may represent some of the important therapeutic applications of electrochemical information transfer in vivo.

Acknowledgments

The author wishes to sincerely thank G. Cerf for her significant contributions to the computer modeling of adaptive cellular impedance changes. This work was supported partially by NSF Grant No. NSF–APR–76–19469 and Electro-Biology, Inc.

Literature Cited

1. *Ann. N.Y. Acad. Sci.* **1974,** 238.
2. *Clin. Orthop.* **1977,** 124.
3. Presman, A. S. "Electromagnetic Fields and Life"; Plenum: New York, 1970.
4. Fukada, E.; Yasuda, I. *J. Phys. Soc. Jpn.* **1957,** *12,* 1158.
5. Bassett, C. A. L. *Biochem. Physiol. Bone, 2nd Ed.* **1971.**
6. Reinish, G. G.; Norwick, A. S. *Nature (London)* **1975,** *253,* 626.
7. Rodan, G. A.; Feinstein, M. B. *Proc. Natl. Acad. Sci. USA* **1976,** *73,* 1829.
8. Brighton, C. T.; Friedenberg, Z. D.; Zemsky, L. M.; Pollis, R. R. *J. Bone Jt. Surg.* **1975,** *58A,* 368.
9. Zengo, A. N.; Bassett, C. A. L.; Prountzos, G.; Pawluk, R. J.; Pilla, A. A. *J. Dent. Res.* **1976,** *55,* 383.
10. Pilla, A. A. *Ann. N.Y. Acad. Sci.* **1974,***238,* 149.
11. Chiabrera, A.; Hinsenkamp, M.; Pilla, A. A.; Ryaby, J. P.; Ponta, D.; Beltrame, F.; Grattarole, M.; Nicolini, C. *J. Histochem. Cytochem.* **1979,** *27,* 375.
12. Vergos, G.; Pilla, A. A.; Bassett, C. A. L. *J. Bone and Jt. Surg.,* Orthopaedic Transactions **1979,** p. 78.
13. Rodan, G. A.; Bourret, L. A.; Norton, L. A. *Science* **1978,** *199,* 690.
14. Pilla, A. A. *Intersoc. Enrgy Convers. Eng. Conf., Conf. Proc., 7th,* Am. Chem. Soc., Washington, D.C., **1972,** p. 761.
15. Pilla, A. A. *J. Bioelectrochem. Bioenerg.* **1974,** *1,* 227.
16. Pilla, A. A. *Ann. N.Y. Acad. Sci.* **1974,** *238,* 149.
17. Pilla, A. A. *J. Bioelectrchem. Bioenerg.* **1976,** *3,* 370.
18. Urry, D. W. *Ann. N.Y. Acad. Sci.* **1978,** *307,* 3.
19. Parsegian, V. A. *Ann. N.Y. Acad. Sci.* **1975,** *264,* 161.
20. Pilla, A. A.; Margules, G. *J. Electrochem. Soc.* **1977,** *124,* 1697.
21. Singer, S. J.; Nicholson, G. L. *Science* **1972,** *175,* 720.
22. Goldstein, L.; Levine, Y.; Katchalsky, E. *Biochemistry* **1964,** *3,* 1913.

23. Whitfield, J. F.; Rixon, R. H.; MacManus, J. P.; Balk, S. D. *In Vitro* **1973,** *8,* 257.
24. Singer, S. J. In "Structure and Function of Biological Membranes"; Rothfield, L. I., Ed.; Academic: New York, 1971.
25. McLaughlin, S. G. A.; Szabo, G.; Eisenmann, G.; Ciani, S. M. *Proc. Natl. Acad. Sci. USA* **1970,** *67,* 1268.
26. Miller, I.; Blank, M. *J. Colloid Interface Sci.* **1968,** *26,* 26.
27. MvLaughlin, S. *J. Membr. Biol.* **1972,** *9,* 361.
28. Britten, J. S.; Blank, M. *J. Bioelectrochem. Bioenerg.* **1977,** *4,* 209.
29. Parsegian, V. A. *Ann. N.Y. Acad. Sci.* **1975,** *264,* 161.
30. Anderson, O.; Feldberg, S.; Nakadomari, H.; Levy, H.; McLaughling, S. *Biophys. J.* **1976,** *16,* 194a.
31. Anderson, O.; Fuchs, M. *Biophys. J.* **1975,** *15,* 795.
32. Pilla, A. A. *Biophys. J.* **1977,** *17* 130a.
33. Margules, G.; Doty, S. B.; Pilla, A. A. Chapter 27 in this book.
34. Pilla, A. A. *J. Electrochem. Soc.* **1969,** *116,* 1105.
35. Pilla, A. A. *J. Electrochem. Soc.* **1970,** *117,* 467.
36. Pilla, A. A. In "Electrochemistry: Calculations, Simulation and Instrumentation"; Mattson, J. S., Mark, Jr., H. B., McDonald, Jr., H. C., Eds.; Marcel Dekker: New York, 1972; p. 139.
37. Bassett, C. A. L.; Pilla, A. A.; Pawluk, R. J. *Clin. Orthop.* **1977,** *124,* 117.
38. Chapman, D.; Cornell, B. A. In "Structure of Biological Membranes"; Abrahamsson, S., Pascher, I. Eds.; Plenum: New York, 1977; p. 85.
39. Bretscher, M. S. *Science* **1972,** *181,* 622.
40. Zwaal, R. F. A.; Roelofsen, B.; Colley, C. M. *Biochim. Biophys. Acta,* **1973,** *300,* 159.
41. Cullins, P.; DeKruijff, B.; McGrath, A. E.; Morgan, C. G.; Radda, G. K. In "Structure of Biological Membranes"; Abrahamsson, S., Pascher, I., Eds.; Plenum: New York, 1977; p. 389.
42. Tsai, K.; Lenard, J. *Nature (London)* **1975,** *253,* 554.
43. Verkleij, A. J.; Zwaal, R. F. A.; Roelofsen, B.; Comfurius, P.; Kastelijn, D.; Van Deenan, L. L. M. *Biochim. Biophys. Acta* **1973,** *323,* 178.
44. Trauble, H.; Overath, P. *Biochim. Biophys. Acta* **1973,** *307,* 491.
45. Muller, J.; Schenk, R.; Willenegger, H. *Helv. Chir. Acta* **1968,** *35,* 301.
46. Cheng, D. K. "Analysis of Linear Systems"; Addison–Wesley: London, 1959.
47. Pilla, A. A. In "Electric Properties of Bone and Cartilage;" Brighton, C. T., Black, J., Pollack, S. R., Eds.; Grune and Stratton: New York, **1979,** p. 455.

RECEIVED November 20, 1978.

22

Nonequilibrium Processes in Binding and Release of Brain Calcium by Low-level Electromagnetic Fields

W. R. ADEY[1] and S. M. BAWIN

Research Service (151), Veterans Administration Hospital, Loma Linda, CA 92357 and Brain Research Institute, University of California, Los Angeles, CA 90024

Calcium efflux from brain tissue exposed to weak oscillating fields shows "windowed" responses to the frequency and intensity characteristic of the imposed fields. They may involve nonequilibrium processes. Low-frequency fields decrease calcium efflux from chick and cat cerebral tissue only between 6 and 20 Hz and between 10 and 100 V/m (approximate tissue gradients, 10^{-7} V/cm). Radiofrequency fields (147 and 450 MHz) increase calcium efflux by about 15% when amplitude modulated between 6 and 20 Hz, but only for tissue gradients of 10–100 mV/cm. Transductive coupling may involve coherene charge states between anionic sites on membrane surface glycoproteins, with long-range cooperative interactions triggered by weak extracellular electric fields. Proton "tunneling" may occur at boundaries between coherent and noncoherent zones.

Concepts of nervous excitation have long been based on a rapid, inward movement of sodium ions and a slower, outward movement of potassium across the cell membrane, generating a brief action potential that propagates down the axon. This picture has been much less clear structurally and functionally for brain tissue. Dendritic branches of cerebral neurons spread widely from a small cell body and occupy a

[1] To whom correspondence should be addressed.

0-8412-0473-X/80/33-188-361$05.00/1

tissue volume orders of magnitude larger than the cell body (*1*). Dendrites of one neuron make close contact with those of adjacent cells, thus providing dendro–dendritic pathways for interaction between cells (*2*). Many of these contacts have synaptoid ultrastructural arrangements. Moreover, a significant proportion of cerebral neurons known as Golgi Type II cells have no long axonal processes. They have a massive proliferation of dendrites making numerous contacts with dendrites of surrounding cells.

Major electrophysiological interactions between cerebral neurons occur without a traffic in impulses. Instead, they involve wavelike processes initiated in dendrites and passed from cell to cell by dendro–dendritic contacts (for reviews of our own and related work, *see* Refs. *3* and *4*). In the retina, for example, sequential activation of receptors, horizontal cells, bipolar cells, and perhaps some amacrine cells occurs without impulse generation (*5*). The term "silent retina" aptly describes its functional transactions. Golgi II cells in the cerebral cortex described above also appear incapable of impulse generation.

There is a sharp disparity between the size of the transmembrane electric gradient and oscillating electric fields that surround brain cells. With a steady-state potential gradient across the membrane of the order of 10^5 V/cm, depolarizations associated with synaptic events that trigger a propagated impulse are of the order of 10^3 V/cm. Associated inward transmembrane current flow in axons during excitation is of the order of 1.0 mA/cm^2, but it may be two orders of magnitude lower in excitation of cerebral neurons. In sharp contrast, intrinsic low-frequency oscillations in fluid around brain cells are weak. Derived in great measure from much larger (by about 200 times) oscillations in dendritic membrane potentials noted above, they give rise to the electroencephalogram (EEG), with electric gradients over the dimensions of a cerebral neuron that average 100 mV/cm.

These biophysical considerations have discouraged serious discussion of either intrinsic or imposed fields with comparable characteristics as factors in cerebral transactional mechanisms. Indeed, a direct influence of extracellular gradients of 100 mV/cm on a transmembrane gradient of 10^5 V/cm would seem quite unlikely. Certainly a change that might be comparable with postsynaptic depolarizing gradients of 1 kV/cm would require efficient amplification of the extracellular field by mechanisms in biological transductive coupling as yet poorly understood (*4, 6*).

There is new evidence that sensitivity of brain tissue to certain weak oscillating electromagnetic fields occurs in the absence of significant tissue heating (less than 0.1°C). Brain-temperature increments in excess of 0.1°C alter cell membrane potentials and synaptic functions and shift hormonal and neurohormonal activity according to classic concepts of nervous excitation, based on ionic equilibrium conditions across cell membranes. On the other hand, brain sensitivity to electromagnetic fields at lower levels may involve a quite different class of nonequilibrium processes. This view is supported by the "windowed" character of sensitivities of calcium binding and electrical activity in brain tissue to low-frequency modulation and intensity characteristics of impressed radio frequency fields (*7–12*).

These interactions may occur at the surface of dendritic branches of cortical neurons where strands of glycoproteins extruded from within the lipid bilayer form a surface polyanionic sheet (*13, 14*). The membrane surface thus functions as a powerful binding zone for cations, particularly calcium and hydrogen (*15, 16*). To better define cerebral calcium binding sites responding to weak electrical stimulation (*17*), we have examined changes in calcium efflux, with and without imposed 450 MHz fields, in relation to the calcium concentration of the cortical bathing medium; calcium–calcium interactions play an important role in the regulation of calcium uptake and release (*18, 19*). We also have tested effects of lanthanum in the bathing solution on calcium efflux from the isolated chick cerebral hemisphere, since the trivalent lanthanum ion blocks calcium flux across excitable membranes (*20, 21, 22, 23*). Electron microscopy of lanthanum-treated barnacle muscle fibers (*24*) and heart cells (*25*) has located the ion in the extracellular compartment, but not in the cytoplasm.

It is only recently that imposition of weak electromagnetic fields at frequencies below 1000 Hz and radio frequency (RF) and microwave fields amplitude modulated at these same low frequencies has proved valuable in studies of the molecular organization of major biological systems, including central nervous functions, immune processes, and hemopoietic sequences. Much of this material is contained in symposia volumes and a limited range of texts. The reader may wish to consult these for further information (*26–35*).

Experimental

Ionic assay techniques and the biophysics of electromagnetic field exposure methods will be summarized briefly. Further details are provided elsewhere (*7, 8, 17, 19, 36*).

Cerebral hemispheres of newborn chicks (4–8 days old) were removed rapidly from the crania following decapitation (*17*). Each hemisphere then was incubated at 37°C in 1 mL of a physiological solution containing 2.16m*M* $CaCl_2$ (the same amount of calcium as in cerebrospinal fluid), 2.4m*M* $NaHCO_3$, and 0.2 μCi radioactive calcium ($^{45}Ca^{2+}$, specific activity 0.735 Ci/m*M* Ca). Following $^{45}Ca^{2+}$ loading for 30 min, the hemispheres were rinsed 3 times in radioactivity-free solution and transferred to 1.0 mL of fresh solution. The $^{45}Ca^{2+}$ efflux was determined by sampling the radioactivity of the exchanging solution in 0.2-mL aliquots diluted in scintillation cocktail (Packard Dimilume).

Calcium concentration was modified in two ways. In one experimental series, calcium was omitted, and in the other 2.0m*M* of calcium was added to the normal solution, which already contained 2.16m*M* $CaCl_2$. Lanthanum concentrations of 0.5 and 2.0m*M* were used, and it was necessary to omit HCO_3^- to avoid precipitation. To avoid the possibility of lanthanum salt formation at the brain surface, samples were rinsed in bicarbonate-free solution before transfer to solutions containing $LaCl_3$. The incubating medium was maintained at a constant osmolarity of 310–313 mosm and at pH 7.6 at 37°C during changes in ionic composition. Decreased pH associated with omitting bicarbonate and adding lanthanum was corrected by adding NaOH. In control experiments the pH was lowered to 6.8 by adding 0.108m*M* HCl to the physiological medium. Exposure to 450-MHz fields was carried out in a TEM mode tapered chamber designed for irradiating biological tissues, cultures and laboratory animals, including monkeys. The air-conditioning system allows regulation of temperature (15°–42°C) and humidity (ambient to 90%). The TE10 mode is excited in a section of a waveguide and drives a pyramidal horn. The horn flare is continued for 4.5 m and terminated by a large, screened wall covered with microwave absorbing material. The 450-MHz generator is housed in a rack console together with modulation and power supplies. The crystal-controlled RF oscillator is a solid-state design with an output of 5.0 W in a 50-ohm load. The driver amplifier (output between 10 and 50 W) and the final amplifier (output up to 500 W) are screen-grid modulated. Sinusoidal or pulsed modulation of the carrier can be adjusted from 0 to 90% between 5 and 35 Hz. The RF output is coupled to the waveguide through RG–8U coaxial cable and a discone probe. The space available for exposure situation at 3 m from the throat of the horn is about 1 m^3. Field distribution in the exposure space is sufficiently uniform so that a "quiet zone" exists in the center of this space. Vertical polarization of the *E* field minimized distortion, owing to, e.g., the introduction of a monkey in the exposure area. In this volume, the field is uniform within narrow limits, and the standing wave ratio is less than 1.2:1 throughout the chamber.

The 450-MHz field incident on the tissue was 0.75 mW/cm^2, sinusoidally amplitude modulated at 16 Hz to a depth of 90%. Comparative studies were made with field powers of 0.375 and 2.0 mW/cm^2. In collaborative studies with Dr. H. Bassen of the Bureau of Radiological Health, we have measured tissue gradients with an implantable dipole probe at 10 to 100 mV/cm with this range of incident fields (37).

Each half brain was tested against the corresponding hemisphere, with left and right sides serving alternately as control and test samples. Efflux of $^{45}Ca^{2+}$ in the normal solution was the control in experiments involving calcium concentrations, pH, and bicarbonate-free solutions. Efflux in the HCO_3^--free medium served as reference in experiments with La^{3+}, since omission of the bicarbonate ion considerably decreased $^{45}Ca^{2+}$ efflux prior to adding lanthanum.

Effects of Changing Extracellular Calcium Levels, Bicarbonate Concentrations, and pH on Cerebral Calcium Efflux

Our previous studies, discussed below, have shown both increased and decreased efflux of calcium from cerebral tissue resulting from interaction with electromagnetic fields of differing intensities, but always having similar low-frequency characteristics in the ELF range from approximately 60–20 Hz, either as an intrinsic field frequency or as an ELF modulation on RF carrier wave. A striking finding in recent studies (*17*) is the reversal in sign in the field effect that follows omission of HCO_3^- and the addition of lanthanum (*see* Figure 1).

Efflux of $^{45}Ca^{2+}$ in calcium-free medium did not differ significantly from control, and adding 2.0m*M* $CaCl_2$ to the testing solution caused only a small increase in efflux. The brain-to-efflux ratio was unchanged at 3.5–4.0, as in previous studies (*10*). Increased efflux induced by field stimulation was the same in high Ca^{2+} solution as in the normal medium. There was also a trend towards increased efflux in calcium-free solution, although the mean 10% increase was not statistically significant. Brain-to-efflux ratios were lower in stimulated samples than in control or ionic treatment alone, reflecting increased $^{45}Ca^{2+}$ efflux by 8.5–11.0%, by comparison with concurrent tests in normal solution. Brain-to-efflux ratios reflected the decrease and were slightly higher than control. Field stimulation had no effect on the reduced $^{45}Ca^{2+}$ efflux.

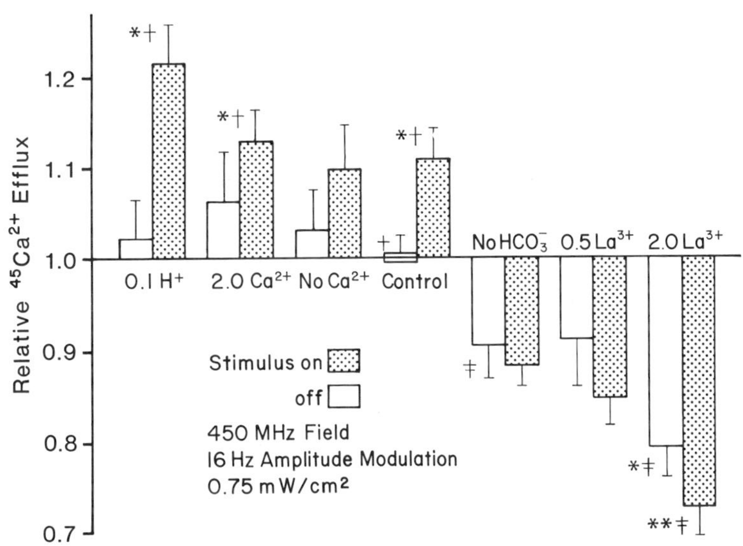

Proceedings of the National Academy of Science USA

Figure 1. Relative $^{45}Ca^{2+}$ effluxes (mean ± SEM) referred to the mean value of control efflux in HCO_3^--containing solution (control) (17). *Statistical comparisons were made either with control efflux or with efflux in HCO_3^--free solution. Open bars, stimulus off; stippled bars, stimulus on. Conditions: A,* mM *H^+; B, 2.0mM Ca^{2+} added; C, no Ca^{2+}; D, control; E, no HCO_3^-; F, 0.5mM* La^{3+}*; G, 2.0mM* La^{3-}. *Stimulus was a 450-MHz field (16-Hz modulation) at 0.75 mW/cm². Statistical analysis: *, P < 0.05; **, P < 0.01; †, compared with control; ‡, compared with no HCO_3^+.*

Lowering the pH of the testing solution to 6.8 with hydrochloric acid (*see* Figure 1, 0.1 H^+) had no major effect on $^{45}Ca^{2+}$ efflux. However, field-stimulated efflux at this pH appeared larger than in normal solution. These findings were confirmed in another series of studies at a lower field level of 0.375 mW/cm^2, where the field-induced increase in low pH medium (12.0%, $n = 30$, $t = 2.609$, $P < 0.05$) was higher than in control solution (8%, $t = 1.756$, NS).

$^{45}Ca^{2+}$ Efflux from Cerebral Tissue in the Presence of Lanthanum Chloride

In the absence of the 450-MHz field, the presence of 0.5m*M* $LaCl_3$ in the bicarbonate-free solution did not change $^{45}Ca^{2+}$ release. However, field stimulation induced a further inhibition of the efflux (7%) and a significant increase in the brain-to-efflux ratio (4.33 ± 0.17) by comparison with control in no-bicarbonate medium (3.75 ± 0.12, $t = 2.747$, $P < 0.05$). With 2.0m*M*/L $LaCl_3$, the nonstimulated efflux was decreased significantly (11–14%). The field-induced decrease in $^{45}Ca^{2+}$ efflux was about the same as in 0.5m*M* La^{3+} solution. This decrease at a field intensity of 0.75 mW/cm^2 did not occur at lower or higher field strengths (0.375 and 2.0 mW/cm^2).

A salient aspect of these experiments is confirmation of our previous findings and those of others that RF fields amplitude modulated at brain wave frequencies increased $^{45}Ca^{2+}$ efflux from isolated chick cerebral tissue (*8, 9, 10*). New observations in this study include an absence of sensitivity of the response to variations of calcium concentration (0–4.16m*M*) in the bathing solution, its enhancement by addition of hydrogen ions (0.108m*M* HCl), and its inhibition in the absence of normal (2.4m*M*) bicarbonate levels (*17*). Addition of lanthanum to the bicarbonate-free solution restored field sensitivity, but the stimulus decreased instead of increasing $^{45}Ca^{2+}$ efflux. We hypothesize that weak, extracellular electric gradients oscillating at low frequencies or occurring as a low-frequency modulation on a RF carrier wave may be transduced in a specific class of extracellular negative binding sites, normally occupied by calcium ions and susceptible to competitive hydrogen-ion binding.

A perspective on these findings and conclusions requires consideration of brain as a tissue. It is also necessary to consider the qualitative and quantitative biophysics of brain-tissue sensitivity to imposed electromagnetic fields.

Cerebral Tissue Models of the Biomolecular Basis of Electromagnetic Field Interaction

Brain tissue has two cellular compartments, neuronal and neuroglial, separated from each other by an extracellular space that forms as much as 20% of the cerebral volume (*38*), but it may change dramatically in size as a response to dehydration, anoxia, and cerebral blood flow. Neuronal and neuroglial cell membranes face each other across narrow channels of this extracellular space, which has typical transverse dimensions of 150 Å and numerous terminal strands of polyanionic glycoproteins described above protrude into it as part of a "greater membrane" (*13*). Electrophysiological evidence supports the concept of interaction between calcium and cell-surface macromolecules in functional modulation of the intercellular space. An apparent reduction in extracellular calcium concentration by as much as 20% occurs during cerebellar stimulation, but the measuring technique would not distinguish between actual removal of calcium from the extracellular space and an alteration in its state of binding to surface polyanions (*39*). There is a sharp increase in the electrical impedance of cerebral tissue associated with increased extracellular calcium levels (*40*). In these measurements, the major part of the impedance-measuring current travels in extracellular channels (*see* Figure 2). Specific resistance of cellular fluid is of the order of 4 ohm cm^{-1}, whereas transmembrane resistance is typically about 5000 ohms cm^{2-}. For this reason, major components of currents induced by imposed fields will be found also in the extracellular space (*41*). They will be cm^{-2}. For this reason, major components of currents induced by imposed directed along the membrane surface, and not transversely through the membrane along channels followed by sodium and potassium ions in the later stages of excitation associated with membrane depolarization.

Therefore we may hypothesize that the first step in transductive coupling of weak extracellular electrical gradients at the cell surface is a longitudinal integrative process. Available data support the view that brain-tissue sensitivity to weak oscillating electromagnetic fields may involve nonequilibrium processes based on long-range atomic interactions.

Windows in Frequency and Intensity Sensitivity of Cerebral Calcium Efflux to Environmental Electromagnetic Fields

In brain tissue, binding of calcium ions is highly sensitive to weak, imposed fields, and has offered a uniquely sensitive and versatile tool for both the site and the mechanisms of transductive coupling. Nonfocal repetitive stimulation of the awake cat brain with pulses at 200/sec producing intracortical gradients of 60 mV/cm, comparable with the EEG, causes a sharp increase in cortical efflux of $^{45}Ca^{2+}$, a phenomenon closely linked to release of synaptic-transmitter amino acids gamma-aminobutyric acid (GABA) and glutamic acid (*19, 42*). Similarly, the $^{45}Ca^{2+}$ efflux from isolated chick cerebral hemisphere was stimulated by

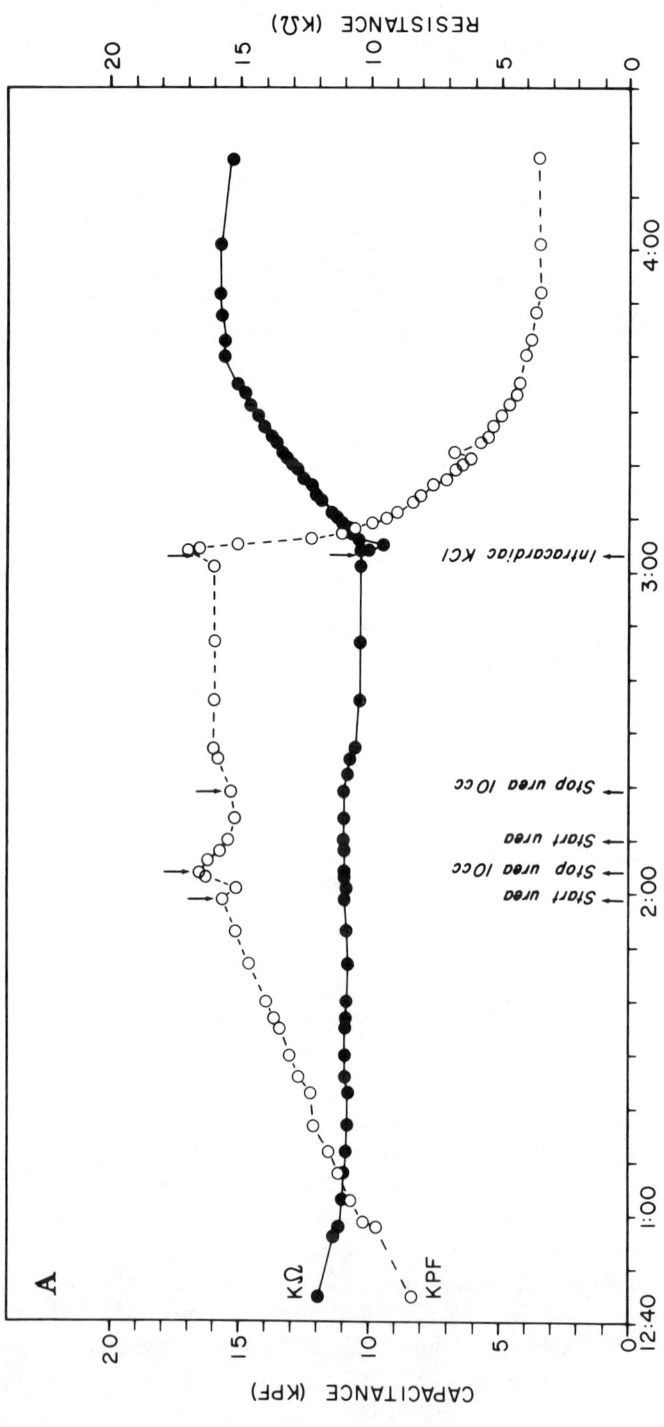

Experimental Neurology

Figure 2. Cerebral impedance measured with stainless steel coaxial bipolar electrodes in cerebral tissue of cat, showing effects of intravenous urea followed by asphyxiation (C). Calcium sharply increases impedance: (●), resistive component of impedance; (○), reactive component. Units on ordinates: kilohms and nanofarads (40).

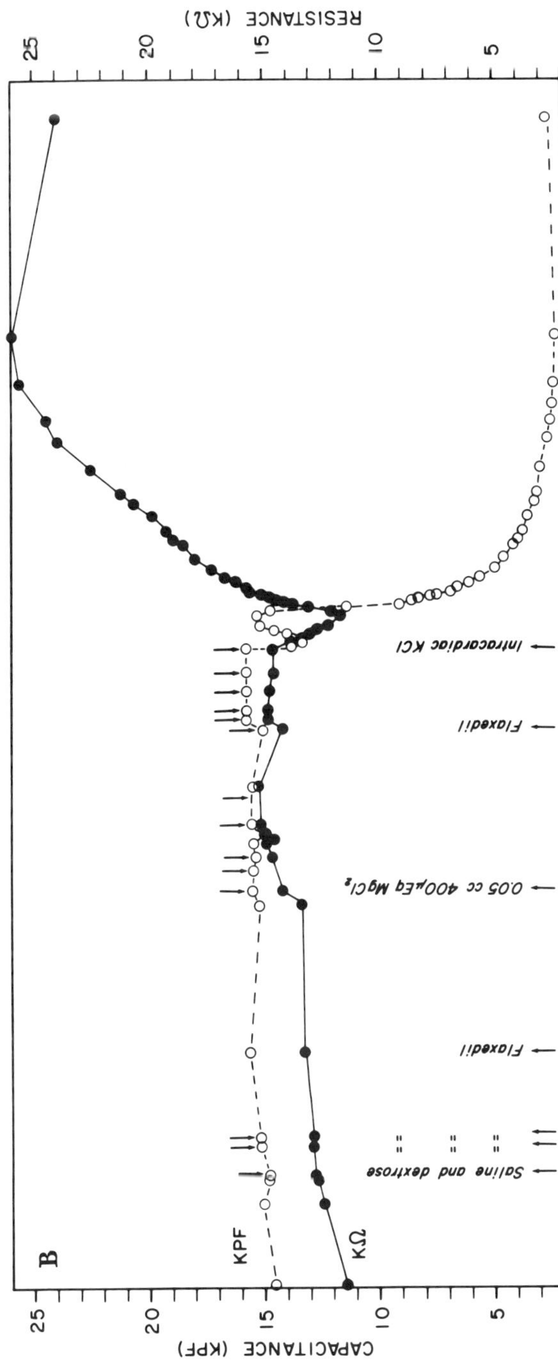

Figure 2. Continued

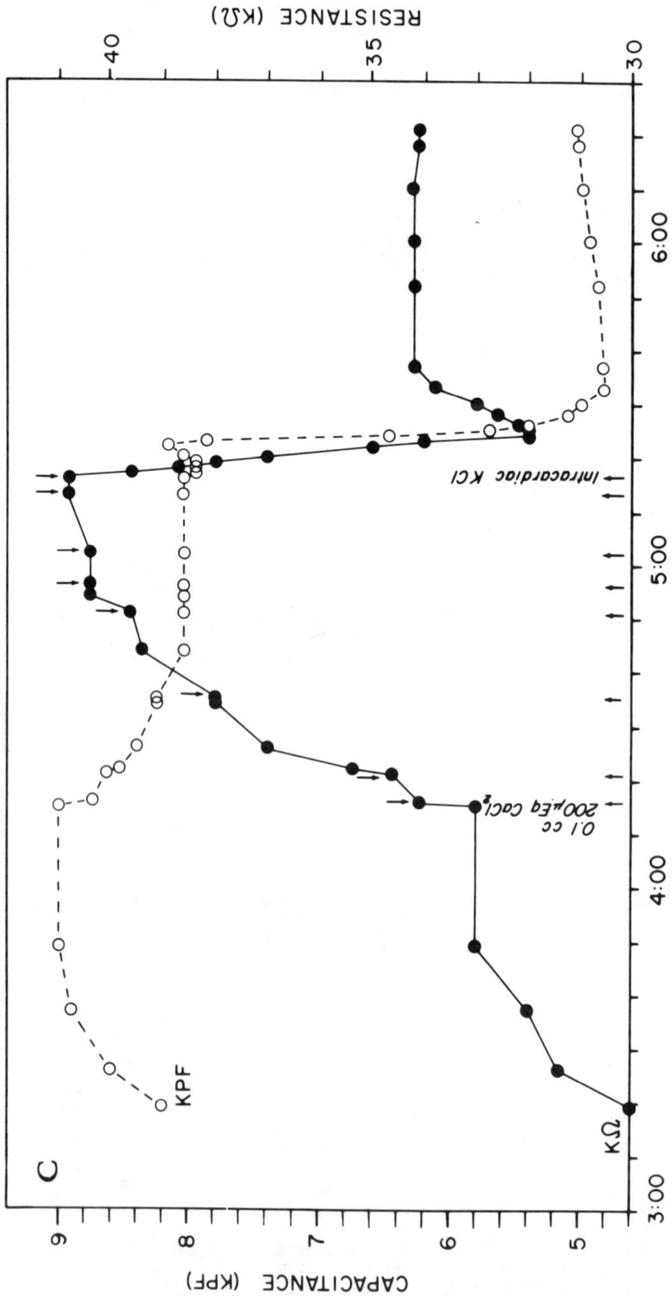

Figure 2. Continued

exposure to a 147-MHz, 0.8 mW/cm^2 field, but only when amplitude modulated at frequencies from 6 to 20 Hz, with maximum effect at 16 Hz (*8*) (*see* Figure 3). This specificity of the tissue response at ELF modulation frequencies has been confirmed by Blackman et al. (*9*), who also noted a narrow field intensity window with maximal effects ca. 0.75 mW/cm^2, and no significant effects at higher or lower field levels. We have confirmed a narrow power window between 0.1 and 1.0 mW/cm^2 for enhanced $^{45}Ca^{2+}$ efflux with a 450-MHz field amplitude modulated at 16 Hz (*10*) (*see* Figure 4). Here, measured tissue RF gradients were of the order of 100 mV/cm for an incident field energy of 1.0 mW/cm^2 (*37*). By contrast, ELF fields with gradients of 10–60 V/m in air decreased $^{45}Ca^{2+}$ efflux. These internal field strengths were several orders of magnitude smaller than the tissue gradients induced by 147 and 450 fields (*7*), but the frequency specificity and the size of the response were the same whether the fields induced an increase or a decrease in the efflux (*see* Figure 5).

The consistency of the magnitude of the effect in all experiments (10–15% in efflux), together with the exquisite sensitivity of the system to narrow windows in frequency and amplitude of the field stimulus, may indicate that a specific class of calcium sites is sensitive to low-frequency electric fields (*17*). Data presented here on field sensitivity of calcium binding after lanthanum treatment may offer the first direct evidence that these field sensitive binding sites are indeed located on the membrane surface and in the extracellular space. A model of nerve membrane excitation proposed by Bass and Moore (*16*) envisages a competitive binding of hydrogen ions to membrane surface sites, with local displacement of calcium ions to adjacent sites on a macromolecular polyanionic substrate. Our data also have shown a sensitivity of calcium binding to both hydrogen and bicarbonate ion concentrations (*17*).

Cooperative Models of Transductive Coupling in Nonequilibrium Ion Binding in Cerebral Tissue

What mechanisms might effect the transductive coupling between a weak, oscillating electromagnetic field and this membrane surface polyanionic sheet? From studies of the biopolymer poly-l-glutamic acid, Schwarz (*43, 44*) concluded that its charge sites may behave coherently, with adjacent elements having the same energy levels. A coherent domain of anionic, fixed-charge sites at the membrane surface requires energy to raise them above the ground state. For some membrane-surface phenomena, energy is supplied by metabolic processes (*45*), but there is as yet no precise knowledge about sources of energy that might contribute to coherent surface charge states. As discussed elsewhere (*10*), it is

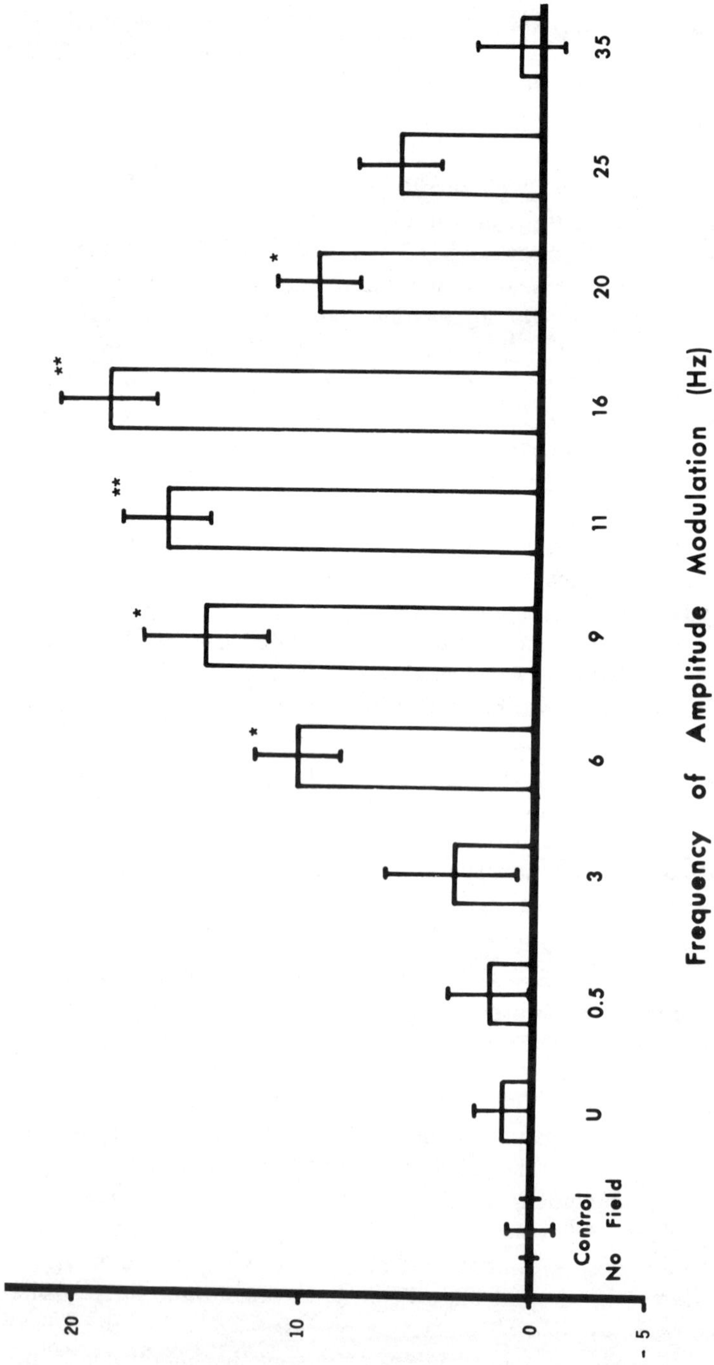

Annals of the New York Academy of Science

Figure 3. Effects of changing frequency of amplitude modulation from 0.5–35 Hz on $^{45}Ca^{2+}$ efflux from chick cerebral tissue: U, unmodulated 147-MHz carrier produced no significant changes. Statistical analysis: *, $P < 0.05$; **, $P < 0.01$ (8).

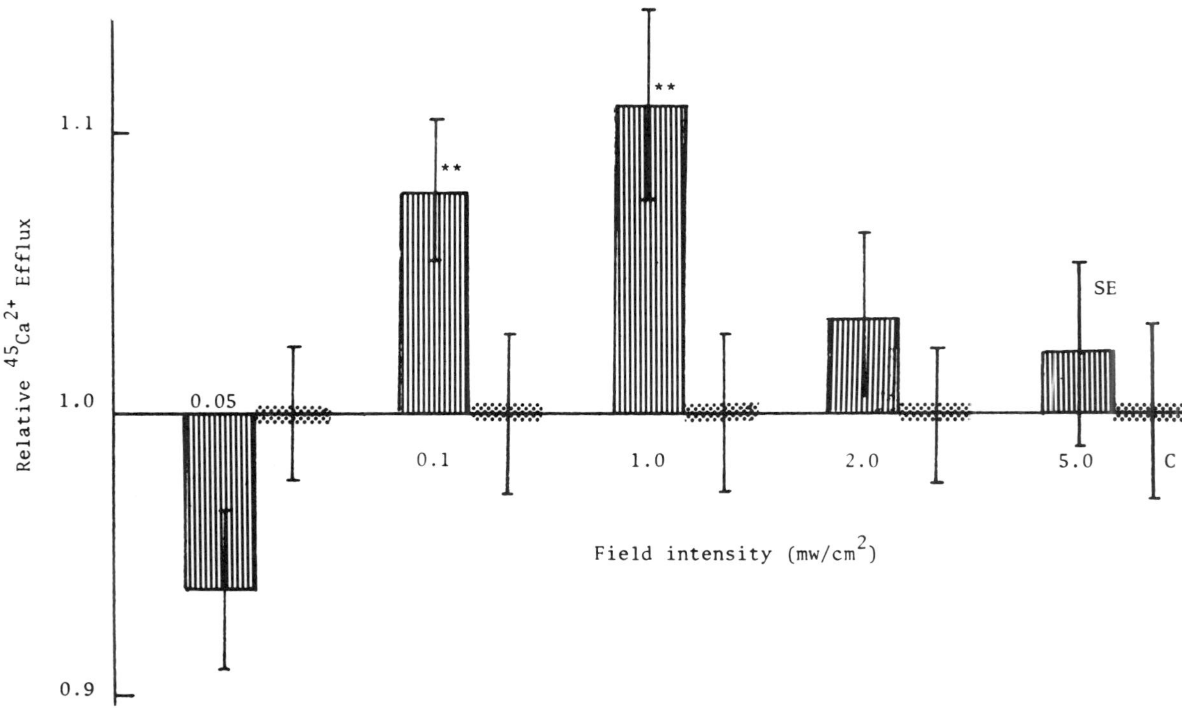

Bioelectrochemistry and Bioenergetics

Figure 4. Effects of changing intensity of 450-MHz field, amplitude modulated at 16-Hz, on efflux of $^{45}Ca^{2+}$ *from chick cerebral hemispheres. Cross-hatched bars show levels of efflux from exposed specimens in relation to control specimens (stippled bars) tested simultaneously in the same experiments. Variance shown as SEMs:* **, $P < 0.05$ (10).

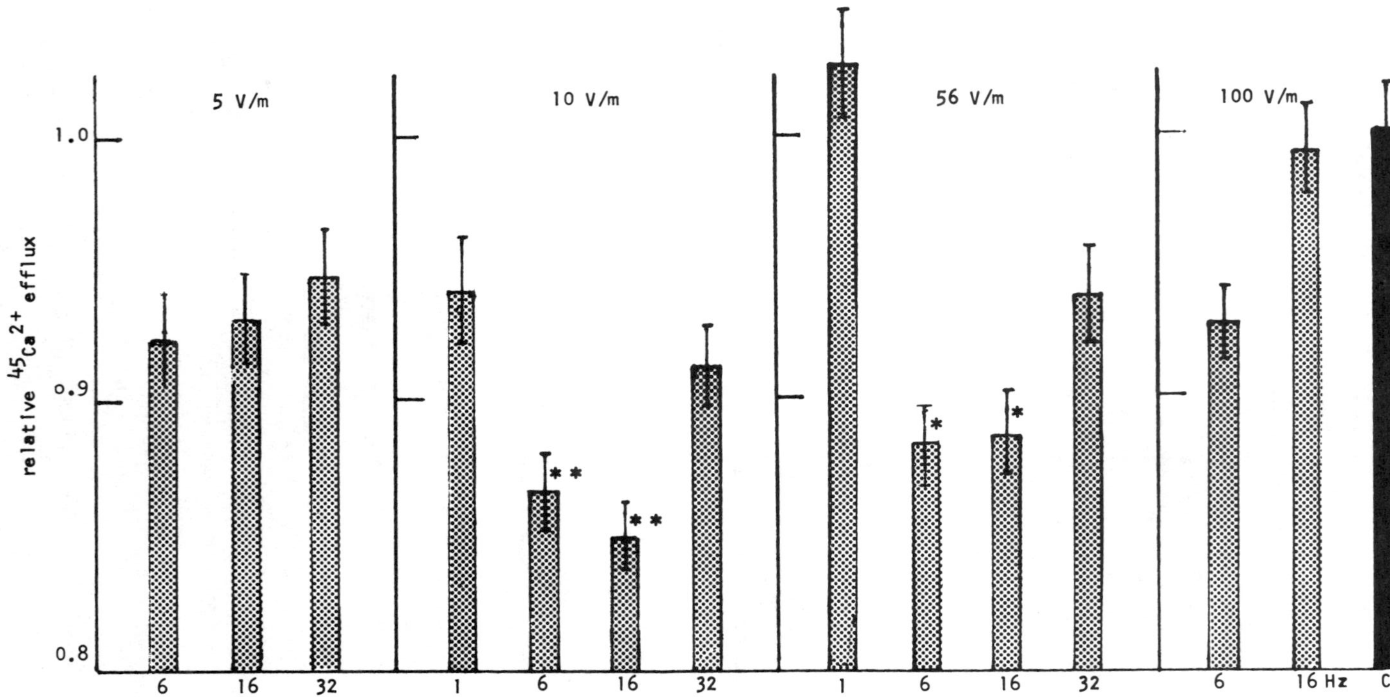

Proceedings of the National Academy of Science USA

*Figure 5. Effects of low-frequency fields on efflux of $^{45}Ca^{2+}$ from freshly isolated chick cerebral hemisphere at four different field intensities (5, 10, 56, and 100 V/m). Field frequencies from 1–32 Hz are shown on the abscissae. Efflux levels were normalized with respect to control efflux levels (C) shown in extreme right bar. Variance plotted as SEMs. Statistical analysis: *, $P < 0.05$; **, $P < 0.01$ (7).*

implicit in the concept of a coherent domain that, for the duration of the coherent state, this patch of membrane would be thermally quiet, isolated from the randomization of energy levels characteristic of the noncoherent condition. Such a system can be restored to the ground state by an extremely weak trigger at the level of Boltzmann (kT) noise or below (*7*), but in a coherent system the resulting effect at a remote site is a far greater release of energy than the trigger itself.

The narrow frequency and power windows so far defined for field interactions with cerebral calcium binding strongly bespeak such a cooperative interaction (*46*) based on long-range interactions in the length of a coherently organized charge system. The calcium–hydrogen-ion interaction described here is consistent with their respective roles proposed in the Bass and Moore model. However, we further hypothesize that the cooperative aspects of the interaction revealed in frequency and power windows may arise in a quite different role for hydrogen ions, with proton tunneling occurring between charge sites that interact across the boundary between coherent and incoherent charge zones (*36*). Such a trigger process would be most effective in the axis of the field electric gradient along this surface biopolymer sheet. As a first transductive step, it would initiate an avalanche or domino effect in the category of quantum amplification (*6, 10*).

Summary

1. Sensitivity of brain tissue to certain weak, oscillating electromagnetic fields occurs in the absence of significant tissue heating (less than 0.1°C), and may involve a class of nonequilibrium processes. This evidence focuses on the windowed character of sensitivities of calcium binding and electrical activity in brain tissue to low-frequency modulation and intensity characteristics of impressed RF fields.

2. ELF fields decrease calcium efflux from chick and cat cerebral tissue by about 15% only in narrow amplitude and frequency windows, between 6 and 20 Hz and between 10 and 100 V/m (approximate tissue gradients, 10^{-7} V/cm).

3. VHF (147-MHz) and UHF (450-MHz) fields increase calcium efflux by about 15% when amplitude is modulated between 6 and 20 Hz, but only for incident fields in the vicinity of 1.0 mW/cm^2.

4. We have now shown that this response to 16-Hz amplitude-modulated, 450-MHz, 1.0 mW/cm^2 fields is insensitive to variations in calcium concentration from 0–4.16m*M* in the bathing solution, but is enhanced by adding hydrogen ions (0.108m*M*, 0.1 *N* HCl) and inhibited in the absence of normal bicarbonate ion levels (2.4m*M*).

5. Lanthanum ions, which block transmembrane movement of calcium, restored field responsiveness in bicarbonate-free solutions, but the field then decreased $^{45}Ca^{2+}$ efflux.

6. Measured tissue gradients were ca. 50 mV/cm, corresponding to EEG gradients, weaker by 10^4 than in synaptic excitation.

7. Low-frequency gradients may be transduced in a specific class of extracellular binding sites, normally occupied by calcium ions and susceptible to competitive hydrogen-ion binding. Transductive coupling may involve coherent charge states between anionic sites on membrane-surface glycoproteins, with long-range cooperative interactions triggered by weak extracellular electric fields. Proton tunneling may occur at boundadies between coherent and noncoherent charge zones.

Acknowledgments

These studies were supported by the Bureau of Radiological Health Grant USPHS 3-R01-FD678-01, National Science Foundation Grant GB-27740, U.S. Air Force Contract F44620-70-0017, and by the Office of Naval Research Contract N-00014-A-200-4037.

Literature Cited

1. Sholl, D. A. "The Organization of the Cerebral Cortex"; Wiley: New York, 1956.
2. "Physiology and Pathology of Dendrites," Kreutzberg, G. W., Ed.; Raven Press: New York, 1975.
3. Elul, R. "The Genesis of the EEG," *Int. Rev. Neurobiol.* **1972**, *15*, 227–272.
4. Schmitt, F. O.; Dev, P.; Smith, B. H. "Electrotonic Processing of Information by Brain Cells," *Science* **1976**, *193*, 114–120.
5. Dowling, J. E.; Ehinger, B.; Hedden, W. L. "The Interplexiform Cell: A New Type of Retinal Neuron," *Invest. Ophthalmol.* **1976**, *15*, 916–926.
6. Adey, W. R. "Models of Membranes of Cerebral Cells as Substrates for Information Storage," *BioSystems* **1977**, *8*, 163–178.
7. Bawin, S. M.; Adey, W. R. "Sensitivity of Calcium Binding in Cerebral Tissue to Weak Environmental Electric Fields Oscillating at Low Frequency," *Proc. Nat. Acad. Sci., U.S.A.* **1976**, *73*, 1999–2003.
8. Bawin, S. M.; Kaczmarek, L. K.; Adey, W. R. "Effects of Modulated VHF Field on the Central Nervous System," *Ann. N. Y. Acad. Sci.* **1975**, *247*, 74–81.
9. Blackman, C. S.; Elder, J. A.; Bennane, S. G.; Weil, C. M.; Eichinger, D. C. "Two Factors Affecting the Radiation-Induced Calcium Efflux from Brain Tissue," URSI Symp. on Bioeffects of Electromagnetic Waves, Washington, D.C., *Radio Sci.*, in press.
10. Bawin, S. M.; Sheppard, A.; Adey, W. R. "Possible Mechanisms of Weak Electromagnetic Field Coupling in Brain Tissue," *Bioelectrochem. Bioenerg.* **1978**, *5*, 67–76.
11. Bawin, S. M.; Gavalas, R. J.; Adey, W. R. "Effects of Modulated VHF Field on the Central System," *Ann. N. Y. Acad. Sci.* **1973**, *247*, 74–80.
12. Takashima, S.; Oronal, B.; Schwan, H. P. "Effects of RF Energy on the EEG of Rabbit," *Proc. IMPI/URSI Symp. on Electromagnetic Fields in Biological Systems, Ottawa, Canada*, 1978, p. 9.
13. "Brain Cell Microenvironment," Schmitt, F. O.; Samson, F. E., Ed.; *Neurosci. Res. Prog. Bull.* **1969**, *7*, 277–417.
14. Singer, S. J.; Nicholson, G. L. "The Fluid Mosaic-Model of the Structure of Cell Membranes," *Science* **1972**, *175*, 720–736.
15. Katchalsky, A. "Polyelectrolytes and Their Biological Interactions" in "Connective Tissue. Intercellular Macromolecules;" New York Heart Association, Ed.; Little, Brown: Boston, 1964; pp. 9–41.

16. Bass, L.; Moore, W. J. "A Model of Nervous Excitation Based on the Wein Dissociation Effect" in "Structural Chemistry and Molecular Biology," Rich, A., Davidson, C. M., Eds.; Freeman: San Francisco, **1968**; pp. 350–368.
17. Bawin, S. M.; Adey, W. R.; Sabbot, I. M. "Ionic Factors in Release of $^{45}Ca^{2+}$ from Chicken Cerebral Tissue by Electromagnetic Fields," *Proc. Nat. Acad. Sci., U.S.A.* **1978,** *75,* 6314–6318.
18. Stahl, W. L.; Swanson, P. D. "Calcium Movements in Brain Slices in Low Sodium and Low Calcium Media," *J. Neurochem.* **1972,** *19,* 2395–2403.
19. Kaczmarek, L. K.; Adey, W. R. "The Efflux of $^{45}Ca^{2+}$ and [^{3}H] Gamma-Aminobutyric Acid from Cat Cerebral Cortex," *Brain Res.* **1973,** *63,* 331–342.
20. Miledi, R. "Lanthanum Ions Abolish Calcium Response of Nerve Terminals, *Nature (London)* **1971,** *229,* 410–411.
21. Van Breemen, C.; deWeer, P. "Lanthanum Inhibition of $^{45}Ca^{2+}$ Efflux from the Squid Giant Axon," *Nature (London)* **1970,** *226,* 760–761.
22. Van Breemen, C.; Farinas, B. R.; Gerba, P.; McNaughton, E. D. "Excitation Contraction Coupling in Rabbit Aorta Studied by the Lanthanum Method for Measuring Cellular Calcium Influx," *Cir. Res.* **1972,** *30,* 44–54.
23. Baker, P. F.; McNaughton, P. A. "Calcium-Dependent Calcium Efflux from Intact Squid Axons: Ca-Ca Exchange or Net Extrusion?" *J. Physiol. (London)* **1976,** *258,* 97P–98P.
24. Henkart, M.; Hagiwara, S. "Localization of Calcium Binding Sites Associated with the Calcium Spike in Barnacle Muscle," *J. Memb. Biol.* **1976,** *27,* 1–20.
25. Langer, G. A.; Frank, J. S. "Lanthanum in Heart Cell Culture. Effect on Calcium Exchange Correlated with Its Localization," *J. Cell Biol.* **1972,** *54,* 441–455.
26. "Brain Interactions with Weak Electric and Magnetic Fields," Adey, W. R., Bawin, S. M., Eds.; *M.I.T. Neurosci. Res. Prog. Bull.* **1977,** *15,* 1–135.
27. "Biologic Effects of Nonionizing Radiation," Tyler, P., Ed.; *Ann. N. Y. Acad. Sci.* **1975,** *247,* 1–543.
28. "Biological Effects and Measurements of Radio Frequency," Hazzard, D. G., Ed.; *HEW Publ. (FDA) 77-8026,* 1977.
29. "The Physical Basis of Electromagnetic Interactions with Biological Systems," Taylor, L. S., Cheung, A. Y., Eds.; University of Maryland: 1977.
30. "Biological Effects of Electromagnetic Waves," Johnson, C. C., Shore, M. L., Eds.; *HEW Publ. (FDA) 77-8010* 1976, Vol. I and II.
31. Baranski, S.; Czerski, P. "Biological Effects of Microwaves"; Dowden, Hutchinson and Ross: Stroudsburg, PA, 1976.
32. "Biologic Effects and Health Hazards of Microwave Radiation," Czerski, P., Ed.; Polish Medical Publishers: Warsaw, 1974.
33. "Biological Effects of Electromagnetic Waves," Justesen, D. R.; Guy, A. W., Eds.; *Radio Sci.* **1977,** *12,* Supplement 6(S).
34. "Biological Effects of Electric and Magnetic Fields Associated with Proposed Project Seafarer"; Report of Committee on Biosphere Effects, National Academy Sciences, U.S.A.: 1977.
35. Biomagnetic Effects Workshop, Lawrence Berkeley Laboratory, Tenforde, T., Ed. Publication LBL-7452, 1978.
36. Adey, W. R. "Experiment and Theory in Long-Range Interactions of Electromagnetic Fields at Brain Cell Surfaces," in "Proceedings Biomagnetic Effects Workshop," Lawrence Berkeley Laboratory, Tenforde, T., Ed.; Publication LBL-7452, 53–78.
37. Bassen, H.; Herchenroeder, P.; Cheung, A.; Neuder, S. "Evaluation of an Implantable Electric-Field Probe Within Finite Simulated Zones," *Radio Sci.* **1977,** *12*(6), 15–25.

38. Van Harreveld, A.; Crowell, J.; Malhotra, S. K. "Extracellular Space in the Cerebral Cortex of the Mouse," *J. Cell. Biol.* **1965,** *25,* 117–137.
39. Nicholson, C.; ten Bruggencate, G.; Steinberg, R.; Stöckle, H. "Calcium Modulated in Brain Extracellular Microenvironment Demonstrated Ion-Selective Micropipette," *Proc. Nat. Acad. Sci., U.S.A.* **1976,** *74,* 114–120.
40. Adey, W. R.; Bystrom, B.; Costin, A.; Kado, R. T.; Tarby, T. J. "Divalent Cations in Cerebral Impedance and Cell Membrane Morphology," *Exp. Neurol.* **1969,** *23,* 29–50.
41. Cole, K. S. "Permeability and Impermeability of Cell Membranes for Ions," *Cold Spring Harbor Symp. Quant. Biol.* **1940,** *4,* 110–122.
42. Kaczmarek, L. K.; Adey, W. R. "Some Chemical and Electrophysiological Effects of Glutamate in Cerebral Cortex," *J. Neurobiol.* **1974,** *5,* 231–241.
43. Schwarz, G. "A Basic Approach to a General Theory for Cooperative Intramolecular Conformation Changes of Linear Biopolymers **1967,** *5,* 321–324.
44. Schwarz, G. "Fundamental Principles of Cooperative Processes; Sharpness and Kinetics of Cooperative Transitions" in "Functional Linkage in Biomolecular Systems;" Schmitt, F. O., Schneider, D. M., Crothers, D. M., Eds.; Raven Press: New York, 1975; pp. 25–32.
45. Yahara, I.; Edelman, G. M. "Restriction of the Mobility of Lymphocyte Immunoglobulin Receptors by Concanavelin A," *Proc. Nat. Acad. Sci., U.S.A.,* **1972,** *69,* 608–612.
46. Adey, W. R. "Evidence for Cooperative Mechanisms in the Susceptibility of Cerebral Tissue to Environmental and Intrinsic Electric Fields" in "Functional Linkage in Biomolecular Systems," Schmitt, F. O., Schneider, D. M., Crothers, D. M., Eds.; Raven Press: New York, 1975; pp. 325–342.

RECEIVED November 18, 1978.

23

Direct Cell-to-Cell Channels: The Basis of Junctional Coupling As Shown by Electrical and Tracer Studies

S. J. SOCOLAR, Y. KANNO[1], and W. R. LOEWENSTEIN

Department of Physiology and Biophysics, University of Miami School of Medicine, P.O. Box 016430, Miami, FL 33101

Cell-to-cell conductance in a nascent cell junction grows by stable quantal steps, indicating that the coupling pathway develops by accretion of direct cell-to-cell unit channels in parallel. This finding emerges from experiments in which transfer impedance is monitored at high resolution in the early stages of junction of a pair of cells isolated from a Xenopus *embryo. It represents the most direct confirmation of the model proposed by Loewenstein in 1966 to characterize the pathway for both electrical coupling and cell-to-cell transmission of small hydrophilic molecules. The main evidence for cell-to-cell channels as well as for the topology of the cell-to-cell pathway is reviewed.*

For a century the idea of the cell as an autonomous enclave held sway in biological thought. Studies over the past few years have shown the existence of specialized membrane passageways that permit the exchange of small hydrophilic molecules between cell interiors. A major implication is that for the smaller molecular components of the cytosol, the functional compartmental unit is not a single cell but, rather, a larger set (*1, 2*). Abundant evidence indicates that such an exchange capability —junctional transmittance—occurs in most kinds of organized metazoan tissues (*see* in Ref. *3*). Here we show that, in the light of recent experiments, the junctional passageways are ensembles of direct cell-to-cell

[1] Present address: University of Hiroshima, Japan.

0-8412-0473-X/80/33-188-379$05.00/1

channels, functioning in parallel, and, at least in early stages of formation, are all alike in their transmittance properties (*see* Figure 2E).

There are considerable grounds, albeit still entirely circumstantial, for the supposition that the intramembrane particles of gap junctions are the sites of the cell-to-cell channels (*4*). In this chapter, however, we focus not on ultrastructural data but on findings regarding cell-to-cell electrical conductance and tracer movements. Significantly, studies of these kinds have provided a firm basis not only for postulating the cell-to-cell passageways but also for characterizing their topology as well as some of their other properties. For background we review briefly, and not entirely chronologically, the underpinnings of the idea of a cell-to-cell channel.

The Evidence for Specialized Junctional Passageways

The first evidence of a cell-to-cell pathway came from studies by neurophysiologists who found at certain muscle (*5*) and nerve (*6*) cell interfaces (without chemical synapses) that a voltage shift in a cell causes a parallel shift in contiguous cells. This indicated that some sort of conductive pathway links the cell interiors, as represented schematically in the equivalent circuit of Figure 1. Findings of such electrical coupling in many other tissues followed. But was the conductive link anything more than just a rather large area of very close apposition between quite ordinary regions of the membranes of neighboring cells? Several kinds of evidence have shown this (*see* Figure 2A) to be unacceptable as a general model.

(a) The area of close membrane apposition in general was not nearly large enough to account for the high cell-to-cell conductance (*7–13*).

(b) Experiments demonstrating cell-to-cell passage of various tracer substances indicated that many of there were far more permeant from cell to cell than outward through nonapposed cell-membrane regions (*7, 14, 15, 16, 17, 18*).

(c) Despite the high flow rates seen in some tracer studies, there was evidence of a moderate molecular size limit for transfer (*16, 17, 18*). (Such a limit, we might add, would not be expected if the cells simply were linked by coarse cytoplasmic bridges, e.g., as may occur in certain embryo cells (*19*).)

(d) Cell-to-cell transmission could be reduced dramatically by experimental intervention without a comparable reduction of nonjunctional membrane conductance (Indeed, the latter is increased in some cases.) (*20*).

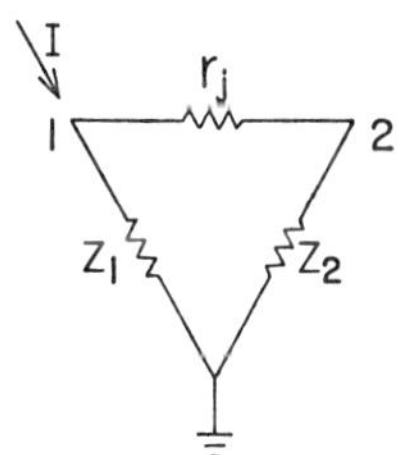

Figure 1. Equivalent electric circuit representation of a coupled cell pair. A current (I) is passed between an electrode inside Cell 1 and another in the extracellular medium (ground). This gives rise to a measurable potential in Cell 2, indicating that the two cell interiors are linked by a conductive pathway, represented here by a junctional element of resistance r_j. *The elements* z_1 *and* z_2 *represent the (lumped) impedances across the nonjunctional surface membranes of the two cells. Each cell is assumed to be isopotential, as is the extracellular medium.*

(e) Cells with a (correctible) genetic incapacity for junctional transmission nevertheless can be apposed closely (*22*).

These features showed that close apposition alone is insufficient to account in general for the observed cell-to-cell transmittance, and that a class of cell interfaces exists (*see* Figure 2B) with junctional membrane specializations affording unusually high permeability. The identification of the electrical (presumably ionic) pathway with the pathway for larger tracer molecules rests on the following: (1) when toe pathway is modified experimentally by elevating cytoplasmic free [Ca^{2+}] (normally

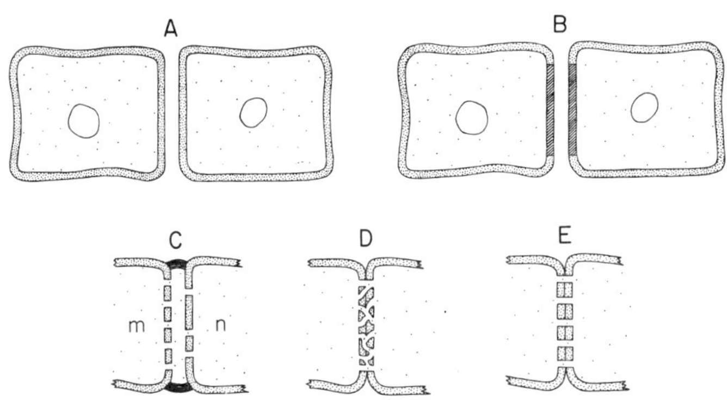

Figure 2. The basis of junctional coupling; some alternative models explored. (A) The cell-to-cell conductive pathway is provided by a large area of close contact between nonspecialized surface membranes of the two cells. (B) In this model, the closely apposed membrane regions are specialized to provide high permeability to small hydrophilic molecules. (C) The junctional membranes are penetrated by uniform aqueous channels, numbering m *in one membrane and* n *in the other; the pathway is mediated by a common intercellular compartment which is insulated by a perijunctional diffusion barrier. (D) The tight cell-to-cell seal provides excellent perijunctional insulation without a median compartment to interrupt the passageways, but they are interrupted by random, nondiscrete interconnections. (E) Current conception: tight cell-to-cell channels, all alike (cf. the 1966 model,* see *Ref.* 1). *See text for comparison of these alternatives in the light of experimental findings.*

$\lesssim 10^{-7}M$) to $\gtrsim 5 \times 10^{-5}M$, both electrical coupling and tracer permeability are abolished; (2) certain cell strains which, because of an inherited defect, display no electrical coupling, are also incapable of junctional transfer of tracer molecules. When, by genetic manipulation, the strains are given the capacity to transfer probe molecules, they also show electrical coupling (reviewed in Ref. *3*). While this evidence is not as direct as one might wish, there is no pressing reason for considering a multipathway model.

Among possible junctional membrane specializations conferring high permeability, pores and carriers come to mind first. The great diversity of molecular species (reviewed in Refs. *23* and *24*) shown to transit the specialized pathway bespeaks a lack of specificity that makes the carrier idea implausible (*15*). To decide whether this means that junctional pores have to be wider than nonjunctional pores, one needs to compare size–selectivity for permeants at the two sites. Otherwise one cannot rule out a difference in surface densities of identical pores as the explanation of permeability difference in junctional and nonjunctional membranes. If permeation occurs via aqueous pores and if there are indeed larger pores in junctional than in nonjunctional membrane, permeability at junctions should diminish by a smaller factor with increasing size of permeant that at a nonjunctional membrane. (Cf. Ref. *25*, where an analogous test, in terms of length constants, is, proposed although the interpretation of results given there is somewhat problematic because nothing is said as to whether the tracer distribution was in a steady state, as would be required for a valid test.) Few reliable paired measurements of junctional and nonjunctional permeabilities exist. However, lower selectivity at a junctional membrane has been shown for tetraethylammonium ion vs. K^+ (*see* Ref. *16*).

Historically, the idea that the pathway might comprise membrane components with unique properties was generated by the discovery that not only small ions but even larger hydrophilic molecules can traverse the junctional pathway (*7*). The stimulus for the experiments leading to this discovery was the knowledge, just then emerged, that junctional transmittance, far from being limited to nerve and muscle, occurs in epithelia (*7*) and glia (*26*), neither of which carries electrical signals. This knowledge led to a re-evaluation of the physiological import of the junctional pathway (*1, 2, 7*), and to a broader search for transmissible molecular species.

Working Out the Topology of the Pathway

The Figure 2B scheme also is deficient. Such a cell-to-cell pathway might exhibit a measure of isolation from the main extracellular space by virtue of the diffusion resistance offered by a very narrow gap. But a theoretical analysis of electrical fields in such a model (*27, 28; see* also Ref. *29*), makes it possible to calculate that the isolation afforded by a 2-nm gap is not in general sufficient to account for the low degree of electrical leakage observed experimentally (cf. *29;* e.g., we have calculated that, given the observed junctional resistance in the *Drosophila* salivary gland (*7*), th gap is too wide by at least three orders to provide the observed junctional insulation). Also, the fluorescent dye Procion yellow, which, upon injection, moves from cell to cell, nevertheless does not enter from outside even when the coupled cells are bathed in it for many hours and the intercellular cleft appears thoroughly infiltrated (*15, 30, 31*). Evidently the cell-to-cell passageway has a rather tight perijunctional diffusion barrier (e.g., Figure 2C–E).

To encompass the junctional transmittance facts known in 1966, it was postulated that direct, aqueous cell-to-cell pathways exist, in the form of "junctional units," each unit combining tightly in series a pair of "junctional-membrane" elements, one derived from each cell, and thus providing a central passageway of high permeance, bounded peripherally by a barrier against leaks to or from extracellular space; many such junctional units in parallel were considered to make up the entire cell-to-cell pathway (*1*). The term "direct" as used here means that no two units are mediated by a common compartment such as that in Figure 2C.

However, several questions remained to be answered experimentally; the answers, when found, could determine collectively whether the elementary cell-to-cell passageway is a well-defined species of membrane channel as channels now are conceived generally:

(a) In a junction, are the passageways, as a population, uniform in conductance, and/or permeance?

(b) Are the passageways direct?

(c) Are the passageways discrete, or do they anastomose, e.g., as in Figure 2D?

Our recently published experimental findings (32) address these questions.

Searching for the Channels

Our experimental study was a test of the proposition that the cell-to-cell passageways are a uniform population of separate, direct cell-to-cell channels functioning in parallel. Such channels should all make equal contributions to the electrical conductance across a cell junction. From the limiting molecular size of permeants (*18; see* also Refs. *16* and *17*), the diameter of the channel bore is estimated as at least 1.5–1.6 nm; hence it seemed that we might be able to resolve experimentally the changes of conductance owing to the opening and/or closing of single channels, and thus test the hypothesis. Measurements of this sort are easiest when made against a background of very small, or zero, junctional conductance, and so we chose to follow with very high resolution (*32*), the junctional conductance changes at the onset of the spontaneous junction formation that occurs when two cells are placed in contact. We used cells isolated from amphibian (*Xenopus*) embryos. (The suitability of such a system had been suggested by experiments with similar cells, where development of junctional conductance had been monitored at low resolution (*33*).) As we passed current from a microelectrode in one cell, we monitored the development of a transfer voltage in the other cell. For a cell pair, the transfer impedance is given by

$$\frac{V_2}{I} = \left(\frac{r_j}{z_1 z_2} + \frac{1}{z_1} + \frac{1}{z_2}\right)^{-1}$$

where V_2 is the transfer voltage, I is the injected current, z_1 and z_2 are the nonjunctional membrane impedances of the two cells, and $r_j = 1/g_j$ is the junctional resistance. Early during junction, when g_j is still quite small, we can say, with negligible error, that

$$\frac{V_2}{I} = z_1 z_2 g_j$$

Thus, under these conditions, changes in the transfer impedance are directly proportional measures of corresponding changes of junctional conductance (if z_1 and z_2 are essentially constant); and, if junctional conductance were to increase by uniform increments, we should expect the transfer impedance to do likewise. That is precisely what we saw in our experiments (*32*).

Quantal Growth of Junctional Conductance

The basic observation was that during junction formation V_2 increased by stable, quantal steps (*see* Figure 3). All steps observed were either of unit size or double that size. We could check and reverse this stepwise growth pattern by microinjection of Ca^{2+} (*32*), a well-known means for blocking the junctional pathways (*34*). Indeed the transfer voltage could be abolished entirely (*see* Figure 3). But within 2 min, (quantal) growth of V_2 would resume spontaneously (*32*).

To test the reliability of the quantal upstep observation, we fitted upstep V_2 values with the linear equation

$$V_2 = N V_2^0$$

where V_2^0, a constant to be evaluated, is the quantum of transfer voltage for the particular experiment and N is an integer variable determined

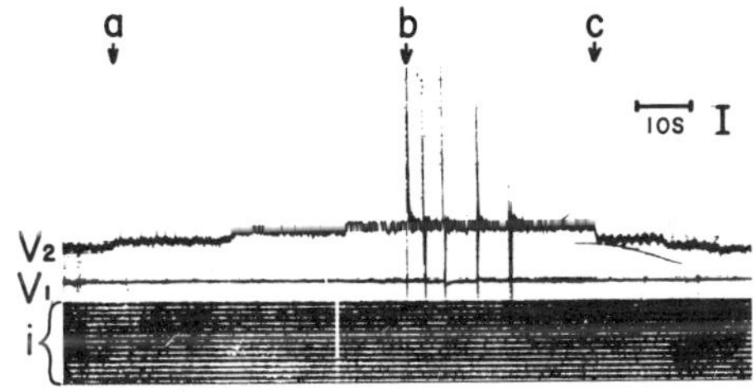

Figure 3. Quantal development of junctional conductance during formation of cell-to-cell channels. Segment of an oscillograph record showing the time course of transfer voltage (V_2) as junction occurs spontaneously in a pair of isolated Xenopus *embryo cells placed in contact* (32)*. At (a) the first of three quantal upsteps occurs. At (b) a series of five pulsatile Ca^{2+} microinjections is delivered into Cell 1 (spikes on V_2 trace are capacitive artifacts caused by solenoid valve controlling pneumatic injection) which leads to (c), a series of quantal down-steps of V_2. No discrete changes are seen here in input voltage (V_1), which is displayed at lower gain than V_2.* Bottom *trace* (I) *shows the sinusoidal 25-Hz current (10-nA rms) injected continuously into Cell 1. Because V_1 and V_2 are rectified by the phase-sensitive detector–amplifier systems used for enhancing their signal/noise ratios, their records appear as D.C. Calibrations: V_1, 100 μV; V_2, 3 μV.*

by inspection: for each recorded V_2, N is the integer that most nearly expresses V_2 as a multiple of the largest common factor. Then, the precision index of the estimated V_2^0 was taken as a measure of the soundness of the quantal hypothesis. As it turned out the standard error of V_2^0 in our experiments ranged from 0.6 to 1.7%, indicating that V_2 indeed grows quantally during cell junction (*32*).

Thus, junctional conductance grows by quantal increments during junction formation; each (spontaneous) upstep on the record reflects a channel opening.

The Channels Are Direct

In the light of these findings, we can pass over as unreasonable topologies like that in Figure 2C, where the passageways are not direct from cell to cell but are mediated by a shared compartment. In such a case there is no reason to believe that channel openings in the cell pair would occur synchronously, in a paired fashion; rather, the numbers of channels in the apposing cell faces should be largely independent. Suppose, then, that one cell face has m channels and the other has n, all alike with unit conductance g_j^0. The integral conductance of the junction then would be

$$g_j = \frac{mng_j^0}{m+n}$$

This value generally would not change quantally as the integers *m* and *n* grow in size. (Formally, an exception occurs if one of the integers is assumed to become quite large, say 50 or more, before channels begin to open on the other face. Although this would manifest itself as quantal growth of transfer voltage within the accuracy of our measurements, it is quite implausible that such extreme asymmetry should be the usual pattern of openings.)

There also is no way that the integral conductance owing to randomly anastomosing passageways (*see* Figure 2D) could develop quantally. Only the schema of Figure 2E is plausible—direct, discrete units in parallel and all alike with respect to permeance. Limited as they were to the earliest stages of junction, however, our results do not preclude the possible later appearance of different channel populations or the later modification of the properties of the early channels.

We have seen that a cell junction forms as if by addition of single-channel units, and it seems natural to associate these units with the "junctional units" proposed in 1966 (*see* Ref. *1*). But it has to be said that we cannot strictly rule out the possibility that each junctional unit comprises a cluster of channels; even interconnection of passageways within such a cluster is admissible if sufficient uniformity prevails. Although for simplicity we think of the junctional unit as housing a single channel, the question of how many channels a junctional unit contains may be easier to answer after the absolute value of the conductance quantum has been measured (a measurement that, for technical reasons, was not achieved in the experiments described (*32*)) and compared with values estimated theoretically for a 1.5–2-nm diameter cylindrical channel (of course, the true geometry of a channel need not be so simple).

Channel Stability

The channels formed in our experiments were unusually stable as compared with other kinds of membrane channels (*32*): once formed, they persisted in the open state for minutes at least. (We observed channel closure only when we injected Ca^{2+}.) The channels certainly do not exist as patent entities in separate cells, but they form and open within minutes when cells establish contact. In terms of current ideas of membrane structure, such aqueous channels appear likely to be bounded by integral membrane proteins. Yet protein synthesis can be inhibited in contacting cells without significantly affecting their capacity to form cell–cell channels (*35*). Together with the speed of channel assembly, this suggests (cf. *36*) that major preformed subunits, or protochannels, move symmetrically via the respective cell membranes to the site of contact, and that when they are properly paired across the interface (i.e., aligned and linked) and stabilized, the channel opens. Thus, the molecular interactions that provide structural stabilization contribute also (though not these alone (cf. *32*)) to patency stabilization.

There are several conceivable sources of the stability. Cytoskeletal components—microtubules or -filaments—might anchor the aligned channels, but this seems unlikely since reagents (colchicine and Cytochalasin B) that dissociate such anchor structures appear not to affect junctional transmittance (*37, 38, 39*) and, indeed, have been used to suppress cell division and cytokinesis during studies of channel formation (*32, 40*). Direct interactions among the channel subunits also are obvious candidates as stabilizers. Yet another possibility to consider is that the link-up of the protochannels may be accompanied by a stabilizing local reorganization of the lipid matrix—perhaps mediated by local changes in lipid composition. Suggestive here is the evidence (reviewed in Ref. *4*) that phospholipids are important in maintaining the structural integrity of gap junctions.

Acknowledgments

We are grateful to B. Rose, I. Simpson, and G. Dahl for discussion, A. Toro for typing the manuscript, I. Armenteros for drawings; and R. Serralta for photography. This work was supported by research grants from the National Science Foundation and the National Cancer Institute—USNIH.

Glossary of Symbols

g_j = junctional conductance; electrical conductance of the direct pathway linking the interiors of two apposed cells; $g_j = 1/r_j$.

g_j^0 = electrical conductance of a single junctional pore in a uniform population of pores mediating junctional passage

I = total electric current passed between interior and exterior of one cell of a pair in a measurement of the pair's junctional conductance (*see* Figure 1)

m, n = respective numbers of uniform elementary junctional pores in two cells hypothetically coupled via an interspace (*see* Figure 2C)

N = the number of quantal units making up the transfer voltage (V_2) if $r_j >> |z_1|, |z_2|$; by inference, the number of junctional channels.

r_j = junctional resistance; electrical resistance of the direct pathway linking the interiors of two apposed cells (*see* Figure 1).

V_1 = input voltage; in measurement of the junctional conductance of a cell pair, the steady-state displacement of potential in Cell 1, that cell containing the current source (*see* Figure 1).

V_2 = transfer voltage; in measurement of the junctional conductance of a cell pair, the steady-state displacement of potential in Cell 2, that member not containing the current source (*see* Figure 1).

V_2^0 = a quantal unit of transfer voltage, discernible as a repeated discrete increment in V_2 during the growth of V_2 in a nascent junction as long as $r_j >> |z_1|, |z_2|$. In a given junction, the magnitude of V_2^0 is proportional to the measuring current I.

z_1, z_2 = nonjunctional membrane impedances of a pair of cells whose junctional conductance is being measured (*see* Figure 1).

1, 2 = denotation of the cell interiors (each presumably an isopotential domain) of a pair of cells whose junctional conductance is being measured (*see* Figure 1).

Literature Cited

1. Loewenstein, W. R. *Ann. N.Y. Acad. Sci.* **1966,** *137,* 441–472.
8. Potter, D. D.; Furshpan, E. J.; Lennox, E. S. *Proc. Nat. Acad. Sci. U.S.A.* **1966,** *55,* 328–336.
3. Loewenstein, W. R. *BBA Revs. Cancer* **1979,** *560,* 1–65.
4. Griepp, E. B.; Revel, J.-P. In: "Intercellular Communication," De Mello, W. C., Ed.; Plenum: New York, 1977; pp. 1–32.
5. Weidmann, S. *J. Physiol. (London)* **1952,** *118,* 348–360.
6. Furshpan, E. J.; Potter, D. D. *J. Physiol. (London)* **1959,** *145,* 289–325.
7. Loewenstein, W. R.; Kanno, Y. *J. Gen. Physiol.* **1964,** *22,* 565–586.
8. Loewenstein, W. R.; Socolar, S. J.; Higashino, S.; Kanno, Y.; Davidson, N. *Science* **1965,** *149,* 295–298.
9. Bennett, M. V. L. *Ann. N.Y. Acad. Sci.* **1966,** *137,* 509–539.
10. Bennett, M. V. L.; Pappas, G. D.; Gimenez, M.; Nakajima, Y. *J. Neurophysiol.* **1967,** *30,* 236–300.
11. Kriebel, M. E. *J. Gen. Physiol.* **1968,** *52,* 46–59.
12. Brink, P.; Barr, L. *J. Gen. Physiol.* **1977,** *69,* 517–536.
13. Sheridan, J. D.; Hammer-Wilson, M.; Preus, D.; Johnson, R. G. *J. Cell Biol.* **1978,** *76,* 532–544.
14. Weidmann, S. *J. Physiol. (London)* **1966,** *187,* 323–342.
15. Payton, B. W.; Bennett, M. V. L.; Pappas, G. D. *Science* **1969,** *166,* 1641–1643.
16. Weingart, R. *J. Physiol. (London)* **1974,** *240,* 741–762.
17. Brink, P. R.; Dewey, M. M. *J. Gen. Physiol.* **1978,** *72,* 67–86.
18. Simpson, I.; Rose, B.; Loewenstein, W. R. *Science* **1977,** *195,* 294–296.
19. Arnold, J. M. *Differentiation* **1974,** *2,* 335–341.
20. Loewenstein, W. R.; Nakas, M.; Socolar, S. J. *J. Gen. Physiol.* **1967,** *50,* 1865–1891.
21. Peracchia, C.; Dulhunty, A. F. *J. Cell. Biol.* **1976,** *70,* 419–439.
22. Azarnia, R.; Larsen, W. J.; Loewenstein, W. R. *Proc. Nat. Acad. Sci. U.S.A.* **1974,** *71,* 880–884.
23. Loewenstein, W. R. *Cold Spring Harbor Symp. on Quant. Biol.* **1976,** *40,* 49–63.

24. Socolar, S. J.; Loewenstein, W. R. In "Methods in Membrane Biology," Korn, E. D., Ed.; Plenum: New York, 1979; Vol. 10, pp. 123–177.
25. Pappas, G. D.; Bennett, M. V. L. *Ann. N.Y. Acad. Sci.* **1966,** *137,* 495–508.
26. Kuffler, S. W.; Potter, D. D. *J. Neurophysiol.* **1964,** *27,* 290–320.
27. Heppner, D. B.; Plonsey, R. *Biophys. J.* **1970,** *10,* 1057–1075.
28. Woodbury, J. W.; Crill, W. E. *Biophys. J.* **1970,** *10,* 1076–1083.
29. Bennett, M. V. L.; Auerbach, A. A. *Anat. Rec.* **1969,** *163,* 152.
30. Bennett, M. V. L. In "Intercellular Staining in Neurobiology," Kater, S. B., Nicholson, C., Eds.; Springer, New York: New York, 1973; pp. 115–134.
31. Imanaga, I. *J. Membr. Biol.* **1974,** *16,* 381–388.
32. Loewenstein, W. R.; Kanno, Y.; Socolar, S. J. *Nature* **1978,** *274,* 133–136.
33. Ito, S.; Sato, E.; Loewenstein, W. R. *J. Membr. Biol.* **1974,** *19,* 339–355.
34. Rose, B.; Loewenstein, W. R. *J. Membr. Biol.* **1976,** *28,* 87–119.
35. Epstein, M. L.; Sheridan, J. D.; Johnson, R. G. *Exp. Cell Res.* **1977,** *104,* 25–30.
36. Loewenstein, W. R.; Kanno, Y.; Socolar, S. J. *Fed. Proc.* **1978,** *37,* 2645–2650.
37. Azarnia, R., 1975 personal communication.
38. Rose, B., 1977 personal communication.
39. Flagg–Newton, J., 1978 personal communication.
40. Ito, S.; Sato, E.; Loewenstein, W. R. *J. Membr. Biol.* **1974,** *19,* 305–337.

RECEIVED November 20, 1978.

24

The Cell-to-Cell Membrane Channel: Permeability and Its Regulation

B. ROSE, I. SIMPSON, and W. R. LOEWENSTEIN

Department of Physiology and Biophysics, University of Miami School of Medicine, Miami, FL 33101

Junctional membrane channels were probed with various fluorescent-labelled peptides and found permeable only to molecules less than 2000 daltons. As determined from the physical dimensions of the largest permeating molecules, a minimal-channel bore with a 15-Å diameter is required to permit their passage through the junction. With the aid of the Ca^{2+}-sensitive light-emitting protein aequorin and the simultaneous measurement of electrical coupling across a cell junction, the junction permeability decreases with increasing $[Ca^{2+}]_i$. As disclosed by coinjection of paired, different-sized tracer molecules, Ca^{2+}-injection caused a graded reduction in the molecular-size limit for channel permeation.

Many kinds of cells are coupled by junctions containing cell-to-cell membrane channels which permit a direct exchange of ions and small molecules between cells (*1, 2*). Since the finding that fluorescein passes through such junctions (*3*), other fluorescent and colorant molecules have been used to probe the permeability of cell-to-cell channels (*4–10*). We have determined an upper-molecular-size limit of about 2000 daltons for channel permeation in cell junctions of the salivary gland of *Chironomus thummi* larvae with the use of linear peptide molecules of known size (*9*). From this, a channel bore of at least 15 Å is computed.

The permeability depends on the cytoplasmic free Ca^{2+} concentration ($[Ca^{2+}]_i$). The original proposal by Loewenstein (*1, 11*) that $[Ca^{2+}]_i$ regulates channel permeability, was supported by much circumstantial evidence (*12–18*). The most direct evidence comes from our experiments

0-8412-0473-X/80/33-188-391$05.00/1

(*19*), in which cytoplasmic Ca^{2+} was monitored with the aid of the Ca^{2+}-dependent luminescence of the photoprotein aequorin (*20*). In addition, the channel permeability was probed simultaneously by measuring electrical cell-to-cell coupling.

We approached the further question of whether this change in channel permeability reflects a reduction in the number of open channels and/or a reduction in channel diameter (aperture) by utilizing pairs of permeability probes of varying size. We found a graded reduction in molecular-size limit of channel permeation, suggesting a graded reduction in channel aperture (*21*).

Materials and Methods

Fluorescent Permeability Probes. The 10^{-2} mol/L solutions of fluorescent probes (*see* Table I; also *see* Refs. *9* and *22* for preparation and purification methods) were injected by pressure into cells of isolated, larval salivary glands of *Chironomus thummi* with the aid of micropipettes. (For composition of the physiological bathing medium, *see* Ref. *14.*) The intracellular and transjunctional spread of fluorescence was observed and photographed through a microscope (Leitz, with Ploem fluorescence system). The individual probes, in cases when several were injected together, could be identified by choosing appropriate optical filters, in accordance with the excitation and emission characteristics of their respective label: (a) λ (excitation): FITC, 450 nm; DANS, 340 nm; and LRB, 540 nm; (b) λ (emission): FITC, 520 nm; DANS, 525 nm; and LRB, 590 nm.

$[Ca^{2+}]_i$ Detection. Purified aequorin was injected into one or two adjacent cells. Light emitted inside the cells owing to the Ca^{2+}–aequorin reaction was recorded with a photomultiplier and, in addition, visualized by a television camera coupled to an image intensifier (*see* Figure 1), which enabled us to determine where within the cell a $[Ca^{2+}]$ change occurred. Sensitivity of the $[Ca^{2+}]_i$ detection was of the order of 5×10^{-7} to 10^{-6} mol/L (*19*).

Electrical Measurements. For simultaneous probing of junctional-channel permeability we monitored electrical coupling between the two adjacent cells with a microelectrode in each cell, recording cell potentials and the voltage changes caused by a current pulse delivered into one of the cells through a third microelectrode (*see* Figure 1). The voltage ratio V_2/V_1 served as an index of junctional permeability (*3*). Cell potentials (E), voltage changes (V), and current (*I*) were displayed on an oscilloscope and videotaped together with the outputs of the photomultiplier and the image intensifier–TV camera, as well as being recorded on a chart recorder.

Results

Molecular-Size Limit of Channel Permeation. The results obtained with the various tracers are summarized in Table I. Amino acids and peptides with sizes less than 2000 daltons passed through the junction.

The tracers generally crossed several cell junctions on either side from the injected cell at rates inversely related to molecular size. Tracers greater than 2000 daltons did not flow detectably through junction, nor did they seem to alter the channels: smaller tracers, injected simultaneously or successively—and with a different fluorescent label for identification—continued to traverse junctions (*see* Figure 2). Such coinjections were undertaken routinely.

The results suggest that for peptides, the cut-off for permeation of the junctional-membrane channels is about 2000 mol wt. (In an earlier

Table I. Cell-to-Cell Passage of Fluorescent Probes

Molecule[a]	*Molecular Weight*	*Cell–Cell Passage*	*Control Molecule*[b]
DANS(SO_3H)	251	+	
DANS(Glu)OH	380	+	
FITC(Glu)OH	536	+	
LRB(SO_3H)	559	+	
DANS$(Gly)_6OH$	593	+	
DANS$(Glu)_3OH$	640	+	
LRB(Glu)OH	688	+	
FITC$(Gly)_6OH$	749	+	
FITC$(Glu)_3OH$	794	+	
FITC(Glu–Tyr–Glu)OH	828	+	
DANS$(Leu)_3(Glu)_2OH$	849	+	
FITC(Glu–Try–Glu)OH	851	+	
LRB$(Gly)_6OH$	901	+	
LRB$(Glu)_3OH$	950	+	
LRB(Glu–Tyr–Glu)OH	982	+	
FITC$(Leu)_3(Glu)_2OH$	1004	+	
FITC$(Lys)_5OH$	1047	+	
LRB$(Leu)_3(Glu)_2OH$	1158	+	
FITC Angiotensin	1444	+	
FITC(Pro–Pro–Gly)$_5$OH	1652	+	
FITC Fibrinopeptide 'A'	1926	—[c]	LRB(Glu)OH LRB$(Leu)_3(Glu)_2OH$
FITC Microperoxidase	2268	—	LRB(Glu)OH
FITC Insulin A chain	2921	—	LRB(Glu)OH
DANS Insulin A chain	3232	—	[d]
FITC Insulin B chain	3897	—	LRB$(Glu)_3OH$
LRB Insulin A chain	4158	—	Fluorescein (330 mol wt)

[a] FITC = fluorescein–isothiocyanate; DANS = dansyl chloride; LRB = lissamine rhodamine B; (Glu)OH = glutamic acid. (Gly)OH = glycine; Leu(OH) = leucine; (Tyr)OH = tyrosine; (Try)OH = tryptophane; (Lys)OH = lysine; (Pro)OH = proline.

[b] Small molecule (passing) used as control for same junction.

[c] R. Wagner in our lab (personal communication) has since found occasional, slight cell–cell passage.

[d] Not tested with control molecule, but junction was coupled electrically.

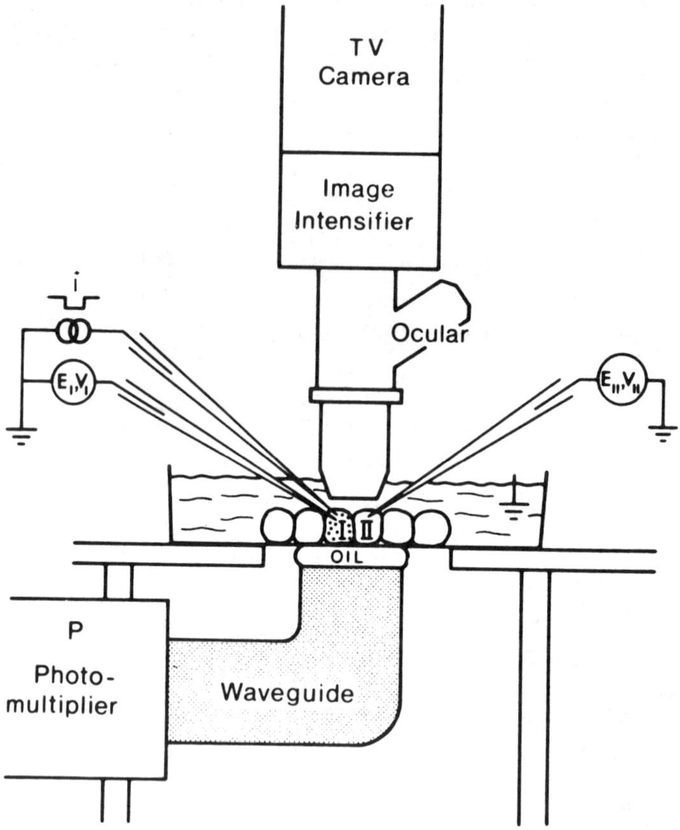

Figure 1. Experiment setup. Television camera coupled to image intensifier by a light guide views the aequorin luminescence of the cells through microscope (darkfield); luminescence also is monitored by photomultiplier. Electrical coupling is measured with the aid of three microelectrodes, by pulsing current (I) between the inside of Cell I and the grounded medium, and measuring the resulting steady-state changes (V) in membrane potentials (E) in Cell I and adjacent Cell II (19).

publication, Ref. 9, we suggested the limit to lie between a molecular-weight range of 1200 and 2000, merely for our lack of tracers in this size range at that time.) From the physical dimensions determined from molecular (CPK) models of the largest permeating probes, we infer a minimal bore diameter of 15–16 Å for the channels to accommodate these molecules (*9, 23*).

It should be mentioned here that in order to serve as a junctional tracer the molecule must be water soluble, nontoxic, not readily taken up by cells from the exterior, and not degraded by the cell for at least the duration of the test. All of our probes met these criteria. In tests for uptake from the exterior, glands were exposed to a 5×10^{-4} mol/L tracer concentration in the bathing solution. Only damaged cells and the gland lumen became and remained fluorescent after such exposure and subsequent washout. All dyes spread rapidly from the injection site throughout the cell and fluorescence was lost readily when those cells were injured, suggesting that cytoplasmic binding was not the limiting cause in the cases of failure of channel permeation. Prolonged incubation of tracers with pooled, mashed cytoplasm of salivary glands and subsequent paper electrophoresis failed to detect any degradation of the tracers (*see* Ref. 9 for the details on these control experiments).

Channel Closure by Ca^{2+}. Ca^{2+} injections into aequorin-loaded cells produced bright luminescent spots, generally restricted to the area of injection. Channel permeability decreased when $[Ca^{2+}]_i$ was elevated at the junctional area (*see* Figure 3). Because of the diffusion restriction in cytoplasm to Ca^{2+} (*24, 25, 26*), injections had to either be rather large or close to junction for the Ca^{2+} to reach junctional area. $[Ca^{2+}]_i$ elevations restricted to only one of a cell's junctions only affected this junction's permeability (*see* Figure 4).

A rise in $[Ca^{2+}]_i$ induced by exposure to metabolic inhibitor CN or to ionophores X537A and A23187—which allow transmembrane entry of

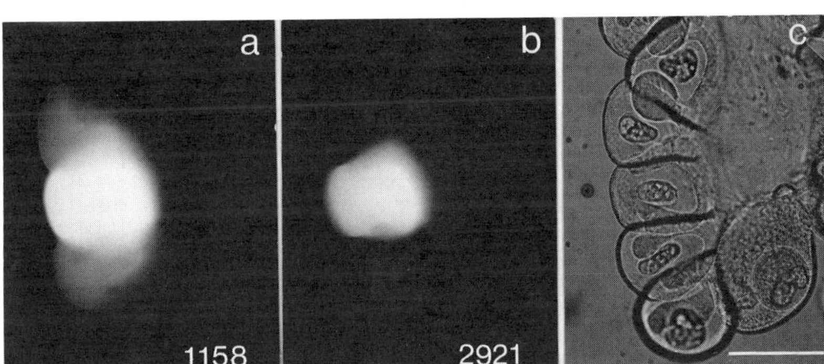

Figure 2. Probing junctional-membrane channels with two molecules of different size and fluorescent label. The red fluorescent tracer LRB-(Leu_3Glu_2)OH(1158 mol wt) is injected into the cell marked with an arrow, together with the yellow–green fluorescent tracer FITC Insulin A chain (2921 mol wt). (a) Distribution of the red tracer (1158 mol wt) photographed (darkfield) in black and white 3 min after injection. (b) Distribution of the yellow–green tracer (2921 mol wt) 35 min after injection. (The two fluorescences are set apart by using different excitation and barrier filters.) The red tracer spread from the injected cell to several neighbors (here only the first-order neighbors are visible for phototechnical reasons). The yellow–green tracer stayed within the injected cell. (c) Brightfield photograph of cells. Calibration, 100 μ.

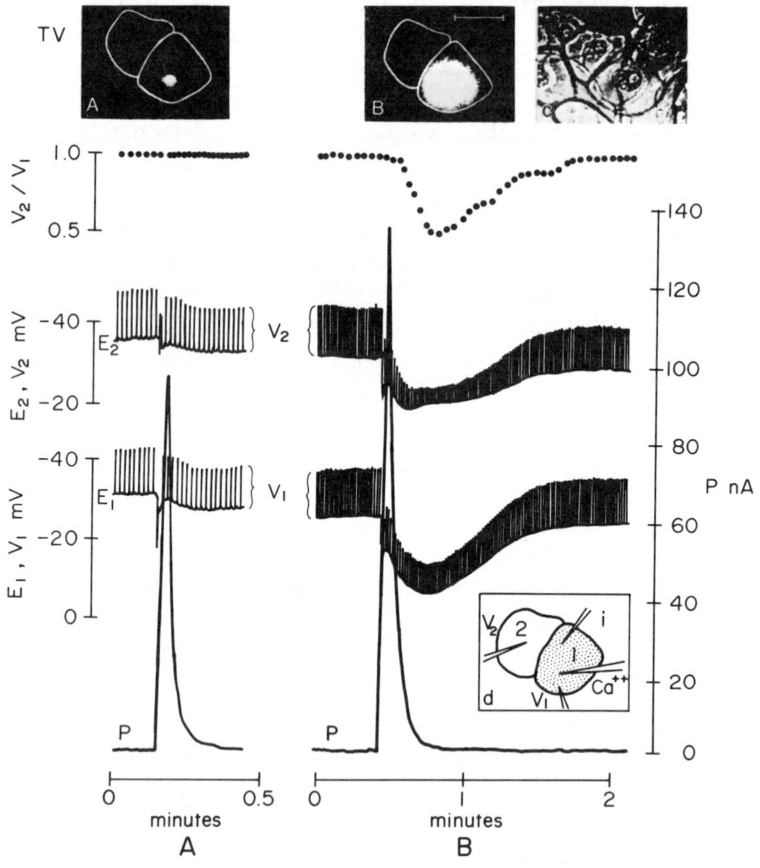

Journal of Membrane Biology

Figure 3. Uncoupling by Ca^{2+} injection. A pulse of Ca^{2+} (0.15 mol/L Ca and 0.1 mol/L EGTA) is injected (pressure) into Cell 1 while electrical coupling between Cells 1 and 2 is monitored. A and B show the respective effects of a small and a large Ca pulse delivered to the same intracellular site. Top: *TV pictures of the resulting aequorin glows (A, B) at time of their maximum spatial spread (corresponding with the peak of P curve); and brightfield TV picture (c) of the cells and micropipettes (diagrammed in (d); dotted cell contains aequorin). Calibration, 100 μ.* Below: *Chart record tracings of the photomultiplier current (P), membrane potentials (E_1, E_2) of the two cells and their displacements (V_1, V_2) produced by the test current pulses ($I = 4 \times 10^{-8}$ A; 1/sec), and plot of V_2/V_1 (coupling ratio). In A the aequorin glow, i.e., the detectable $[Ca^{2+}]_i$ elevation, is restricted to a radius of ca. 15 μ from the injection pipette and does not reach the region of cell junction 1:2. The cells stayed coupled. In B the $[Ca^{2+}]_i$ elevation extends to the region of cell junction. The cells uncouple. Note different time scales in A and B (19).*

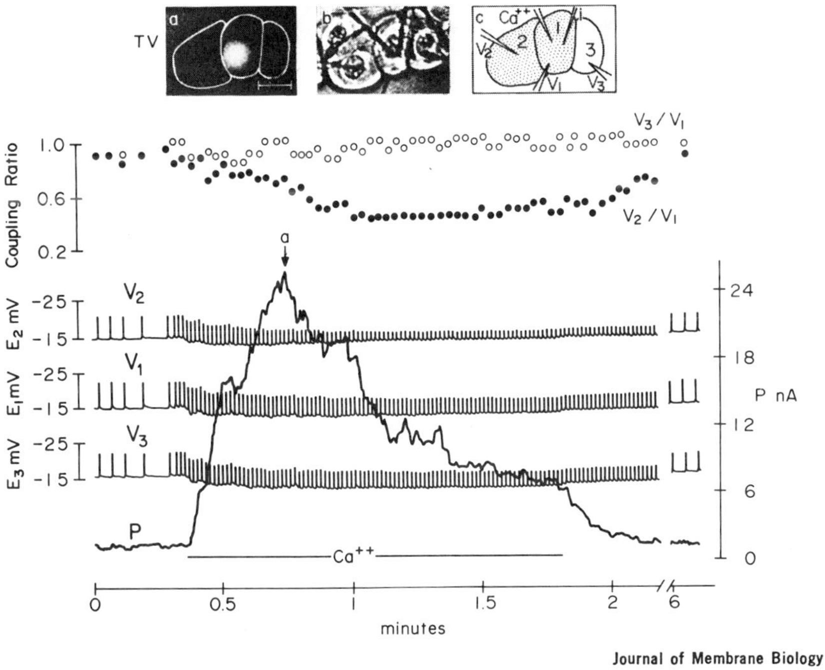

Figure 4a. Single-junction uncoupling by Ca^{2+} injection. Ca^{2+} is injected (iontophoresis) close to one cell junction (1:2) elevating $[Ca^{2+}]_i$ locally in the region of this junction. Coupling is measured across junctions 1:2 and 1:3. TV picture (a) shows the aequorin glow at the time of maximal spatial spread. Below are the chart record tracings of the photomultiplier current (P), the membrane potentials (E) and their displacements (V) in the three cells (test current pulses 4 × 10^{-8} A; 1/sec, and the coupling ratios V_2/V_1 and V_3/V_1).

Ca^{2+} into cells—was accompanied by a fall in channel permeability whenever $[Ca^{2+}]_i$ was elevated at the junction (*see* Figures 5 and 6).

Ca^{2+} itself was never seen to cross junctions, as disclosed by experiments in which two adjacent cells contained aequorin, and Ca^{2+} was injected into one of them close to the cell's common junction. This lack of channel permeation of Ca^{2+} sharply contrasts the permeation of the much larger, permeant peptide molecules. It might suggest the following mechanism of permeability regulation: Ca^{2+} binds to the channel, changes the channel-fixed charge configuration or molecular conformation effecting closure, and hence limiting its own flux and that of other molecules.

Recently, Meech and Thomas (*27*) found that Ca^{2+} injections into snail neuron caused a fall in intracellular pH. The authors suggested that many actions, hitherfore ascribed to Ca^{2+}, in effect may be caused by H^+ (*see* also Ref. *28*). H^+ release during Ca^{2+} accumulation is a well-known phenomenon in biological systems (e.g., mitochondria; *29, 30, 31*),

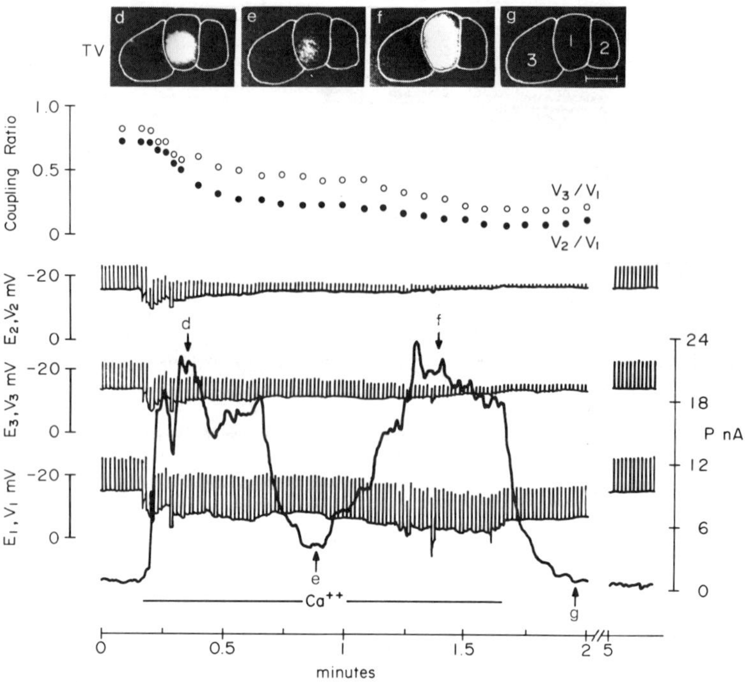

Figure 4b. Multiple junction uncoupling by Ca injection. The rate of Ca^{2+} injection is increased causing the $[Ca^{2+}]_i$ elevation to spread farther reaching both junctions 1:2 and 1:3 (TV picture d). The coupling now becomes depressed at both junctions. During a second peak of the Ca^{2+} injection, the Ca^{2+} swamps the cell (TV picture f) depressing the coupling further. The bar marks the duration of the Ca current, the arrows on the P*-curve indicate the time correspondence of TV pictures, and the calibration bar on the TV pictures equals 100 μ* (19).

and H^+ is known to compete with Ca^{2+} for binding on cell membranes (*32*). With respect to the channel-permeability mechanism the question now naturally arose whether Ca^{2+} or H^+ is the effector, or whether they both are. Lowering intracellular pH in *Xenopus* embryonic cells by exposure to CO_2 indeed lowered cell-to-cell channel permeability (*33*). The question remained, however, whether a decrease in pH will not in turn elevate $[Ca^{2+}]_i$. Lea and Ashley (*34*) demonstrated such an elevation of $[Ca^{2+}]_i$ during lowering of the pH_i by exposure to CO_2 of barnacle muscle; and Rose and Rick (*35*) showed such an elevation in *Chironomus* salivary gland cell when lowering pH_i by CO_2 exposure or by HCl injection.

Which of the two ions is sufficient for channel closure? We now have an answer that Ca^{2+} is: $[Ca^{2+}]_i$ elevation in *Chrionomus* salivary gland cells causes channel closure at constant intracellular pH (buffered or even at elevated intracellular pH (*35*).

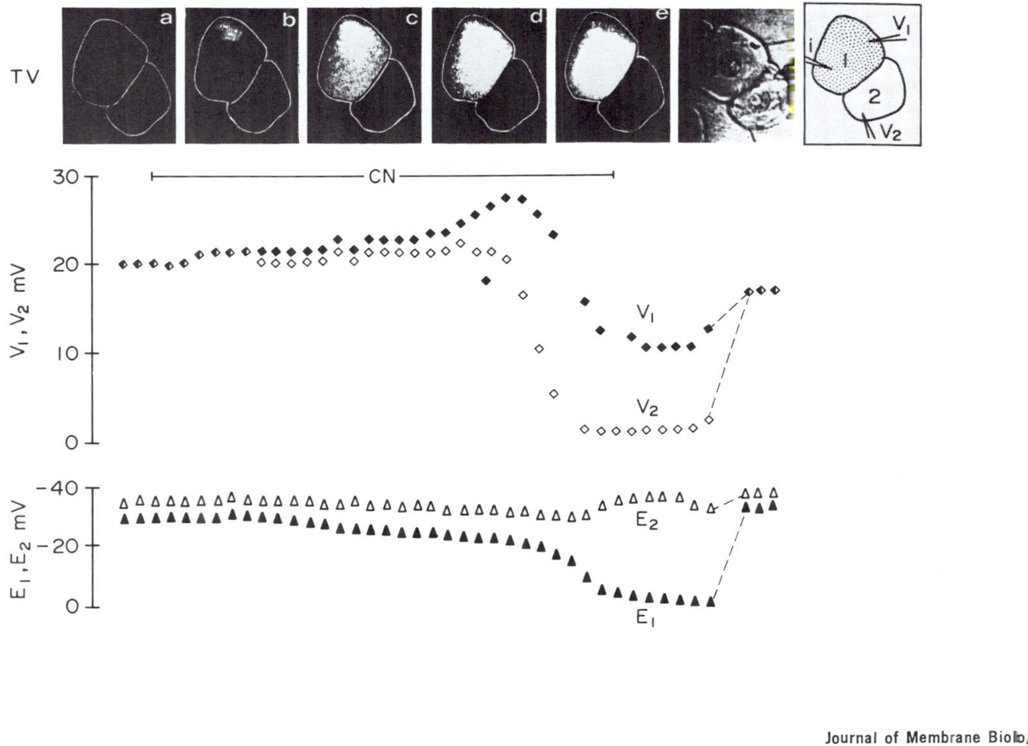

Figure 5. Uncoupling by cyanide. Cells in Ca-free medium are exposed to 5 × 10^{-3} mol/L sodium cyanide (bar). $[Ca^{2+}]_i$ rises at first locally (TV picture b) and then diffusely (TV pictures c, d, and e), associated with uncoupling. Time correspondence of TV pictures a, b, c, d, and e is marked on the P-curve. I = 4 × 10^{-8} *amp* (19).

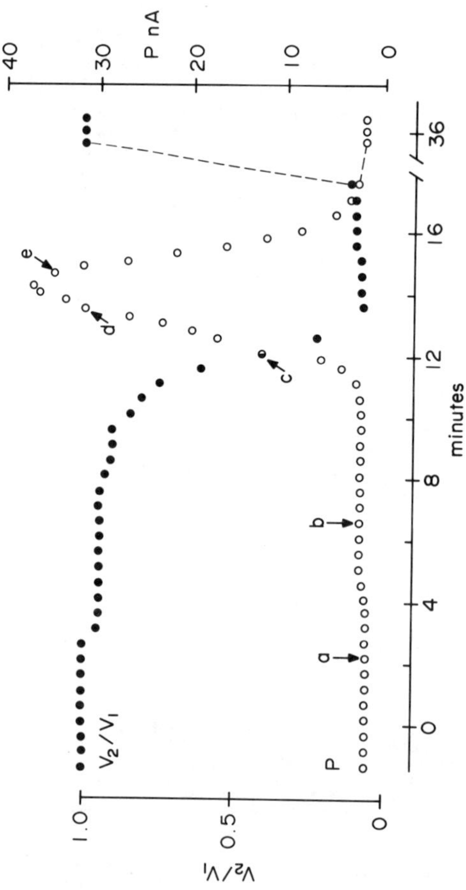

Figure 5. Continued

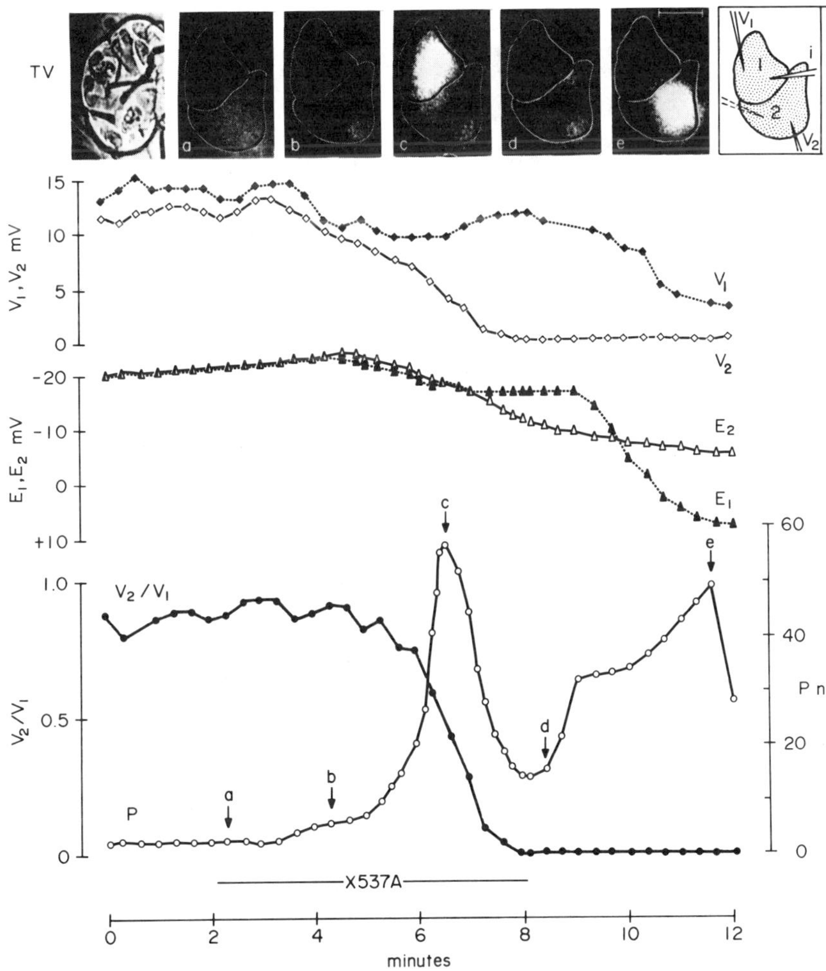

Journal of Membrane Biology

Figure 6. Uncoupling by ionophore X537A. Cells in Ca-containing medium are exposed to 1×10^{-5} *mol/L ionophore X537A* (bar). *Cells 1 and 2 are injected with aequorin. Appearance of luminescence in Cell 1 (signalling* $[Ca^{2+}]_i$ *elevation) correlates with uncoupling. A general* $[Ca^{2+}]_i$ *elevation in Cell 2 begins several minutes later than in Cell 1; a local elevation around electrode* V_2 *is seen early on. The decline in luminescence in each cell is the result of aequorin depletion.* $I = 4 \times 10^{-8}$ *amp.*

In principle, a decrease in junctional permeability—as measured electrically—may be the reflection of a reduction in the number of open channels or in the aperture of the individual channels. The latter alternative opens the exciting possibility that the permeability change may be selective, i.e., the permeability may fall for the larger molecular species, but not for the smaller, inorganic molecule carrying the electrical current.

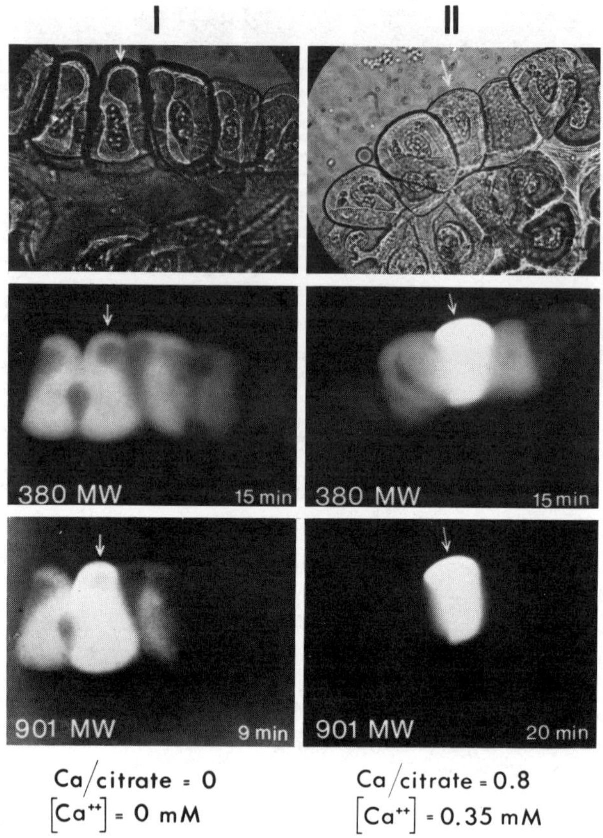

Figure 7. Selective block of junctional transit.

Permeability of junctional membrane channels is probed with pairs of fluorescent-labelled tracers. The fluorescence of one tracer in each pair is yellow–green and the other is red. The pairs are microinjected into a cell (arrow) together with Ca-buffers and the tracer fluorescence is photographed. Each series of photographs gives, from top to bottom, *a view of the cells; the intracellular fluorescence of the smaller tracer molecule; that of the larger molecule (set apart by excitation and barrier filters). The molecular weights on the photographs identify the tracer pairs: (I) control—both molecules of the pair*

The first suggestion that this may be so came from work in which the junctional passage of the 330 mol wt fluorescein was blocked after small Ca^{2+} injections, with little or no detectable reduction of electrical coupling (*36*).

The development of the fluorescent junctional permeability probes permitted distinction between these alternatives for the first time. We tested the junction simultaneously with two probes of different molecular size and fluorescent color (Table II) while raising $[Ca^{2+}]_i$. In each test we used a pair of molecular probes and compared their transit times

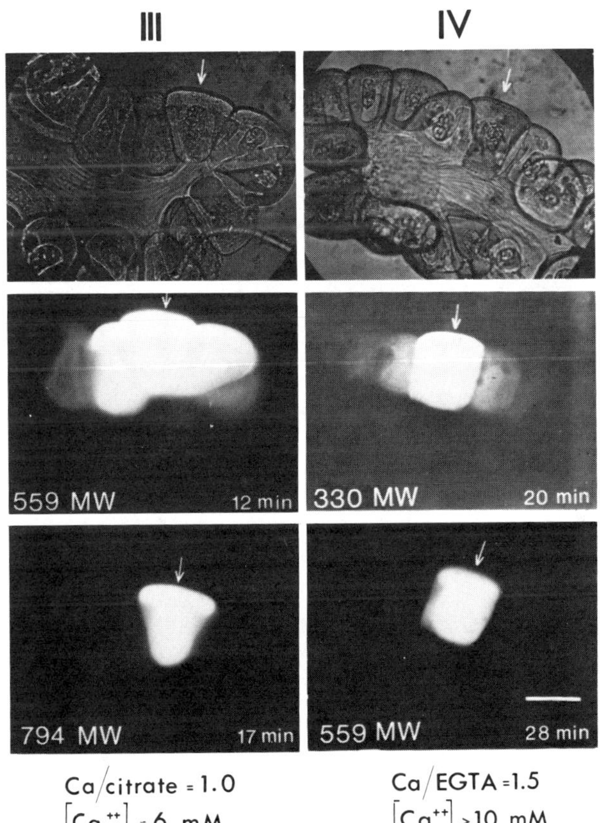

have passed from cell to cell; (II–IV) the molecules are injected together with Ca^{2+}; passage of the larger molecule is blocked. The Ca–citrate ratios were: a, 0; b, 0.8; c, 1.0; citrate$_{total}$, 0.03 mol/L. The photographs of the blocked tracers (II–IV) were taken 5 min after those of their respective smaller companions; the injection-to-photography interval was 35-fold the control transit time in II, III, and IV. In (I) the photograph of the larger tracer was taken before that of the smaller tracer; the injection-to-photography interval for the larger tracer was 15-fold the control transit time. Calibration, 100 μm.

during simultaneous diffusion in the same direction across a junction. If and only if the nonselective mechanism (a decrease in the number of open channels) governs the Ca^{2+}-induced permeability reduction, both probes should be retarded equally irrespective of their size or other structural differences.

The result was that with moderate $[Ca^{2}]^{+}_{i}$ elevations, the transjunctional flow of the larger molecules was slowed more than that of the small ones; in some cases it was even sensibly blocked, while the small one continued to pass (*see* Figure 7). Figure 8 presents the data of all the tests for four tracer pairs. Here, junctional permeability is expressed in terms of junctional transit speed. To normalize for variations in $[Ca^{2+}]_i$ in the junctional area in the various experiments, the transit speed of a probe is given relative to the mean transit speed of the probe in Ca^{2+}-free control injections (determined at the same concentration ratio for a given tracer pair). The relative transit speeds of the smaller molecule (V_S) versus that of the larger one (V_L) in each pair are plotted. Included are

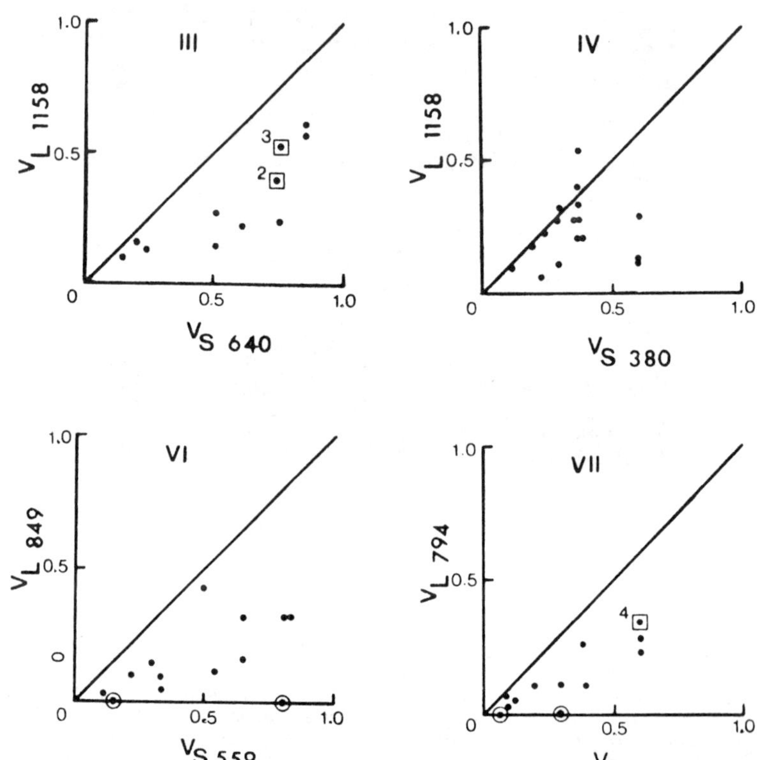

Figure 8. Selective retardation of junctional transit by Ca^{2+}. Plotted are the relative junctional transit speeds (v) modified by Ca^{2+} for the small (S, abscissae) and large members (L, ordinates) of tracer pairs III, IV, VI, and VII. Each value of v is the ratio of the mean junctional transit time in Ca-free controls to a single measurement of transit time in elevated $[Ca^{2+}]_i$. The subscripts give the molecular weights of the tracers. Each graph presents all data for a given tracer pair, including those of selective block (circled; $V_L = 0$). Squares denote repeated data points, with the multiplicity indicated. The line represents $V_S = V_L$, that is, equal (nonselective) retardation of junctional transit. The resultant confidence levels for $V_L < V_S$ are (III) 0.5, (IV) 5, (VI) < 0.5, and (VII) $< 0.1\%$. (The numbers of controls and two other underlying statistics, the ratios of mean transit times (L/S) of the controls ± s.e. were: (III) 5, 1.29 ± 0.27; (IV) 8, 2.67 ± 0.58; (VI) 5, 2.36 ± 0.54; (VII) 11, 1.18 ± 0.16, respectively.)

Table II. Tracer Pairs

Pair No.		*Molecular Wt*		*Molecular Wt*
I	LRB(Leu)$_3$(Glu)$_2$OH	1,158	+ DANS (Leu)$_3$(Glu)$_2$OH	849
II	LRB(Leu)$_3$(Glu)$_2$OH	1,158	+ FITC(Glu)$_3$OH	794
III	LRB(Leu)$_3$(Glu)$_2$OH	1,158	+ DANS(Glu)$_3$OH	640
IV	LRB(Leu)$_3$(Glu)$_2$OH	1,158	+ DANS(Glu)OH	380
V	LRB(Gly)$_6$OH	901	+ DANS(Glu)OH	380
VI	DANS(Leu)$_3$(Glu)$_2$OH	849	+ LRB SO_3H	559
VII	FITC(Glu)$_3$OH	794	+ LRB SO_3H	559

also those cases with selective block to one molecule in a pair, invariably the larger one ($V_L \approx 0$). The line $V_L/V_S = 1$ is drawn in each graph and represents equal retardation of both tracers, namely nonselective reduction of junctional permeability. Nearly all points fall below the line. Application of the two-sample *t*-test to the aggregate data of each graph, for comparing the retarded junctional tracer movement with the control tracer movement shows that for all four tracer pairs the permeability reduction is selective for the larger molecule, implying that the molecular-weight limit for channel permeation decreases in a graded fashion.

Viewed in terms of the idea (*1*) and evidence (*37, 38*) that the permeable junction is made of channels, this graded decrease of molecular-size limit may be attributed to a graded diminution of effective channel size upon $[Ca^{2+}]_i$ elevation. Alternatively, the junction might contain channels of different size and Ca^{2+}-sensitivity. The results then would be accounted for by an all-or-none closure of channels with Ca^{2+} sensitivities directly related to channel size. However, the recent findings by Loewenstein, Kanno, and Socolar (*37*) of quantal increases during de novo formation of cell-to-cell channels support the notion that the channels are of unitary nature at least when nascent.

Acknowledgment

The work in the authors' laboratory was supported by research grants from the National Institute of Health and the National Science Foundation.

Glossary of Symbols

Å = Ångstrøm (10^{-10} m)
I = current
A = Ampere
nA = nanoampere (10^{-9} A)

mV = millivolt (10^{-3} V)
μ = micrometer (10^{-6} m)
nm = nanometer (10^{-9} m)
λ = wavelength
mol wt = molecular weight
P = photomultiplier current
E_1, E_2 = membrane potential of Cells 1 and 2, respectively
V_1, V_2 = steady-state change in membrane potential (of Cell 1 and 2, respectively) in response to current test pulse passed between interior of Cell 1 and grounded bath
pH_i = intracellular (cytoplasmic) pH
$[Ca^{2+}]_i$ = intracellular (cytoplasmic) concentration of ionic free Ca
X537A
A23187 = ionophores' identification number assigned by manufacturer
FITC = fluorescein isothiocyanate
LRB = lissamine rhodamine B
DANS = dansyl = 5-dimethylamino-1-naphthalene sulfonyl
leuOH = leucine
gluOH = glutamic acid
tryOH = tryptophane
lysOH = lysine
proOH = proline
v = ratio of mean junctional transit time (control injections) to a single measurement of transit time (test)

Literature Cited

1. Loewenstein, W. R. "Permeability of Membrane Junctions," *Ann. N.Y. Acad. Sci.* **1976,** *137,* 441.
2. Loewenstein, W. R. "Permeable Junctions," *Cold Spring Harbor Symp. Quant. Biol.* **1975,** *40,* 49.
3. Loewenstein, W. R.; Kanno, Y. "Studies on an Epithelial (Gland) Cell Junction. I. Modifications of Surface Membrane Permeability," *J. Cell Biol.* **1964,** *22,* 587.
4. Kanno, Y.; Loewenstein, W. R. "Low-Resistance Coupling Between Gland Cells. Some Observations on Intercellular Contact Membranes and Intercellular Space," *Nature* **1964,** *201,* 194.
5. Potter, D. D.; Furshpan, E. J.; Lennox, E. S. "Connections Between Cells of the Developing Squid as Revealed by Electrophysiological Methods," *Proc. Nat. Acad. Sci.* **1966,** *55,* 328.
6. Payton, B. W.; Bennett, M. V. L.; Pappas, G. D. "Permeability and Structure of Junctional Membrane at the Electrotonic Synapse," *Science* **1969,** *166,* 1641.
7. Rose, B. "Intercellular Communication and Some Structural Aspects of Membrane Junctions in a Simple Cell System," *J. Membrane Biol.* **1971,** 5, 1.
8. Johnson, R.; Sheridan, J. D. "Junctions Between Cancer Cells in Culture: Ultrastructure and Permeability," *Science* **1971,** *174,* 717.

9. Simpson, I.; Rose, B.; Loewenstein, W. R. "Size Limit of Molecules Permeating the Junctional Membrane Channels," *Science* **1977**, *195*, 294.
10. Brink, P. R.; Dewey, M. M. "Nexal Membrane Permeability to Anions," *J. Gen. Physiol.* **1978**, *72*, 67.
11. Loewenstein, W. R. "Cell Surface Membranes in Close Contact. Role of Calcium and Magnesium Ions," *J. Colloid Interface Sci.* **1967**, *15*, 34.
12. Loewenstein, W. R.; Nakas, M.; Socolar, S. J. "Junctional Membrane Uncoupling. Permeability Transformations at a Cell Membrane Junction," *J. Gen. Physiol.* **1967**, *50*, 1865.
13. Loewenstein, W. R.; Penn, R. D. "Intercellular Communication and Tissue Growth. II. Tissue Regeneration," *J. Cell. Biol.* **1967**, *33*, 235.
14. Politoff, A. L.; Socolar, S. J.; Loewenstein, W. R. "Permeability of a Cell Membrane Junction. Dependence on Energy Metabolism," *J. Gen. Physiol.* **1969**, *53*, 498.
15. Rose, B.; Loewenstein W. R. "Junctional Membrane Permeability. Depression by Substitution of Li for Extracellular Na, and by Long-Term Lack of Ca and Mg; Restoration by Cell Repolarization," *J. Membrane Biol.* **1971**, *5*, 20.
16. Oliveira-Castro, G. M.; Loewenstein, W. R. "Junctional Membrane Permeability: Effects of Divalent Cations," *J. Membrane Biol.* **1971**, *5*, 51.
17. Délèze, J. "The Recovery of Resting Potential and Input Resistance in Sheep Heart Injured by Knife or Laser," *J. Physiol. (London)* **1970**, *208*, 547.
18. DeMello, W. C. "Effects of Intracellular Injection of Calcium and Strontium on Cell Communication in Heart," *J. Physiol. (London)* **1975**, *250*, 231.
19. Rose, B.; Loewenstein, W. R. "Permeability of Cell Junction Depends on Local Cytoplasmic Calcium Activity," *Nature* **1975**, *254*, 250.
20. Shimomura, O.; Johnson, F. H.; Saiga, Y. "Extraction, Purification and Properties of Aequorin, a Bioluminescent Protein," *J. Cell. Comp. Physiol.* **1962**, *59*, 223.
21. Rose, B.; Simpson, I.; Loewenstein, W. R. "Calcium Ion Produces Graded Changes in Permeability of Membrane Channels in Cell Junction," *Nature*, **1977**, *267*, 625.
22. Simpson, I. "Labeling of Small Molecules with Fluorescein," *Anal. Biochem.* **1978**, *89*, 304.
23. Flagg-Newton, J.; Simpson, I.; Loewenstein, W. R. "Permeability of the Cell-to-Cell Membrane Channels in Mammalian Cell Junction," *Science* **1979**, *205*, 404.
24. Hodgkin, A. L.; Keynes, R. D. "Movement of Labelled Calcium in Squid Giant Axon," *J. Physiol. (London)* **1957**, *138*. 253.
25. Podolsky, R. J.; Costantin, L. L. "Regulation by Calcium of the Contraction and Relaxation of Muscle Fibers," *Fed. Proc., Fed. Am. Soc. Exp. Biol.* **1964**, *23*, 933.
26. Rose, B.; Loewenstein, W. R. "Calcium Ion Distribution in Cytoplasm Visualized by Aequorin: Diffusion in Cytosol Restricted Eenrgized Sequestering," *Science* **1975**, *190*, 1204.
27. Meech, R. W.; Thomas, R. C. "The Effect of Calcium Injection on the Intracellular Sodium and pH of Snail Neurones," *J. Physiol. (London)* **1977**, *265*, 867.
28. DeWeer, P. "Aspects of the Recovery Processes in Nerve," In "Int. Rev. Science, Physiol. Series One, Vol. 3, Neurophysiol."; C. C. Hunt, Ed.; Butterworths: London, 1975; p. 231.
29. Bartley, W.; Amoore, J. E. "The Effects of Manganese on the Solute Content of Rat Liver Mitochondria," *Biochem. J.* **1958**, *69*, 348.
30. Chappel, J. B.; Greville, G. D.; Bioknell, K. E. "Stimulation of Respiration of Isolated Mitochondria by Manganese Ions," *Biochem. J.* **1962**, *84*, 61p.

31. Saris, N. E. "Studies on Active Calcium Transport in Mitochondria," *Acta Chem. Scand.* **1963,** *17,* 882.
32. Carvalho, A.; Sanui, H.; Pace, N. "Calcium and Magnesium Binding Properties of Membranes," *J. Cell. Comp. Physiol.* **1963,** *62,* 311.
33. Turin, L.; Warner, A. "Carbon Dioxide Reversibly Abolishes Ionic Communication Between Cells of Early Amphibian Embryo," *Nature* **1977,** *270,* 56.
34. Lea, T. Y.; Ashley, C. C. "Increase in Free Ca^{2+} in Muscle After Exposure to Co_2," *Nature* **1978,** *275,* 236.
35. Rose, B.; Rick, R. "Intracellular pH, Intracellular Ca^{2+}, and Junctional Cell–Cell Coupling," *J. Membrane Biol.* **1979,** *44,* 377.
36. Délèze, J.; Loewenstein, W. R. "Permeability of a Cell Junction During Intracellular Injection of Divalent Cations," *J. Membrane Biol.* **1976,** *28,* 71.
37. Loewenstein, W. R.; Kanno, Y.; Socolar, S. J. "Quantum Jumps of Conductance During Formation of Membrane Channels at Cell–Cell Junction," *Nature* **1978,** *274,* 133.
38. Socolar, S. J.; Kanno, Y.; Loewenstein, W. R. "Direct Cell-to-Cell Channels: The Basis of Junctional Coupling as Shown by Electrical and Tracer Studies," In "Advances in Chemistry. Bioelectrochemistry: Ions, Surface Membranes"; M. Blank, Ed.; **1980,** *188,* p. 000.

RECEIVED November 27, 1978.

25

The Influence of the Composition of the Bathing Solutions on Frog Skin Membrane Potential

MAX BENDER, K. HILLMAN, L. TIRRI, M. GRABEN, D. FENSTER, and M. ROSENSAFT

Chemistry Department, Fairleigh Dickinson University, Teaneck, NJ 07666

Potential is extremely dependent on bathing solutions, effects being different one side vs. the other. Ionic concentration gradient applications in either direction appeared inconsequential. Polyvalent ions (outside only) caused potential surging. Effects including a marked resistivity decrease with increased voltage indicate electrokinetic phenomena. Utilizing potential profile data, a mechanism includes the skin as a series of electrogenic Gouy–Chapman electrokinetic barrier layers. Barriers control separation of charge, permeability, and sign and magnitude of the potential. Barriers near skin surfaces are influenced strongly by bathing solution composition which also has catalytic effects on the rate of pumping. Pumping current (not reversible to applied emf) is in a different path than shunt current (reversible). Passive flow (electroneutral via thermodynamic gradient) is not observed. Indirect effects of the solutions are explained through leaky capacitance.

Considering that the measured electrical potential of frog skin is so highly dependent on the composition of the solutions with which it is in contact and that this dependency is so different on the inside of the skin compared with the outside (*1*), a detailed study (*2, 3, 4, 5, 6*) is presented showing specific effects associated with the presence and concentration of particular solution constituents. Many polyvalent cations were examined in addition to Na, K, and Ca. Thermodynamic gradients

0-8412-0473-X/80/33-188-409$08.75/1

were applied and consideration was given to ion transport and electrical resistivity. Observations are interpreted with the literature for insight into the mechanism of the electrical activity. Literature searches by Sheldon (*7*) and Neuschwanter (*8*) were helpful in this connection.

Experimental

Distilled deionized water was used throughout the work. Chemicals were Analytical Reagent grade. The frogs were both sexes of Rana pipiens. They were obtained from the Alburg area of Vermont. Upon receipt of shipment and afterward, any frog was discarded which at any time appeared to have "red leg," a viral infection attacking the small blood vessels of the skin. The frogs were kept at 5°C in a refrigerator supplied with air vents. They were contained in a plastic tub in 1 in. of solution the composition of which was the standard reference Ringer Solution I (Horowitz–Lockwood (*9*)) used throughout this work diluted 1:5. The pH of this solution was 7.4–7.6.

Solution	*mM dm*$^{-3}$
NaCl	111.30
KCl	3.36
$CaCl_2 \cdot 2H_2O$	2.04
$NaHCO_3$	1.19
Dextrose	5.55

A frog was allowed to warm up in water at room temperature for 1 hr prior to having its skin mounted for measurements. The animal was pithed and the abdominal skin removed. Only one membrane was taken from any frog, this being a circular patch from the essentially unpigmented center area.

The cell in which the skin was mounted was made of lucite. It consisted of a pair of mirror-image-matched, cylindrically shaped chambers each having a 1-cm diameter circular opening at one end. The inside diameter of the chamber cylinder was 2.5×10^{-2} m and its length was 2.85×10^{-2} m. Flange thickness, i.e. the thickness of the 1-cm opening for each compartment was 0.3 cm. The capacity of each chamber was 11.5 mL. In assembling the cell, the chambers were aligned opening to opening with the membrane in between covering these openings. Bolts and wing nuts passing through collars around each chamber served to hold the cell together. Clamping was sufficiently tight to render the chamber–membrane area leak-proof while not crushing the frog skin. The collars also served as a rest for the cell keeping it from tipping during measurements. Each chamber had three openings at the top, one for oxygen supply, one for electrical accommodation, and the third for adding and removing the solutions with which the membrane was treated.

Upon mounting, Ringer Solution I was added immediately to both compartments and surgical-grade oxygen was bubbled through each using hypodermic syringes. A reduction valve regulated the flow so that with a pressure slightly above atmospheric, there was sufficient agitation to stir the solutions in the compartments in addition to furnishing an oxygen supply. The oxygen was bubbled first through water to insure no evaporation of the solutions in the chambers. Oxygen was used throughout each experiment unless otherwise specified.

Measurements were made at the ambient temperature of the laboratory which ranged between 296 and 302 K. Series of data obtained within this range indicated that any variation owing to temperature could be neglected compared with the variations experienced from one frog to another at the same temperature or the same skin at different times at the same temperature. (For studies on temperature effects *see* Snell and Leeman (*10*)). Controls were used in each experiment. All experiments began under standard Solution I conditions and then the solutions were changed, the usual run taking place at essentially constant temperature.

The electrical potential (E) across the membrane was measured with a high resistance voltmeter, i.e. the Keithley 610 A or the 602 solid-state electrometer. The electrometer was connected across a pair of saturated KCl calomel reference electrodes that were in contact with the bathing solutions by means of salt bridges. The salt bridges consisted of surgical tubing filled with a 3% KCl–agar solution. Junction potentials for the salt bridges were less than 0.5 mV in all cases. Likewise, the calomel electrodes differed by no more than 0.5 mV when compared with Ag–AgCl reference electrodes. Potentials in this work are cited on the basis of the compartment solution facing the inner side of the skin being positive. A minus sign in front of any given reported value indicates that there was a reversal in this charge under the experimental conditions.

The fresh membrane potential in Solution I (both sides) varied greatly from one frog to another. Thus, in all of the work a given series of measurements were repeated, each series being based on a sufficient number of skins for verification of the results.

Na^{22} was used in connection with the study on Na transport under a sodium thermodynamic gradient. It was administered from a 5-mL teaching sample obtained from Nuclear—Chicago. The whole sample contained 10 μg of Na^{22}, the carrier solution being 0.1N in HCl. Radiation counts were made with a Geiger–Mueller tube which was also from Nuclear—Chicago. The Na^{22} was utilized in the given experiments where the same solution was applied to both sides of the skin, where an Na gradient was applied from the membrane inside to the outside, and where the same Na gradient was reversed so that it acted from outside to inside, respectively. For each of these three types of experiments the Na^{22} was used in the membrane inside solution, the radioactive pick-up measurements being made on the opposite side. Then another set of three experiments was made the same as before except that the Na^{22} was placed in the outside solutions, the inside solutions being examined for radioactive pick-up.

First the skin was mounted with dextrose-free Solution I on both sides and allowed to stand for 30 min. Then these solutions were replaced with the test solution according to the experiment and 0.10 mL of the Na^{22} solution were added to one side or the other according to the experiment.

Every 30 min 0.05 mL was taken from the chamber opposite that of the Na^{22} inclusion and E was measured. Addition of the Na^{22} and removal of samples were made with disposable hypodermic syringes. Two sheets of microscope lens paper cut into a 1.8×10^{-2}-m circle were used to absorb the 0.05 mL samples in the center of a planchette and the radiation count was taken. For reference, background radiation evaluations were made before each experiment, taking an average of three 1-min counts. These amounted to 13–22 cpm.

In the electrical resistivity measurements the cell used for the frog skin potentials was fitted with platinum electrodes each approximately 6 cm^2 in area and 1.5 cm away on opposite sides from the 1 cm^2 opening that holds the membrane. Ringer's bathing solution was the liquid medium and measurements were made with and without frog skin membrane in place.

Three different measurement techniques were considered. One utilized the electrometer, applying 1 mV direct current and reading the resistance directly. In the second, frog skin potential was nulled with a calibrated Leeds and Northrup student potentiometer while 5–10 mV were impressed on the membrane and the current flow was measured with the electrometer. The resistance was calculated using Ohm's Law. In the third technique the Industrial Instruments (Beckman) conductivity bridge RC-16-B-2 was used with 1000 cycles of alternate current (ac) at 2 V being applied.

Results

The Same Solution on Both Sides of the Membrane. Table I shows the effect on E as the concentration of Na is varied in Solution I, and the content of the other solution components is unchanged. Before applying the test solution the membrane was exposed to Solution I for 1 hr, the potential remaining essentially constant in this connection. After the given period in the test solution the membrane was re-exposed to Solution I for 1 hr, the potential remaining constant except where otherwise indicated.

In Table II, M_{KCl} is varied. The concentrations 0–33.6mM were introduced after 60 min in Solution I, 50mM after 40 min, and 275 and 320mM concentrations after 30 min. The period in Solution I after the test was 30 min.

Table III pertains to Ca being varied in Solution I. Concentrations of Ca ranging from 0 through 20.4mM were placed in the compartment after 60 min in Solution I; 55.0mM $CaCl_2$ was introduced after 50 min in Solution I; and all other concentrations were introduced after 30 min in Solution I. The period after testing was 30 min.

Different Solutions on Opposite Sides of the Membrane. An NaCl thermodynamic potential was applied using Solution I on one side and an alteration of Solution I where the NaCl was replaced by an osmotic equivalent of sucrose (222.60mM). E measurements were made with the gradient passing from the inside to outside of the frog skin membrane and vice versa. Results are plotted in Figure 1. Each frog skin first was equilibrated with Solution I on both sides for 60 min before replacement with the test solutions. The difference between inside to outside vs. outside to inside is marked.

There was little difference in this effect whether Ca was present or not (in Solution I) with and without the inclusion of 0.15% EDTA.

Table I. The Same Solution on Both Sides of Membrane. Effect of NaCl Concentration Change in Solution I

		E, mV				Total Time in Test Solution (min)
$M_{NaCl}/10^{-3}$ (mol dm^{-3})	I	Solution I Pretest	in Test Solution 45 min	in Test Solution 90 min	Solution I After-Test	
0	0.011	60	25	7	50[a]	120
11.0	0.022	44	42	42	48	90
55.6	0.066	68	60	60	68	110
111.3[b]	0.122	95	—	—	95[c]	—
222.3	0.233	90	40	40[d]	87	80
556.5	0.567	44	0	−0.25[e]	[f]	135
1113.0	1.124	87	0	−0.30[e]	[g]	135

[a] Not very stable.
[b] Control.
[c] Steady plateau over 4-hr period.
[d] In test solution for 80 min.
[e] In test solution for 110 min.
[f] 25% Recovery.
[g] No recovery; *E* became −0.5 mV, then −0.3 mV after 100 min in Solution I (final).

Table II. The Same Solution on Both Sides of Membrane. Effect of KCl Concentration Change in Solution I

		E, mV			Time in Test Solution (min)
$M_{KCl}/10^{-3}$ (mol dm^{-3})	I	Solution I Pretest	In Test Solution 60 min	Solution I After-Test	
0	0.119	83	108	88	65
0.34	0.119	65	78	58	65
1.68	0.120	81	93	85	75
3.36[a]	0.122	95	—	95[b]	—
6.72	0.125	40	20	41	90
16.8	0.135	60	45	58	80
33.6	0.152	95	69	80	85
50.0	0.168	28	11	12	185
275.0	0.393	28	5	—	—
320.0	0.438	41	3	—	—

[a] Control.
[b] Steady plateau over 4-hr period.

Table III. The Same Solution on Both Sides of Membrane. Effect of $CaCl_2$ Concentration Change in Solution I

		E, *mV*			
$M_{CaCl_2}/10^{-3}$ *(mol dm^{-3})*	I	*Solution I Pretest*	*In Test Solution 60 min*	*Solution I After-Test*	*Time in Test Solution (min)*
0	0.116	50	43	—	150
0.204	0.116	110	95	112	90
1.02	0.119	67	66	69	105
2.04[a]	0.122	95	—	95[b]	—
4.08	0.128	76	52	64[c]	85
10.2	0.146	70	4	13	65
20.4	0.177	116	0	70[c]	55
55.0	0.281	46	1	—	—
65.0	0.311	33	21	—	—
75.0	0.341	29	3	—	—
85.0	0.371	43	17	—	—
90.0	0.386	58	12	—	—
200.0	0.716	50	−11	15	90

[a] Control.
[b] Steady plateau over 4-hr period.
[c] 90 min.

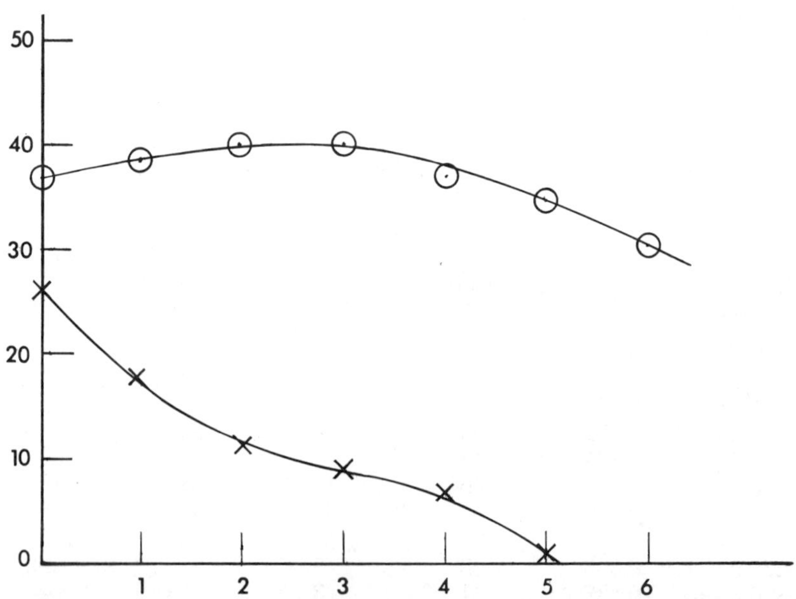

Figure 1. Frog skin potential under NaCl gradient: (⊙), high Na at inside; (×), high Na at outside. Abscissa = time (hours); ordinate = potential E *(millivolts).*

Also, replacing the sucrose with an equivalent amount of choline chloride, which raises the ionic strength (I) to that for Solution I, made no difference. The removal of dextrose also brought about no change.

Table IV represents a series of experiments showing the effect on E of increasing the concentration in Na, K, and Ca by 0.1M in Solution I, the opposing solutions being Solution I and Solution I with 0.1M choline chloride. Increased Na, K, and Ca solutions were applied to the inside of the membrane in one set of experiments and to the outside in another set. All membranes were treated first with Solution I for 1 hr before exposure to the experimental solutions. The values given are the percent decrease in E as originally measured in Solution I. In all cases membrane potential is decreased.

Na Transport—Na Thermodynamic Gradient. Na tracer measurements were made looking for Na transfer with Solution I (but without the dextrose) on both sides of the membrane. Also, measurements were made under the application of the Na thermodynamic potential according to the experiments delineated in Figure 1 (except that no dextrose was present in these solutions). These Na gradient experiments were repeated after adding 0.13% EDTA. Solution counts for all runs were essentially in the same range indicating that there was no measurable transport of Na from one compartment to the other. The total time for each run was 3.5 hr. E readings came out somewhat reduced evidently owing to the presence of the acidic Na^{22} addition which brought the pH to about 5.7. The reduction in potential was greater on the outside of the membrane than on the inside. But the trends of E vs. time matched very well with the previous experimental observations where the Na^{22} had not been used.

The Influence of pH on *E*. With the application of Solution I (pH = 7.4) to both sides of the frog skin, acidity developed with time. This effect was more pronounced in the outer solution. After 24 hr the pH of the solution at the inside membrane dropped to 7.2 and that of the solution on the outside of the membrane dropped to 6.0.

While studies of the effect of pH on E are available in the literature, particular information was required both for the experiments on polyvalent cations (vide post) and for the comparison of inside solution effects with outside solution effects. Experimental results are summarized in Table V. These are consistent with the results of Amberson and Klein (*11*) and with the acidic Na^{22} observations of this work.

Experimentally, the composition of the test solutions was the same as for Solution I except that the appropriate buffer mixtures giving the desired pH were substituted for the $NaHCO_3$, and the buffer concentration in the Ringer solution was now 2mM/L. For pHs of 3.00, 3.50, 4.00, and 4.50, mixtures of sodium hydrogen phthalate and HCl were used. For pHs of 5.00, 5.50, 6.00, 6.50, and 7.00, mixtures of sodium dihydrogen phosphate and NaOH were used (*12*).

Table IV. Different Solutions on Opposite Sides to the Increased Ionic Strength of

Time in Test Solution (hr)	*Increased Na I = 0.222*			
	Inside vs.		*Outside vs.*	
	Solution I	*Choline Cl*	*Solution I*	*Choline Cl*
1	50	75	84	55
5	56[a]	76[a]	94.4	77

[a] 4.5 hr.

Skins were brought first to steady state in the presence of Solution I as shown by E not varying more than 1% from the average of three successive measurements 5 min apart. This usually required 30 to 60 min. Then the compartment solutions were removed, the cell was rinsed twice with Solution I and filled with test solution on one side and with fresh Solution I on the other side. E measurements were taken every 5 min for 90 min. After this the solutions were removed, the cell

Table V. Effect of pH on

Compartment	*pH*	*Steady State*		*0-min Test*	
		Test	*Control*	*Test*	*Control*
outside	3.0	37.4	29.5	23.3	29.5
inside	3.0	28.8	22.6	19.7	22.8
outside	3.5	32.1	24.6	35.1	24.8
inside	3.5	29.3	35.8	28.6	35.4
outside	4.0	51.7	28.0	52.7	27.8
inside	4.0	35.6	32.1	35.0	31.8
outside	4.5	55.5	37.8	54.2	38.0
inside	4.5	28.2	31.2	28.9	31.3
outside	5.0	45.3	29.7	43.7	29.9
inside	5.0	27.7	31.6	27.5	31.4
outside	5.5	54.9	28.7	56.1	28.9
inside	5.5	29.1	23.6	29.3	23.6
outside	6.0	43.0	35.5	42.8	35.1
inside	6.0	29.8	35.0	30.1	35.0
outside	6.5	46.1	31.8	46.5	30.0
inside	6.5	44.9	31.4	44.2	31.5
outside	7.0	29.7	22.5	30.0	22.6
inside	7.0	34.9	26.4	35.0	26.4

of the Membrane. The Percent Decrease in *E* Owing Na, K, Ca, and Choline Chloride

Increased K I = 0.222				*Increased Ca* I = 0.242				
Inside vs.		*Outside vs.*		*Inside vs.*		*Outside vs.*		*Standard Solution I*
Solution I	*Choline Cl*	*Solution I*	*Choline Cl*	*Solution I*	*Choline Cl*	*Solution I*	*Choline Cl*	*on Both sides*
75	95	86	56	47	41	71	55	19
76[a]	95[a]	92.2	59	50[a]	79[a]	89.5	73	55[b]

[b] 48% for 4 hr.

was double rinsed in and filled with fresh Solution I, and *E* readings were taken every 5 min for 30 min.

Sign of Charge Reversal of the Membrane Potential. In Table I high Na^{+} concentrations reverse the sign of the potential although values for *E* were very low. High Ca^{++} concentrations (*see* Table III) likewise brought about reversal with the effect being much stronger than that of Na.

Membrane Potential (mV)

90-min Test		*0-min Solution I*		*30-min Solution I*	
Test	*Control*	*Test*	*Control*	*Test*	*Control*
6.3	27.8	9.9	28.1	27.2	28.2
12.4	21.8	13.1	22.4	13.9	24.4
22.5	25.6	24.0	26.0	37.3	26.2
28.3	33.4	28.9	33.8	30.6	33.5
39.9	27.5	39.3	29.5	38.7	25.3
32.5	25.4	32.5	25.2	33.9	26.5
49.1	35.8	51.3	36.6	50.4	36.6
28.0	30.1	28.4	30.2	25.7	30.6
44.1	29.0	44.9	29.1	37.9	28.6
27.6	30.5	27.2	31.0	28.8	29.2
56.0	27.9	57.3	28.1	45.5	27.8
28.9	22.8	29.8	22.6	30.9	23.0
42.2	35.8	42.3	36.0	37.4	37.1
30.0	31.2	31.0	33.0	31.8	28.5
43.9	26.4	45.1	26.0	43.3	25.0
44.2	32.1	44.5	34.0	42.6	35.1
27.4	20.0	27.7	20.4	26.8	19.4
32.2	22.7	32.3	23.0	31.5	21.6

Table VI. Effect of Distilled Water on Frog Skin Membrane Potential (mV)

Min.	*Exp. 1* Steady State in Solution I 45 min; Then Distilled H_2O on Both Sides	*Exp. 2* Direct Application of Distilled H_2O on Both Sides	*Exp. 3*[a] Solution II[b] on Inside Distilled H_2O on Outside	*Exp. 4*[a] Solution II[b] on Outside, Distilled H_2O on Inside	*Exp. 5* Solution II[b] on Both Sides
0	+35.0	−13.5	−6.5	−15.5	+44.6
15	—	−26.0	−8.5	−16.2	+40.7
30	+31.5	—	−6.5	−16.8	+39.7
45	—	−35.0	−4.0	−18.0	+39.5
60	−10.0	−37.0	—	−17.4	+40.0
90	−13.5	—	—	—	+38.7
105	−16.5	−35.5	—	—	+37.5
120	−19.0	—	—	—	+36.3
145	−20.0	—	—	—	—
150	—	−32.5	—	—	+34.5
200	—	−32.0	—	—	+32.1
210	—	—	—	—	+31.8
225	—	—	—	—	+31.3
240	—	−30.5	—	—	—
255	—	—	—	—	+30.9
285	—	−24.5	—	—	+30.7
315	—	—	—	—	+30.5
330	—	−23.5	—	—	+31.2
360	—	−20.0	−14.5	−22.0	+31.8
390	—	−19.5	—	—	+32.8
420	—	−19.0	—	—	+34.0
450	—	−19.0	—	—	+35.2
480	—	−16.5	—	—	+36.1
525	—	−15.5	—	—	+37.6
575	—	−13.5	—	—	—
590	—	−12.0	—	—	—
600	—	—	—	—	+39.3

[a] Direct application.
[b] Solution II: 56.74mM NaCl, 0.33 mM KCl, 1.02mM $CaCl_2$, 1.00mM $NaHCO_3$; $I = 0.061$.

In the other concentration extreme, reversal takes place when I is low. This is evident from the results with distilled water in the bathing compartments as shown in Table VI.

The membrane is more sensitive to reversal in distilled water if it is not bathed previously in Ringer's solution. Also it is more sensitive to reversal with distilled water in the inside compartment.

Polyvalent Ions. Table VII show the effect of Al^{+3} as a function of concentration with $AlCl_3$ being added to Solution I. First the membrane was exposed to Solution I on both sides from 35 to 65 min, then to the test solution on both sides. There is a drop in potential and this depression of *E* increases for the most part with increased concentration. In contradistinction to the Al (and Ca), the introduction of $LaCl_3$ perks up the potential markedly. Then it descends (*see* Figure 2). Experimentation methodology was the same as for Al.

La and Al are compared with 19 other polyvalent cations in Tables VIII and IX. In this study the concentration of the cation in each case is 1m*M*. This is sufficiently small so that none of the bathing solutions had a pH below 5, which is where membrane potential begins to suffer (*see* Table V). The vehicle is Solution I except that it is buffered with 2.0m*M* Trizma to avoid precipitation of given cations with the bicarbonate. In one series the cation was applied to the inside of the membrane only and in the other to the outside. Solution I was used across the membrane from the test. As a control to a particular run, a test solution with the particular cation omitted but with sufficient HCl to adjust the pH to that of the test solution was used on the same side of the membrane as the test solution (Solution I was on the opposite side). The measurement technique and procedure was the same as for the pH experiments outlined in Table V.

Resistivity Measurements. Average values obtained in the case of each method are outlined in Table X. The frog skin thickness used for computation was 0.2 mm, which was an average for the various skins measured. Thus the ratio of solution length to membrane thickness amounted to 150:1 and the ratio of the resistivities of solution to membrane could be estimated using the difference of the cell resistance with and without membrane.

Table VII. Effect of Including Aluminum in Solution I

		E, mV				
$M_{AlCl_3}/10^{-3}$ *(mol dm^{-3})*	I	*Solution I Beginning*	*Solution I Just Before Test*	*Beginning of Test Solution*	*60 min of Test Solution*	*150 min of Test Solution*
0.5	0.125	48.0	32.0	25.5	18.0	21.0[a]
1.0	0.128	42.0	27.0	18.0	10.0	7.0
9.0	0.176	23.5	14.0	4.0	1.5	2.0
13.0	0.200	42.0	27.5	18.5	7.5	3.0
19.0	0.236	59.0	38.5	19.0	7.5	4.0[b]

[a] 29.0 mV after 310 min in test solution.
[b] 1.0 mV after 260 min in test solution.

Note that in terms of the dimensions of the apparatus, the resistance of 187 Ω measured for the Ringer's solution ($I = 0.122$) with the ac bridge amounts to a conductivity of 0.0161 Ω^{-1} cm^{-1} which is comparable with that for 0.1*M* KCl at 25°C, i.e. 0.012856 Ω^{-1} cm^{-1}. This is a check on the methodology. Frog skin resistivity decreases with applied voltage (*see* Figure 3). The interpolated resistivity at 50 mV allows an Ohm's Law estimation of a short-circuit current of 50 μA which compares with values reported in the literature (*13, 14*). The corresponding skin resistance (1000 Ω) is in range with values reported by Ussing and Windhager (*15*) and Ussing (*16*).

Discussion

The various effects observed and studied in terms of the literature are grouped here for convenience according to particular subject areas. These areas are interrelated. Given literature references apply to more than one of these areas.

Specific Ionic (and Compound) Effects: Ionic Strength—Inside vs. Outside. Increasing *I* in reference to Solution I through the addition of more Na, K, or Ca depresses the potential (*see* Tables I–IV). There was greater sensitivity to this effect on the outside than on the inside of the membrane.

When K is removed from Solution I (on both sides of the membrane) the potential is increased (*see* Table II). This is in contradistinction to Koefoed–Johnsen and Ussing (*17*) who in developing their concepts, quote Fukuda (*18*) in that the potential across the skin is reduced greatly or abolished if there is no K in the inside bathing solution, i.e. that there is no K to exchange for Na by the nonelectrogenic pump. Further, in terms of the function of K, Klahr and Bricker's observations (*19*) are not in agreement with this original model since they found that they could maintain a potential on exposure of frog skins to K concentrations that exceeded estimated intracellular values. Also not in agreement with Koefoed–Johnsen and Ussing are Steinbach's observations (*20*) of a loss of K from the frog skin to the outside solution.

With the present results the potential is reduced drastically when there is not enough Na in the inside solution, but reduction of Na in the outside solution makes little difference as was the situation with K. Meanwhile the Koefoed–Johnsen and Ussing model requires sufficient Na in the outside solution (*17, 21, 22*), while the Na content of the inside solution is not supposed to matter. Cereijido and Curran (*14, 23*) find that *E* remains unaffected despite a fortyfold change in the outside bathing solution Na concentration and a fivefold change in the short-circuit current.

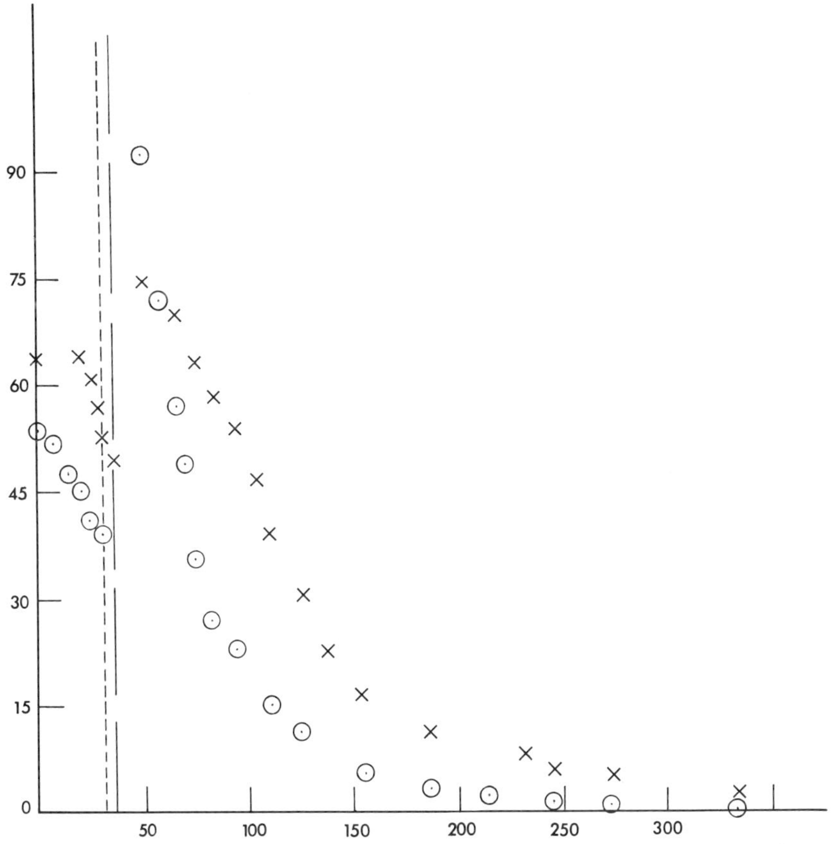

Figure 2. Influence of La on frog skin membrane potential: (×), 20mM La; (⊙), 35mM La. Abscissa = time (min); ordinate = potential E *(millivolts). (— — —), solution change to test for 20mM La; (- - -), solution change to test for 35mMLa.*

Absence of Ca made little difference (*see* Table III). This parallels Lindley and Hoshiko (*1*) who noticed no difference in the skin potential in Ca-free solutions as opposed to that in solutions containing 1–2 mequiv/L Ca.

With respect to large cations, choline chloride added to outside solution (as increased *I, see* Table IV) contributes to depressing the potential. However when it is added as increased *I* over Ringer's solution to the inside compartment (against the added Na, K, or Ca on the outside) it served to alleviate some of the depression. Its use as an osmotic equivalent in place of sucrose according to the NaCl thermodynamic gradient experiments matters little regarding the depressed potential when there is insufficient Na on the inside.

Table VIII. Effect of Polyvalent Cations in Outside

Cation Salt	E (mV)	
	Solution pH	Steady State
$BeSO_4 \cdot H_2O$	5.0	36.6
$MgSO_4$	7.5	42.0
$Sr(NO_3)_2$	7.5	35.2
$BaCl_2 \cdot 2H_2O$	7.5	29.3
$ScCl_3 \cdot 6H_2O$	5.0	45.3
$YCl_3 \cdot 6H_2O$	6.0	49.2
$La(NO_3)_3 \cdot 6H_2O$	6.5	44.8
$Ce(NO_3)_3 \cdot 6H_2O$	6.0	43.5
$Cr(NO_3)_3 \cdot 9H_2O$	5.0	40.8
$MnCl_2 \cdot 4H_2O$	6.5	35.2
$FeSO_4 \cdot 7H_2O$	5.0	36.6
$CoCl_2 \cdot 6H_2O$	6.5	38.8
$NiCl_2 \cdot 6H_2O$	7.5	64.1
$CuSO_4 \cdot 5H_2O$	7.5	49.3
$Zn(NO_3)_2 \cdot 6H_2O$	6.5	57.2
$Cd(NO_3)_2 \cdot 4H_2O$	7.5	30.5
$HgCl_2$	7.5	32.8
$Al(NO_3)_3 \cdot 9H_2O$	5.0	30.7
$Pb(C_2H_3O_2)_2 \cdot 3H_2O$	5.5	23.1
$Th(NO_3)_4 \cdot 4H_2O$	5.0	71.9
$UO_2(C_2H_3O_2)_2 \cdot 2H_2O$	5.0	44.5

[a] Percentage decrease in *E* 120 min after steady state.

With Solution I originally on both sides of the membrane, acetylcholine included in the outside raised *E* by 50% while on the inside it lowered *E* by about 50% (*24*). This is in keeping with the findings of Barnes (*25*). Meanwhile the (cationic) surfactant cetyl trimethyl ammonium bromide lowers *E* drastically when applied to either the outside or the inside, with the effect being greater on the outside (*26, 27*). Anionic surfactants behaved similarly although not to the same degree, while nonionics had little effect. The potential decreased with the concentration of the cationics and anionics, and the decay rate increased sharply beyond the critical micelle concentration (*26, 27*).

Recovery. The experiments of Rennie et al. (*26, 27*) with cationics and anionics show partial recovery of the potential when the skin is washed with surfactant-free medium. More recovery takes place if the original exposure time to surfactant is less, if surfactant concentration is less, and when the action of the detergent is milder. Ussing and Zerahn (*13*) report recovery of the potential from CO_2 on reexposure to air. In like manner reversibility in varying degrees was observed throughout the present results.

Solution on Frog Skin Membrane Potential

E (mV)				Control[a]
Test 0 min	*Test 90 min*	*Solution I 0 min*	*Solution I 30 min*	*% Decrease*
33.1	21.4	28.8	50.3	9.1
41.8	40.9	44.3	42.5	9.6
35.3	32.7	32.9	32.1	8.1
29.5	27.3	27.6	26.1	20.0
45.1	29.1	36.6	37.7	19.8
59.3	46.7	47.3	47.3	1.1
55.2	43.9	46.3	46.2	6.6
63.0	56.0	55.1	57.5	8.9
48.5	21.5	35.9	40.4	−20.9
41.4	39.4	37.9	34.7	8.6
38.6	33.2	35.4	33.4	3.9
55.2	53.8	52.3	43.9	1.1
78.8	76.3	72.7	67.4	0.0
83.1	68.1	68.4	69.6	2.7
71.3	65.0	69.4	62.3	−1.8
45.8	43.2	37.3	26.7	6.7
5.1	2.1	2.0	1.6	1.4
40.3	24.9	32.3	38.8	10.3
30.7	25.5	25.5	25.8	14.1
73.2	50.3	54.5	70.4	−0.3
18.1	12.1	16.9	23.5	1.8

In Tables VIII and IX the Be and Al experiments undergo "super" recovery. (Hg^{++}, however, permanently poisons the membrane). When *E* is decreased owing to insufficient Na on the inside, replacement with Solution I partly brings back the potential (*see* Table I). There is partial recovery in the cases of Na, K, and Ca when applied in excess (*see* Tables I, II, and III) although highest concentrations of Na and K showed no recovery. With respect to the study of the effect of decreased pH (*see* Table V), there is also better recovery when the pH is not lowered quite as much, the outside showing better response than the inside.

Production of Surges in Potential. According to Tables VIII and IX many polyvalent cations bring about a surge in *E*. Al in the form of $Al(NO_3)_3$ shows it under the particular conditions of these experiments. The surge effect happens only with respect to the outside solution. Most of the cations are transition elements. Cations which produced an increase in *E* are compared with each other in Table XI, where changes are shown as a percentage of the original steady-state potential values.

Table IX. Effect of Polyvalent Cations in Inside

Cation Salt	*Solution pH*	E (mV) *Steady State*	*Test 0 min*
$BeSO_4 \cdot H_2O$	5.0	29.7	29.2
$MgSO_4$	7.5	32.0	31.8
$Sr(NO_3)_2$	7.5	31.5	31.7
$Bacl_2 \cdot 2H_2O$	7.5	27.8	28.2
$ScCl_3 \cdot 6H_2O$	5.0	30.8	30.7
$YCl_3 \cdot 6H_2O$	6.0	30.7	30.7
$La(NO_3)_3 \cdot 6H_2O$	6.5	48.3	49.0
$Ce(NO_3)_3 \cdot 6H_2O$	6.0	24.6	25.2
$Cr(NO_3)_3 \cdot 9H_2O$	5.0	25.5	27.7
$MnCl_2 \cdot 4H_2O$	6.5	30.3	30.3
$FeSO_4 \cdot 7H_2O$	5.0	34.7	34.3
$CoCl_2 \cdot 6H_2O$	6.5	28.4	31.8
$NiCl_2 \cdot 6H_2O$	7.5	39.3	39.4
$CuSO_4 \cdot 5H_2O$	7.5	63.0	63.2
$Zn(NO_3)_2 \cdot 6H_2O$	6.5	42.6	43.2
$Cd(NO_3)_2 \cdot 4H_2O$	7.5	28.9	29.6
$HgCl_2$	7.5	37.7	37.0
$Al(NO_3)_3 \cdot 9H_2O$	5.0	27.5	27.5
$Pb(C_2H_3O_2)_2 \cdot 3H_2O$	5.5	35.0	34.5
$Th(NO_3)_4 \cdot 4H_2O$	5.0	42.6	43.0
$UO_2(C_2H_3O_2)_2 \cdot 2H_2O$	5.0	32.9	31.9

[a] Percentage decrease in E 120 min after steady state.

Table X. Electrical Resistance of Frog Skin

Method	*Measured Resistance of Cell (Ω)*	*Resistivity (Ω cm)*	*Ratio of Resistivities Membrane/Solution*
Electrometer: direct reading 1 mV dc	no membrane = 7.23×10^6 with membrane = 7.34×10^6	solution = 2.4×10^6 membrane = 5.5×10^6	2.3
Potentiometer electrometer 5–10 mV dc	no membrane = 1.00×10^6 with membrane = 1.00×10^6	solution = 3.3×10^5 membrane = 3.3×10^5	1.0
Bridge: 1000 cycles 2 V ac	no membrane = 187 with membrane = 245	solution = 62.3 membrane = 2890.0	46.5

Solution on Frog Skin Membrane Potential

E (mV)			Control[a]
Test 90 min	*Solution I 0 min*	*Solution I 30 min*	*% Decrease*
25.8	27.2	27.3	4.7
30.7	30.4	30.6	11.0
29.4	29.4	28.7	6.5
25.7	26.5	26.1	11.3
30.7	31.3	31.3	3.4
30.2	30.5	32.7	−3.8
46.5	42.8	45.9	8.8
23.6	25.9	25.8	9.8
23.7	25.9	25.8	−14.7
30.1	30.9	28.9	5.2
32.0	32.6	31.3	0.8
28.9	31.6	27.1	6.6
35.9	36.0	36.5	3.0
62.0	60.4	58.0	2.4
41.6	42.1	46.4	0.7
30.3	31.5	29.0	7.7
8.1	7.9	3.8	6.9
26.0	26.0	28.2	10.7
31.1	32.0	32.8	14.7
39.6	42.0	45.6	5.0
23.4	30.1	31.9	11.3

Co, Cu, Cd, Pb, Ce, and Al bring on a spike representing more than a 30% increase. Except for Al, *E* for these cations remained above the steady state after 90 min in testing. With Co, Cu, Pb, and Ce, the potential remained greater than the steady-state potential when replaced with standard Ringer's solution. The membranes in Al recovered, retaining a greater *E* than the steady state after reexposure to Solution I for 30 min. In Cd the membranes suffered a decrease in *E* after 30 min. in Solution I following test solution. When the surges were less than 30% the membranes, except in the case of Cr and Th solutions, progressively decreased to the steady state level and remained thus on reexposure to Solution I. The Cr and Th experiments showed a large drop in *E* with time but this recovered to the steady-state value after 30 min of being in Solution I.

The polyvalents which brought about little change in *E* are Mg, Sr, and Ba, elements of the IIA group of the periodic table. This is the case for Ca (*see* Table III). Be, Hg, and UO_2 which are toxic (*28, 29, 30*) depress the potential. This is from the outside solution in all cases. On the inside Be and UO_2 had little effect. Hg, which was the most poison-

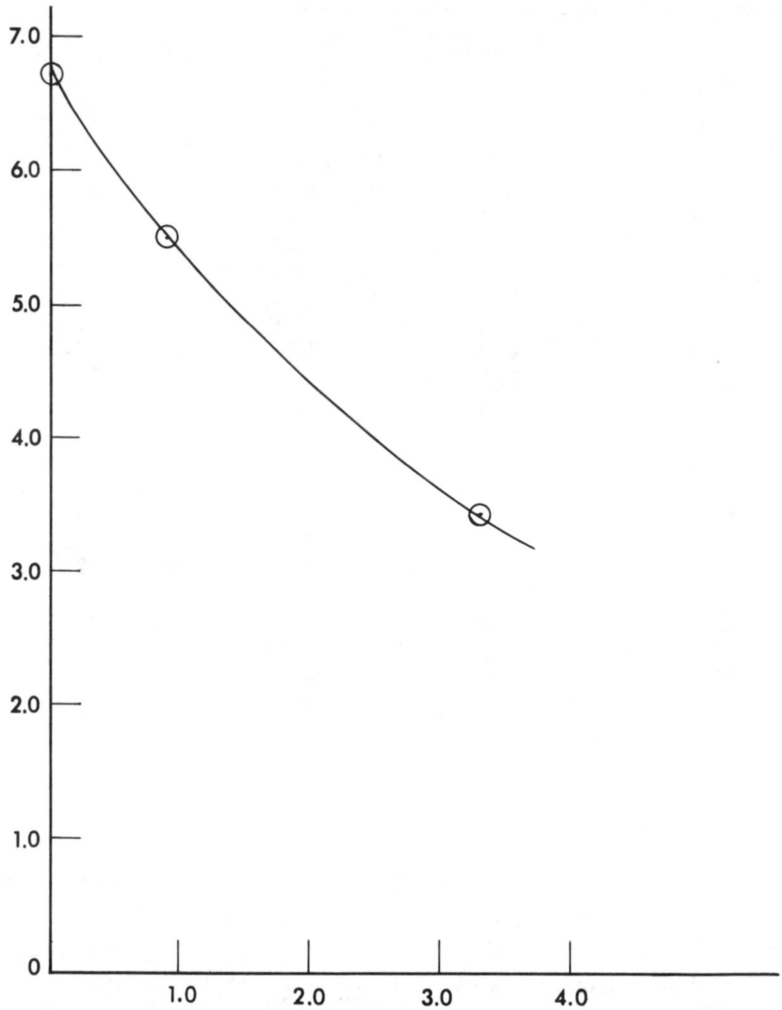

Figure 3. Electrical resistivity of frog skin: abscissa = log millivolts (millivolts); ordinate = log resistivity (ohm centimeters).

ous, depressed *E* from the inside although not to the same degree as from the outside. While Hg tests showed no recovery, the Be test membranes came back when Ringer's solution was applied and even surpassed the original steady state. After UO_2 exposure there is recovery but nowhere up to the steady state. Sc appears to fall between these depressants and the "no-change" category. While the membrane does not respond immediately to it, there is a degree of poisoning. Some recovery takes place when Solution I is applied.

Part of the explanation in the literature of the stimulation of the potential by $CuSo_4$ (*13, 21*) is that the Cu^{++} changes the surface so that passive Cl^- is inhibited from shunting through the skin. In the above reported work with surface active agents (*26, 27*), an initial increase in *E* was observed often, owing to nonionic Tergitol 15-s-9 which amounts to 18%.

E is increased through electrical stimulation (*see* Barnes (*25*), Finkelstein (*32, 33*), and Candia (*34*)).

Anions. The difference in behavior of outside Al in the experiments summarized in Table VII vs. those in Tables VIII and IX might be the result of NO_3^- vs. Cl^-. In connection with inside solution experiments, it mattered little according to Table VIII whether the test cations were added as the nitrate, the sulphate, or the acetate to the chloride Ringer's solution.

Since Cl^- was the anion for Na, i.e., NaCl, used when applying Na thermodynamic gradients across the membrane, it follows that a Cl^- gradient is also ineffective in one direction or the other. Also with regard to these experiments, the greatly reduced Cl^- concentration on the outside made little difference compared with the Ringer's Solution I Cl^- concentration. Considering that the effect on *E* was much the same in the parallel set of thermodynamic gradient experiments (briefly mentioned earlier in the chapter) where the NaCl was replaced osmotically by choline chloride instead of by sucrose in the inside solution, it makes

Table XI. Comparison of Cations That Stimulate Frog Skin Potential[a]

Cation	*Spike 0 Min*	*90-Min Test*	*30-min in Solution I After Test*	*Control (Over Period of Experiment)*
Mn	17.7	11.9	−1.4	−8.6
Co	42.3	38.6	13.2	−1.1
Ni	22.9	19.0	5.1	0.0
Cu	68.6	38.1	41.2	−2.7
Zn	24.6	13.6	8.9	1.8
Cd	50.0	41.6	−12.4	−6.7
Pb	33.0	10.4	11.7	−14.1
Fe	5.5	−9.3	−8.7	−3.9
Ce	44.8	28.7	32.2	−8.9
La	23.2	−2.0	3.1	−6.6
Cr	18.8	−17.2	−1.0	20.9
Al	31.3	−18.9	26.3	−10.3
Y	20.5	−5.1	−3.9	−1.1
Th	1.8	−30.0	−2.1	0.3

[a] Percent change in steady-state *E* (+ = increase; − = decrease).

little difference whether the Cl^- concentration on the inside is low or high. Thus the membrane on either side is not too sensitive to the Cl^- content being reduced to 0.1113*M*.

In the literature (*13, 15, 17, 26, 27, 35, 36*) $SO_4^=$ Ringer's solution gives higher potentials than Cl^-, the reasoning being that the frog skin is less permeable to $SO_4^=$ and thus shunting is minimized. In this connection Ferreira (*37*) shows that the short-circuit current (identified with Na tarnsport) is appreciably less when $SO_4^=$ is used in place of Cl^-. Br^- and I^- also lower this current but not to the same degree as $SO_4^=$. Mullins (*38*) observed a rate movement of radioactive Cl^- which was 2.5 times that of I^- when both sides were bathed in Cl Ringer's Solution. If I^- Ringer's solution was used, the transport of I^- improved. However Cl^-, Br^-, and I^- have about the same equivalent conductance.

Lindley and Hoshiko (*1*) and Singer and Civan (*39*) compare anions in terms of their effect on the potential. It has already been mentioned that surfactants with large anions decrease *E* and that this occurs more when the application is on the outside than when on the inside (*26, 27*).

Electrokinetic Aspects. A variety of observations appears to point to Gouy–Chapman electrokinetic diffuse double-layer effects. According to Tables I, III, and VI collectively it is apparent that there is an optimum range of *I* where the potential is positive. Thus the frog skin can have zero potential (i.e. reach an isoelectric point) at given high and low I. The high sensitivity of the membrane to change in solution concentration and its reversal of sign of potential at low ionic concentrations is typical of electrokinetic phenomena (*40*). Also, at concentrations higher than the Ringer Solution I concentrations, Ca^{++} brings about the decrease in *E*, reversing it to a low negative value much more effectively than Na^+. The Schulze–Hardy principle could apply here (*40*). Concerning the effect of divalent cation note that Ussing and Windhager (*15*), when using $MgSO_4$ in place of Na_2SO_4 on the outside (Na_2SO_4 Ringer's solution being maintained inside) lowered and even reversed *E*.

The depressing of *E* by increased *I* has its electrokinetic explanation in the swamping of the diffuse double layer. Reversibility of *E* recovery on reexposure to Ringer's solution after the test solutions had been applied (as already discussed) parallels electrokinetic phenomena where in many cases of having altered a given system, the restoration to original conditions is less likely to bring about the original state the more drastic the alteration. As an example, this is found in flocculation–deflocculation phenomena (*40*).

In connection with the lowering of *E* in decreased pH, the high sensitivity to H^+ may have to do with its being a potential determining ion (*42, 43*).

Depending on the way they are adsorbed into the diffuse double layer the organic cations in acetyl choline, choline chloride, and cetyl trimethyl ammonium bromide (*26, 27*) could influence *E*. Their respective anions OH^-, Cl^-, and Br^-, play a role as well. This should be the case for the polyvalent cations and the anionic surfactants (*26, 27*). Nonionic Tergitol (*26, 27*) possibly expands the double layer and *E* increases.

The role of $CuSO_4$ discussed above regarding surge in potential and anions could be in its affecting double layers at the interfaces of skin and bathing solutions. Electrokinetics would be at the heart of Ferreira's observed changes in the short-circuit current as a function of the presence of SO_4 vs. Cl vs. Br vs. I (*37*); Mullins' findings of Cl and I transport being so different despite their similar equivalent conductances (*38*); the conclusions of Curran et al. (*44*) that Ca decreases active Na transport and hormone increases it through their effects on the permeability of the outward facing membrane of the potential generating cells; and the experiments of Cereijido et al. (*45*) where the measured Na permeability for different outside concentrations did not conform with rates expected on a passive basis.

The experiments of Lindley and Hoshiko (*1*) where different cations and also anions are arranged in a selectivity order depending on how effectively they replace Na, K, and Cl in maintaining frog skin potential, recall the "lyotropic" (or Hofmeister) series in electrokinetics. Likewise Singer and Civan (*39*) in their experiments with toad urinary bladders, show an anionic "lyotropic" series according to the effect on transepithelial potential, short-circuit current, and mucosal-to-serosal flux. They consider this effect to be the local interaction of anion and skin. Kidder et al. (*46*) also refers to a selectivity order for Li, Na, and K.

Reversal of the potential in distilled water (*see* the Results Section) is understandable in terms of diffuse double-layer theory, the drastic reduction in *I* bringing about a radical change in the sorption of potential determining ions. References in the literature to reversal of the sign of the potential as a result of bathing solution ionic changes include Zadunaisky (*47*) and Martin (*48*). Zadunaisky found that when the Na^+ of the outside Ringer's solution is substituted by the choline ion, *E* is reversed in sign. Martin observed a negative potential when he maintained the outside solution with choline sulfate Ringer's solution plus 2m*M* KCl. When the skin was treated with $10^{-5}M$ $CuSO_4$ or the KCl was replaced by K_2SO_4, the negativity of *E* disappeared. Furthermore, reference has been made already to Ussing and Windhager (*15*) with respect to reversing *E* when they replaced Na with Mg. They noted an increase in the electrical resistance of the skin. The electrical gradient they obtained for frog skin through microelectrode impalement was reversed in direction using Mg in place of Na.

The influx–outflux shunt experiments of Mandel and Curran (*49, 50*) have possibilities of interpretation via electrocapillarity–electrokinetics theory. They observed that different permeabilities of Cl, urea, K, Na, and mannitol in ouabain-poisoned skin (minimal active transport) were dependent on applied electrical potential difference and external salts concentration, with the effects being more or less reversible. Major changes were observed when the potential was depolarized (depolarization potential depending on outside concentrations). In other experiments (*51*) aimed at finding the dependency of Na (active) current on applied emf, the observed independence of (higher) outside Na concentration (swamping?) and the saturation (rather than decrease) at higher depolarization voltages also points to electrokinetic–electrocapillary effects.

Also such effects are shown in Candia's research (*34*) where "passive unidirectional" Na outflux and the resistance of the frog skin changed with applied pd. When alternating current passed through the skin there was an increase in the dc E generated by skin, the explanation being the existence of a double lattice of fixed charges as rectifying structure in which Na was mobile. The presently observed decrease in resistivity with increased applied voltage follows. Ions more fixed in the double layer begin to respond at higher voltage (*42, 43*).

Sorption–Desorption at Inside and Outside Skin Surfaces. Many measurements were made utilizing flame photometry and atomic absorption to look for ion transport of Na, K, and Ca between the bathing solutions across the skin. Evaluation of any through transport was clouded over by the observed absorption–desorption. This varied so greatly from one frog to another that data are not reported in detail but generalizations are made as follows.

Absorption–desorption between the outside of the skin and its bathing solution seemed to be independent of that between the inside of the skin and its bathing solution. The magnitude was independent of the potential and any existing concentration gradient. In low Na bathing solutions (approximately 1mM) whether inside or outside, Na almost always left the membrane. The concentrations of the solutions changed in the range from -0.1 to $+28.1$mM for the various skins tested over 4–6 hours. Krogh (*51*) reports that frogs take up Na and Cl through their skin even from solutions as dilute as $10^{-5}M$ in NaCl.

When the Na in the outside solution was at a high concentration (111.3mM), it went into or came out of the membrane, while the range of solution concentration change was between -22.9 and $+33.1$mM. However with high Na on the inside it was absorbed in most of the runs; e.g. for six frogs the average decrease in bathing solution cencentration was 31mM. One skin showed $+4.6$mM and another -1.0mM.

Changes in K concentration varied from − 2.74 to + 6.5m*M* and for Ca the range was − 1.6 to 1.0m*M* with absorption in some cases and desorption in other cases regardless of the side.

Effect of Thermodynamic (Concentration) Gradients Across the Frog Skin. The frog skin membrane *E* does not seem to be affected by concentration differences across the membrane whether from outside to inside or vice versa. For instance the lower curve of Figure 1 is essentially the same when 1m*M* is on both sides and the upper curve is the same with NaCl Ringer's solution on both sides. Observations by Mandel and Curran (*51*) are interesting in this respect. They found that the short-circuit current actually increased by some 70% when a steep Na gradient was applied from the inside to the outside solution, opposing inward Na transport. Candia (*34*) finds Na outflux behaving as if an important fraction of it is independent of the electrochemical gradient.

In the present measurements, as a result of NaCl having been used it follows that a Cl^- gradient in either direction makes no difference. Further, gradients associated with H^+, the various polyvalent cations considered, and the large cations (and anions for that matter) have little if any effect on the potential. Rather, any change in the potential is a matter of local interaction of bathing solution and frog skin according to ionic species and concentration. On one hand on the inside, and on the other hand on the outside, these interactions are essentially separate from each other.

Transport–Electrogenics. The "zero" Na transport found in the present results regardless of direction of Na gradient or of no gradient, contradict the requirement of Na transport in the Koefoed–Johnson and Ussing model and modifications. Various experimenters (*23, 53, 54, 55*) have found that the epithelial membrane maintains its electrochemical balance without Na being transported. Actually McAfee (*56*), with frogs living in freshly kept deionized water, found that their Na levels essentially did not decrease. This contrasts with the model that calls for Na permeability between the outside bathing solution and the interior of the cell(s). He explains that tracer experiments may give the impression that frog skin is permeable to Na by virtue of exchange reactions and/or exchange diffusion taking place, or by leaks produced through skin edge damage on clamping between plastic chambers for measurements.

There is much evidence of Na transport being at least partially active rather than passive as called for in the model. For instance in the McAfee experiments, where only a few Na ions were in the outside solution and the concentration was too low to be measured, the outside solution could only be transported actively into the cell against its Na content. Krogh (52) observed that frogs take up NaCl from the surrounding medium even if the Na concentration in the latter is as low as 10^{-5}*M*.

Measurements of epithelial cell Na content were made of the skin being bathed in a relatively low Na outside solution by Rotunno, Pouchan, and Cereijido (*57*) and Cereijido and Rotunno (*23*). After allowing for bound Na in the cell, the free Na was still appreciably greater than that in the outside solution. Suggestions in behalf of active Na pumping are deduced also from the research of Cereijido et al. (*45*), Klahr and Bricker (*19*), Ussing (*58*), and Bricker et al. (*59*).

At the same time various experiments described in the literature indicate the transport in frog skin to be much more complex than is accounted for by the original model. Ferreira's results (*37*) suggest that at least some of the Na transport he observed is anion dependent in a nonspecific way. According to Ussing (*60*), depending on the bathing solutions, active transport is not limited to Na. Sucrose (sugars), proteins, Cl^-, and $SO_4^=$ etc. can be involved. Cereijido and Rotunno (*23*) look at the movement of substances such as these (in the absence of an electrochemical potential gradient) as a coupling with active Na transport (*60, 61, 62, 63*). Mandel and Curran (*51*) and others (*64, 65, 66, 67*) point to the intial entry of Na into the skin as not being simple (passive) diffusion. Rather this Na flux is mediated by Na concentration and also, by applied electrical potential. It is inhibited by Li (*65*) and K (*64*) and is sensitive to metabolic inhibition and ouabain (*67*). Curran et al. (*44*) find that Ca or hormones greatly affect active Na transport.

A net transfer of ions across the skin between bathing solutions is minimal, the current being in the range of microamperes. Thus in the light of the magnitude of the absorption–desorption reported here and the (slow) release of H^+ from the skin, "through" transport does not add much to changing the composition of the bathing solutions over the nominal period of experimentation. In the same vein the separate influence on the outside solution on E together with that of the inside solution, points to minimal travel of ionic species through the membrane. Otherwise, effects on E would be similar no matter from which side the given ionic solution concentrations were applied.

The present open-circuit measurements of the frog skin membrane potential are such that with the resistivity of the frog skin being so much lower than that of the electrometer ($10^{14}\ \Omega$), practically all of the current generated must shunt back through the skin.

H^+ needs must be considered in transport. It has a high specific conductance explained according to making and breaking H bonds (*68, 69, 70*). Fleming's results (*71*) indicate that production of acid in the bathing solution (as was observed and reported above) is connected with the transport process.

While it was proposed originally that a single-cell layer in the stratum germinativum was responsible for active transport (*71, 72*), the measurement of potentials as a function of depth of penetration of the skin with microelectrodes furnishes evidence that centers active in the generation of the overall potential exist practically across the whole skin (*14, 15, 60, 73, 74, 75*). In this light the many instances observed of a lack of connection between potential and existing thermodynamic gradients is understood (*23, 60, 61, 62*).

With respect to the paths of ionic currents, the original Koefoed–Johnson and Ussing model (*17*) has passive Na entering cells and active Na leaving. Later work, which involved electron microscopy (*76, 77*) and which pointed to the need to consider shunt circuits, resulted in modifications to this approach, including the concept of ions passing between cells rather than through them (*15, 16, 23, 49, 50, 51, 60, 74, 78, 79*).

Metabolism. The observed need for an adequate amount of Na on the inside could be enzymatic. Ussing (*60*) refers to an active transport pump being able to handle almost anything that gets into it but in which Na plays some particular role so that in its absence the pump is inhibited. The observed poisonous action of Be, Sc, Hg(II), and UO_2 could fit into this category of affecting the potential via the metabolism. Be is particularly interesting in this respect because on replacement with Ringer's solution the potential increased considerably beyond what it was before application of the Be. A similar effect was observed for Al. At least to a degree, the surges in potential observed for given surfactants, i.e. Tergitol and the various polyvalent cations already mentioned, may be explainable this way. H^+ certainly must have its impact.

The literature is replete with information on the specific effects of particular ionic species on enzyme activity. For example, Papahadjopoulos (*80*) finds that acidic phospholipids interact strongly with bivalent metals in the presence of univalent salts. This was indicated by an increase in surface potential and decreases in the surface pressure of monomolecular films and in the zeta potential of liquid crystalline vesicles (there being a sharp change in the rate of diffusion of K^+ and Na^+ ions). Blank and Britten (*81*) observed that Co^{++}, Ni^{++}, and Zn^{++} activated Na–K ATPase in the presence of excess ATP in a similar way as do Mg^{++} and Mn^{++}. Na–K ATPase causes the transport of Na^+ and K^+ across the membrane when ATP is hydrolyzed.

Mechanism. The present experimentation and many results reported in the literature are seen to be difficult to explain in terms of Koefoed–Johnsen and Ussing (*15, 17, 60*). In addition there is a strong indication of electrokinetic effects but no allowance has been made for them in their model.

A thermodynamic gradient of given ionic species requires appropriate electrode reactions to be converted to emf. The lack of emf response obtained in the various applications of concentration gradients between bathing solutions across the skin indicates that the necessary electrode mechanism for these particular passive gradients is absent in the frog skin.

There is no allowance in the model(s) for the negative wells occurring in potential profiles such as those obtained by Engbaek and Hoshiko (*73*); Ussing and Windhager (*15*); Cereijido and Curran (*14*); Whittembury (*82*); and Scheer and Mumbach (*83*). Actually it is difficult to reconcile the various different whole profiles observed with the Koefoed–Johnsen and Ussing model, considering the position of plateaus and wells and the multiplicity of these plateaus and wells, using distance of penetration of the microelectrode as a yardstick. This is also true for Snell and Chowdhury (*75*) who while not showing wells according to their particular technique of frog skin puncture with microelectrodes, describe the electrical potential to be a smooth and monotonically increasing function of the depth of penetration.

Considering these difficulties with the Koefoed–Johnson and Ussing model, the following mechanistic viewpoints are offered. Beginning with potential profiles experimentation (*14, 15, 73, 75, 82, 83*), reference is made to Figure 4 which contains two examples profiles. This is a copy of detail graphs from Engbaek and Hoshiko's work (*73*). The plotted potential is that between the outside solution and the microelectrode in the skin. Plateaus are evident and there are jumps and drops in potential. Given plateau areas have their peaks and crevices.

A literal approach of interpretation of a profile being the actual representation of the potential inside the skin at various depths of penetration vs. the outside solution requires that each jump represents a particular area or layer of electrogenic pumping in the direction of the overall potential. Each drop also represents an area of electrogenic pumping but in the opposite direction to the overall E. Any straight line plateau represents a zone in the skin where no pumping is taking place and the potential remains constant.

Thus the depressions in the profiles are wells, i.e. Figure 4a has six wells and Figure 4b has 5. The negative wells found for many profiles are understandable in this light. Larger jumps correspond to more energetic pumping layers. Plateaus are the in-between of pumping layers (*15, 23, 53, 59*) and can represent the tread of the step up when the pumping of two adjacent layers is in the same direction or summit or well when pumping of the two layers is in opposite directions.

While there may be barrier layers (*14*) not particularly associated with pumping layers, the pumping layers must have built-in barriers i.e. valve arrangements of sufficient resistance to accomplish separation of charge as chemical energy is (metabolically) converted to electrical energy. The voltage generated by each pump–barrier is positive or

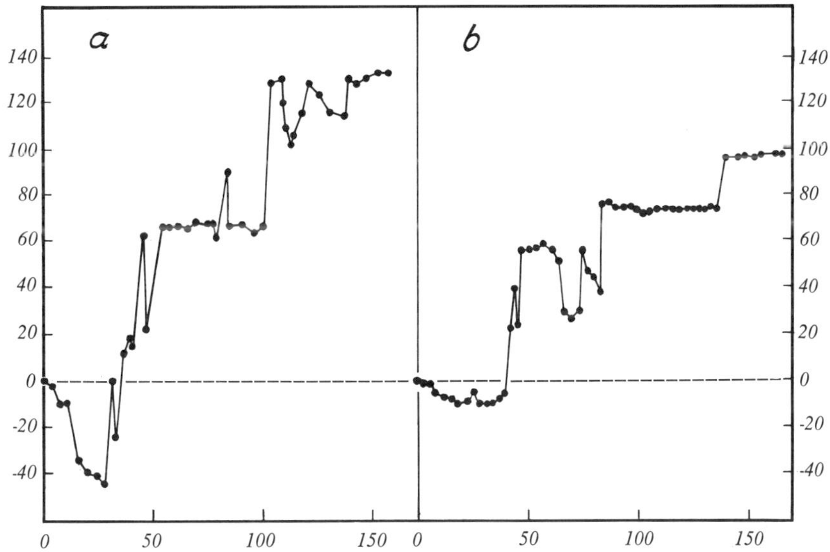

Figure 4. Frog skin potential profiles: abscissa = distance from outside surface (units of meters × 10^{-6}); ordinate = potential (units of volts × 10^{-3}) (73).

negative with respect to the overall E, depending on the sign of charge of the ion being pumped, the location of the pump compared with its barrier, and the sign of charge of this barrier. This succession of electrogenic pump–barrier layers is in the sense of a voltaic pile. Figure 5 is a diagram of such an arrangement. E is the algebraic sum of all of the electrogenic pumping layer voltages generated.

The barriers are electrostatic, being made up or electrokinetic diffuse Gouy–Chapman layers radiating out from the molecular chains, cells, cell membranes, and/or phase interfaces etc. of the skin as cores. Figure 6 shows this schematically, ψ_0 arbitrarily being called negative for the purpose of illustration. They depend on the particular skin region's chemical nature with its content of ions, water molecules, and other molecules for magnitude and sign.

At the skin surfaces the electrokinetic double layers of the barriers extend into the solutions, ψ_0 being based on skin together with solution. The arbitrary graph of ψ vs. distance in the lower left corner of Figure 6 applies to solution–skin interface and barriers in the skin interior. These solution–skin barriers are very meaningful to the overall E as evidenced by the strong effects which depend on bathing solution composition already described. Solution interactions are not only electrokinetic; they must be in the nature of a composite of catalytic action, which influences the metabolism per se and the electrokinetic interplay. (DeSimone (*84*)

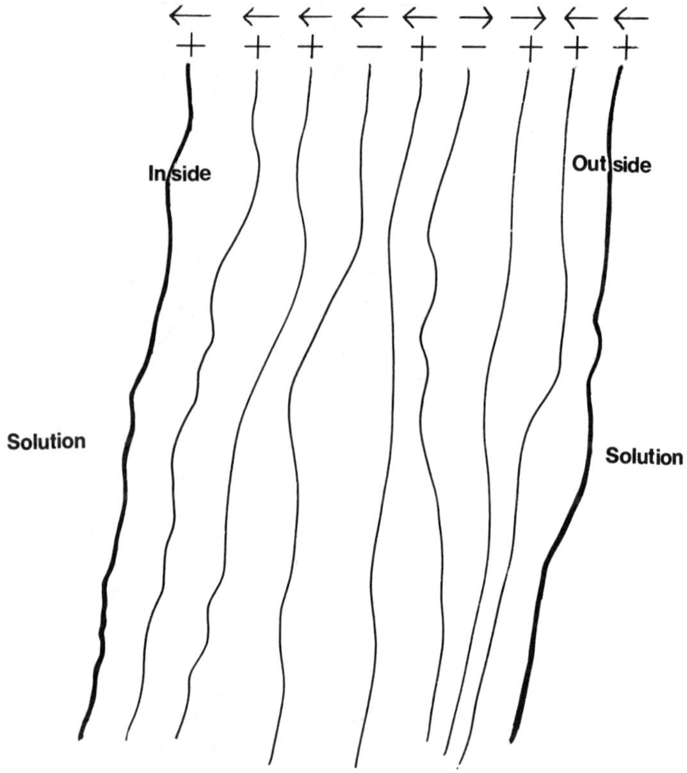

Figure 5. Hypothetical series of electrogenic pump–barrier layers in frog skin

investigated the influence of enzymic surface reactions on the structure of the diffuse double layer, and the influence of potential on concentration profiles and reaction rates.)

Electrokinetically, solution–skin barriers are affected by solution composition insofar as the sign of their potential, the magnitude of this potential, their density of charge, and their thickness. Thus, according to their sign of potential, E is positive or negative depending on the orientation of the electric current generated by the pumps.

Assuming that there is sufficient metabolic energy, E should depend on the permeability of the barrier to shunting, E building up with lower permeability. Permeability should be lower when the potential of the barrier, the double-layer thickness, and the charge density (+ or − as the case may be) are greater. Likewise the degree of passive electroneutral transport would depend on the state of the double layer (*42, 43*).

It already has been brought out that for the present open-circuit methodology of measuring E, using E = 50 mV as an arbitrary basis,

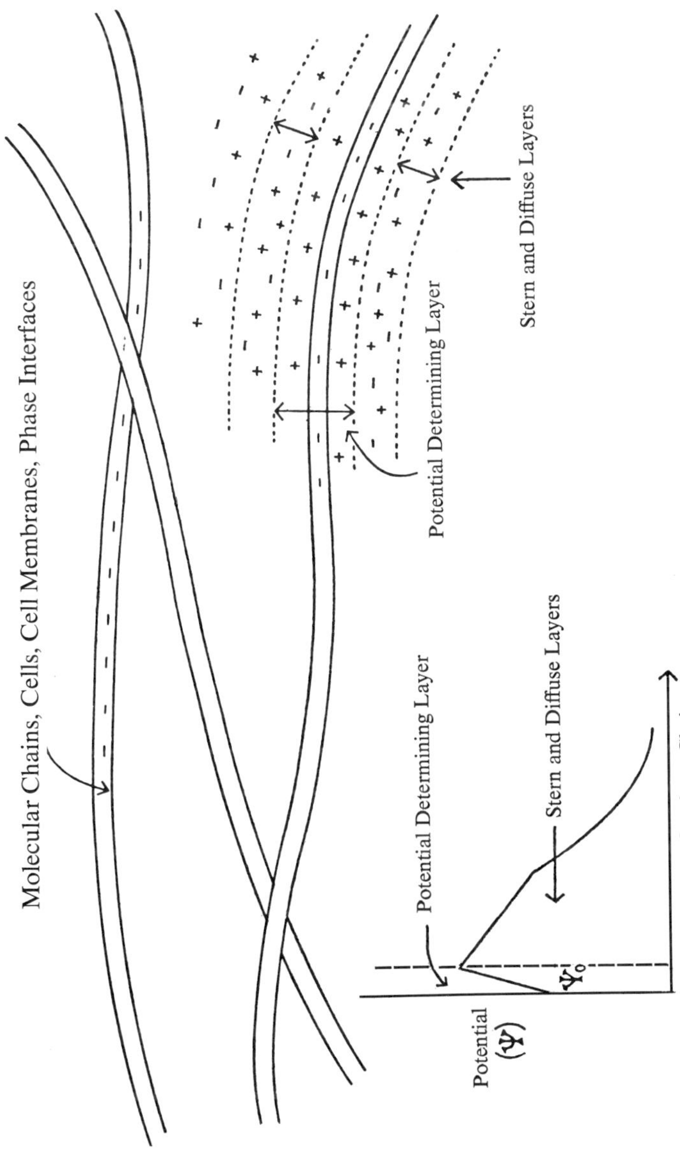

Figure 6. Gouy–Chapman double layers in skin interior (42)

that, considering the resistance of the skin (1000 Ω), practically all of the current generated by the skin shunts back through it. This shunt is in a pathway separate from the pumping system current. Otherwise there would be electroneutralization and no potential would develop.

In this light, the pump–generator–voltaic pile part of the circuit is not reversible, i.e. when back emf is applied. According to the barrier system and the interaction with the bathing solutions, the pile is generating positivity in the solution on one side of it and negativity in the solution on the other side.

There is no need for the shunt path to be irreversible although there actually might be a situation of asymmetry and tunneling according to direction of flow. The shunt path is presumably the path travelled by electric current when emf is applied (from either direction). The permeability of the barrier system to transport already has been referred to in terms of the electrokinetics of the barriers. With respect to actual conductance this should be a function of those electrically charged particles which are sufficiently unbound so that they move and carry the current. Thus those particles in a given zone that are present in the largest concentration with the greatest mobility should be carrying greater shares of the current. Electric current carriers, according to species and their proportionate number, may not necessarily be the same in different parts of the skin since there is the possibility of particular charged species carrying current on one side of a barrier giving their electrical energy to charged species on the other side. Proton pumping and proton transport are good possibilities. Nagle and Morowitz (*68*) and Stoeckenius (*85*) consider such mechanism in detail. They refer to localized chains of hydrogen bonds that can function as proton wires. In being localized, pump flow remains separated from shunt flow.

Presumably passive transport, here defined as that of electrically neutral molecules or electroneutrally matched cations and anions according to the make-up of the acting thermodynamic concentration gradient, does not go through the pumping part of the circuit, where transport other than that generated by the pumps is forbidden. Passive transport does have to pass through electrokinetic barriers as does the shunt or current under applied emf. In this connection Bender et al. (*42, 43*) have studied the diffusion and sorption of salts in what appears to be electrokinetically charged membrane systems. As observed in the present systems of study, the absence of Na transport under a passive gradient, the lack of response of E to concentration gradients in general, and the lack of direct effects of the inside solution on the outside part of the skin and the outside solution on the inside part, indicate little or no actual passive transport of electrolyte taking place.

Note that the junctions of the pumping part of the circuit and the shunt part involve the bathing solutions with the potentials being measured across the solutions. The same should apply for any point of penetration in the potential profile (the potential being that which was measured across the outside bathing solution and the solution of the microelectrode). The potential that now applies is presumably only from that part of the voltaic pile between the microelectrode and the outside bathing solution. The shunt is also across this section of the skin.

With the potential generated being so dependent on bathing solution contents, attention should be paid to the microelectrode. Its solution generally has been 3*M* KCl, to which the skin is exposed in contradistinction to a Ringer's solution makeup. Considering the extreme thinness of the walls such that the glass has membrane properties of intimacy with solution, permeability, and large surface area, there should be much reactivity. Namely there will be some dissolving of the glass, sorption of chemicals, hydration, and ion exchange, the Ca in the glass being more active (*see* Ussing and Windhager (*15*) regarding the effects of divalent Mg). Thus there is a skin–solution–glass Gouy–Chapman diffuse double layer involving the microelectrode, and the measured potential should be affected accordingly.

Microelectrode dimensions and therefore the ratio of surface area to bulk can differ from one electrode to another for different series of measurements. Also chemical differences would be expected in the microelectrodes of different investigators. Probably the negative wells observed at the beginning of various profiles are explainable through this state of the microelectrode solution, especially since there is so little of the voltaic pile in the pump–shunt circuit at this point of penetration. Also the microelectrode electrokinetics can differ in the different zones of penetration owing to differences in chemical nature in the different parts of the skin.

Since the barriers are electrostatic and a small amount of current passes through, the skin should have leaky capacitance properties. This is represented by a series of capacitors with the barriers acting as plates. The dielectric constant of the medium between is a composite based on dipole moments and polarizability of the various molecules according to their structure and the structure of their interassociation. It is interesting in this connection of composite dielectric constant that preferential absorption of K over Na was observed to increase with lower dielectric constant in work with cellulose and cellulose acetate membranes (Bender et al.(*42, 43*).

One might think in terms of an overall capacitor across the skin with the plates being the skin–solution interface barriers. These interact across the dielectric medium of the skin influencing each other's polarity and magnitude of charge and that of the barriers in between. This concept (*see* Figure 7) could apply also to that part of the skin between

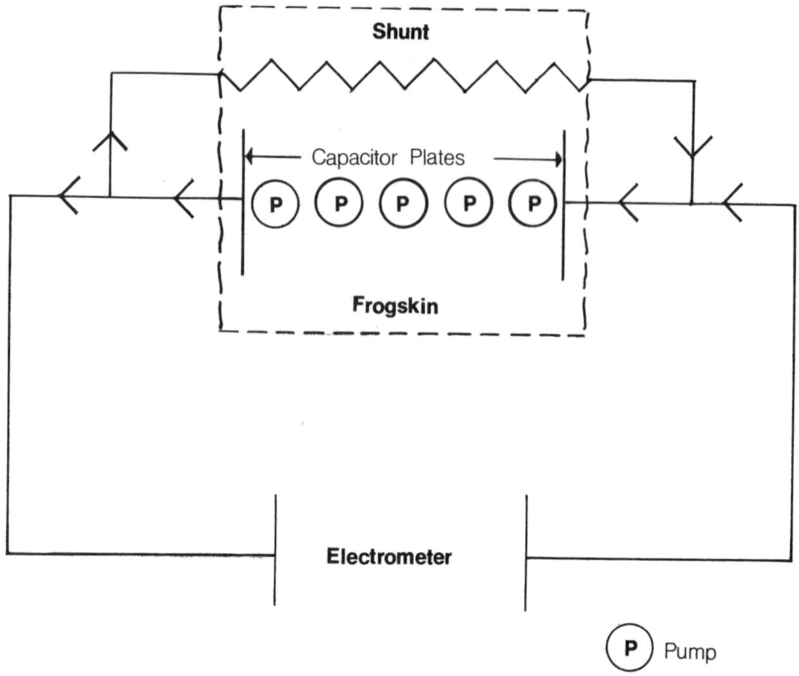

Figure 7. Capacitor—circuit arrangement

the bathing solution and the microelectrode which is positioned for measuring a point in electrical gradient profile measurements.

Capacitance effects as well as electrokinetic interactions could be a strong reason behind the increases in potential observed resulting from electrical stimulation and also from the introduction of the polyvalent cations and other species. Even though as has been shown, each of the bathing solutions has its direct influence only on the skin side with which is is in contact and not on the opposite side, there are important indirect effects. Presumably these are of a capacitance nature. It is indeed evident throughout that E is not the simple algebraic sum of potentials corresponding to the outside of the skin and the inside. For instance, there are the results using distilled water as the bathing solution where E is reversed in sign or reduced to zero by applying a given bathing solution makeup only on one side. Measurements with microelectrodes imbedded in the skin to a given depts (*14, 75*) clearly point to these indirect effects with changes in K concentration in the inside bathing solution being observed to influence strongly the potential measured across the microelectrode and the outside solution. The frog skin should be capable of electrical storage in a similar vein, although to a lower degree, as other animal systems (electric eel) which deliver effective electrical shocks.

Although a capacitance response per se is extremely rapid, the overall effect resulting from a change in bathing solution composition takes time considering the phsyical and chemical reactions and diffusional processes also occurring. But there also can be rapid communication originating with the solutions and passing on from cell to cell as in nervous system effects. For instance, Steggerda and Ponder (*86*) saw depression in the potential upon destruction of the nervous system; Shoffeniels (*87*) observed the permeability to be under the influence of a nervous control.

Literature Cited

1. Lindley, B. D.; Hoshiko, T. *J. Gen. Physiol.* **1964,** *47,* 749.
2. Hillman, K. M.Sc Thesis, Physiology Dept., Fairleigh Dickinson University, Teaneck, N.J., 1968.
3. Tirri, L. B.Sc. Thesis, Fairleigh Dickinson University, Teaneck, N.J., May 1968.
4. Graben, M. B.Sc. Thesis, Fairleigh Dickinson University, Teaneck, N.J., May 1969.
5. Fenster, D. Research under NSF COSIP Grant, Fairleigh Dickinson University, Teaneck, N.J., Summer 1969.
6. Rosensaft, M. M.Sc. Thesis, Fairleigh Dickinson University, Teaneck, N.J., May 1972.
7. Sheldon, J. A. "A Literature Survey of the Frog Skin Potential," M.S. Thesis, Physiology Dept., Fairleigh Dickinson University, Teaneck, N.J., Sept. 1966.
8. Neuschwanter, C. "The Influence of a Magnetic Field on the Potential Difference in Isolated Frog Skin," B.Sc. Thesis, Fairleigh Dickinson University, Teaneck, N.J., May 1972.
9. Lockwood, A. P. M. *Comp. Biochem. Physiol.* **1961,** *2,* 241.
10. Snell, F. M.; Leeman, C. P. *Biochim. Biophys. Acta* **1957,** *25,* 311.
11. Amberson, W.; Klein, H. J. *Gen. Physiol.* **1928,** *11,* 823.
12. "Handbook of Chemistry and Physics," 44th ed.; Chemical Rubber: Cleveland, 1963; p. 1719.
13. Ussing, H. H.; Zerahn, K. *Acta Physiol. Scand.* **1951,** *23,* 110.
14. Cereijido, M.; Curran, P. F. *J. Gen. Physiol.* **1965,** *48*(4), 543.
15. Ussing, H. H.; Windhager, E. E. *Acta Physiol. Scand.* **1964,** *61,* 484.
16. Ussing, H. H. *Acta Physiol. Scand.* **1965,** *63,* 141.
17. Koefoed–Johnsen, V.; Ussing, H. H. *Acta Physiol. Scand.* **1958,** *42,* 298.
18. Fukuda, T. R. *Jpn. J. Med. Sci.* 3 **1942,** *8,* 123.
19. Klahr, S.; Bricker, N. S. *J. Clin. Invest.* **1964,** *43*(5), 922.
20. Steinbach, H. B. *J. Cell. Comp. Physiol.* **1937,** *10,* 51.
21. Hashida, K. *J. Biochem. (Tokyo)* **1922,** *1,* 21.
22. Linderholm, H. *Acta Physiol. Scand.* **1952,** *27,* supplement 97.
23. Cereijido, M.; Rotunno, C. A. *J. Gen. Physiol.* **1968,** *51,* 280S.
24. Landes, F. B.Sc. Thesis, Fairleigh Dickinson University, Teaneck, N.J., May 1970.
25. Barnes, T. C. *Am. J. Physiol.* **1940,** *130,* 557.
26. Rennie, G. K.; Hill, J. C.; Goddard, E. D.; Kulkarni, R. D. *J. Pharm. Pharmacol.* **1975,** *27,* 363.
27. Kulkarni, R. D.; Goddard, E. D. "Abstract E9," 49th National Colloid Symposium, June 1975.
28. Schubert, J. *Sci. Am.* **1958,** *199,* 27.
29. Harrington, C. D. "Uranium Production Technology"; Van Nostrand: New York, 1959.

30. Merck Index, 8th ed., 1968.
31. Ussing, H. H. *Acta Physiol. Scan.* **1949** B, *19*, 43.
33. Finkelstein, A. *J. Gen. Physiol.* **1964**, *47*(3), 545.
33. Finkelstein, A. *J. Gen. Physiol.* **1964**, *43*(3), 545.
34. Candia, O. A. "Abstracts," Eleventh Annual Meeting, Biophysical Society **1967**, *52*, TB10, p. 52.
35. Koefoed–Johnsen, V.; Ussing, H. H.; Zerahn, K. *Acta Physiol. Scand.* **1952**, *27*, 38.
36. Ussing, H. H. *J. Gen. Physiol.* **1960**, *43*(5), 135.
37. Ferreira, K. T. G. *Biochim. Biophys. Acta* **1968**, *150*, 587.
38. Mullins, L. J. *Am. J. Physiol.* **1958**, *194*, 369.
39. Singer, I.; Civan, M. M. *Am. J. Physiol.* **1971**, *221*, 1019.
40. Bender, M. *J. Phys. Chem.* **1953**, *57*, 466.
41. Bender, M. *J. Colloid Sci.* **1954**, 9(4), 359.
42. Bender, M.; Moon, J. K.; Stine, J.; Fried, A.; Klein, R.; Bonjouklian, R. *J. Chem. Soc., Faraday Trans. 1* **1975**, *71*, 491.
43. Bender, M.; Khazai, B.; Dougherty, T. E. *J. Colloid Interface Sci.* **1978**, *63*(2), 346.
44. Curran, P. F.; Herrera, F. C.; Flanigan, W. J. *J. Gen. Physiol.* **1963**, *46*, 1011.
45. Cereijido, M.; Herrera, F. C.; Flanigan, W. J.; Curran, P. F. *J. Gen. Physiol.* **1964**, *47*, 879.
46. Kidder, G. W.; Cereijido, M.; Curran, P. F. *Am. J. Physiol.* **1964**, *207*, 935.
47. Zadunaisky, J. A. *Am. J. Physiol.* **1964**, *207*(5), 1010.
48. Martin, D. W. *J. Cell. Comp. Physiol.* **1964**, *63*, 245.
49. Mandel, L. J.; Curran, P. F. *J. Gen. Physiol.* **1972**, *59*, 503.
50. Mandel, L. J.; Curran, P. F. *Biochim. Biophys. Acta.* **1972**, *282*, 258.
51. Mandel, L. J.; Curran, P. F. *J. Gen. Physiol.* **1973**, *62*(1), 1.
52. Krogh, A. *Skand. Arch. Physiol.* **1937**, *76*, 60.
53. Curran, P. F.; Cereijido, M. *J. Gen. Physiol.* **1965**, *48*, 1011.
54. Huf, E. G.; Doss, N. S.; Wills, J. P. *J. Gen. Physiol.* **1957**, *41*, 397.
55. Levinsky, N.; Sawyer, W. *J. Gen. Physiol.* **1953**, *36*, 607.
56. McAfee, R. D. *Science* **1972**, *178*, 183.
57. Rotunno, C. A.; Pouchan, M. I.; Cereijido, M. *Nature* **1966**, *210*, 597.
58. Ussing, H. H. *Acta Physiol. Scand.* **1949**, *17*, 1.
59. Bricker, N. S.; Biber, T. U. L.; Ussing, H. H. *J. Clin. Invest.* **1963**, *42*, 88.
60. Ussing, H. H. *Ann. NY. Acad. Sci.* **1966**, *137*(2), 543.
61. Crane, R. K. *Fed. Proc., Fed. Am. Soc. Exp. Biol.* **1965**, *24*, 1000.
62. Curran, P. F. *Fed. Proc., Fed. Am. Soc. Exp. Biol.* **1965**, *24*, 993.
63. Franz, T. J.; Van Bruggen, J. T. *J. Gen. Physiol.* **1967**, *50*, 933.
64. Rotunno, C. A.; Vilallonga, F. A.; Fernandez, M.; Cereijido, M. *J. Gen. Physiol.* **1970**, *55*, 716.
65. Biber, T. U. L.; Curran, P. F. *J. Gen. Physiol.* **1970**, *56*, 83.
66. Biber, T. U. L.; Sanders, M. L. *J. Gen. Physiol* **1973**, *61*, 529.
67. Biber, T. U. L. *J. Gen. Physiol.* **1971**, *58*, 131.
68. Nagle, J. F.; Morowitz, J. J. *Proc. Natl. Acad. Sci.* **1978**, *75*, 298.
69. Brattain, W. M., personal communication, Aug. 22–26, 1977 (Gordon Conference).
70. Moore, W. J. "Physical Chemistry," 4th ed., 1972; p. 436.
71. Fleming, W. R. *J. Cell. Comp. Physiol.* **1957**, *49*, 129.
72. Ussing, H. H. "The Use of Tracers in the Study of Active Ion Transport Across Animal Membranes," *Cold Spring Harbor Symp. Quant. Biol.* **1948**, *8*, 193.
73. Engbaek, L.; Hoshiko, T. *Acta Physiol. Scand.* **1957**, *39*, 348.
74. Ussing, H. H. "Transport of Electrolytes and Water Across Epithelia"; The Harvey Lectures, Ser. 59, Academic: New York, 1965.
75. Snell, F. M.; Showdhury, T. K. "Intracellular Transport," "Symp. of the International Soc. for Cell Biol.," Warren, K. B., Ed.; Academic: New York, 1966; Vol. 5, p. 141.

76. Farquhar, M. G.; Palade, G. E. "Functional Organization of Amphibian Skin," *Proc. Natl. Acad. Sci. (U.S.A)* **1964**, *51*, 569.
77. Farquhar, M. G.; Palade, G. E. *Cell Biol.* **1965**, *25*, 263.
78. Ussing, H. H. *Ber. Bunsenges. Phys. Chem.* **1967**, *71*, 807.
79. Kedem, O.; Essig, A. *J. Gen. Physiol.* **1967**, *48*, 1047.
80. Papahadjopoulous, D. "Surface Properties and Permeability Characteristics of Acidic Phospholipids: Interaction of Monolayers and Hydrated Liquid Crystals with Bivalent Metals," 156th National Meeting, ACS, Sept. 1968; COLL 57.
81. Blank, M.; Britten, J. S. In "Monolayers," *Adv. Chem. Ser.* **1975**, *144*.
82. Whittembury, G. *J. Gen. Physiol.* **1964**, *47*, 795.
83. Scheer, B. T.; Mumbach, M. W. *J. Cell. Comp. Physiol.* **1960**, *55*, 259.
84. DeSimone, J. A. *J. Theor. Biol.* **1977**, *68*, 225
85. Stoeckenius, W., submitted for publication in *Soc. Gen. Physiol.*
86. "Physiology of the Amphibia"; Moore, J., Ed.; Academic: New York, 1964; p. 261.
87. Schoffeniels, E. "Cellular Aspects of Membrane Permeability"; Pergamon: N.Y., 1967; p. 88.

RECEIVED October 17, 1978.

26

Destruction of the Electrophysiological Potential of Excised Frog Skin by Surfactants

R. D. KULKARNI and E. D. GODDARD

Union Carbide Corporation, Tarrytown Technical Center,
Tarrytown, NY 10591

The effect of various surfactants on the electrophysiological potential of freshly excised frog abdominal skins bathed in Ringer's solution was evaluated using sodium–linear alkyl benzene sulfonate (LAS, C_{12}), sodium lauryl sulfate (SLS), sodium alkylethoxy sulfate (Neodol 25-3EOS, AES), cetyltrimethylammonium bromide (CTAB), and the nonionic Neodol 25-7EO. Surfactants can cause a rapid, irreversible decay of the frog skin potential, the nature and the rate of the potential decay being dependent on the type of surfactant used. In decreasing order of potential-destroying effectiveness, the grading of the surfactants was CTAB > SLS > LAS > AES > Neodol 25-7EO. A marked increase in potential decay rate was observed above their CMC for both CTAB and SLS. On the other hand, addition of Neodol 25-7EO caused a marked reduction in the decay rate produced by SLS, and this is ascribed in part to mixed micelle formation.

When a freshly excised "still living" frog abdominal skin is bathed in aerated Ringer's solution, a steady-state potential is developed (*1*) under open circuit conditions. The magnitude of this potential, with the innerside of the skin electrically positive in relation to the outside, can be as high as 100–150 mV (*1, 2*). The origin of such potential is thought to be metabolic sodium and potassium transport reactions associated with active sodium transport or the "sodium pump" process (*3, 4, 5*), resulting in spontaneous transport of sodium ions from the outside to the inside of the skin.

[1] Work carried out at Lever Brothers Research Center, Edgewater, New Jersey.

0-8412-0473-X/80/33-188-445$05.00/1

The existence of such potentials is vital to the functioning of animal systems (*6*) and is associated with a number of coupled transport processes (*4*), control of cell volume (*3*), and so on. Introduction of surfactant molecules, which are thought to interfere with metabolic reactions responsible for the sodium pump, can destroy the generated potential. In the case of skin, the rate of potential destruction is likely to be an indication of the aggressiveness of the surfactant towards the skin.

This chapter deals with the effect of various cationic, anionic, and nonionic surfactants on the decay of frog skin potential.

Experimental

Materials. The source of skin was the common American frog, Rana pipiens. The frogs, 3–5 in. in length, were obtained from Texas through The Connecticut Valley Biological Supply Company. They were stored in the refrigerator until use. The steady-state skin potential for these frogs was strongly dependent on their condition: sluggish frogs gave significantly lower potential than vigorous specimens. Just before use the frogs were taken out of the refrigerator, washed, and kept in room-temperature water until they became active. They then were sacrificed and the abdominal skin was removed, washed, and stored in normal Ringer's solution. The skins were used within 1 hr of their dissection.

Sodium lauryl sulfate (SLS) was obtained from BDH Chemicals, Ltd. and was > 99% pure. Cetyltrimethylammonium bromide (CTAB) was supplied by Fine Organics and was twice recrystallized before use. Sodium linear alkyl benzene (LAS) sulfonate with an average 12-carbon chain length was a Lever Brothers product purified by desalting in hot 80% ethanol and extraction with petroleum ether. Nonionic surfactant Neodol 25-7EO, an ethoxylate based on a 12–15-carbon chain alcohol, was supplied by Shell, as was the Neodol 25-3EO sulfate (AES).

Double-distilled water with conductivity less than 1.5×10^{-6} mhos/cm at 25°C was used throughout.

Apparatus. The apparatus used for the electrometric measurement is similar to that developed by Ussing (*1, 7*) and is shown in Figure 1. The cell consists of two conical compartments, each of 20-mL capacity, separated and sealed from each other by the membrane under investigation. These compartments are filled with sulfate Ringer's solution (111.2m*M* Na_2SO_4, 2.0m*M* K_2SO_4, 1.0m*M* $CaSO_4$, and 2.6m*M* $NaHCO_3$) to an appropriate level. The solution in each compartment is circulated and simultaneously aerated by the rising purified air bubbles in the bubbling tubes as shown.

The electrical potential measuring device consists of two calomel sleeve junction reference electrodes which are coupled to each of the half-cell compartments by agar bridges. These reference electrodes in turn are connected to a Keithley 610 A electrometer coupled to a Honeywell Electronic 19 chart recorder.

The experiments were conducted by first mounting the membrane carefully in between the two half-cell compartments, avoiding wrinkles, strains, and leakages. This was followed by the addition of 25-mL sulfate

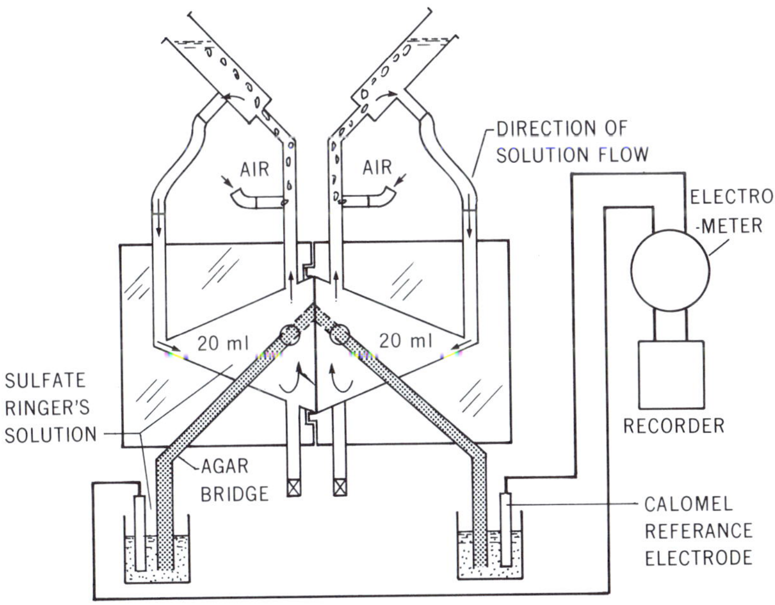

Figure 1. Apparatus for electrometric measurements

Ringer's solution to each half cell and aeration of these cell compartments at 180 bubbles/min until a steady-state potential was recorded. Next, an aliquot of surfactant stock solution (generally 1–5 mL) was added to the outer compartment to obtain the required surfactant concentration. To avoid any hydrostatic disturbances across the membrane, the surfactant solution addition was accompanied by the simultaneous addition of an equal volume of sulfate Ringer's solution to the inner compartment. All the surfactant stock solutions were prepared in the sulfate Ringer's solution.

Throughout the experimentation care was taken to avoid any hydrostatic pressure head across the membrane. It was observed that even a small pressure difference not only altered the observed potential, but also changed the rate of its decay following surfactant insult. In general, application of positive pressure from outside to inside increased the skin potential and its decay after surfactant insult, while the reverse effect was observed with negative pressure. Accordingly, the hydrostatic head across the membrane was kept at minimum. This precaution was necessary to obtain reproducible results, especially when using low surfactant concentration and for certain mixed-surfactant systems. Figure 2 illustrates the reproducibility obtained when the above precautions were taken.

Results and Discussion

In the present study, sulfate Ringer's solution was chosen over conventional chloride Ringer's solution. This was because sulfate ions, which

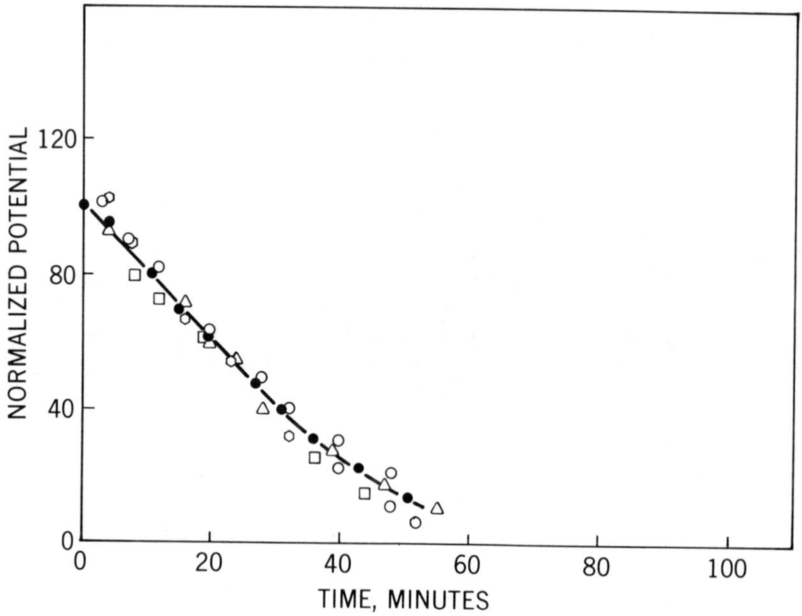

Figure 2. Potential decay pattern of the membrane in the presence of a representative alkyl benzene sulfonate surfactant. Reproducibility study.

have a lower permeability through the membrane, generate a higher membrane potential. Using sulfate also gave lower susceptibility to the hydrostatic pressure gradient across the membrane.

The effect of different surfactants, all used at the same concentrations, on skin potential is shown in Figure 3. In this figure, the normalized skin potential is plotted as a function of time following surfactant insult. Using normalized potential, in which the starting potential is set at 100%, facilitates comparison of surfactant effects. Actual starting potentials varied appreciably from skin to skin, but good reproducibility for a particular treatment was obtained when normalized potentials were used (*see* Figure 2). Figure 3 shows that after the first 10–20 min, the skin potential steadily decreases with time with 0.005% surfactant in the solution. The rate of potential decrease is higher for the cationic surfactant CTAB than for the anionic surfactants SLS and LAS. This observation is consistent with the well-known fact that cationic surfactants such as CTAB are much more aggressive to biological membranes than anionic or nonionic surfactants. In the absence of added surfactant, the skin potential maintained its value to within 10% even after 8 hr; addition of 0.005% CTAB completely destroyed the potential in about 60 min. Once the potential is destroyed in this way, repeated washing of the skin with

Ringer's solution will not restore it. In other words, the surfactants irreversibly impair the active metabolic sodium transport reactions which generate the potential. The distinct differences among various types of surfactant seen in Figure 3 disappear at higher concentration levels. This is shown in Figure 4 where normalized potential–time plots are given for 0.05% surfactant concentration. The effects for the different ionic surfactants are clearly much larger than for the nonionic Neodol 25-7EO. The above results clearly underline the importance of a charged head group on the surfactant molecule in its attack on the membrane.

Because of the wide disparity of decay rates caused by ionic and nonionic surfactants, it was interesting to examine the effect of mixing such species. Figure 5 demonstrates that a 0.05% SLS solution causes a more rapid decay of skin potential than a similar solution to which 0.05% Neodol 25-7EO is added, i.e. an intermediate effect (In work to be published elsewhere (*8*), it is shown that preapplication to frog skin of certain protectants such as polyoxethylenes or cationically substituted cellulose derivatives also mitigates the SLS effect.), rather than an additive effect, on the potential decay rate is obtained. The most likely cause of this is the formation of mixed micelles, which effectively decreases the concentration of free surfactant anion (LS^-) monomer. However, reduction of the LS^- monomer concentration does not provide a full explanation for the observed effect. The critical micelle concentration (cmc) of

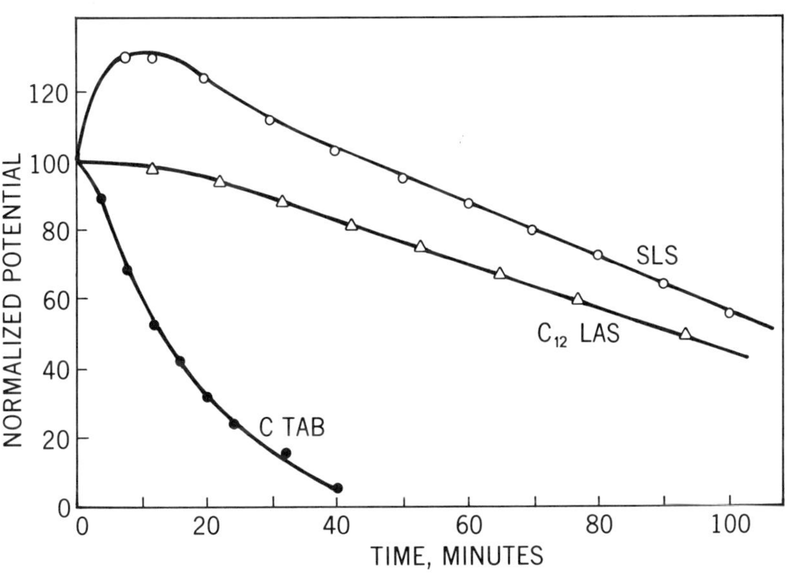

Figure 3. Effect of various surfactants at 0.005% concentration on the membrane potential

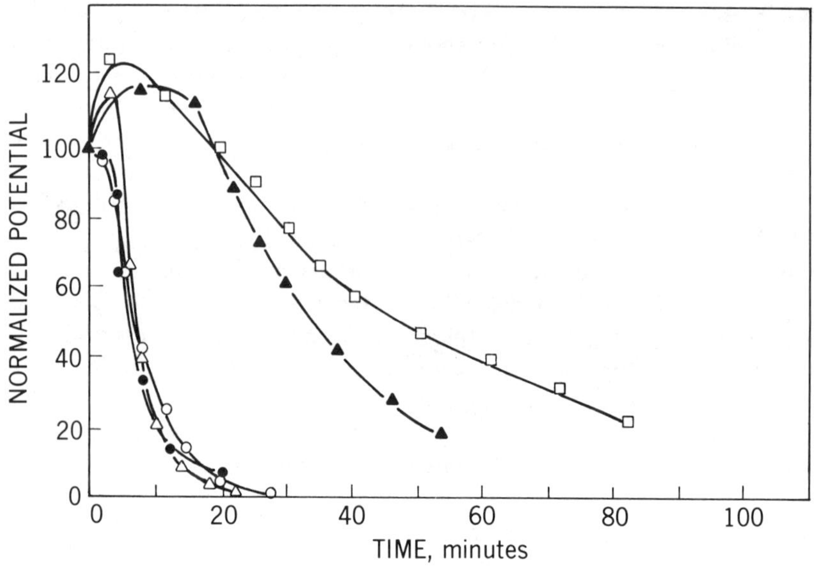

Figure 4. Effect of various surfactants at 0.05% concentration on the membrane potential: (●), C TAB; (○), C_{12} LAS; (△), SLS; (□), N 25-7EO; and (▲), AES (3EO).

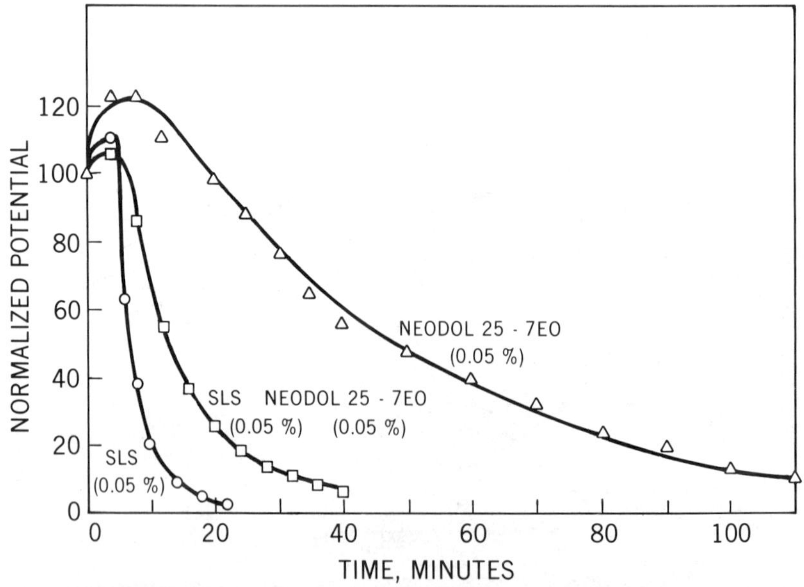

Figure 5. Mitigating effect of Neodol 25-7EO on SLS action

SLS in sulfate Ringer's solution is 0.04% or $1.3 \times 10^{-3}M$. From tabulated data of cmcs and salt effects (*9, 10*) in the literature one can estimate the cmc of an equiweight mixture of SLS and Neodol 25-7EO to be at most one tenth of this concentration. Reference to Figures 5 and 6 indicates that the observed decay rate for the mixture clearly exceeds the additive effect of 0.05% Neodol 25-7EO and free LS^- at a level of $< 0.005\%$. One concludes either that the mixed micelles of SLS and Neodol 25-7EO have an effect greater than that of the nonionic micelles alone, or that the membrane activity of the free SLS (at its reduced level) is greater in the presence of the nonionic surfactant. That the activity of the mixed micelles is less than that of SLS micelles is evident from results given later in the chapter (*see* Figures 5, 10, and 11). It is also interesting that the potential destroying activity of AES lies between that of an alkyl sulfate, for example, and a nonionic surfactant (*see* Figure 4). This might well result from shielding of the charge of the anionic head group by the neighboring ethoxy groups.

Figures 6–11 demonstrate the effect of surfactant concentration on the decay of skin potential. Several features can be noted from these figures.

1. An increase in the surfactant concentration increases the potential decay rate. Figures 10 and 11, discussed later, show that the decay rates observed with SLS and CTAB show a pronounced increase above their respective cmc's.
2. While a continuous potential decay with time is obtained most frequently, SLS, except at high concentration, shows an initial sharp rise in potential followed by a steady decrease. The initial rise in potential may be the result of physical adsorption of SLS onto the membrane and an increase of its charge, which will influence its permeability to sulfate ions prior to the eventual potential destruction process. The higher the surfactant concentration, the lower the height of the maximum, and the quicker it is achieved.
3. An exponential or nonlinear potential decay pattern is observed with surfactant concentrations higher than 0.005%, and a linear potential decay is obtained at lower concentrations.

Analysis of the Data. Surface-active agents, especially the ionic and, more especially, the cationic type, have pronounced biological activity. Early experiments (*11*), for example, demonstrated the pronounced lytic activity of soaps and alkyl sulfates towards erythrocytes. Other experiments have demonstrated the interfacial activity of ionic surfactants towards various lipid materials such as cholesterol (*12*), and also towards protein (*13*). While such information is well known, knowledge of how surfactants impair cell structure, and, more particularly, interfere with the metabolism of living cellular systems, is still

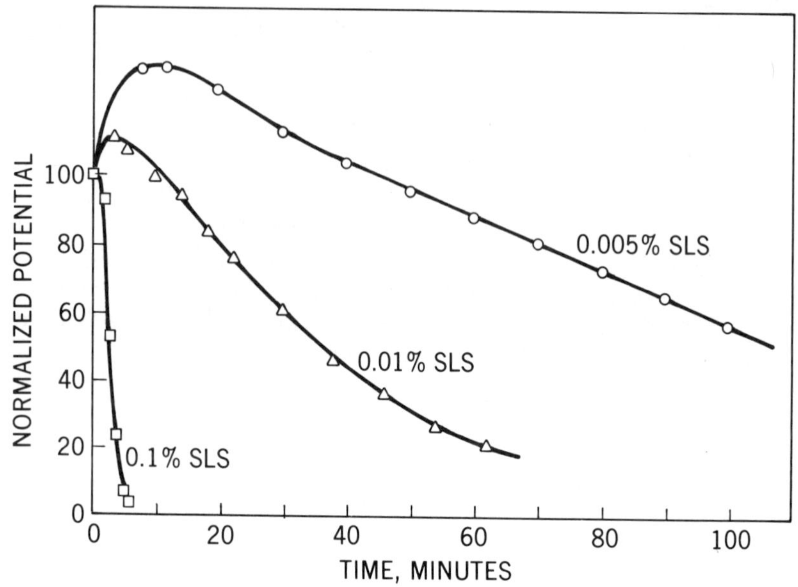

Figure 6. Effect of SLS concentration on the membrane potential decay

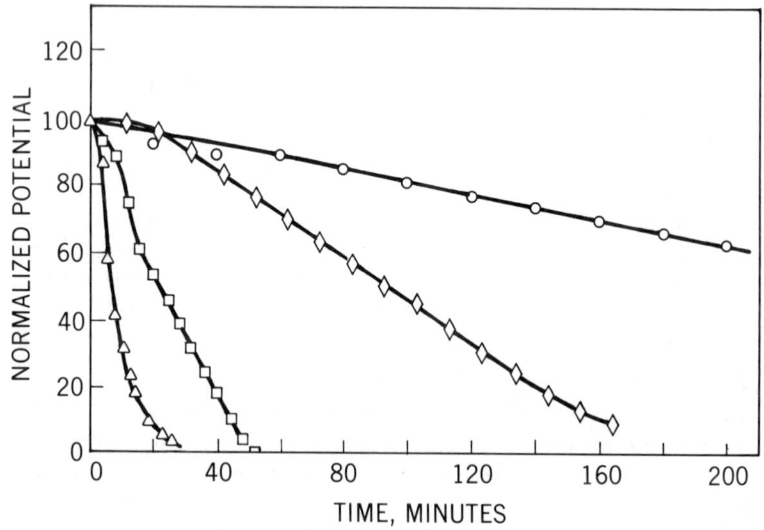

Figure 7. Effect of C_{12} LAS on the membrane potential: (○), 0.001% C_{12} LAS; (◇), 0.005%; (□), 0.02%; and (△), 0.05%.

fragmentary. Accordingly, we are not concerned here with mechanistics but rather present a mathematical and phenomenological description of the observed effects.

The linear potential decay observed at lower surfactant concentrations can be expressed as:

$$\frac{d\Delta\Psi}{dt} = -k \tag{1}$$

where k is a zero-order rate constant. Integration of Equation 1 under the limits $t = 0$ and $\Delta\Psi = \Delta\Psi°$, gives:

$$\Delta\Psi = \Delta\Psi° - kt \tag{2}$$

As mentioned earlier, such a decrease in the skin potential following surfactant addition results from interference with the metabolic reactions which generate the membrane potential through active sodium transport.

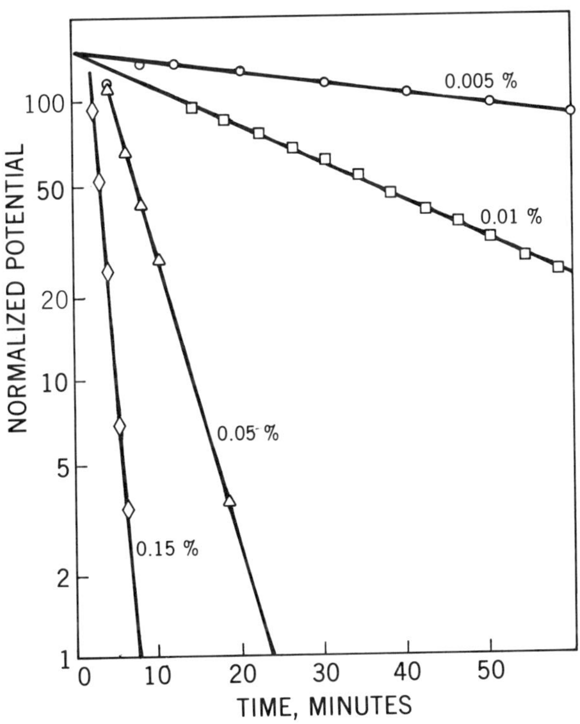

Figure 8. Logarithmic potential decay pattern of the membrane in SLS solutions

The ionic flux, J_i, of Species i across the membrane is expressed by a relationship similar to that developed by Kedem (*14*), viz.

$$J_i = \frac{ZF\Delta\Psi}{R_i} - \frac{R_r}{R_i} J_r \tag{3}$$

where Z is the valence of the ion, F is Faraday's constant, $\Delta\Psi$ is the membrane potential, and R represents resistance. J_r is tthe contribution to J_i arising from coupling with the flow of the metabolic membrane reaction; no other coupling reactions are assumed. The first term on the right-hand side of Equation 3 is the contribution to J_i resulting from a difference in the electrochemical potential of i across the membrane, i.e., it represents passive transport of i.

In the present case Na^+ and SO_4^{--} are the main transporting ionic species and their flux are written as

$$J_{Na^+} = -\frac{F\Delta\Psi}{R_{Na^+}} + \frac{R_r}{R_{Na^+}} J_r \tag{4}$$

$$J_{SO_4^{--}} = \frac{2F\Delta\Psi}{R_{SO_4^{--}}} \tag{5}$$

Under open-circuit and steady-state conditions, electrical neutrality across the membrane will be maintained. Thus

$$J_{Na^+} = 2J_{SO_4^{--}} \tag{6}$$

From Equations 4, 5, and 6 we have

$$\Delta\Psi = \left[\frac{R_{Na^+}R_{SO_4^{--}}}{4R_{Na^+} + R_{SO_4^{--}}} \cdot \frac{R_r}{FR_{Na^+}}\right] J_r \tag{7}$$

The expression in parentheses is an index of permeability and, hence, Equation 7 is expressed as

$$\Delta\Psi = \lambda J_r \tag{8}$$

At constant membrane permeability

$$\Delta\Psi \propto J_r \tag{9}$$

i.e., the potential generated across the skin is directly proportional to the flow of the metabolic reaction responsible for active sodium transport.

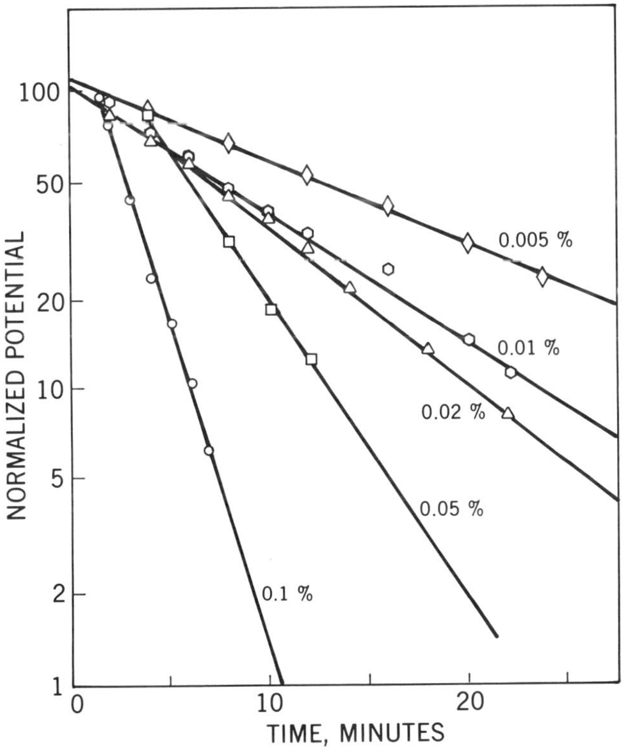

Figure 9. Logarithmic potential decay pattern of the membrane in CTAB solutions

The effect of added surfactant on J_r is seen by combining Equations 2 and 8 as follows

$$J_r = \frac{1}{\lambda} [\Delta\Psi^\circ - kt] \tag{10}$$

Equation 10 indicates how the metabolic reaction rate decreases linearly with time after surfactant insult, assuming constant membrane permeability. This assumption was checked by determining tritium transport before and during exposure of a membrane to 0.5% SLS surfactant solution. These experiments, conducted using a diffusion cell modeled after Loveday (*15*), showed that the water flux across the membrane did not change noticeably in the 1-hr period following exposure to the surfactant. Actual values obtained were 4.3×10^{-5} and 4.5×10^{-5} g/cm^2/sec, with and without 0.5% SLS in Ringer's solution, respectively.

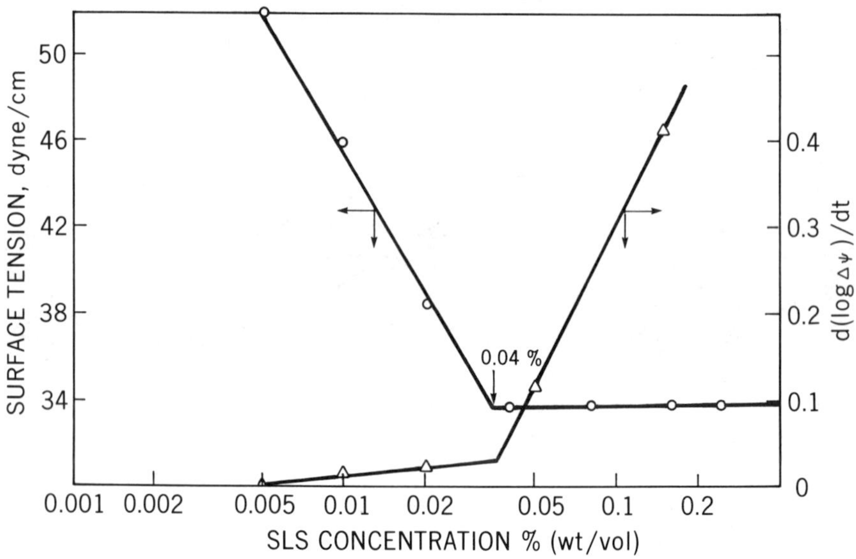

Figure 10. Effect of SLS concentration on surface tension and the logarithmic potential decay rate of the membrane

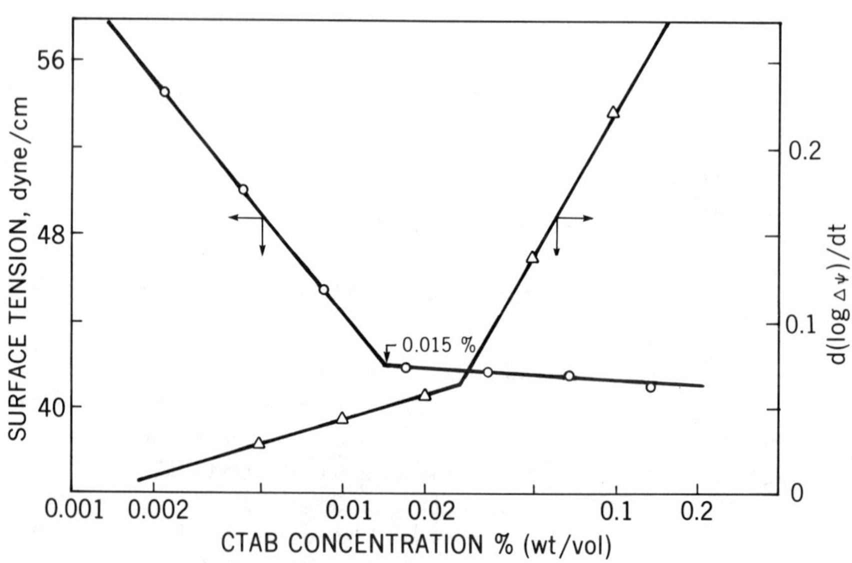

Figure 11. Effect of CTAB concentration on surface tension and the logarithmic potential decay rate of the membrane

The curves exhibiting nonlinear potential decay, i.e., at higher surfactant concentration, are replotted in Figures 8 and 9, in semilogarithmic form. The plots reveal linearity over most of the time course indicating that the potential decay follows first-order kinetics. Under these conditions the rate of destruction of potential is proportional to that potential itself. Then

$$\frac{d\Delta\Psi}{dt} = k'\Delta\Psi \tag{11}$$

where k' is the first-order rate constant. Integration of Equation 10 under the limits $t = 0$ to t and $\Delta\Psi = \Delta\Psi^\circ$ to $\Delta\Psi$ gives

$$\ln \Delta\Psi = -k't + ln\ \Delta\Psi^\circ$$

or

$$\Delta\Psi = \Delta\Psi^\circ e^{-k't} \tag{12}$$

From Equations 8 and 12

$$J_r = \frac{1}{\lambda}\ \Delta\Psi^\circ e^{-k't} \tag{13}$$

Hence, sodium transport by the metabolic reaction will be inhibited exponentially with time, again assuming λ is constant.

The first-order rate constant k' is a function of surfactant concentration. Its relationship to concentration is evident from Figures 10 and 11 in which k' is plotted against SLS and CTAB concentration, respectively. A sharp increase in the first-order decay rate is obtained near the cmc of the surfactant, as determined from surface tension and concentration measurements. These plots show that these surfactants are much more effective in destroying the potential of the membrane when they are in micelle form. This observation is interesting in light of work reported by Choman (*16*) in which he observed a significant increase in the swelling and water uptake of calf skin when the surfactant concentration of the solution contacting it exceeded the cmc. The effects of surfactant concentration reported here are consistent with those reported earlier by Webb (*17*) who obtained evidence of the active sodium pumping's being inhibited at higher surfactant concentration.

Finally, comments are in order concerning the high potency of the surfactants above their cmc. Although no change in permeability was detected during the time span of the potential tests, it is possible that certain structural changes were initiated. Micellar solutions of surfactants are known to solubilize and complex with components of cells, such as lipids and proteins. Qualitatively, over long periods of exposure, e.g. overnight, the more concentrated surfactant solutions brought about changes in the texture of the membranes. Exposure over a weekend substantially reduced the strength of the membranes which were torn easily. More work is required to establish what structural changes may be involved in the first few minutes of exposure.

Conclusions

1. Surfactants irreversibly destroy the natural potential across frog skin membranes and the effect increases with concentration. At lower surfactant concentration, potential decays linearly with time following zero-order kinetics; at higher surfactant concentrations, potential decays logarithmically with time following first-order kinetics.
2. A sharp increase in the potential decay rate is observed above the cmc of ionic surfactants.
3. The cationic surfactant CTAB causes the most rapid destruction of potential while the nonionic surfactant exhibits the least effect on skin potential with the effect of anionic surfactants being intermediate.
4. Mixtures of ionic and nonionic surfactants cause slower potential decay than the ionic surfactant alone.

Glossary of Symbols

Ψ = potential across membrane of time t
Ψ° = potential at $t = 0$
J_i = ionic flux of Species i across membrane
R_i = resistance to Species i
Z = valence of Ion i
F = Faraday constant
J_r = contribution to J_i from coupling
R_r = the corresponding resistance term
λ = permeability index of the membrane

Literature Cited

1. Ussing, H. H.; Zerahn, K. *Acta Physiol. Scand.* **1950**, *23*, 110–127.
2. Koefoed–Johnsen, V.; Ussing, H. H. *Acta Physiol. Scand.* **1958**, *42*, 298–308.
3. Brittar, E. E. In "Membranes and Ion Transport"; Wiley Interscience: London, 1970; Vol. 1.
4. Schultz, S. G.; Curran, P. F. *Physiol. Rev.* **1970**, *50*(4), 637–718.
5. Linderholm, H. *Acta Physiol. Scand.* **1952**, *27*, 1–143.
6. Hoffman, J. F. In "The Cellular Functions of Membrane Transport"; Prentice–Hall: Englewood Cliffs, NJ, 1969.
7. Rennie, G. K.; Hill, J. C.; Goddard, E. D.; Kulkarni, R. D. *J. Pharm. Pharmacol.* **1975**, *27*, 363–366.
8. ———, unpublished data.

9. Schick, M. J.; Manning, D. J. *J. Am. Oil Chem. Soc.* **1966**, *43*, 133.
10. Schick, M. J. *J. Phys. Chem.* **1963**, *67*, 1796.
11. Schulman, J. H.; Rideal, E. K. *Proc. R. Soc. London, Ser. B* **1937**, 122, 29.
12. Goddard, E. D.; Schulman, J. H. *J. Colloid Sci.* **1953**, *8*, 309.
13. Schulman, J. H.; Rideal, E. K. *Proc. R. Soc. London, Ser. B* **1937**, 122, 46.
14. Schultz, S. G. "Mechanisms of Adsorption," In "Biological Membranes"; Dowben, R. M., Ed.; Little, Brown: Boston, 1969.
15. Loveday, D. E. *J. Soc. Cosmet. Chem.* **1961**, *12*, 224.
16. Choman, B. R. *Investig. Dermatology* **1961**, *37*(4), 263–271.
17. Webb, G. D. *Acta Physiol. Scand.* **1965**, *63*, 377–384.

RECEIVED October 17, 1978.

27

Impedance of Living Cell Membranes in the Presence of Chemical Tissue Fixative

G. S. MARGULES[1], S. B. DOTY, and A. A. PILLA[2]

Bioelectrochemistry Laboratory, Orthopedic Research Laboratories and Department of Applied Chemistry and Chemical Engineering, Columbia University, 630 W. 168th Street, New York, NY 10032

The role of electrochemical information transfer in cell regulation has been studied by evaluating the transient impedance of the toad bladder membrane. These studies have isolated the relaxation times associated with dielectric, specific adsorption, phase transfer, and membrane transport processes for both the living and glutaraldehyde-fixed membrane. The results for the living system show that passive Na^+ transport is rate limited by the phase transfer step and that specific adsorption may play a "gating" role in this process. Both active and fixed membranes exhibit similar dielectric and specific adsorption behavior, but the transport pathway is significantly different. The active membrane appears to have an adsorption layer of Na^+ in the membrane phase which, along with phase transfer, regulates transport. In comparison, fixed-membrane transport can be interpreted by a diffusion layer approach wherein membrane concentration is distributed uniformly and does not appear selective.

Recent studies have shown that the concept of electrochemical information transfer at living cell membranes (*1–9*) can be useful to describe the criteria necessary to modify the structure at a cell's electrified surfaces and junctions in a functional manner. This approach was inspired by the

[1] Present address: Cordis Corporation, P.O. Box 370428, Miami, FL 33137.
[2] To whom correspondence should be addressed.

0-8412-0473-X/80/33-188-461$06.00/1

early pioneering work of Yasuda and Fukada (*10, 11*) followed by that of Bassett and Becker (*12, 13, 14*), who proposed that changes in the electrical environment of a living cell could beneficially modify function. In order to illustrate the existence of potential dependent surface processes at living-cell membranes, experiments were designed involving the determination of the impedance of the toad urinary bladder epithelial system. Comparison of experimental results with membrane models, which took into account surface processes such as specific adsorption (binding) and phase transfer (partitioning) coupled with membrane and aqueous transport, showed that the electrochemical information transfer concept could be quantitated (*5*). In other words, the number of surface and surface-coupled transport processes present in this system could be detected. In addition, the value of each relaxation time (or time constant) was established. This indicated that such studies could provide a meaningful approach to the elucidation and, therefore, the possibility of modification of living-cell function. Both in vitro studies involving amphibian red blood cell dedifferentiation (*6, 8*), Ca^{2+} kinetics in embryonic chick limb rudiments (*7*), and in vivo clinical applications wherein otherwise incurable bone fractures have been consolidated (*15*) show that selective chronic modification of cell surface and membrane transport phenomena, by the application of specific pulsating current, can lead to beneficial modification of cell function in tissue growth and repair.

One of the most important results from the toad bladder impedance studies is the ability to isolate and quantitate potential dependent cell surface processes from membrane and aqueous transport. In this manner, relaxation times of the various interfacial and transport processes, excitable by low-level (microampere per square centimeters), pulsating current, could be evaluated. Furthermore, results obtained using various tissue ionic environments (*16*) shows that the regulating step for passive Na^+ transport in this system appears to be a phase transfer (penetration) step of finite (relatively slow) rate. Experiments such as this provide the ability to quantitate the change in relaxation time of one or several processes affected by changes in cellular environment which may occur during various portions of a given cell's functional state or cycle.

The methodology used for living-cell impedance studies was designed to allow surface and transport (aqueous or membrane) to be uncoupled in time (frequency) if the relaxation processes involved had sufficiently different kinetics. This was achieved by obtaining impedance data over a very wide frequency range in order to allow mechanistic-step uncoupling. Using this basic electrochemical relaxation methodology, this chapter studys the effect of a modification of membrane structure using tissue fixatives.

Initially, using a substance such as glutaraldehyde was considered to mimic cell death and therefore the passive sodium transport pathway would be isolated further from any possible coupling with metabolic pathways. In fact, this study shows that not only does the latter occur, but the Na^+ membrane transport step is altered while surface parameters such as specific adsorption appear unchanged.

Chemical fixatives such as glutaraldehyde often are used to stabilize large molecular structures so that three-dimensional structural analyses can be carried out. When a living cell membrane is "fixed", some bulk membrane components are stabilized structurally. If these entities are involved in the transport pathway detectable by impedance measurements, changes in model parameters and even the basic models can be expected. In juxtaposition to classical studies using radioactive tracers in which steady-state membrane transport is evaluated, the impedance technique used in this study allows the determination of associated kinetic parameters. Therefore, this type of study is viewed as complementary and perhaps synergistic to steady-state evaluations which provide information of a thermodynamic nature.

Theory

In order to quantitate the response of a living membrane to the passage of dynamic current, it has been necessary to consider both interfacial and transport steps (*3, 5, 9, 17, 19–25*). The original analysis considered that the total current was composed of a charging portion not involving membrane transport and a phase-transfer (partitioning) portion wherein the passage of current carrying species across the membrane can occur. This led to a description of membrane impedance (*5*) which could be used to explain the transient electrochemical response of the toad urinary bladder. Central to this model is the consideration that membrane transport is driven by both concentration (diffusion) and electric field gradient (migration). When glutaraldehyde is used on a living-cell system, one of the primary results of molecular stabilization is a disappearance of resting transmembrane potential differences. Since these potentials usually are associated with ionic transport, it is clear that migration effects in passive Na^+ transport are not the same as those in unfixed preparations. Because of this it is useful to briefly review the characteristics of the general membrane model which are relevant for the active and fixed toad bladder.

Recall that the total membrane current $i(s)$ (s is the complex frequency variable of the Laplace transformation) is the sum of a charging current, $i_c(s)$, and a phase-transfer current, $i_p(s)$, and that both components have been detected in the unfixed bladder system. Under these conditions the following may be written

$$i_c(s) = q_E s_\eta(s) + q_i s \Gamma_i \Delta\Gamma_i(s) \tag{1}$$

which shows that the charging current is dependent both on a change in potential, $\eta(s)$, and a change in concentration, $\Delta\Gamma_i(s)$, of a specifically adsorbed (bound) species at the membrane surface. In Equation 1 the coefficients q_E and q_i represent the variation of charge q_i with potential (E) and aqueous concentration C_i of the adsorbing species, respectively; Γ_i is the resting surface concentration of the adsorbed entity. The kinetic expression of adsorption, represented by $\Delta\Gamma_i(s)$ in Equation 1 is given by (*18*)

$$\Delta\Gamma_i(s) = \frac{a}{1 + \Gamma_i s/v_i}\,\eta(s) \tag{2}$$

where a is a coefficient representing the potential dependence of, and v_i is the exchange rate constant for, adsorption. Equation 2 is written for the condition that the adsorbing species does not develop a concentration gradient, a situation actually observed using an unfixed bladder membrane preparation (*5*). When Equation 2 is combined with Equation 1, the admittance $Y_c(s)$ corresponding to the charging current pathway is

$$Y_c(s) = q_E s + q_i\Gamma_i s/(1 + \Gamma_i s/v_i) \tag{3}$$

which shows by inspection that the impedance $Z_c(s) = 1/Y_c(s)$ for this portion of the membrane relaxation (or excitation) can be described exactly by an aperiodic equivalent electric circuit consisting of a capacitor, C_D, in parallel with a series resistance/capacitance, R_A/C_A, combination (*see* Figure 1). The former element represents the dielectric behavior of the membrane, related essentially to its overall lipoprotein structure, and the latter *RC* combination represents the specific adsorption process. Since the toad urinary bladder exhibits passive Na^+ transport, it is necessary to provide for a phase-transfer current pathway. This may be written for an unfixed membrane as (*3, 5, 17*)

$$i_P(s) = I_E\eta(s) + I_M C_m \Delta C_m(o, s) \tag{4}$$

in which I_E and I_M are coefficients representing the potential dependence and membrane concentration dependence of phase transfer, respectively. Note that aqueous concentration gradients are not considered (either intra- or extracellular), since experimental data shows that the rate-limiting step appears to be the penetration or partitioning of Na^+ into the membrane phase, not its transport through, up to, or away from, the membrane system. The quantity $\Delta C_m(o, s)$ represents the variation in membrane concentration, C_m, of the transporting species as a result of the current perturbation and is written for the plane $x = 0$ (e.g. extracellular fluid/membrane interface).

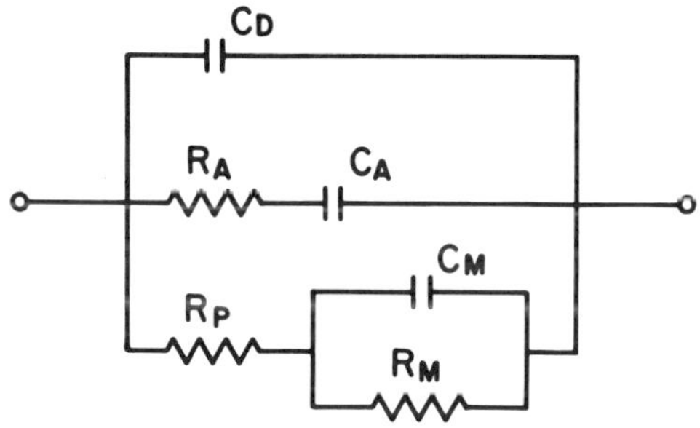

$C_D = 0.385 \pm 0.123 \mu F$ $R_P = 3525 \pm 1111 \Omega$

$R_A = 242 \pm 90 \Omega$ $R_M = 449 \pm 418 \Omega$

$C_A = 0.459 \pm 0.174 \mu F$ $C_M = 99 \pm 87 \mu F$

Figure 1. Electrical equivalent circuit representing the transient linear response of the toad urinary bladder membrane system exposed to normal amphibian Ringer's solution on both mucosal and serosal surfaces. Three relaxation processes are shown corresponding to dielectric (C_D) *specific adsorption* R_a, C_a, *and membrane transport* (R_P, R_M, C_M) *phenomena. The values for these parameters were obtained for 38 bladder preparations illustrating that the functional response for each membrane is identical.*

The current pathway for transmembrane transport can be written using Equation 4 provided that $\Delta C_M(o, s)$ is expressed in explicit form. This has been derived previously in a general manner using the Nernst–Plank equation with no restriction on the value of the voltage field within the membrane phase. Thus, the membrane transport impedance, $Z_P(s)$, is

$$Z_P(s) = \frac{1}{I_E}$$

$$\left[1 + \frac{1}{nFC_mD_M[FV/2RT + 1/\delta_M + \delta_M/3(F^2V^2/R^2T^2 + s/D_M]}\right] \quad (5)$$

in which D_M is the membrane phase diffusion coefficient, V is the voltage field developed across the membrane thickness, δ_M, and the other terms have been defined previously, or have their usual significance. Inspection of Equation 5 shows a series configuration of both phase-transfer (partitioning) and mass-transport steps. The former is represented by a pure resistor, R_P, and the latter by a parallel resistor/capacitor combination, R_M/C_M (*see* Figure 1).

The total impedance of the unfixed toad bladder membrane system can be defined now by Equations 3 and 5. Inspection of these equations shows that, under linear perturbation conditions, this membrane can be represented exactly by the aperiodic equivalent electric circuit shown in Figure 1. The physical significance of all of the elements has been obtained (*5*) by comparison with Equations 3 and 5. For a bladder membrane exposed on both mucosal (inner) and serosal (external) surfaces with normal amphibian Ringer solution the transient response is described by the model just discussed. Of importance is the fact that two surface processes of specific nature have been isolated. Thus a specific adsorption relaxation (R_A, C_A; (*see* Figure 1)) not involving the transporting species is present. This process appears to be slow enough (i.e. R_A sufficiently large, *see* Figure 1) to prevent the build-up of aqueous concentration gradients of the adsorbing species, the identity of which is not clear at the present time. The second surface process isolated is the phase-transfer step (R_p, Figure 1) of Na^+ membrane transport (R_M, C_M (*see* Figure 1)). All of the experiments to date using unfixed membrane preparations (*5, 16*) exposed to normal Ringer solution show that the phase-transfer step is rate limiting (i.e. $R_p >> R_M$; *see* Figure 1). In addition, as the mucosal Na^+ concentration is decreased, R_P increases, i.e. the penetration step becomes even more rate limiting. This result shows that less Na^+ passively enters the epithelial cell, which ultimately returns it to the circulation and tissue via the Na^+/K^+ active transport systems, as the mucosal Na^+ concentration decreases. It thus seems that Na^+ may regulate its own passive transport via a specific membrane surface event which is perturbable by a transient change in interfacial potential.

As mentioned above, the derivation of the model used to explain the transient impedance of the active bladder membrane system made no restrictions on the value of the voltage field present in the membrane phase. In addition the absence of aqueous transport polarization allowed the specific contribution of membrane transport in Equation 5 to be derived. The expression obtained for membrane transport indicates that there should be a resting membrane concentration, C_m, of Na^+ which is given by

$$C_M = I_E/I_M \, nFC_m\delta_M/3 \tag{6}$$

This equation shows that C_M is proportional to the membrane thickness δ_M and is coupled to both phase-transfer kinetics, I_E, and membrane transport kinetics via I_M.

Since Equation 6 has been evaluated for an active membrane, it appears that Na^+ transport kinetics may involve a certain membrane residence time which plays some regulatory role. It would be expected, therefore, that chemical fixation may allow further elucidation of this transport phenomena if the interactions of Na^+ with membrane structure can be modified.

Experimental and Results

For this study the membrane system of the isolated toad (Bufo marinus) urinary bladder is used. The isolation procedure developed for this work is given below. The toad is pithed so that no further brain or sensory activity is present. Under these conditions the bladder may be excised rapidly while maintaining an isotonic saline environment. The bladder lobes are tied with suture material and placed in a container of aerated amphibian Ringer solution (*see* Figure 2). While in this condition, the bladder is opened, care being taken not to disturb the mucosal (inner) layer, uniformly stretched on a cork template, and placed in the experimental chamber (*see* Figure 3). For the initial studies, the solution in each compartment of the experimental chamber (i.e. on both the mucosal and serosal sides of the membrane) is, in mM/1: NaCl, 109; $CaCl_2$, 0.9; $NaHCO_3$, 2.4; KCL, 2.5; and $C_6H_{12}O_6$, 5.5. For chemical fixation of the membrane a 1% glutaraldehyde solution is made using the above Ringer's solution. In this manner the preparation is fixed under the desired initial active condition. The chamber is constructed such that 1 cm^2 of bladder area is exposed. It has provision for rapid and simultaneous filling of each compartment via syringe entry points to avoid membrane deterioration and hydrostatic pressure differences (*26, 27, 28, 29*). Once mounted, the bathing solutions are kept aerated and

Figure 2. A typical isolated toad bladder membrane system. The bi-lobed structure is removed while in a filled condition.

Figure 3. The chamber used for transient impedance studies. The stretched bladder is shown pinned on the cork template which is in position prior to apposition of the two chamber compartments. The syringes were used for simultaneous compartment filling.

the resting transepithelial potential difference (TEPD) is monitored using SCE reference electrodes (Tacussel Type C-10) via an electrometer (Keithley model 602). Bladders having a 20–50-mV TEPD were chosen for study since this potential range indicated good cellular integrity (*30, 31*). Note that when glutaraldehyde is added to the membrane the TEPD drops to zero within 2 min, indicating that function related to Na–K-activated ATPase enzyme activity ceases. An electron micrograph of the fixed bladder is shown in Figure 4. The standard organelles, surface coat, and tight junctions are visible, indicating normal epithelial cell ultrastructure. In the fixed state, bulk structural integrity is maintained, rendering the preparation appropriate for impedance analysis.

A four-electrode system is used for all impedance measurements. The perturbation signal is applied via two high-surface-area carbon rods to reduce faradaic polarization effects. Membrane response is obtained via two sintered tantalum electrodes which are placed within 0.5 mm of each membrane surface to minimize ohmic drop contributions. The experimental impedance system is shown in schematic form in Figure 5. All measurements are performed using low-amplitude input perturbation such that the membrane experiences only a small fraction ($< 10\%$) of the TEPD. In this manner the linear conditions required for model verification are fulfilled.

In order to obtain the wide frequency range necessary to evaluate the multiparameter models appropriate for this membrane, a measurement system capable of 100-nsec time resolution has been used (*16*). Since the desire is to freeze the membrane behavior in time, it is necessary to apply as few perturbations in as short a time as possible. For this reason, signal-averaging techniques cannot be used not withstanding the requirement for millivolt-level perturbation to satisfy linearity conditions. This has prompted the development of a new membrane impedance technique wherein the shape of the input perturbation greatly increases the signal:noise ratio. In essence, this approach involves allowing the shape of the input waveform to be determined by the impedance of the membrane itself (*32*). This allows the signal amplitude to be greatly increased at the critical short time range when the membrane response is lowest. For this, the output impedance of the step function generator (Monsanto model 300A) is made much lower than that of the membrane (*see* Figure 5).

The input current waveform, $I(t)$, thus established is measured across a resistor (*see* Figure 5). Both input and response, $V(t)$, wave-

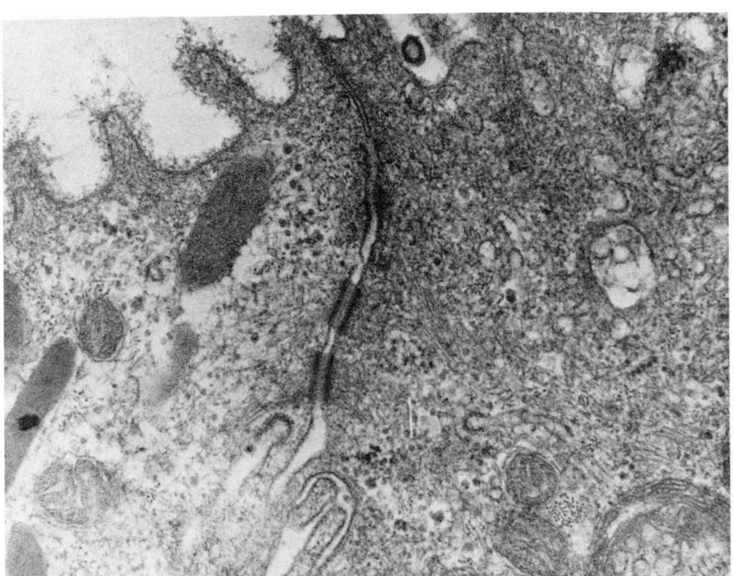

Figure 4. Electron micrograph (30,000×) illustrating the apical mucosal surface of two adjacent epithelial cells of the isolated toad bladder. The unit cell membrane and the villus projections into the bladder lumen are covered with a fibrous glycocalyx-like material. The lateral cell membranes are connected by tight junctions and desmosomes, and followed by extensive interdigitation of the membranes to the basal surface. The cell cytoplasm contains mitochondria, dense granules, free ribosomes, and numerous microfilaments.

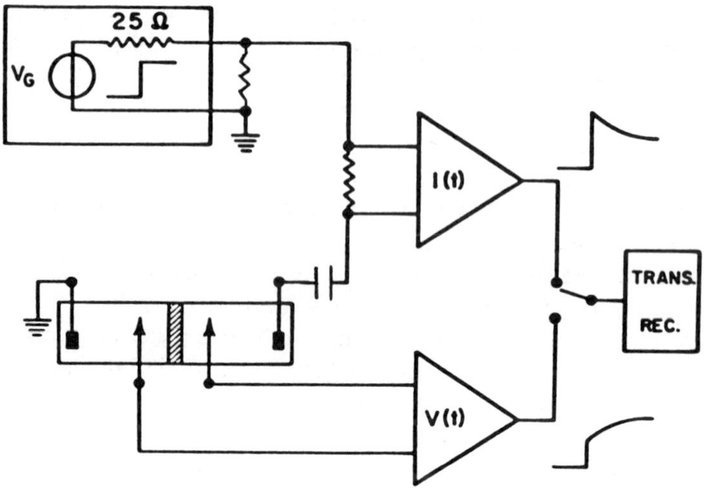

Figure 5. Schematic of the experimental set-up used for transient impedance studies of the toad bladder membrane system. The generator, V_g, has relatively low output impedance so that the input waveform (measured via the resistor across the inputs of amplifier I(t)*;* top curve*) is determined by the membrane impedance. Digital transient recording allows a time resolution of 100 μsec to be achieved. Membrane response is obtained via amplifier* V(t) (bottom curve*) which is driven by measuring electrodes independent of those via which* I(t) *is applied.*

forms are detected using a differential amplifier system (Tektronix model P6046 and Electro-Biology model 100033) and recorded using a transient recorder (Nicolet model 1090A) having a 100-nsec time resolution. Each signal is stored on cassette tape (Nicolet model NIC-193) for future processing. The raw experimental data stored on cassette tape is displayed on an XY recorder (Esterline Angus model XY530), smoothed, and points are chosen for digital Laplace transformation (*33, 34, 35*). This is performed along the real axis (σ) of the complex frequency plane (s). The real impedance function, $Z(\sigma)$, describing the transient behavior of the bladder membrane system, thus obtained is evaluated over the frequency range, 10^{-1}–10^{7} rad/sec. This wide range allows the isolation of relaxation steps owing to surface processes from those involving transport phenomena, as has been shown already for an active membrane system (*5*).

Once $Z(\sigma)$ is obtained the possible relaxation pathways are separated for analysis using a graphical de-embedding technique. The individual time constants are evaluated starting at the highest available frequency range and continuing by functional analysis to steady-state (dc) response. Based on the original impedance model the following function should be observed over the highest frequency range ($\sigma \to \infty$)

$$Z(\sigma) = R_e + 1/C_D\sigma \tag{7}$$

This allows isolation of the electrolyte resistance, R_e (present owing to the electrolyte pathway between the membrane surfaces and the monitoring electrodes, (*see* Figure 5) and the dielectric capacitance of the membrane, C_D. A representative plot of this type obtained for an active membrane is shown in Figure 6. Linearity obtains only over the highest frequency range, illustrating that a second relaxation process is present. Because $Z(\sigma)$ is a real function, the nature of this second time constant may be tested merely by subtracting the values of R_e and C_D previously obtained. If this process involves specific adsorption not involving membrane or aqueous transport (as was suspected in previous studies (5)) the following can be written

$$\frac{1}{B(\sigma)} = \frac{1}{1/(Z(\sigma) - R_e) - C_D\sigma} = R_A + \frac{1}{C_A\sigma} \tag{8}$$

where R_A and C_A represent the kinetics and surface concentration of the specifically adsorbed species, respectively. Their physical significance is given by Equation 3. A plot of this type for an active membrane is shown in Figure 7 where linearity is observed over a sufficiently wide

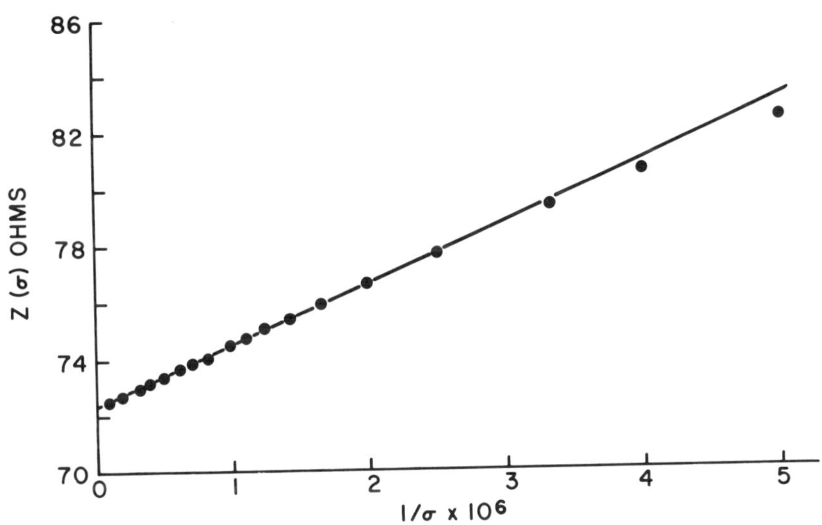

Figure 6. Plot showing functional analysis of the real axis impedance function for the toad bladder membrane. This behavior is observed for all preparations, active or fixed, and represents the highest frequency range attainable (up to 10^7 rad/sec) in this study. This plot allows the dielectric capacitance, C_D, of the membrane to be evaluated. This quantity is identical before and after glutaraldehyde fixation.

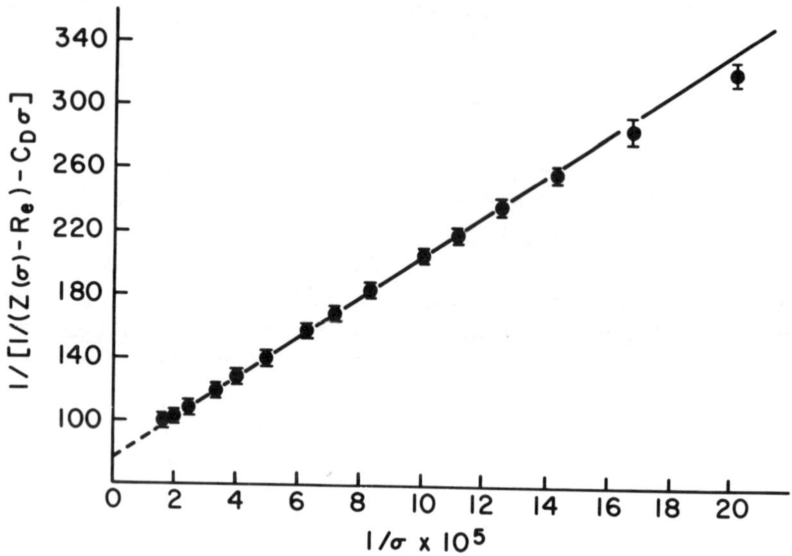

Figure 7. This plot shows the graphical de-embedding procedure which isolates the specific adsorption relaxation pathway for the toad bladder membrane. The quantities R_A *and* C_A *thus obtained are present for both active and fixed preparations.*

frequency range to adequately isolate these parameters. Note that in a previous study (5) the specific adsorption pathway was not able to be isolated because of a lack of sufficiently high frequency data. (The required time resolution of 100 nsec was achieved in this study primarily as a result of the addition of the transient recorder). Once again, deviation from linearity for this function as σ decreases shows the presence of a third relaxation process. Progressive isolation has revealed that the third time constant appears related to the membrane transport pathway. Under these conditions, for $\sigma \rightarrow$ o:

$$\frac{1}{F(\sigma)} = \frac{1}{B(\sigma) - C_A\sigma/(1 + R_AC_A\sigma} = R_P + R_M - R_M{}^2C_M\sigma \quad (9)$$

which is sufficient to diagnose the membrane transport pathway shown in Figure 1. Here R_P is the phase-transfer resistance and R_M and C_M represent membrane transport kinetics as given in Equation 5. A plot of this type is shown in Figure 8. Note that the behavior shown in Equation 9 always is observed for an active membrane attesting to the presence of migration—diffusion with a nonnegligible voltage field in the membrane phase-transport pathway. The final separate evaluation of R_P, R_M, and

C_M requires using an additional functional relationship which, since the sum $R_P + R_M$ is known, is written as:

$$\frac{1}{1/F(\sigma) - (R_P + R_M)} = -\frac{1}{R_M} - \frac{1}{R_M{}^2 C_M \sigma} \qquad (10)$$

This expression allows the isolation of R_M and combined with Equation 9 the independent evaluation of R_P and C_M. Note that Equation 9 and 10 are not absolutely necessary to isolate the membrane transport parameters, but they are useful as a cross check in diagnosis. No further time constants can be detected.

The above analysis applies to an active membrane in normal Ringer's solution. It clearly depicts three relaxation processes, two of which are related solely to surface processes and a final one corresponding to a coupled surface and membrane transport pathway. The form of the impedance as analyzed here is unique for this physical system because of the characteristics imposed by the model. In the context of normal circuit synthesis procedures, the equivalent electric circuit obtained is

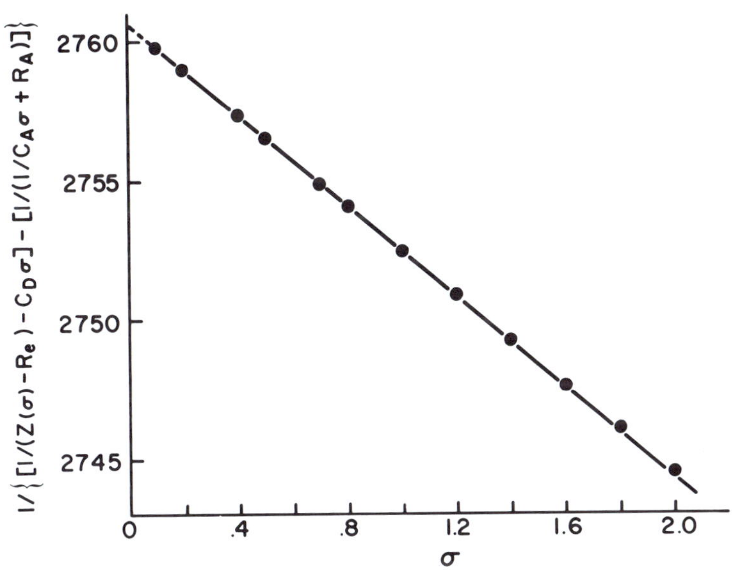

Figure 8. Low-frequency real-axis impedance behavior of an active membrane allowing isolation of the membrane transport parameters R_P, R_M, *and* C_M. *As shown in Figure 1,* $R_p >> R_m$ *illustrating that phase transfer (partitioning) appears to be the rate-determining step for passive* Na^+ *transport in this system. The mass transport pathway for a fixed membrane is purely resistive* (see *Figure 9) exhibiting diffusion layer-type behavior.*

not unique. However, the physical restraints given by the model (*see* Equation 5) confine the circuit configuration to that shown in Figure 1, which applies for an active bladder.

Utilizing the above approach, 49 bladder preparations were analyzed. For this series the TEPD was 28 ± 9 mV. When normal Ringer's solution was used in mucosal and serosal compartments, the behavior shown in Figure 1 always was obtained. The parameter values shown illustrate that there are quantitative differences from bladder to bladder but the qualitative behavior was observed consistently. In addition, the relative values of the observed parameters always were obtained. Examination of these shows that the value of C_D is well within the expected range (cf. lipid bilayers). Note that the value of C_D previously observed (*5*), when the R_A, C_A pathway was not isolated, was, as anticipated, too high and contained a contribution from C_A. Isolation of what appears to be a specific adsorption pathway shows that this surface phenomenon has a relatively rapid time constant (69 ± 18 usec). Interestingly, this value shows that the perturbation of a specific surface process in a living cell probably would be functionally significant for chronologically applied pulses having a 100-μsec duration. At this time it is not clear what specific entities are involved in this process.

The parameters involved in the membrane transport pathway show, as anticipated from the model, that there is a non-negligible membrane phase concentration of the transporting species (C_M). This is consistent with approaches involving the presence of energy barriers in a transport pathway (*24, 25, 36*) which would result in planes of varying concentration in the pathway. The relative value of R_P and R_M is highly significant, since it clearly shows that phase transfer (partitioning) is the rate-limiting step for the transport process. As in the earlier study (*5*) a shunt-resistive pathway, often observed in classical short-circuit techniques (*37, 38*), was not detected in this work. Clearly the linear perturbation conditions present here do not elicit this type of response, but rather test the membrane preparation without modifying its structural or functional baseline characteristics. Further evidence of this is seen in the electron micrograph of Figure 4 which shows that the tight junction and membrane fusion regions are intact after exposure to normal Ringer's solution and after impedance determination.

The measurement of impedance of a fixed bladder was carried out by first evaluating this function in the active state. Once this was accomplished, the mounted bladder was subjected to 1% glutaraldehyde in a bathing solution identical to that for the active state. As mentioned earlier, the TEPD falls to zero within approximately 2 min and does so independently of how the glutaraldehyde was added (mucosal or serosal compartments only or both simultaneously). At this point, a linear perturbation was applied causing approximately the same transepithelial potential change as for the active bladder. Observation of the response just after TEPD = 0 showed impedance behavior identical to that obtained preglutaraldehyde. The transient response then slowly varied for approximately 1–1.5 hr before becoming constant, indicating that fixation was complete with respect to electrochemical dynamics. Only those results with respect to this condition will be reported here. The observed variation in impedance until steady-state fixation was achieved will be reported elsewhere.

The impedance of the fixed bladder membrane was evaluated according to the same procedure as described above for the active membrane. In every case, the functional behavior for $\sigma \to \infty$ was identical to that for the unfixed state (*see* Figure 6 and Equation 7). Once the dielectric capacitance, C_D, had been isolated, it was again possible to detect the second relaxation given by Equation 8. Functional behavior identical to that illustrated in Figure 7 (isolating R_A and C_A) was obtained, indicating the presence of a specific adsorption (binding) process. Up to this point the impedance behavior of the active and fixed bladder were functionally

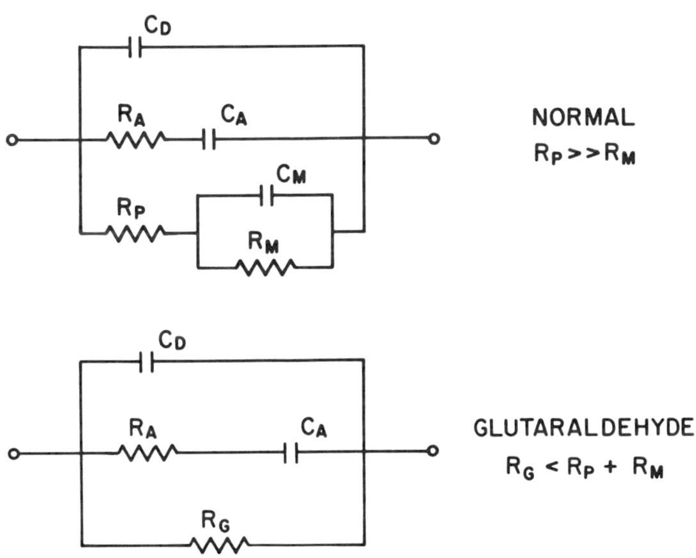

Figure 9. Comparison of the transient impedance behavior of a normal (active) and glutaraldehyde-fixed toad bladder membrane system. Dielectric and specific adsorption response are functionally identical; however, the transmembrane transport pathway is significantly different. The active membrane appears to have a specifically adsorbed layer of Na^+ (R_M, C_M) *in the membrane phase which, along with* R_p*, regulates transport behavior. Fixed-membrane transport appears to be interpretable by a diffusion layer approach.*

identical. However, when an attempt was made to isolate the third relaxation process (related to membrane transport), Equations 9 and 10 were not obeyed. The following response was observed (after isolation of C_D, R_A, and C_A)

$$\frac{1}{B(\sigma) - C_A\sigma/(1 + R_A C_A\sigma)} = R_G \tag{11}$$

where R_G represents the membrane transport pathway (resistive only) for a glutaraldehyde fixed preparation. Typical behavior for a fixed membrane is given in Figure 9, shown in comparison with the electrical equivalent circuit for an unfixed membrane. A substantial change in the third relaxation process, apparently indicating a structural modification in the membrane phase-transport pathway is evident. Typical relative values for an active and the same fixed membrane are shown in Table I. Here C_D is essentially unchanged after fixation. The specific adsorption process is still present with little change in C_A but a relatively large modification of R_A. Finally, R_M, C_M, and R_P have been replaced by R_G for the fixed bladder, with R_G substantially lower than R_P, but larger than R_M.

The above results were obtained by comparing the transient response of an active vs. a fixed membrane in the presence of normal amphibian Ringer's solution. Some preliminary results have been generated using low (by 80X) Na^+ concentration in the mucosal compartment. Once again, the primary modification after fixation has been in the membrane transport pathway via the disappearance of C_M and an increase in conductivity. These results will be reported elsewhere.

It thus appears that the effect of glutaraldehyde fixation is similar whatever the initial conditions of the membrane preparation. In all cases the membrane appears to lose its ability to maintain a resting concentration of the transporting species.

Table I. Comparison of Transient Impedance Parameters for a Typical Active and Glutaraldehyde-Fixed Toad Bladder Membrane Preparation[a]

	Active	*Fixed*
C_D (μF)	0.348	0.345
R_A (Ω)	575.0	252.5
C_A (μF)	0.288	0.352
R_M (Ω)	432.0	—
C_M (μF)	82.1	—
R_P (Ω)	4530	—
R_G (Ω)	—	1655

[a] Terms are defined in text.

Discussion

The initial study of the impedance of the toad bladder system reported elsewhere (*5*) demonstrated that it was possible to isolate potential dependent membrane surface phenomena from transmembrane and aqueous transport processes. The present results extend this analysis for the active membrane through the detection of a specific adsorption relaxation. This was possible by a modification in the experimental design such that a 100-nsec time resolution could be obtained. Also, using transient recording techniques allows a considerable increase in accuracy because the time-dependent impedance variations in a living system are minimized.

Examination of the data obtained in this study for the active membrane shows that the dielectric capacitance of the membrane system is well within the range expected for lipoprotein structures having 70–100-Å thicknesses. The value of 0.340 ± 0.082 uF obtained for C_D obviously relates to bulk membrane structure and does not appear to include a contribution from the compact double layers present at the membrane interfaces. This result is expected, provided that the charged interfaces involve, in large part, electrostatic interactions. Under these conditions the response of these structures to charge injection would appear, on the time scale of these experiments, to be infinitely rapid and therefore inseparable from the dielectric capacitance.

The form of the second relaxation pathway (R_A, C_A) clearly indicates that it is a surface process apparently representing a specific adsorption phenomenon. Studies in progress involving modification of Na^+ and Ca^{2+} concentration in the mucosal compartment do not yet allow distinction to be made between two suspected mechanisms. The first involves the possible gating mechanism of Ca^{2+} for Na^+ transport which is suspected for other cellular systems (*39, 40, 41*). The second possibility is that the specific adsorption process involves Na^+ ion which, in a sense, gates itself. Placed in the context of the observed modification of the Na^+ transport pathway (studies to be reported elsewhere), wherein partitioning appears to become more difficult as (Na^+) decreases, the latter interpretation may be closer to physical reality.

As previously found, the actual membrane transport pathway for an active system appears very rapid ($\tau \approx 120\ \mu$sec; *see* Figure 1) tending to indicate a membrane phase barrier mechanism rather than one which involves, e.g. a carrier. Indeed, the value of τ_M is very similar to that of a heterogenous process. This leads to the suspicions that the resting membrane concentration of Na^+ is on a plane perhaps just inside the mucosal membrane–extracellular fluid interface.

Under these conditions the passive Na^+ transport mechanism might be conceived as its displacement from this plane by a gated entry of Na^+

upon application of a transient potential difference. The membrane phase entry or gating kinetics are represented by R_P, which, by analogy to charge transfer kinetics at the electrode–electrolyte interface, does not appear to involve specific adsorption for this step. In addition, studies (to be reported elsewhere) show that for an active membrane subjected to low mucosal Na^+, the TEPD is relatively unchanged. This may indicate that the membrane phase barrier model is still operative under these conditions and that only the gating process becomes more difficult.

When glutaraldehyde is added to the active membrane, the results indicate that a major modification in transmembrane transport kinetics occurs. The disappearance of τ_M may mean that the membrane phase barrier model is no longer valid because of structural modifications both at the gating site ($R_G << R_P$) and in the membrane transport pathway.

The primary use of chemical cross-linking or fixation in membrane studies has been to stabilize the compound for structural investigation, particularly in the 5–20-Å range (*see* Ref. *42* for a review). In many studies it is possible to use very specific cross-linking reagents such as bis-imidates which can, e.g., cross-link the proteins associated with ribosomes (*43*). This has been performed so specifically that ribosomal protein synthesis is not altered, but the ensuing cross-linked structure is stable enough for such analyses as gel electrophoresis. Glutaraldehyde is a very reactive and quick-acting agent. For example, rat liver was penetrated initially by glutaraldehyde at the rate of 0.1 mm in 15 min (*44*). However, its reactivity is not specific enough to be able to identify, molecularly, the entities involved in a given membrane (*45*). Fixation alters cell membrane permeability (*46*) even though glutaraldehyde may not cause any obvious protein conformational changes (*45*). Glutaraldehyde appears to achieve its major cross-linking with amino acids (*47*). The major reactive amino acids are: arginine; cysteine; glutamine; histidine; lysine; tryptophan; and tyrosine. Therefore, it might be expected that proteins rather than mucopolysaccharides or lipids are among those entities preferentially cross-linked in the course of fixation for, e.g., electron microscopy studies.

Based on the above information and the kinetic results of the present study it may be possible that passive Na^+ transport in the toad bladder epithelial cell system involves a molecular pathway which at least has peptide properties. Since the principal steady-state electrochemical modification is a disappearance of the TEPD, it is clear that for this transport pathway the fixation has altered a fundamental bulk membrane property which is related to the resting membrane concentration of the transporting species. It is entirely conceivable that this could occur by a protein or protein–lipid structural modification which through cross-linking competitively occupies the Na^+ sites related to the barrier defined in τ_M.

It is interesting to assess how the absence of TEPD could affect the membrane transport pathway irrespective of (or coupled with) structural changes. To do this it is necessary once again to consider the phase-transfer current, $i_p(s)$, as defined in Equation 4. It will be remembered that the variation in concentration just inside the membrane phase, $\Delta C_m(o, s)$, was obtained in explicit form with no restriction on the voltage field, V, present in the transport pathway. If V becomes very small or nonexistant, the solution of the transport equation used for the final derivation of Equation 5 is no longer valid. The primary reason for this is that, as $V \rightarrow 0$, the membrane concentration profile for membrane thickness δ_M, tends to be such that $\Delta C_m(o, s) = \Delta C_m(-\delta_M, s)$. In other words, concentration variations at both membrane interfaces tend to follow each other (*9, 17*), whereas when V is non-negligible the situation is equivalent to the energy barrier model, and $\Delta C_m(-\delta_M, s = 0$. Using the above conditions, the phase-transfer impedance $Z_p(s)$ for $V \rightarrow 0$ is given by

$$Z_P(s) = \frac{1}{I_E}\left[1 + \frac{I_M}{\eta F C_m D_M\,[FV/2RT + 2/\delta_M]}\right] \tag{12}$$

Comparison of Equation 12 with Equation 5 shows that the current pathway for membrane transport still contains a phase-transfer kinetic term, R_P, related to $1/I_E$, but membrane phase transport is now purely resistive in nature. The term containing frequency, s, in the denominator of Equation 5 is no longer present in Equation 12. In fact, this quantity remains purely resistive, even if $V = 0$.

If membrane transport obeys Equation 12, then the diagnostic criteria followed by an active membrane after isolation of the dielectric and specific adsorption behavior would be identical to that given by Equation 11. Examination of Equation 12 shows that it is impossible to separate phase transfer from membrane transport. The results of this study show only that the value of this resistive response, R_G, is not related simply to R_P and/or R_M found for an active membrane. If, as appears likely, the entry point and membrane barrier are modified structurally by glutaraldehyde, then Na^+ selectivity is affected most probably and the membrane mass transport reverts to diffusion-layer-type behavior. Under these conditions membrane thickness (δ_M) would be expected to be the major variable as indeed is shown by Equation 12. However, it is not yet clear if the fixed membrane is merely acting like a thin, porous structure through which only ions of a certain size may transport. Results to be reported elsewhere concerning the change in membrane impedance during the first hour after addition of glutaraldehyde may shed further light on this.

In contrast to the major modification in transport kinetics after fixation, relatively minor changes were observed for dielectric capacitance, C_D, and specific adsorption, R_A/C_A, kinetics. This result is indeed interesting since it shows that even relatively nonspecific cross-linking can be used to differentiate the kinetics of potential dependent heterogeneous processes from those for transport phenomena. The least changed quantity is C_D, which was originally thought to be related to nonspecific charge interactions associated primarily with membrane lipids. This measurement alone may provide some kinetic evidence that glutaraldehyde primarily cross-links proteins. In contrast, the quantitative changes observed in specific adsorption kinetics are somewhat interesting. The concentration of specifically adsorbed species related to C_A remains nearly constant after fixation. However, adsorption kinetics (R_A) are significantly more rapid. Whether or not this is related to the change in phase-transfer kinetics (which has either become very rapid or disappeared altogether), is unclear at present. It is interesting that the kinetics of both specific adsorption and membrane transport apparently have become more rapid. This would seem to indicate that the structural modifications engendered by glutaraldehyde have their primary influence on specific kinetic parameters. It is intriguing to speculate that this may mean that potential dependent kinetics of primarily heterogenous (surface) nature are a principal step in living cell regulatory processes.

Conclusions

When the concept of electrochemical information transfer in vivo was formulated originally it was based on the central idea that the kinetics of potential dependent interfacial phenomena could in some way be involved in cell regulation. The first step in establishing the relevance of this approach to tissue growth and repair modification required the detection and quantitative assessment of electrochemical phenomena at living-cell membranes. This was accomplished for the active toad urinary bladder membrane system using a transient impedance technique (5). The present work extends this study for the active membrane by providing a substantial improvement in the high-frequency range by using transient recording techniques. This allows the isolation of a specific adsorption (binding) relaxation and a clear determination of the surface process of phase transfer involved in Na^+ transport.

With these results in hand it was possible to gain further insight concerning the active membrane transport mechanism. The present results once again demonstrate that the rate-determining step is related to partitioning kinetics ($R_p >> R_m$). In addition, the transport time constant, τ_M, is small enough to lead to the suspicion that there is a plane in the membrane phase at which the majority of Na^+ resides, i.e. membrane transport behaves like a heterogeneous process. For this particular membrane system it is interesting to speculate that passive Na^+ transport occurs by a potential dependent entry which allows the ion to adsorb onto a membrane-phase plane. If the Na^+ concentration in this plane is relatively independent of its mucosal concentration (as observed in a study to be reported elsewhere) then phase transfer would be regulatory and each ion entry should provoke an equivalent exit of Na^+ into the intracellular fluid. In other words, the amount of membrane-phase Na^+ is established by the number of sites available at specific molecular entities, and further Na^+ entry is followed by displacement of membrane-phase Na^+, which is a relatively rapid process compared with phase entry. Obviously for the active membrane preparation this is a highly tentative picture. The types of data which are generated by these transient impedance studies do, however, allow the separation of kinetic processes related to one of the principal transport functions of this system.

When glutaraldehyde is added to the membrane preparation, the impedance results obtained are significant when compared with those of the active control. Since a functional modification in the membrane transport pathway occurred upon fixation, it was necessary to re-examine the proposed transport model. As originally derived, membrane transport was considered to occur under the influence of a substantial voltage field and that membrane-phase concentration changes could be described by considering the inside mucosal surface only. The value of the time constant for membrane transport (after entry or partitioning) for the active system shows that this may be a reasonable picture. After fixation, membrane transport appears to be represented best by considering that membrane-phase concentration changes occur uniformly across the membrane and that the voltage field may be substantially lower. If indeed membrane transport occurs as just described for a fixed preparation, it is interesting to note that this behavior is analogous to that which occurs in a thin diffusion layer. In other words, the cross-linked membrane may be responding in much the same manner as a thin, porous layer in so far as transport is concerned. Experiments are presently in progress to assess this by subjecting the preparation to ionic entities of substantially different size. This may allow the geometry of the stabilized pore to be established. Interestingly, cross-linking does not appear to strongly affect the membrane surface or the basic lipid structure. Thus the dielectric capacitance, which is a reflection of lipid asymmetry and the surface charge interactions, is nearly identical to that obtained for the active control preparation. It is possible that this information may be useful in assessing the mechanism of glutaraldehyde fixation for both light and electron microscopy studies. The fact that the specific adsorption process does not disappear after fixation appears to indicate that the number of binding sites is not altered substantially. However, adsorption kinetics are changed sufficiently to lead to the suspicion that some structural change has occurred. In addition, since both adsorption and transport kinetics were altered by fixation, it may be that the same or linked molecular entities are involved. It is possible, for example, that the specific adsorption process is linked to phase transfer, as has been suspected already (*5*). Thus, if phase transfer involves a gating phenomenon associated with a conformational change, then the major functional effect fixation has on transport also could be reflected at the mucosal surface entry point. Comparison of the effect of fixation on both of these processes as a function of different initial active conditions may provide further evidence of the extent to which they are coupled.

The studies presented here represent an attempt at the quantitative evaluation of dynamic interfacial electrochemical phenomena at living-cell membranes. The modern bioelectrochemical techniques developed for this study allow the isolation of kinetic phenomena which appear related to a principal transport function of this cell. Chemical fixation causes functional changes which can be detected easily via this approach and allow certain preliminary mechanistic conclusions to be reached. The application of these results to the use of electrochemical information for the modification of tissue growth and repair processes is significant. The relative values of surface and transport time constants have been assessed. They fall within the ranges originally predicted, and are in fact, useful for the proper encoding of electromagnetically injected current waveforms to obtain functional modification in in vitro and in vivo applications. One essential aspect of electrochemical information is to excite only surface or surface and membrane transport processes for a given cell or tissue. Based on the detected relaxation times it is evident that waveform duration could play a significant role once amplitude and exposure (repetition rates) were approximately established. This has been proved in studies on the rate of chromatin–DNA structure modification as a function of the pulse duration and the repetition rate of injected current. For this system if the induced current waveform has sequential segments in which surface only, followed by surface and membrane transport phenomena are excited, there is a repetition rate window in which DNA uncoiling is accelerated vs. controls (*6, 8*). Further evidence for waveform selectivity has been illustrated in Ca^{2+} uptake studies in embryonic bone explants. These studies show that membrane transport must be affected before Ca^{2+} uptake is modified (*7*).

Studies such as those presented here provide evidence on a mechanistic front that it is possible to use low-level current injection ($\mu A/cm^2$) to selectively modify cellular processes. The therapeutic implications of this are substantial. In fact, specific electromagnetic current induction can be used to heal otherwise incurable bone fractures (*15*). As the theoretical and experimental approach expands it will be possible to evaluate the cellular response to pulsating current at each stage of its cycle or development. As this occurs it is hoped that fundamental cellular regulatory mechanisms can be understood better and that new therapeutic applications will be forthcoming.

Acknowledgment

This work was supported partially by NSF Grant No. NSF–APR–76–19469 and Electro-Biology, Inc.

Literature Cited

1. Pilla, A. A. *Proc. Intersoc. Energy Conver. Eng. Conf., 7th, San Diego,* Am. Chem. Soc., Washington, D.C., *1972*, p. 761.
2. Pilla, A. A. *J. Bioelectrochem. and Bioenergetics* **1974,** *1,* 227.
3. Pilla, A. A. *Ann. N. Y. Acad. Sci.* **1974,** *238,* 149.
4. Pilla, A. A. *J. Bioelectrochem. and Bioenergetics* **1976,** *3,* 370.
5. Pilla, A. A.; Margules, G. *J. Electrochem. Soc.* **1977,** *124,* 1697.
6. Chiabrera, A.; Hinsenkamp, M.; Pilla, A. A.; Ryaby, J. T.; Ponta, D.; Belmont, A.; Beltrame, F.; Grattarola, M.; Nicolini, C. *J. Histochem. Cytochem.* **1979,** *27,* 375.
7. Vergos, G. Pilla, A. A.; Bassett, C. A. L. *J. Bone Jt. Surg.* Orthopaedic Transactions, **1979,** p. 78.
8. Chiabrera, A.; Hinsenkamp, M.; Pilla, A. A.; Nicolini, C. In "Chromatin Structure and Function"; Nicolini, C., Ed.; Plenum: New York, 1979, in press.
9. Pilla, A. A. *Proc. Int. Soc. Electrochem., 29th, Budapest, Hungary;* Budapest Press: Budapest, 1978; p. 39.
10. Yasuda, I. *J. Kyoto Med. Soc.* **1953,** *4,* 395.
11. Fukada, E.; Yasuda, I. *J. Phys. Soc. Jpn.* **1957,** *12,* 1158.
12. Bassett, C. A. L.; Becker, R. O. *Science* **1962,** *137,* 1063.
13. Bassett, C. A. L.; Pawluk, R. J.; Becker, R. O. *Nature (London)* **1964,** *204,* 652.
14. Becker, R. O.; Murray, D. G. *Clin. Orthop.* **1970,** *73,* 169.
15. Bassett, C. A. L.; Pilla, A. A.; Pawluk, R. J. *Clin. Orthop.* **1977,** *124,* 117.
16. Margules, G.; Pilla, A. A. *J. Electrochem Soc.* **1978,** *125,* 171C
17. Pilla, A. A. *Biophys. J.* **1977,** *17,* 130a.
18. Lorenz, W.; Mockel, F. *Z. Elektrochem.* **1956,** *60,* 507.
19. deLevie, R.; Seidah, N. G.; Moreira, H. *J. Membr. Biol.* **1974,** *16,* 17.
20. Miller, I.; Blank, M. *J. Colloid Interface Sci.* **1968,** *26,* 26.
21. MvLaughlin, S. *J. Membr. Biol.* **1972,** *9,* 361.
22. Parsegian, V. A. *Ann. N.Y. Acad. Sci.* **1975,** *264,* 161.
23. Britten, J. S.; Blank, M. *J. Bioelectrochem. and Bioenergetics* **1977,** *4,* 209.
24. Anderson, O.; Feldberg, S.; Nakadomari, H.; Levy, H.; McLaughlin, S. *Biophys. J.* **1976,** *16,* 194a.
25. Anderson, O.; Fuchs, M. *Biophys. J.* **1975,** *15,* 795.
26. Walser, M. *J. Clin. Invest.* **1969,** *48,* 1714.
27. Crocker, B. P.; Tisher, C. C. *Kidney Int.* **1972,** *5,* 145.
28. Erlij, D. *Pfluegers Arch.* **1976,** *364,* 91.
29. Lief, P. D.; Mutz, B. F.; Banks, W. *Am. J. Physiol.* **1976,** *230,* 1722.

30. Finn, A. L. *Nature (London)* **1974,** *250,* 495.
31. Leaf, A.; Anderson, J.; Page, L. B. *J. Gen. Physiol.* **1958,** *41,* 657.
32. Chiabrera, A.; Margules, G. S.; Pilla, A. A. *Proc. Int. Soc. Electrochem., 29th, Budapest, Hungary;* Sandor, L., Ed.; University of Budapest Press: Budapest, 1978, p. 13.
33. Pilla, A. A. *J. Electrochem. Soc.* **1969,** *116,* 1105.
34. Pilla, A. A. *J. Electrochem. Soc.* **1970,** *117,* 467.
35. Pilla, A. A. In "Electrochemistry: Calculations, Simulation and Instrumentation"; Mattson, J. S.; Mark, H. B., Jr.; McDonald, H. C., Jr. Marcel Dekker: New York, 1972; p. 139.
36. Hille, B. *J. Electrochem. Soc.* **1978,** *125,* 171c.
37. Goefoed–Johnsen, V.; Ussing, H. H. *Acta Physiol. Scand.* **1958,** *42,* 298.
38. Reuss, L.; Finn, A. L. *J. Membr. Biol.* **1975,** *20,* 191.
39. Rasmussen, H. *Science* **1970,** *170,* 404.
40. Rodan, G. A.; Feinstein, M. B. *Proc. Natl. Acad. Sci. U.S.A.* **1976,** *73,* 1829.
41. Rodan, G. A.; Bourret, L. A.; Norton, L. A. *Science* **1978,** *199,* 690.
42. Peters, K.; Richards, F. M. *Annu. Rev. Biochem.* **1977,** *46,* 523.
43. Bickle, T. A.; Hershey, J. W.; Traut, R. R. *Proc. Natl. Acad. Sci. U.S.A.* **1972,** *69,* 1327.
44. Hopwood, D. *J. Anat.* **1967,** *101,* 83.
45. Quiocho, F. N.; Bishop, W. H.; Richards, F. M. *Proc. Natl. Acad. Sci. U.S.A.* **1967,** *57,* 525.
46. Krames, B.; Page, E. *Biochem. Biophys. Acta* **1968,** *150,* 24.
47. Sabatini, D. D.; Bensch, K.; Bavinett, R. J. *Anat. Rec.* **1962,** *142,* 274.

RECEIVED November 20, 1978.

28

A Role for Water in Growth, Metabolism, and Intracellular Organization[1]

F. M. ETZLER and W. DROST-HANSEN[2]

Laboratory for Water Research, Department of Chemistry, University of Miami, Coral Gables, FL 33124

Evidence for the existence of vicinal (interfacial) water is reviewed briefly and its role in cellular functioning is outlined. In the present study, data are reported for the growth of a thermophilic alga, Cyanidium caldarium, *as a function of temperature. Several growth optima and minima were observed; the growth minima generally occurring at the temperatures where the vicinal water undergoes structural changes. When cultured on a minimal medium in the presence of NH_4^+ as the nitrogen source the organism exhibited a growth peak above 45°C. However, when the NH_4^+ was replaced by NO_3^-, no growth occurred above 45°C. This result is discussed in terms of our theory of multiple metabolic pathways, determined in part by the properties of vicinal water.*

For some microorganisms cultured over wide temperature ranges, the occurrence of multiple growth optima has been reported sporadically in the literature. A typical example of such multiple growth optima is shown in Figure 1 from a study by Mitchell and Houlahan (*1*). These authors measured the growth of a mutant of *Neurospora crassa* which required riboflavin for optimum growth. It is seen from Figure 1 that at reduced levels of riboflavin, a distinct minimum in the growth curve occurs near 30°C. In previous studies (*2, 3, 4*) the senior author has

[1] Contribution No. 27, LWR.
[2] To whom correspondence should be addressed.

0-8412-0473-X/80/33-188-485$05.00/1

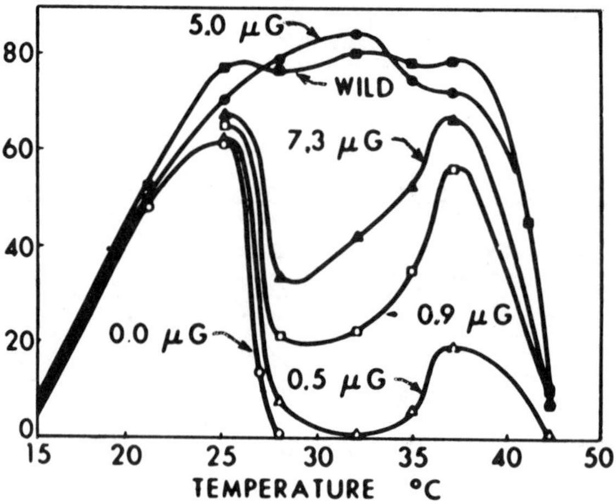

Figure 1. Multiple growth optima for a mold (a mutant of Neurospora crassa*) as a function of temperature with different amounts of riboflavin present* (1)

proposed that minima in growth curves reflect structural transitions of the intracellular water. In this chapter, we discuss some aspects of multiple temperature optima and particularly the suggestion that maxima and minima in the growth processes may reflect different metabolic pathways, the selection of which depends on the structure of the vicinal, intracellular water.

Before proceeding, we wish to summarize some of the observations which have been reported in the literature on vicinal water and to illustrate the likely role of such water in cellular functioning through a number of examples, mostly taken from the literature.

Summary of Vicinal Water Properties

Water and aqueous solutions adjacent to most (solid) interfaces possess notably different properties from those of the respective bulk systems. Such interfacial water is referred to as vicinal water. Because the properties of vicinal water differ from the properties of bulk aqueous systems, it must be concluded that the structure of vicinal water differs from the bulk structure (*5, 6, 7, 8, 9*). The evidence presently available also suggests that the modification of the structure extends over considerable distances, apparently as much as 100 to 1000 Å or roughly 30 to 300 molecular diameters. The evidence for extensive structure modification derives in part from the following types of measurements: (a) disjoining pressure (*10, 11*); (b) viscosity (*11, 12, 14*); (c) ion selectivity (*14, 15*); (d) ultrasonic absorption and velocities (*16, 17*); (e) dielectric data

(*18, 19, 20, 21, 22*); (f) conductance data (*23, 24*); and (g) ultraslow mechanical relaxation (shear) (*23, 24, 25*). Additional evidence is reviewed in Refs. *6* and *7*, while Ref. *8* should be consulted for a collection of recent contributions towards the elucidation of the role of vicinal water in cellular systems.

In this chapter we discuss some specific aspects of vicinal water as it affects intracellular organization and, in particular, the effects on metabolic pathways and growth rates.

Thermal Anomalies: Paradoxical Effect. The properties of vicinal water exhibit thermal anomalies, manifested as more or less abrupt changes in the temperature coefficient of those properties over at least four different temperature intervals, specifically near 14° to 16°C, 29° to 32°C, 44° to 46°C, and 59° to 62°C (*5, 6, 7*). Another unique aspect of vicinal water is the so-called "paradoxical effect" (*7*): it appears, to a first approximation, that the temperatures of the thermal transitions (T_k) are independent of the chemical nature of the solid surfaces and relatively independent of the concentration and nature of the solutes in the aqueous phase.

Structural Aspects. The structure of vicinal water is not known. In fact, it is not even clear if vicinal water is "more structured" or "less structured" than bulk water. Possible types of vicinal water structures are clathrates and high-pressure ice polymorphs (*6, 7*). However, it is possible also that no definite new types of structures occur in vicinal water although the existence of multiple thermal transitions strongly suggests that definite, relatively large-scale, cooperative effects must play a role—something which is difficult to envision if identifiable structured entities do not exist.

Molecular-Weight Effects. Vicinal water appears to occur at most (or all) aqueous phase/solid surfaces. Hence, the effects of vicinal water must be expected to be manifested in cellular systems because of such interfaces as the cytoplasm/cell membrane or the interface at various organelles. However, it is important to note that vicinal water also occurs in aqueous solutions of macromolecules (*7, 8*). Thus, there is further reason to expect that cellular systems should reveal the effects of vicinal water.

Effects of Vicinal Water on Growth Processes

Illustrative Examples. The evidence for the occurrence of—and a role for—vicinal water in cells is derived from a variety of observations on living systems. One of the most useful approaches for demonstrating the existence of vicinal water has been through studies of thermal effects: the frequent occurrence in cellular systems of thermal anomalies (at the transition temperatures of vicinal water) strongly implicates vicinal water in some rate-determining processes in cells. A few examples of thermal effects in living systems are reviewed in this section.

Figure 2 shows a typical example of the abruptness with which a large number of cellular systems respond to temperatures near the vicinal water transition temperatures. The mean germination period (measured in hours) for freshly harvested as well as "after-ripened" seeds of barley (var. Golden Promise) as reported by Roberts and Smith (*26*) is plotted in this graph. A remarkable and sudden retardation of the germination

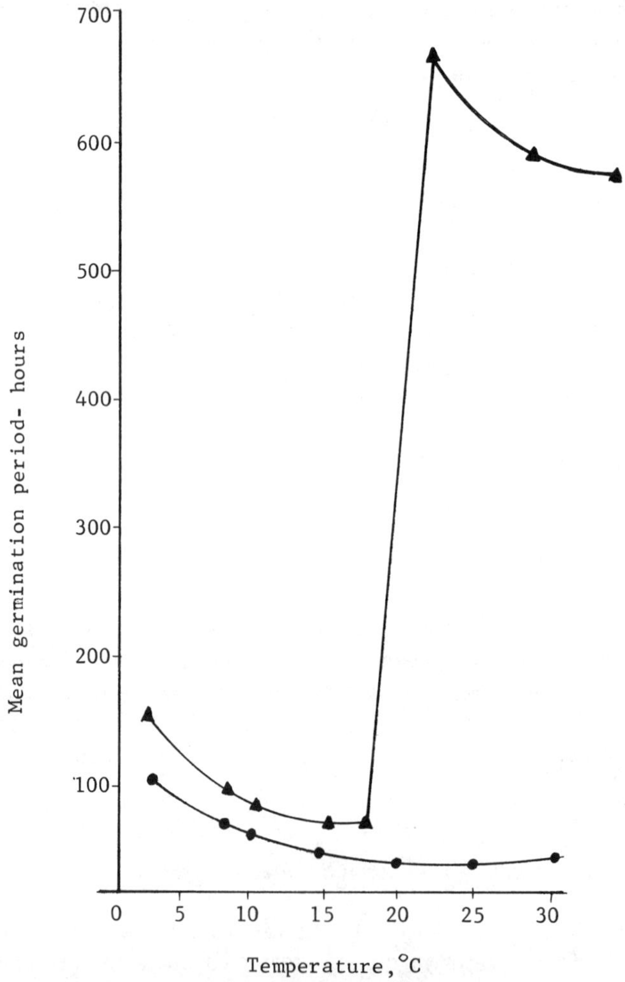

Figure 2. Mean germination period of freshly harvested (▲) and after-ripened (●) barley seeds as a function of temperature (21)

occurs above approximately 16°C. (*See* also Ref. *6* for a discussion of the possible role of vicinal water in vernalization.)

Figure 3 shows the effects of temperature on the survival of the green photosynthesizing macroalga *Valonia macrophysa* (*27*). The domain of survival (for 3-days exposure) is limited sharply by the temperatures of two of the vicinal water transition temperatures, namely 15° and 30°C. Thorhaug (*27*) has provided many additional examples of such thermal boundaries (*see also* Ref. *28*).

The example shown in Figure 3 clearly suggests that a temperature "domain" exists, with sharply defined boundaries, in which the organism can function. For other organisms, a different domain might be between 30° and 45°C. At each of these boundaries rapid changes with temperature are expected in a variety of cell functions; as examples, membrane permeabilities and membrane resistances may change (*9*) and a large number of enzymatic reactions may be affected (*6, 7, 29*), etc. This in turn has led to various theories to explain a number of physiological phenomena. As an example, Drost–Hansen (*6, 30, 31*) proposed earlier that body temperatures of mammals and some birds have been chosen through evolutionary processes to be as far away from either of these T_k's (30° and 45°C) as possible—i.e., body temperatures are near 37°C (98.6°F) which is indeed close to the average body temperature of about 150 mammals and those birds which do not fly (ostrich, penguin, kiwi, etc.). Other implications are discussed in Refs. *6, 8,* and *31.*

Finally, Figure 4 shows the effects of temperature on the rate of germination of turnip seeds. The ordinate in this figure is the logarithm of the reciprocal of the time (in hours) for 50% germination, plotted as a function of the reciprocal absolute temperature. Anomalous changes in slope occur near 15° and 30°C. A detailed discussion of these data can be found in Refs. *32* and *33*.

Many examples can be added to demonstrate the occurrence of remarkably abrupt changes in physiological parameters (including growth) at the vicinal water transition temperatures (*6, 7, 8, 28, 29, 33, 34*). The available data are not restricted to plants and algae which have been used here only to demonstrate the nature of the phenomenon.

Mechanism of Control by Vicinal Water in Cellular Functioning. Vicinal water appears to exist adjacent to (most) solid/aqueous interfaces (*4*). These solid/water interfaces include membrane systems (*9, 34*) and the interface of dissolved biomacromolecules with the aqueous phase (*29*). The effects of vicinal water are also frequently apparent in enzyme kinetic data in bulk solution (i.e., in vitro; away from any biological surfaces (*29, 32*); and in enzyme data in living cellular systems).

In the case of isolated membrane systems, it is not difficult to imagine a direct functional role for vicinal water. Changes in permeability and/or

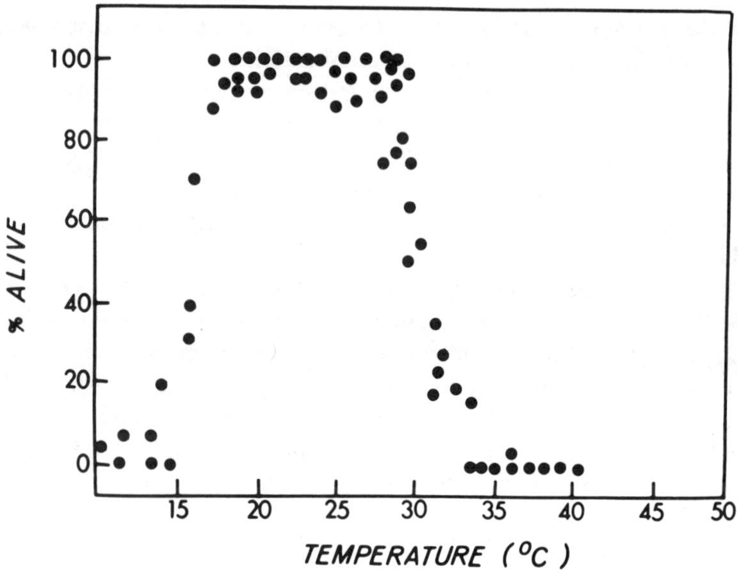

Figure 3. Survival of Valonia macrophysa *(a green macroalga) as a function of temperature after 3 days. Each point represents between 25 to 100 cells (22).*

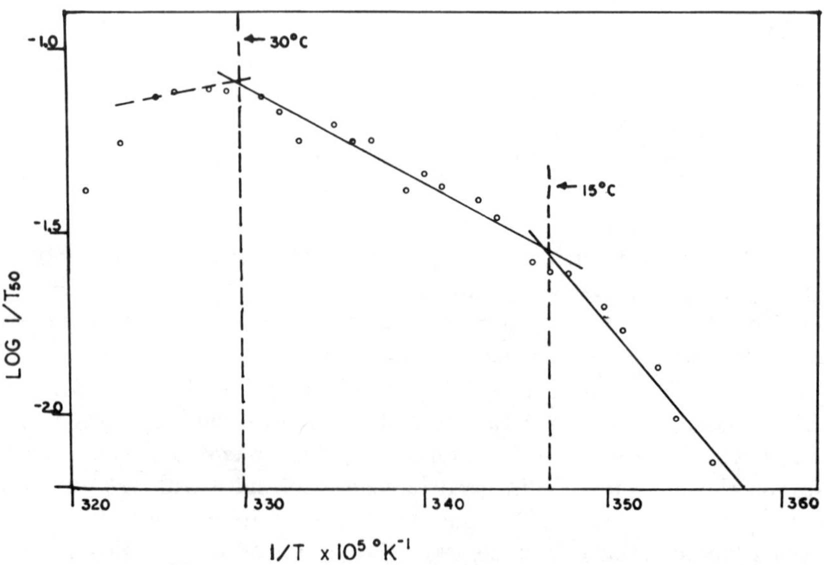

Figure 4. Effects of temperture on germination of turnip seeds (Arrhenius plot). Log 1/time to 50% germination vs. 1/temperature. See text for details.

ionic activities have been discussed (*8, 9*). However, for the occurrence of thermal anomalies in many processes of intact cells, it is far more difficult to identify a site of action. Nonetheless, one important suggestion has been made recently by Clegg (*35*) who consider the possibility that the majority of the cellular macromolecules function primarily in the vicinally ordered water of various intracellular interfaces rather than in a bulk aqueous phase. This chapter should be consulted for some stimulating thoughts on the subject.

Multiple Growth Optima and Multiple Metabolic Pathways

Multiple Growth Optima. Earlier in this chapter, a few examples were discussed which illustrate the abrupt changes sometimes observed in cellular functioning at the vicinal water transition temperatures (T_k). In many cases, the T_k's correspond to absolute upper thermal limits for organisms (*see* especially Refs. *6, 28,* and *29*). However, in some cases, particularly for single-cell organisms, multiple growth optima have been observed. In these instances, the region around one (or more) T_k is associated with partial or complete suppression of growth, while at both higher and lower temperatures, growth does occur. A few illustrative examples are discussed below.

Figure 5 shows a typical example of multiple growth optima. The ordinate is the amount of growth (after 16 hr) of *Streptococcus faecalis,* as a function of temperature (*36, 37, 38*). A distinct growth minimum occurs near 29°C.

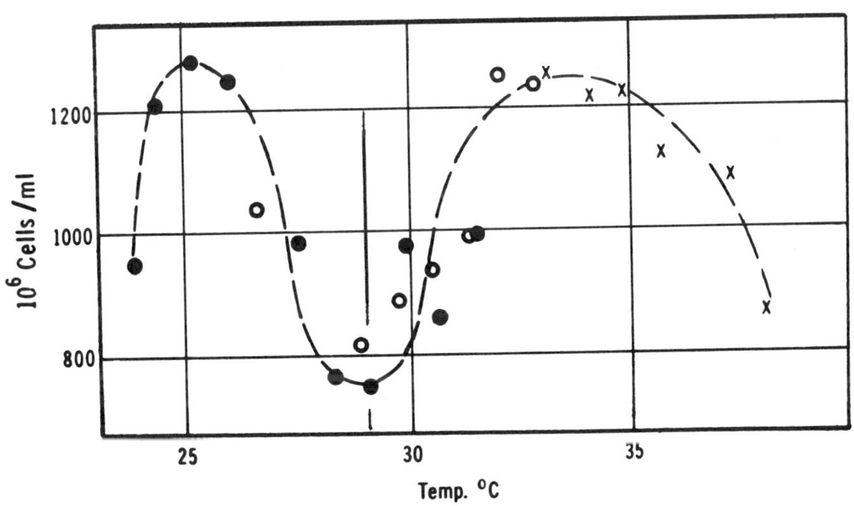

Figure 5. Number of cells of Streptococcus faecalis *as a function of temperature* (40)

Another example is shown in Figure 6 in which minima in growth rate are reported for two varieties of beets (39). Again, distinct minima are observed in the growth of the two varieties, namely between 11° and 14°C and ca. 14° to 16°C, respectively—in other words, near T_k's ca. 14° to 16°C. Note also the cessation of growth in the range between 28° to 31°C.

Other examples of multiple growth optima have been discussed in some of the literature referred to earlier. In the next section we discuss a tentative hypothesis to explain the occurrence of bimodal (and multimodal) growth optima.

Multiple Metabolic Pathways. In dicussing the occurrence of multiple growth optima, Oppenheimer and Drost–Hansen (2) proposed that different metabolic pathways were utilized in different temperature regions (i.e., in the regions between T_k's) owing to some structure-dependent processes related to water-structure transitions.

An attempt to elucidate such processes was made by Schmidt and Drost–Hansen (*3, 4*) who measured pH changes accompanying the growth of *E. coli* over a wide temperature range. Depending on the

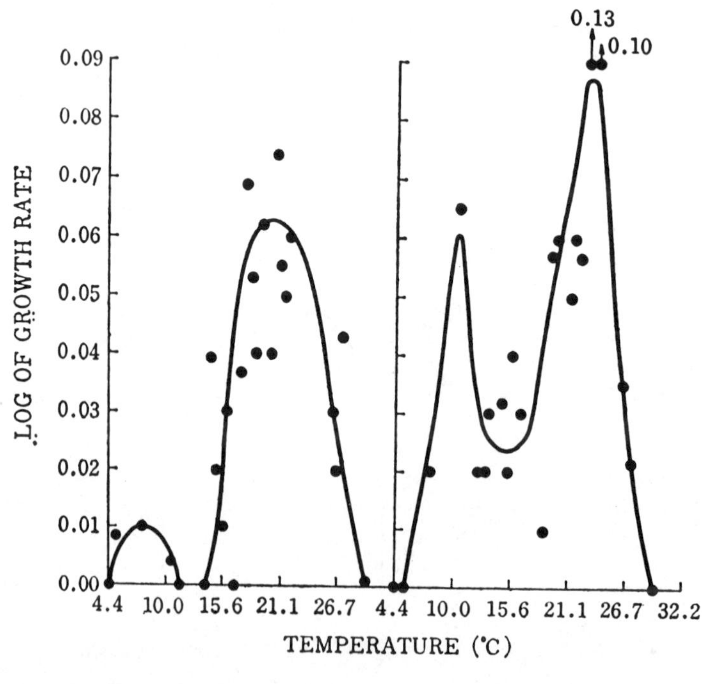

Figure 6. Plot of logarithm of growth rate for two varieties of beets as a function of temperature (41)

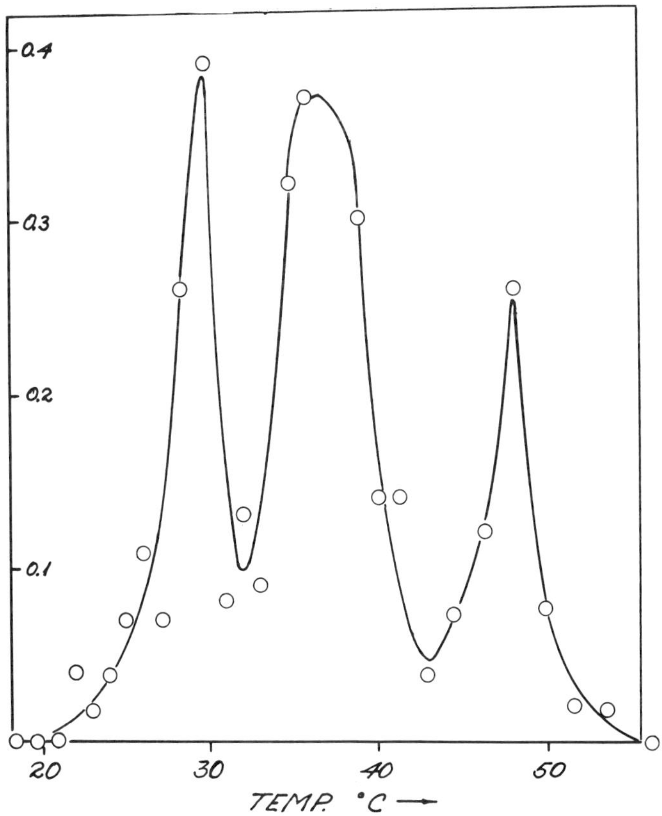

Figure 7. Growth of Cyanidium caldarium *(grown on NH_4^+-containing merium) as a function of temperature.* See *text for details.*

temperature, either increases or decreases in pH were observed. However, while the results obtained were entirely consistent with the proposal by Oppenheimer and Drost–Hansen (*2*), the experiments did not prove a cause–effect relationship between (vicinal) water structure changes and the selection of one specific metabolic pathway over another.

Recently, Etzler and Drost–Hansen (*32*) have re-examined the question of possible mechanisms for multiple growth optima. In this study, the rate of growth of a green thermophilic photosynthesizing alga (*Cyanidium caldarium*) was measured (at closely spaced temperatures—*see* Ref. *40*). Figure 7 shows the amount of growth observed after 21 days as a function of temperature. One observes no less than three clearly separated growth optima with distinct minima near 32° and 43°C—in good agreement with expectations from the theory, according to which temperature regions around T_k lead to impaired cell functioning.

An Example of Metabolic Pathway Selection

Multiple growth optima are observed most frequently for growth on minimal media (*2, 3, 6; see* also Ref. *29*). In the example discussed above, the growth of *Cyanidium caldarium* did indeed take place in a minimal medium: nitrogen was provided in the form of $(NH_4)_2SO_4$; in addition, the medium contained KH_2PO_4, $MgSO_4$, $CaCl_2$, $FeSO_4$, H_3BO_3, and trace amounts of Zn, Mn, Ni, Co, and H_2SO_4 (pH $\sim$ 2). However, no other solutes, organic or inorganic, were added to the medium. Atmospheric CO_2 served as the only carbon source!

In a more recent series of experiments, the alga (*Cyanidium caldarium*) was grown in the same medium, only modified to contain NO_3^- instead of NH_4^+ as the nitrogen source. The result is shown in Figure 8.

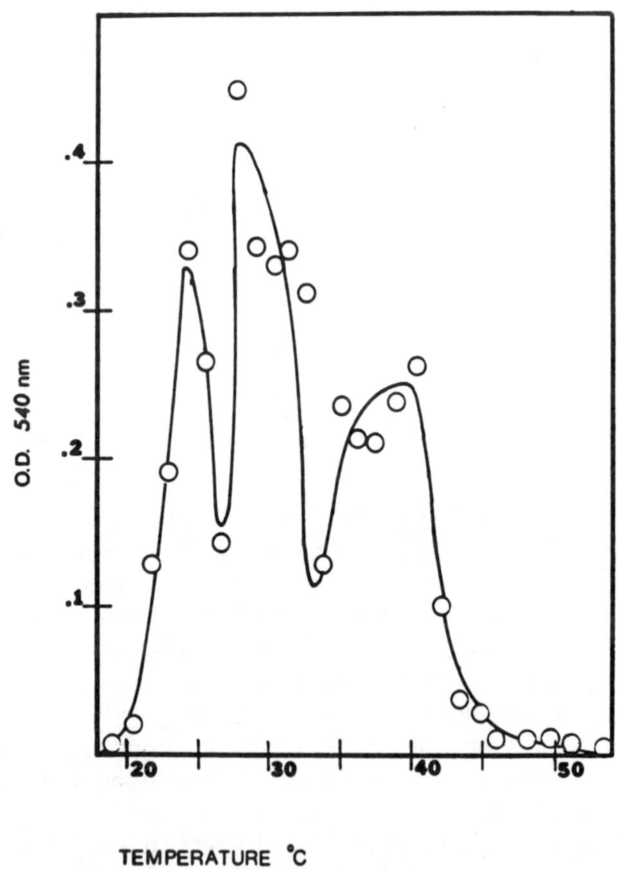

Figure 8. Growth of Cyanidium caldarium *(grown on NO_3^--containing medium) as a function of temperature.* See *text for details.*

It is apparent from these data that growth did not occur above 45°C—as opposed to the results obtained with the NH_4^+-containing medium in which a growth peak occurred at a higher temperature, namely ca. 50°C. It appears that a change has taken place in the choice of metabolic pathways. NO_3^- apparently cannot be utilized above 45°C—i.e., above T_k (the temperature of the vicinal water transition). As in the experiments by Schmidt and Drost–Hansen, the result is consistent with the hypothesis of a controlling role for vicinal water in cellular functioning but again it does not prove this point. On the other hand, we are unable to propose any alternative explanation, particularly considering the diversity of systems in which different metabolic requirements appear modified upon crossing a vicinal water transition temperature. We stress that it may be possible to construct other mechanisms to explain multiple growth optima and that it is difficult—or impossible—at this time to identify specific molecular processes, involving vicinal water, which readily can explain the phenomenon. On the other hand, the abruptness of change of growth rates which often is observed and the conspicuous role played by temperatures near T_k do lend credence to the notion of a likely role of vicinal water in multiple growth optima. On this basis then it seems reasonable to propose that the mechanism may be explained in terms of a choice between different, allowable metabolic pathways.

Summary

Vicinal water occurs near most interfaces between a solid and an aqueous phase. The vicinal water appears to exert profound effects on some regulatory processes in cellular systems. The examples discussed included various rate processes such as growth, body temperatures, upper (lethal) thermal limits, and rates of germination. In addition, vicinal hydration effects appear to influence other cell processes, including the choice of metabolic pathways. This is suggested strongly by the results reported in this chapter, showing a thermophilic alga (*Cyanidium caldarium*) able to grow above 45°C in the presence of NH_4^+ but not with NO_3^- as the only nitrogen source.

Acknowledgments

The senior author wishes to express his gratitude to NSF (Grant No. GB 40199) and EPA (Grant No. R80-38-86-01) for the support by these agencies of his research on the role of vicinal water. Our sincere thanks are extended also to J. S. Clegg for his continued interest, encouragement, and advice.

Literature Cited

1. Mitchell, H. K.; Houlahan, M. B. *Am. J. Bot.* **1946,** *33,* 31.
2. Oppenheimer, C. H.; Drost-Hansen, W. *J. Bacteriol.* **1960,** *80*(1), 21.
3. Schmidt, M. G.; Drost-Hansen, W. "Abstracts of Papers," ACS Meeting, Chicago, Ill., September 1961.
4. Drost-Hansen, W. *Ann. N. Y. Acad. Sci.* **1965,** *123* (Art. 2), 471.
5. Drost-Hansen, W. *Ind. Eng. Chem.* **1969,** *61*(11), 10.
6. Drost-Hansen, W. In "Chemistry of the Cell Interface"; Brown, H. D., Ed.; Academic: New York, 1971; pp. 1–184.
7. Drost-Hansen, W. *Phys. Chem. Liq.* **1978,** *7*(3, 4), 243–348.
8. Drost-Hansen, W.; Clegg, J. S., Eds. "Cell-Associated Water"; Academic: New York, 1979.
9. Drost-Hansen, W. In "Physical Principles of Biological Membranes"; Snell, Wolken, Iverson, Lam, Eds.; Gordon & Breach Science Publishers: New York, 1970; p. 243.
10. Peschel, G.; Adlfinger, K. H. *Z. Naturforsch., Teil A* **1971,** *26,* 707.
11. Peschel, G.; Belouschek, P. In "Cell-Associated Water"; Academic: New York, 1979; Chapter 1.
12. Peschel, G.; Adlfinger, K. H. *J. Colloid Interface Sci.* **1970,** *34,* 505.
13. Dordick, R.; Korson, L.; Drost-Hansen, W. "Precision Viscosity Measurements on Aqueous Solutions," *J. Colloid Interface Sci.* **1979,** *72*(2), 206–214.
14. Wiggins, P. M. *Clin. Exp. Pharmacol. Physiol.* **1975,** *2,* 171.
15. Hurtado, R.; Drost-Hansen, W. In "Cell-Associated Water"; Academic: New York, 1979; Chapter 4.
16. Brun, S. G.; Graae Sørensen, P.; Drost-Hansen, W. "Ultrasonic Absorption Measurements," submitted for publication.
17. Bruun, S. G.; Graae Sørensen, P.; Drost-Hansen, W. "Ultrasonic Velocity Measurements," submitted for publication.
18. Ballario, C.; Bonincontro, A.; Cametti, C.; D'Agostino, S. *Nature (London), Phys. Sci.* **1971,** *234,* 179.
19. Ballario, C.; Bonincontro, A.; Cametti, C. *J. Colloid Interface Sci.* **1978,** *63*(3), 567.
20. Ballario, C.; Bonincontro, A.; Cametti, C. *Nuovo Cimento* **1973,** *6*(15), 611.
21. Ballario, C.; Bonincontro, A.; Cametti, C. *J. Colloid Interface Sci.* **1975,** *54*(3), 415.
22. Clegg, J. S.; Drost-Hansen, W. In "The Physical Basis of Electromagnetic Interactions with Biological Systems"; Taylor, L. S., Cheung, A. Y., Eds.; Institute for Physical Sciences and Technology and School of Medicine: University of Maryland, 1977; p. 121.
23. Kerr, J. E. D. Ph.D. Dissertation, University of Miami, 1970.
24. Drost-Hansen, W. *J. Geophys. Res.* **1972,** *77*(27), 5132.
25. Schufle, J. A.; Huang, C. T.; Drost-Hansen, W. *J. Colloid Interface Sci.* **1976,** *54,* 184.
26. Roberts, E. H.; Smith, R. D. In "The Physiology and Biochemistry of Seed Dormancy and Germination"; Kahn, E., Ed., Elsevier/North-Holland Biomedical Press; 1977; p. 385.
27. Thorhaug, A. Part II of "Biologically Allowable Thermal Limits"; Report, EPA-660/3-74-003 (U.S. Environmental Protection Agency) by W. Drost-Hansen and A. Thorhaug; U.S. Government Printing Office; Washington, D.C., 1974.
28. Drost-Hansen, W. *Chesapeake Sci.* **1969,** *10*(3, 4), 281.
29. Drost-Hansen, W. In "Effects and Methods of Control of Thermal Discharges," Serial No. 93-14 (Senate Committee on Public Works), EPA Publication, Part 2, pp. 999–1140; U.S. Government Printing Office: Washington, D. C., 1973.

30. Drost-Hansen, W. *Naturwissenschaften* **1956,** *43,* 512.
31. Drost-Hansen, W. *Ann. N. Y. Acad. Sci.* **1973,** *204,* 100.
32. Etzler, F. M.; Drost-Hansen, W. In "Cell-Associated Water"; Academic: New York, 1979; Chapter 5.
33. Etzler, F. M.; Drost-Hansen, W. In "Colloid and Interface Science"; Kerker, M., Ed.; Academic: New York, 1976; Vol. III, p. 533.
34. Drost-Hansen, W. In "L'eau et les Systemes Biologiques"; Alfsen, A., Berteaud, A.-J., Eds.; Edition du Centre National de la Recherche Scientifique: Paris, 1977; p. 177.
35. Clegg, J. S. In "Cell-Associated Water"; Academic: New York, 1979; Chapter 10.
36. Davey, C. B.; Miller, R. J. *Nature* **1966,** *209,* 638.
37. Davey, C. B.; Miller, R. J. *Soil Sci. Soc. Am., Proc.* **1964,** *28,* 9.
38. Davey, C. B.; Miller, R. J. *J. Bact.* **1966,** *91,* 1827.
39. Nishiyama, I. *Res. Bulletin No. 102,* Hokkaido National Agricultural Experimental Station: Hokkaido, Japan, 1972; p. 125.
40. Drost-Hansen, W. "Temperature Gradient Incubators," to be published in Proceedings, EPA Mini-Symposium on Modeling (Boston, November, 1977); Fisher, J., Ed.; U.S. Government Printing Office: Washington, D.C., 1979.

RECEIVED November 20, 1978.

INDEX

INDEX

D

E

F

I

J

K

L

M

N

O

P

R

S

T

Jacket design by Carol Conway.
Editing and production by Candace A. Deren.

The book was composed by Service Composition Co., Baltimore, MD, printed and bound by The Maple Press Co., York, PA.